Advances in Robot Path Planning, Volume II

Advances in Robot Path Planning, Volume II

Guest Editor

Jonghoek Kim

MDPI

Basel • Beijing • Wuhan • Barcelona • Belgrade • Novi Sad • Cluj • Manchester

Guest Editor
Jonghoek Kim
Defense System Engineering
Sejong University
Seoul
Republic of Korea

Editorial Office
MDPI AG
Grosspeteranlage 5
4052 Basel, Switzerland

This is a reprint of the Special Issue, published open access by the journal *Applied Sciences* (ISSN 2076-3417), freely accessible at: https://www.mdpi.com/journal/applsci/special_issues/87FX7VXW55.

For citation purposes, cite each article independently as indicated on the article page online and as indicated below:

Lastname, A.A.; Lastname, B.B. Article Title. *Journal Name* **Year**, *Volume Number*, Page Range.

ISBN 978-3-7258-3757-1 (Hbk)
ISBN 978-3-7258-3758-8 (PDF)
https://doi.org/10.3390/books978-3-7258-3758-8

Contents

applied sciences

MDPI

Article

Non-Parametric Calibration of the Inverse Kinematic Matrix of a Three-Wheeled Omnidirectional Mobile Robot Based on Genetic Algorithms

Jordi Palacín *, Elena Rubies, Ricard Bitrià and Eduard Clotet

Robotics Laboratory, Universitat de Lleida, Jaume II, 69, 25001 Lleida, Spain
* Correspondence: jordi.palacin@udl.cat

Featured Application: Odometry calibration of a three-wheeled omnidirectional mobile robot.

Abstract: Odometry is a computation method that provides a periodic estimation of the relative displacements performed by a mobile robot based on its inverse kinematic matrix, its previous orientation and position, and the estimation of the angular rotational velocity of its driving wheels. Odometry is cumulatively updated from tens to hundreds of times per second, so any inaccuracy in the definition of the inverse kinematic matrix of a robot leads to systematic trajectory errors. This paper proposes a non-parametric calibration of the inverse kinematic (IK) matrix of a three-wheeled omnidirectional mobile robot based on the use of genetic algorithms (GA) to minimize the positioning error registered in a set of calibration trajectories. The application of this non-parametric procedure has provided an average improvement of 82% in the estimation of the final position and orientation of the mobile robot. This is similar to the improvement achieved with analogous parametric methods. The advantage of this non-parametric approach is that it covers a larger search space because it eliminates the need to define feasible physical limits to the search performed to calibrate the inverse kinematic matrix of the mobile robot.

Keywords: odometry; odometry calibration; inverse kinematic; omnidirectional mobile robot

Citation: Palacín, J.; Rubies, E.; Bitrià, R.; Clotet, E. Non-Parametric Calibration of the Inverse Kinematic Matrix of a Three-Wheeled Omnidirectional Mobile Robot Based on Genetic Algorithms. *Appl. Sci.* **2023**, *13*, 1053. https://doi.org/10.3390/app13021053

Academic Editor: Jonghoek Kim

Received: 16 December 2022
Revised: 4 January 2023
Accepted: 10 January 2023
Published: 12 January 2023

1. Introduction

Odometry is a direct computation method that provides a periodic estimation of the relative displacement of a mobile robot through the use of its inverse kinematic matrix, its previous orientation and position, and the estimation of the angular rotational velocity of its driving wheels. Odometry is cumulatively updated from tens to hundreds of times per second so any inaccuracy in the definition of the inverse kinematic matrix of the robot causes systematic trajectory estimation errors. Borenstein et al. [1] proposed the University of Michigan Benchmark (UMBmark) test to estimate and correct systematic odometry errors in differential drive mobile robots. The existence of systematic odometry errors in a differential drive mobile robot is usually evidenced when a straight trajectory becomes curved. The UMBmark defines a motion experiment in which a differential drive mobile robot follows a square-shaped path in either a clockwise or counterclockwise direction in order to provide unbiased error compensations in both directions. The UMBmark test assumes that systematic odometry errors are caused by an inaccurate definition of the distance from the wheels to the center of rotation of the mobile robot and by an inaccurate definition of the diameters of the wheels. However, the trajectory of the mobile robot may also be conditioned by such other parameters as the controllers driving its motors [2–4]. The effects of all the error sources in the odometry of the mobile robot are usually summarized in the computation of its effective inverse kinematics [5].

Compared with non-holonomic robots, odometry estimation in omnidirectional mobile robots is far more complex due to their enhanced motion capabilities. The additional degrees of freedom offered by omnidirectional robots, along with a larger number of parameters involved in the definition of their kinematic models, can accentuate the effects of systematic errors [6,7]. In the case of a three-wheeled omnidirectional motion system, Maddahi et al. [8] proposed a method to reduce odometry errors based on the application of two corrective indices to the inverse kinematic matrix of the robot. In summary, the application of these two corrective indices has the effect of correcting the angular velocities of the wheels, which can be considered equivalent to a parametric calibration of the radii of the wheels. Maddahi et al. [8] concluded that the application of these corrective indices provides better error reduction in the case of the mobile robot performing straight trajectories rather than curved trajectories. Alternatively, Lin et al. [9] proposed a method to reduce odometry errors based on the direct correction of the kinematic model of an omnidirectional mobile robot using different calibration trajectories. Similarly, the inverse kinematics and trajectory performance of four-wheeled omnidirectional mobile robots have been specifically analyzed by different authors. Maulana et al. [10] analyzed the inverse kinematics in a mecanum mobile robot using stepper motors. Jia et al. [11] analyzed the kinematic model of a mecanum wheeled robot with four wheels. Xu et al. [12] analyzed the tracking performance of a four-wheeled omnidirectional mobile robot using sliding mode control. Li et al. [13] proposed a procedure to reduce odometry errors by considering the problem of wheel slippage due to the motion constraints originated by wheel redundancy in four-wheeled omnidirectional mobile robots. In this case, Li et al. [13] proposed the determination of a non-parametric velocity compensation matrix that is applied to the kinematic model of the mobile robot with the objective of reducing the specific odometry error caused by wheel slippage. In a similar direction, Savaee et al. [14] proposed a non-parametric approach to calibrate the effective kinematic matrix of an omnidirectional mobile robot. This proposal was based on the comparison of trajectory simulations, performed with a virtual mobile robot, with experimental trajectory measurements performed by the real mobile robot. Under such conditions, the assumption was that offline calibration of the kinematic matrix of the mobile robot was less prone to local minima effects [15]. Prados et al. [16] analyzed the motion performance of a four-wheeled omnidirectional mobile robot tailored for surveillance application in limited spaces.

Finally, in a previous work [17], we performed the parametric optimization of the inverse kinematic matrix of a real three-wheeled omnidirectional mobile robot. This approach was based on the definition of a set of benchmark omnidirectional trajectories and an offline parametric optimization based on odometry computation. The parameters analyzed as possible error sources were: the radius of the three omnidirectional wheels, the distance from the wheels to the center of rotation of the mobile robot and the angular orientation of the wheels. These nine parameters (three error sources for each of the three wheels) were iteratively calibrated in order to optimize the inverse kinematic matrix that defines the odometry of the mobile robot. The underlying hypothesis of this parametric optimization was the possibility of directly identifying assembling imprecisions originated during the construction of the robot and thus being able to correct the cause of systematic errors in the trajectory of the robot. However, the analysis of the results obtained in [17] showed that the calibrated values of the parameters have no direct feasible interpretation in the mobile robot. For example, the diameter of the wheels that improves the odometry does not correspond to the wheel diameters obtained with accurate measurements.

New Contribution

This paper proposes a non-parametric calibration of the inverse kinematic (IK) matrix of a three-wheeled omnidirectional mobile robot. This calibration procedure is based on the multidimensional search capabilities of genetic algorithms (GA), which are used to iteratively fit a numerical description of the inverse kinematic matrix that improves the odometry of a three-wheeled omnidirectional mobile robot.

Genetic algorithm optimization has many practical applications in robotics [18–22]. In this paper, genetic optimization is applied in an offline procedure that recomputes the odometry from the mobile robot data registered in 36 representative straight and curved trajectories. These trajectories were inspired by the proposal of Maddahi et al. [8] and Batlle et al. [23] and were already proposed in [17] as a benchmark for offline omnidirectional mobile robot calibration.

The performance of the experimental application of this non-parametric calibration procedure in a real three-wheeled omnidirectional mobile robot has been evaluated in terms of odometry improvement. The main advantage of this non-parametric calibration procedure, relative to a parametric calibration [17], is that it does not require the definition of feasible physical-interpretable limits during the iterative search performed by genetic algorithms. As a result, the search space is larger but the iterative calibration procedure converges faster. The main theoretical disadvantage of this non-parametric calibration procedure is the leak of feasible physical interpretation of the fitted inverse kinematic matrix (such as wheel diameter, distance of the wheel, etc.), although a previous work [17] showed that the result of a parametric calibration cannot be physically interpreted.

2. Materials and Methods

The materials and methods used in this paper are: the APR-02 three-wheeled omnidirectional mobile robot, the method used to compute the odometry of the mobile robot, the representative calibration trajectories, and the dataset of training and validation trajectories used to calibrate the inverse kinematic matrix of this mobile robot.

2.1. APR-02 Three-Wheeled Omnidirectional Mobile Robot

The robot assessed in this paper is the autonomous APR-02 three-wheeled omnidirectional robot, which was initially devised as a mobile telepresence platform [24]. The inclusion of an onboard high-end computer in the following prototypes allowed the development of autonomous applications, e.g., supporting and guiding older people with mobility impairments [25], and an early gas leak detection system [26]. The main characteristic of this assistive mobile robot [27] is the use of an omnidirectional motion system based on three optimal omnidirectional wheels [28] driven by three brushed direct current motors (BDCM) [29]. The odometry of the robot is calculated from the information provided by the encoders of the motors. The advantage of a three-wheeled against a four-wheeled omnidirectional motion system is the avoidance of wheel slippage [17,30].

Figure 1 shows some images of the APR-02 mobile robot completing a displacement. The trajectory of the mobile robot is controlled by a target motion command (v, α, ω, t_r) [30] that defines: the translational velocity of the motion (v); the angular orientation of the planned motion (α), defined in a complete range from 0° to 360°; the angular rotational speed of the base of the mobile robot (ω) while performing a displacement; and the time duration of this motion command (t_r), which can be replaced by a target distance displacement. The path planning algorithm of the mobile robot periodically updates the target motion command every 300 ms. The maximum value of the time duration (t_r) is always lower than 500 ms as a kind of watchdog security measure to stop the mobile robot automatically in case of malfunction of its central processing unit. This omnidirectional mobile robot is able to move in any angular orientation (α) without having to perform any previous maneuver, also being able to rotate over itself while moving [30].

The determination of the ground truth trajectory of the mobile robot is based on the application of SLAM [31] to the precise information gathered from its onboard Hokuyo UTM-30LX 2D LIDAR, which is placed horizontally or tilted down [32] depending on the expected surrounding environment. The practical disadvantage of using a high-precision LIDAR is its high cost. There are also other cheaper positioning alternatives that have been proven useful for kinematic calibration [33] and in the application of mobile robots intended to operate in the presence of dynamic obstacles, such as people [34].

(a) (b) (c) (d)

Figure 1. APR-02 mobile robot performing a transversal (or lateral) displacement: (**a**) starting point; (**b**,**c**) intermediate points; and (**d**) destination point.

2.2. Odometry Estimation

Odometry is a computation method that provides a periodic estimation of the relative displacement of a mobile robot. This computation requires an accurate and precise definition of its inverse kinematic matrix, its previous position and orientation, and the estimation of the angular rotational velocities of its driving wheels. The consequence of any inaccuracy or imprecision is the generation of systematic odometry errors. Additionally, odometry interprets the angular rotational speeds of the wheels of the mobile robot as linear displacements, so it requires non-slippage wheel conditions throughout the trajectory. In the case of a three-wheeled omnidirectional mobile robot, the wheels have no motion constraints and no slippage is originated while performing any continuous motion [30].

The kinematics and the odometry of a three-wheeled omnidirectional mobile robot such as the APR-02, was described previously in [17,30]. This practical odometry estimation is challenging because of the degree of freedom provided by its omnidirectional motion system. Figure 2 presents the definition of the omnidirectional motion system of the APR-02 mobile robot: (x, y, θ) is the position of the mobile robot referred to the absolute or fixed world frame (X_W, Y_W), (X_R, Y_R) is the relative mobile robot frame, (v, α, ω) is the motion command that specifies the target trajectory planned for the mobile robot, $(\omega_a, \omega_b, \omega_c)$

are the angular velocities of the wheels required to implement the motion command, and (V_a, V_b, V_c) are the deduced linear velocities of the wheels.

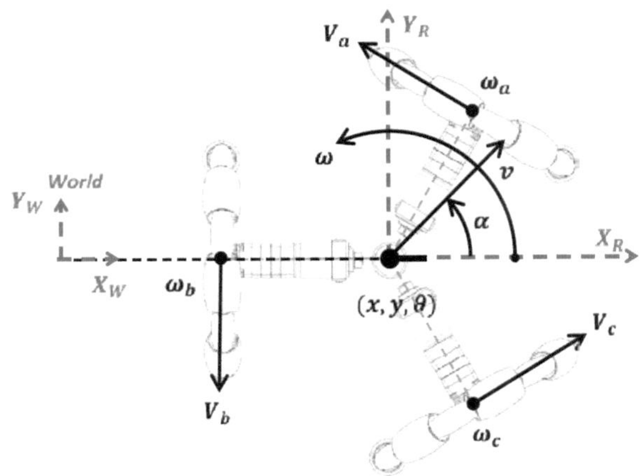

Figure 2. Detail of the omnidirectional motion system of the APR-02 mobile robot. (X_R, Y_R) is the mobile robot frame in which X_R represents the front and forward direction of the robot.

Odometry uses the instantaneous estimate of the current angular velocities of the three wheels, a, b, c, given as $(\omega_a(k), \omega_b(k), \omega_c(k))$, and defined in rad/s; and the previous position of the mobile robot $(x(k-1), y(k-1), \theta(k-1))$, defined in the absolute world frame (X_W, Y_W). Both values are used to update the current position of the mobile robot $(x(k), y(k), \theta(k))$ by using the following expression [17]:

$$\begin{bmatrix} x(k) \\ y(k) \\ \theta(k) \end{bmatrix}_{World} = \begin{bmatrix} x(k-1) \\ y(k-1) \\ \theta(k-1) \end{bmatrix}_{World} + \Delta t \cdot R(\theta(k-1))^{-1} \cdot M^{-1} \cdot \begin{bmatrix} \omega_a(k) \\ \omega_b(k) \\ \omega_c(k) \end{bmatrix} \quad (1)$$

where Δt is the sampling time at which the angular velocities of the wheels of the robot are updated and k is the current discrete sample number that describes a time elapsed $t(k)$ (where $t(k) = \Delta t \cdot k$) since the initialization of the robot. In the case of the APR-02, this sampling time coincides with the sampling time used internally by the three proportional, integral, and derivative (PID) controllers ($\Delta t = 10$ ms) [29] of the three brushed direct current motors (BDCM) driving its three omnidirectional wheels.

In Equation (1), $R(\theta(k-1))^{-1}$ is the inverse of the rotation matrix computed from the previous instantaneous angular orientation of the mobile robot $\theta(k-1)$:

$$R(\theta(k-1))^{-1} = \begin{bmatrix} \cos(\theta(k-1)) & -\sin(\theta(k-1)) & 0 \\ \sin(\theta(k-1)) & \cos(\theta(k-1)) & 0 \\ 0 & 0 & 1 \end{bmatrix} \quad (2)$$

and M^{-1} is the inverse of the compact kinematic matrix of the mobile robot that can be computed analytically as [17]:

$$M^{-1} = \frac{1}{R_a \sin(\delta_b - \delta_c) - R_b \sin(\delta_a - \delta_c) + R_c \sin(\delta_a - \delta_b)} \cdot \begin{bmatrix} r_a(R_b \cos(\delta_c) - R_c \cos(\delta_b)) & -r_b(R_a \cos(\delta_c) - R_c \cos(\delta_a)) & r_c(R_a \cos(\delta_b) - R_b \cos(\delta_a)) \\ r_a(R_b \sin(\delta_c) - R_c \sin(\delta_b)) & -r_b(R_a \sin(\delta_c) - R_c \sin(\delta_a)) & r_c(R_a \sin(\delta_b) - R_b \sin(\delta_a)) \\ r_a \sin(\delta_b - \delta_c) & -r_b \sin(\delta_a - \delta_c) & r_c \sin(\delta_a - \delta_b) \end{bmatrix} \quad (3)$$

where (r_a, r_b, r_c) are the radii of the wheels, (R_a, R_b, R_c) are the distances between the center of the robot and the wheels, and $(\delta_a, \delta_b, \delta_c)$ are the angular orientations of the wheels defined in the mobile robot frame (X_R, Y_R). Figure 3 presents the definition of all these parameters involved in the omnidirectional motion system of the APR-02. These have the following design values: $r_{a,b,c} = 0.148$ m, $R_{a,b,c} = 0.195$ m, $\delta_a = 60°$, $\delta_b = 180°$, $\delta_c = 300°$.

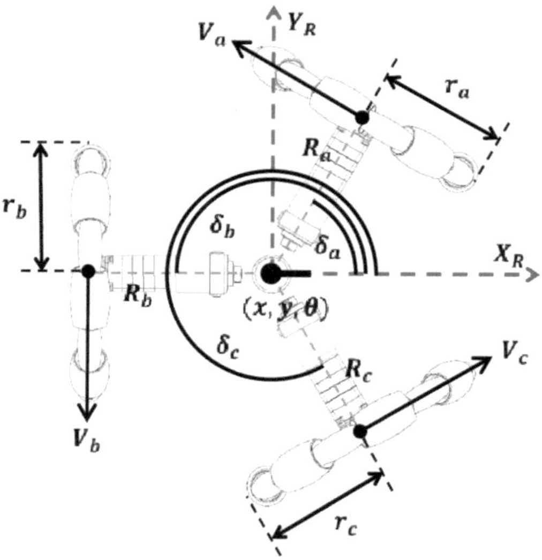

Figure 3. Parametric definition of the omnidirectional motion system of the APR-02 mobile robot. (X_R, Y_R) represents the mobile robot frame in which X_R is the front of the mobile robot.

2.3. Calibration Trajectories

The calibration trajectories used in this paper are the same 36 representative benchmark trajectories proposed in [17]. These trajectories were proven useful for improving the odometry of a three-wheeled omnidirectional mobile robot [17]. The motion commands required to generate the calibration trajectories are listed in Table A1 of [17]. These benchmark trajectories define a characteristic flower-shaped figure that is representative of the motion capabilities of an omnidirectional mobile robot.

Table 1 lists the information registered by the APR-02 mobile robot while performing a calibration trajectory defined by a motion command (v, α, ω), executed over a predefined distance or time. The path-planning algorithm of the mobile robot converts this target motion into the target angular rotational velocities $(\omega_{Ma}, \omega_{Mb}, \omega_{Mc})$ to be applied to the three PID controllers [29] that supervise the motors of the robot. The main information registered in Table 1 is the complete sequence of angular rotational velocities of the wheels measured during a displacement, $\omega_{a,b,c}$. This information is used by the mobile robot to compute the odometry in real-time, but the registration of this information allows future offline re-computation of the odometry with a different inverse kinematic matrix (allowing its calibration). The APR-02 also estimates its position by applying SLAM to the 2D scans provided by its onboard LIDAR. This real-time information is stored as the reference ground truth trajectory followed by the mobile robot during the experiment, along with the scans provided by the LIDAR sensor for future offline analysis.

Table 1. Information registered by the mobile robot while performing a calibration trajectory.

Parameter	Description
(v, α, ω, d)	*Target motion command for the robot* v: translational velocity of the displacement of the robot, (m/s) α: angular orientation of the displacement, (°) ω: angular rotational speed of the robot during the displacement, (rad/s) d: linear distance to be achieved during the displacement, (m)
$(\omega_{Ma}, \omega_{Mb}, \omega_{Mc})$	*Target angular velocities for the wheels* ω_{Mi}: target angular velocity defined for the motor of the wheel i, computed when receiving the motion command, (rpm)
$\omega_{a,b,c} = \begin{bmatrix} t(k=1\ldots F) \\ \omega_a(k=1\ldots F) \\ \omega_b(k=1\ldots F) \\ \omega_c(k=1\ldots F) \end{bmatrix}$	*Angular velocities of the wheels provided by the encoders* $t(k)$: time the update k was received, (s) ω_i: instantaneous angular velocity measured by the encoder of the motor i, updated periodically at a frame rate of 10 ms, (rpm)
$(x, y, \theta) = \begin{bmatrix} x(k=1\ldots F) \\ y(k=1\ldots F) \\ \theta(k=1\ldots F) \end{bmatrix}$	*Trajectory of the mobile robot estimated with the odometry* x, y: location of the robot, (m) θ: angular orientation of the robot, (°)
$LIDAR = \begin{bmatrix} t_{LIDAR}(p=1) & & t_{LIDAR}(p=N) \\ \begin{bmatrix} d(1) \\ \ldots \\ d(1080) \end{bmatrix}^1 & \ldots & \begin{bmatrix} d(1) \\ \ldots \\ d(1080) \end{bmatrix}^N \end{bmatrix}$	*Scans provided by the onboard LIDAR* $t_{LIDAR}(p)$: time the scan p was received, (s) $d(r)$: distance scan corresponding to the angular orientation r, updated periodically at a frame rate from 200 to 300 ms, (mm)
$(x, y, \theta)_{LIDAR} = \begin{bmatrix} x_{LIDAR}(p=1\ldots N) \\ y_{LIDAR}(p=1\ldots N) \\ \theta_{LIDAR}(p=1\ldots N) \end{bmatrix}$	*Ground Truth trajectory of the mobile robot estimated with SLAM* x_{LIDAR}, y_{LIDAR}: location of the robot, (m) θ_{LIDAR}: angular orientation of the robot, (°)

2.4. Dataset of Training and Validation Trajectories

This paper uses two datasets composed of training and validation trajectories. The training dataset is used to calibrate the inverse kinematic matrix of the APR-02 mobile robot. The training dataset is the same one used in [17] to obtain comparable calibration results. The training dataset is composed of 36 benchmark calibration trajectories, each repeated 5 times, with a total of 180 trajectories registered for training. The validation dataset is used to assess the final performance of the non-parametric inverse kinematic matrix fitted. The validation dataset is composed of 36 new benchmark calibration trajectories. Once again, each trajectory is repeated 5 times, making up a total of 180 new trajectories registered for the testing phase. Figure 1 shows the mobile robot APR-02 completing a calibration trajectory.

3. Procedure for Genetic Algorithm Calibration of the Inverse Kinematic Matrix

The non-parametric procedure used in this paper to calibrate the inverse kinematic matrix of the APR-02 mobile robot is based on the multivariate search capabilities of genetic algorithms [35]. This nature-inspired search is used to recompute the odometry of the mobile robot offline (Equation (1)) with the information registered while completing a trajectory. The application of a genetic search iteratively finds a local unconstrained minimum of an objective cost function in a multivariate search space. Figure 4 shows a diagram describing the genetic algorithm search, which is started from an initial population that defines a starting point in the search space and then performs a bounded search.

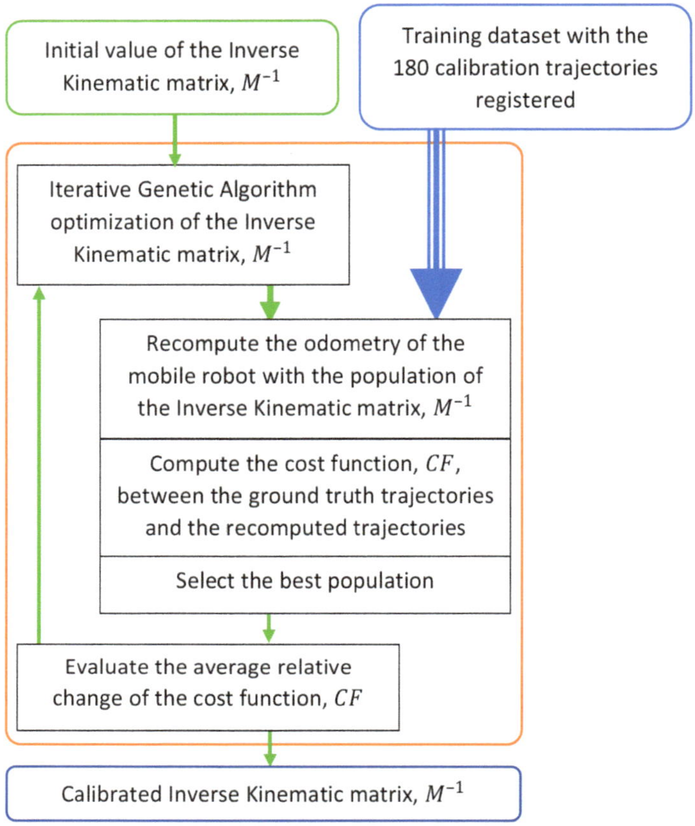

Figure 4. Diagram of the Genetic Algorithm search performed to calibrate the inverse kinematic matrix of the APR-02 mobile robot.

The initial population of the search is the theoretical value of the inverse kinematic matrix M^{-1} defined in Equation (3). This is interpreted by the genetic algorithm as a population vector, V, that defines the ninth-dimensional search space:

$$M^{-1} = \begin{bmatrix} m_{1,1} & m_{1,2} & m_{1,3} \\ m_{2,1} & m_{2,2} & m_{2,3} \\ m_{3,1} & m_{3,2} & m_{3,3} \end{bmatrix} \tag{4}$$

$$V = \begin{bmatrix} m_{1,1} & m_{1,2} & m_{1,3} & m_{2,1} & m_{2,2} & m_{2,3} & m_{3,1} & m_{3,2} & m_{3,3} \end{bmatrix} \tag{5}$$

The upper and lower bounds (V_{UB} and V_{LB}) of the GA search were configured to allow a $\pm 10\%$ variation of the values defined in the initial population V.

The iterative search performed by genetic algorithms requires the computation of a cost function CF to evaluate the performance of each population proposed during the search. This cost function is the same one used in [17] to compare the previous reference parametric calibration results [17] with the non-parametric calibration presented in this paper. This function is defined as:

$$CF = \frac{1}{Z} \cdot \sum_{i=1}^{Z} \sqrt{\left(x^i_{LIDAR}(N) - x^i(F)\right)^2 + \left(y^i_{LIDAR}(N) - y^i(F)\right)^2 + \left(\theta^i_{LIDAR}(N) - \theta^i(F)\right)^2} \tag{6}$$

where Z is the number of trajectories used to compute the cost function (the 180 trajectories included in the training dataset); i represents an individual trajectory completed by the mobile robot; $x^i_{LIDAR}(p)$ is the x location of the mobile robot estimated by applying SLAM to the p scan provided by the onboard LIDAR while completing the trajectory i; N identifies the last scan provided by the LIDAR when the robot reaches the end of the trajectory; $x^i(k)$ is the x location of the mobile robot estimated with the internal odometry using the current values of the inverse kinematic matrix M^{-1}; F identifies the last odometry evaluation obtained when the robot reaches the end of the trajectory; θ is the final angular orientation of the mobile robot.

This cost function formulation was used to calibrate the effective kinematic parameters of an omnidirectional mobile robot by Savaee et al. [14]. This cost function apparently applies the same weights to the distance and angular error obtained at the end of the displacement. However, this perception is incorrect because an error in a specific angular orientation of the robot affects the estimation of the subsequent trajectory cumulatively (see Equation (1)). Therefore, any error in the intermediate estimation of the angular orientation has a huge potential cumulated weight in the computation of the cost function.

At this point, note that this cost function is computed using only the ending position of the mobile robot. Therefore, the evaluation of this cost function only requires: (1) the last position of the ground truth trajectory estimated with the SLAM procedure (computed while completing the trajectory) and (2) the last location estimated with the odometry using the inverse kinematic matrix defined by the current population analyzed in the iteration. This offline computation of the odometry requires all the information provided by the encoders during the displacement; this is why each trajectory registered by the mobile robot also includes the information provided by the encoders.

Finally, the iterative search performed by the genetic algorithm stops if the average relative change in the best population found of a cost function CF is less than or equal to 10^{-6}, which is the default value used in standard GA searches.

4. Results

This section compares the performance of the inverse kinematic matrices of the APR-02 mobile robot. This section presents: (1) the reference theoretical value of the inverse kinematic matrix of the mobile robot, (2) the reference parametric calibration of the inverse kinematic matrix obtained previously in [17], and (3) the result of the non-parametric calibration performed in this paper.

4.1. Reference Theoretical Value of the Inverse Kinematic Matrix

The exact or theoretical value of the inverse kinematic matrix (M^{-1}) of the APR-02 mobile robot can be computed analytically from Equation (1) as [17]:

$$M^{-1} = \begin{bmatrix} -0.0854478398400646 & 0.0000000000000000 & 0.0854478398400646 \\ 0.0493333333333333 & -0.0986666666666667 & 0.0493333333333333 \\ 0.2529914529914529 & 0.2529914529914529 & 0.2529914529914529 \end{bmatrix} \quad (7)$$

Figure 5 shows the evolution of a sample trajectory followed by the APR-02 mobile robot: the red line displays the ground truth trajectory estimated with SLAM [31] and the blue line the odometry computed using the theoretical value of the inverse kinematic matrix. As could be expected, Figure 5 shows differences between the ground truth trajectory and the odometry due to the existence of systematic odometry errors. At this point, note that the inverse kinematic matrix of a three-wheeled omnidirectional robot is very sensitive to any inaccuracy in the implementation of the parameters of the motion system [17]: radii of the wheels (r_a, r_b, r_c), the distance between the center of the robot and each wheel $(R_a,\ R_b,\ R_c)$, and the angular orientation of each wheel in the mobile robot frame $(\delta_a,\ \delta_b,\ \delta_c)$. In any case, the effects of the systematic errors are best evidenced in a mobile robot performing curved trajectories (Figure 5) rather than straight trajectories [17]. In the case of an omnidirectional mobile robot, the difference between a curved and a straight trajectory depends only on the target angular rotational speed, which is the ω parameter in the motion command, (v, α, ω).

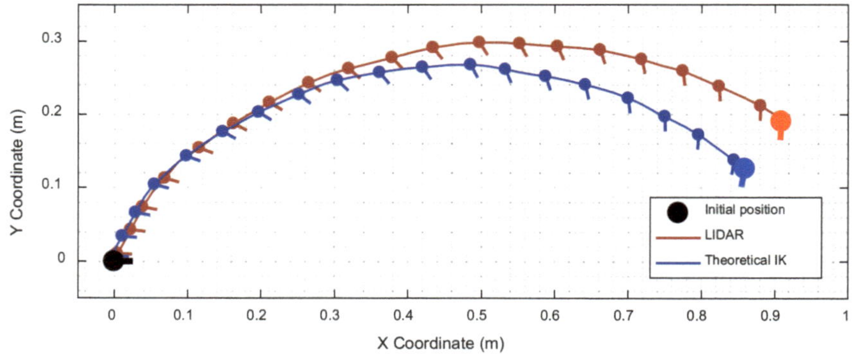

Figure 5. Sample trajectory followed by the APR-02 mobile robot for the motion command $(v, \alpha, \omega \neq 0)$ = (0.2 m/s, 30°, 0.35 rad/s). The ground truth trajectory was estimated from the information of the LIDAR (red line) while the odometry was estimated with the theoretical value of the inverse kinematic matrix of the mobile robot (blue line).

4.2. Reference Parametric Optimization of the Inverse Kinematic Matrix (from [17])

The parametric optimization presented in [17] was focused on calibrating the value of the parameters of the APR-02 mobile robot and the computation of the inverse kinematic matrix of the robot from these parameters. Nine specific mobile robot parameters were calibrated in [17]: the radii of the wheels (r_a, r_b, r_c), the distance between the center of the robot and each wheel (R_a, R_b, R_c), and the angular orientation of each wheel $(\delta_a, \delta_b, \delta_c)$. The computation of this parametric optimization required 903 s in a high-performance workstation.

The exact value of the inverse kinematic matrix $(M_{PARAMETRIC}{}^{-1})$ obtained from the parametric calibration presented in [17] is:

$$M_{PARAMETRIC}{}^{-1} = \begin{bmatrix} -0.0892930568372762 & -0.0005606566978203 & 0.0867466159313470 \\ 0.0520955668635434 & -0.1010213675626970 & 0.0533117820461363 \\ 0.2364093797047320 & 0.2341358647406650 & 0.2353792947863630 \end{bmatrix} \quad (8)$$

This parametric inverse kinematic matrix represents the following errors in the determination of the parameters of the mobile robot: 4.7% in the size of the radius of the wheel a (r_a), 2.4% in r_b, 3.2% in r_c, 12% in the distance from the wheel a to the center of rotation of the robot (R_a), 10% in R_b, 11% in R_c, 0.4% in the angular orientation of the wheel a (δ_a), 0.2% in δ_b, and 0.5% in δ_c. These huge errors obtained in the parametric calibration, higher than 10% in some cases, were too high to be justified as being caused by assembling or manufacturing errors. Therefore, the drawback of the parametric calibration was that the results could not be interpreted as an improved description of the physical parameters of the mobile robot: the radii of the wheels $(r_a,\ r_b,\ r_c)$, the distances from the wheels to the center of rotation of the mobile robot $(R_a,\ R_b,\ R_c)$, and the angular orientations of the wheels $(\delta_a,\ \delta_b,\ \delta_c)$. Consequently, these results suggested the development of a comparative non-parametric calibration of the inverse kinematic matrix of the mobile robot.

4.3. Non-Parametic Inverse Kinematic Matrix Calibrated with Genetic Algorithms

As stated above, the difficult physical interpretation of the parametric optimization results obtained in [17] combined with the difficulty of establishing feasible limits to the parametric search suggested the implementation of an alternative non-parametric calibration of the inverse kinematic matrix of the mobile robot. The complete procedure used in this non-parametric calibration is described in Section 3. The optimization is based on the multivariate search capabilities of the genetic algorithms, which are less prone to local minima than other gradient search algorithms. The computation of this non-parametric optimization required 637 s in the same high-performance workstation used in the reference parametric optimization [17] (29% less time).

The best non-parametric inverse kinematic matrix (${M_{NON-PARAMETRIC}}^{-1}$) obtained with the application of the calibration procedure proposed in this paper (described in Section 3) is:

$$ {M_{NON-PARAMETRIC}}^{-1} = \begin{bmatrix} -0.0883286476963123 & 0.0000000000000000 & 0.0873330402667908 \\ 0.0522829495809178 & -0.1008246738469390 & 0.0536189007203205 \\ 0.2357044251338890 & 0.2338948216993490 & 0.2348793373133590 \end{bmatrix} \quad (9) $$

The following expression is used to compare the inverse kinematic matrix obtained with the parametric (Equation (8)) and non-parametric calibration (Equation (9)). This expression performs the Hadamart product (element-wise product) [36] of the difference between these two matrices.

$$ \left({M_{NON-PARAMETRIC}}^{-1}\right)_{ij} \cdot (1 + {M_{DIFFERENCE}}^{-1})_{ij} = \left({M_{PARAMETRIC}}^{-1}\right)_{ij} \quad (10) $$

$$ {M_{DIFFERENCE}}^{-1} = \begin{bmatrix} 1.091\% & 1.505 \times 10^{15}\% & -0.671\% \\ -0.358\% & 0.195\% & -0.572\% \\ 0.299\% & 0.103\% & 0.212\% \end{bmatrix} \quad (11) $$

The difference matrix ${M_{DIFFERENCE}}^{-1}$ shows that the results of both calibrations (parametric and non-parametric) are very similar, with differences lower than 0.7% except in the case of the coefficients $m_{1,1}$ (1.091%) and $m_{1,2}$ (1.505×10^{15}%). The huge difference in the coefficient $m_{1,2}$ is because the non-parametric calibration maintained the original value defined in the theoretical inverse kinematic matrix (Equations (3) and (7)), with a value very close to zero: -1.208×10^{-17}. Alternatively, the result of the parametric calibration performed in [17] (displayed in Equation (8)) showed $m_{1,2}$ to reach the value of -0.00056, which is very difficult to interpret as a parametric inaccuracy originated during the assembling the robot. The differences between the parametric and non-parametric calibration of the inverse kinematic matrices are small but cannot be neglected because the odometry is cumulatively updated ten times per second with the displacement information provided by the encoders.

Figure 6 compares the offline computation of the odometry of four randomly chosen validation trajectories (from the 36 × 5 available in the validation dataset). The ground truth trajectory computed from the LIDAR information is labeled in red and the odometry is computed using the different inverse kinematic matrices evaluated in this paper: theoretical IK (green), parametric IK (brown), and non-parametric IK (magenta). Figure 7a–d shows in detail the final position of the robot achieved in the trajectories depicted in Figure 6. Figure 7a–d shows small differences between the ground truth trajectory of the mobile robot (red line) and the odometry computed with the parametric (brown line) and non-parametric (magenta line) inverse kinematic matrices obtained for the APR-02.

Finally, Tables 2 and 3 compare the performance of the non-parametric calibration performed in this paper. Table 2 presents the average values of the cost function (CF) obtained with the 180 benchmark trajectories contained in the training dataset. The reference cost function, $CF_{Theoretical\ IK}$, was computed using the theoretical inverse kinematic matrix which is described in Equations (3) and (7). The values of the parametric inverse kinematic matrix (Parametric IK) were obtained in [17] and its values are presented in Equation (8). The numerical values of the non-parametric optimization of the inverse kinematic matrix performed in this paper with genetic algorithms (Non-parametric IK) are described in Equation (9). Table 2 shows that the best cost function computed with the training trajectories is practically the same for the parametric and non-parametric calibrations. Please note that the column indicating $CF_{TRAINING}$ presents the cost function computed with the calibration trajectories included in the training dataset.

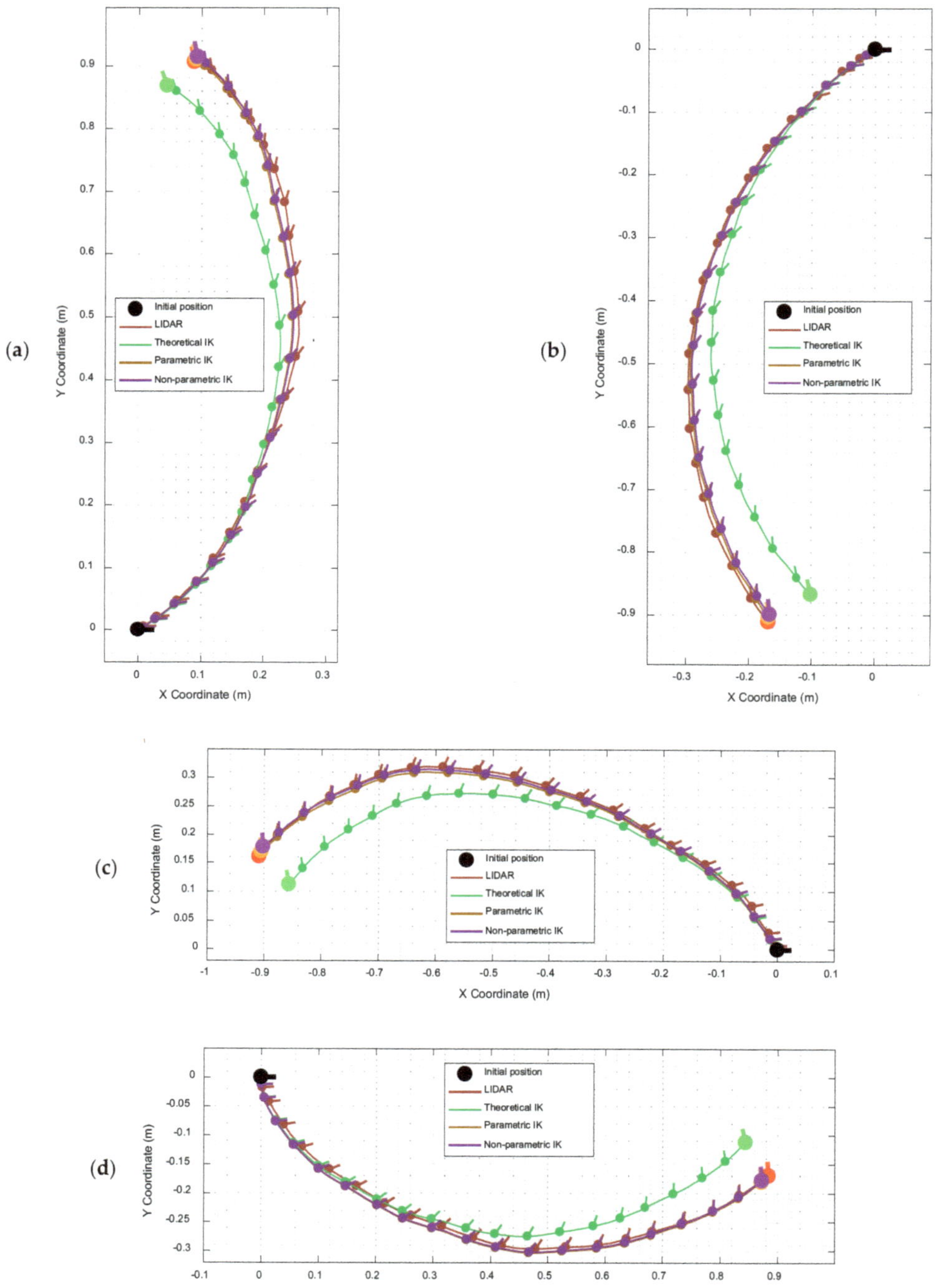

Figure 6. Comparison between the ground truth trajectory followed by the mobile robot (red line) and the odometry estimated with: the theoretical IK (green line), the parametric IK (brown line) and the non-parametric IK (magenta line). Trajectories originated by the following motion commands, (v, α, ω): (**a**) (0.2 m/s, 30°, 0.35 rad/s); (**b**) (0.2 m/s, 210°, 0.35 rad/s); (**c**) (0.2 m/s, 120°, 0.35 rad/s); and (**d**) (0.2 m/s, 300°, 0.35 rad/s).

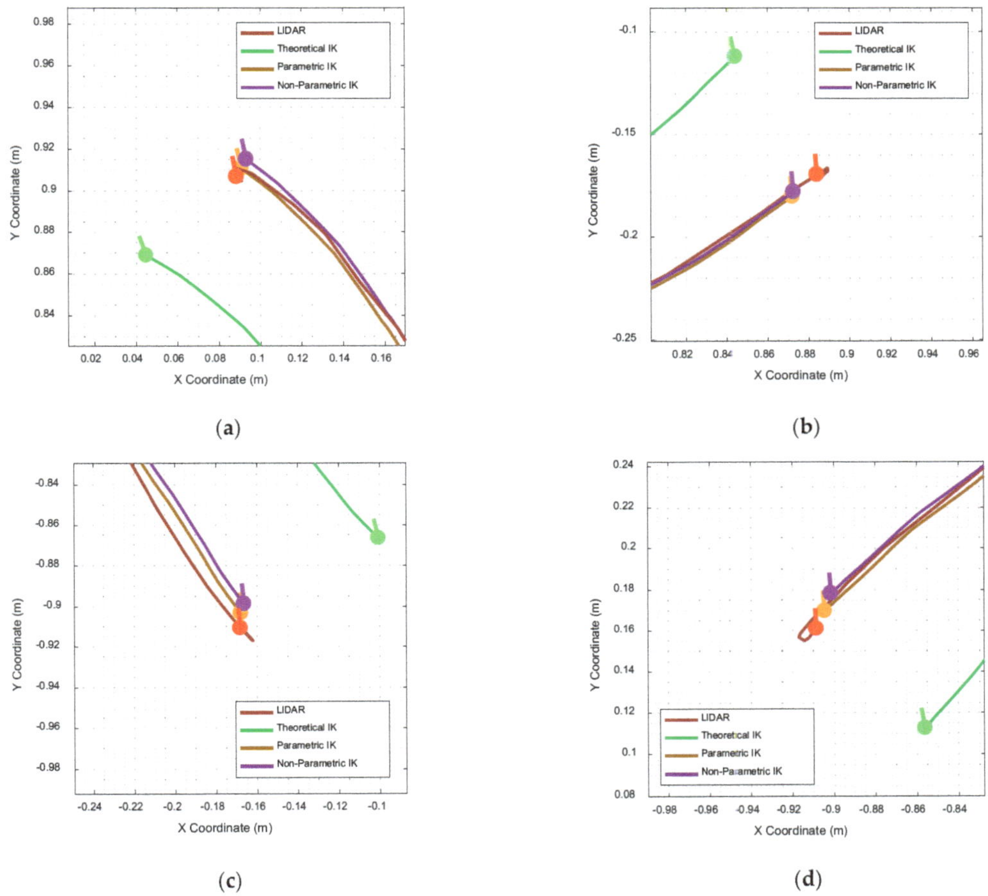

Figure 7. Detail of the final position of the robot corresponding to the full trajectories displayed in Figure 6. Ground truth trajectory of the mobile robot (red) and odometry of the mobile robot computed using the theoretical IK (green), the parametric IK [17] (brown) and the noon-parametric IK (magenta).

Table 2. Values of the cost function obtained when computing the odometry of the 180 trajectories included in the training dataset with the theoretical and calibrated inverse kinematic matrices.

M^{-1}	$CF_{Theoretical\ IK}$	$CF_{TRAINING}$	**Improvement**
Theoretical IK	0.1234	-	-
Parametric IK [17]	0.1234	0.0215	82.61%
Non-parametric IK *	0.1234	0.0215 *	82.60% *

* Calibration result obtained in this paper.

Table 3. Values of the cost function obtained when computing the odometry of the 180 trajectories included in the validation dataset with the theoretical and calibrated inverse kinematic matrices.

M^{-1}	$CF_{Theoretical\ IK}$	$CF_{VALIDATION}$	**Improvement**
Theoretical IK	0.1251	-	-
Parametric IK [17]	0.1251	0.0229	81.68%
Non-parametric IK *	0.1251	0.0227 *	81.81% *

* Calibration result obtained in this paper.

Table 3 shows the value of the cost function evaluated with the trajectories included in the validation dataset. These validation trajectories were registered several months after the registration of the trajectories included in the training dataset. Table 3 shows that the offline computation of the odometry of the 180 trajectories included in the validation dataset offers practically the same cost function and the same improvement with both the parametric and non-parametric inverse kinematic matrices. Additionally, the similarity between the values of the cost function obtained with the training dataset (Table 2) and the validation dataset (Table 3) seems to indicate no over-time changes in the kinematics of the mobile robot APR-02.

5. Discussion and Conclusions

This paper proposes a non-parametric calibration procedure of the effective inverse kinematic matrix of a three-wheeled omnidirectional mobile robot based on the search capabilities of genetic algorithms. The calibration procedure used a training dataset composed of five repetitions of 36 representative benchmark trajectories (180 trajectories in total). These benchmark trajectories summarize the motion performances of a three-wheeled mobile robot [17]. The calibration procedure is based on genetic algorithms because the mutation and combination performed during the search is less prone to local minima during the multivariate searches required to optimize the kinematics of a robot [14,17].

The application of this non-parametric calibration procedure to the offline computation of the odometry of the training trajectories reduced the value of the cost function from 0.1234 to 0.0215, representing an improvement of the cost function of 82.6% (see Table 2). This improvement is practically the same as that (82.6%) obtained in the parametric calibration procedure conducted previously in [17] with the same training trajectories. The analysis implemented with the validation trajectories confirmed an average improvement in the cost function of 81.8% (see Table 3) which is visually observable in the estimated trajectory of the mobile robot (see Figures 6 and 7). The calibration results obtained in this paper have a similar order of magnitude as the improvements presented by Maddahi et al. [8] and Batlle et al. [23], who highlighted the importance of using curved trajectories for calibrating the kinematics of an omnidirectional mobile robot. Nevertheless, the use of different calibration trajectories precludes direct comparison of the achievements.

As a summary, the non-parametric calibration of the inverse kinematic matrix of a three-wheeled omnidirectional mobile robot has two main advantages: (1) it avoids the problem of defining feasible physical-interpretable limits applied in the parametric calibration; and (2), as a consequence, this non-parametric calibration is much easier to implement and converges faster than the parametric calibration. The offline implementation of the non-parametric calibration in a high-performance workstation required an average of 637 s, while the parametric calibration conducted previously in [17] required 903 s (41% more computational time). The definition of feasible, physically interpretable limits in a parametric calibration of the kinematic matrix is a problem that cannot be neglected because of the complex numerical relationship between the parameters that define the inverse kinematic matrix of a mobile robot. The comparative analysis of the results obtained in this paper demonstrates that the improvement in odometry that can be obtained with a non-parametric calibration is practically the same as that which can be obtained with a parametric calibration. Finally, the disadvantage of performing a non-parametric calibration of the inverse kinematic matrix of a mobile robot is that the calibration result is a numerical matrix that cannot be physically interpreted. However, this disadvantage does not really exist because the parameter variation obtained in the parametric calibration performed in [17] could not be physically interpreted either.

Future work will analyze the application of this non-parametric procedure in several three-wheeled omnidirectional mobile robots and the application of alternative methods based on artificial neural networks [37] to estimate the kinematics of an omnidirectional mobile robot.

Author Contributions: Formal analysis, E.R. and E.C.; investigation, J.P. and E.R.; methodology, J.P.; resources, R.B.; software, R.B. and E.C.; writing—original draft, E.R. and E.C.; writing—review & editing, J.P. All authors have read and agreed to the published version of the manuscript.

Funding: This research has been partially funded by the Departament de Recerca i Universitats de la Generalitat de Catalunya: FI SDUR 2022 grant.

Conflicts of Interest: The authors declare no conflict of interest. The funders had no role in the design of the study, in the collection, analyses, or interpretation of data, in the writing of the manuscript or in the decision to publish the results.

References

1. Borenstein, J.; Feng, L. Measurement and correction of systematic odometry errors in mobile robots. *IEEE Trans. Robot. Autom.* **1996**, *12*, 869–880. [CrossRef]
2. Štefek, A.; Pham, V.T.; Krivanek, V.; Pham, K.L. Optimization of Fuzzy Logic Controller Used for a Differential Drive Wheeled Mobile Robot. *Appl. Sci.* **2021**, *11*, 6023. [CrossRef]
3. Ding, T.; Zhang, Y.; Ma, G.; Cao, Z.; Zhao, X.; Tao, B. Trajectory tracking of redundantly actuated mobile robot by MPC velocity control under steering strategy constraint. *Mechatronics* **2022**, *84*, 102779. [CrossRef]
4. Thai, N.H.; Ly, T.T.K.; Dzung, L.Q. Trajectory tracking control for differential-drive mobile robot by a variable parameter PID controller. *Int. J. Mech. Eng. Robot. Res.* **2022**, *11*. [CrossRef]
5. de Jesús Rubio, J.; Aquino, V.; Figueroa, M. Inverse kinematics of a mobile robot. *Neural Comput. Appl.* **2013**, *23*, 187–194. [CrossRef]
6. Sousa, R.B.; Petry, M.R.; Moreira, A.P. Evolution of Odometry Calibration Methods for Ground Mobile Robots. In Proceedings of the IEEE International Conference on Autonomous Robot Systems and Competitions (ICARSC), Ponta Delgada, Portugal, 15–17 April 2020; pp. 294–299. [CrossRef]
7. Hijikata, M.; Miyagusuku, R.; Ozaki, K. Wheel Arrangement of Four Omni Wheel Mobile Robot for Compactness. *Appl. Sci.* **2022**, *12*, 5798. [CrossRef]
8. Maddahi, Y.; Maddahi, A.; Sepehri, N. Calibration of omnidirectional wheeled mobile robots: Method and experiments. *Robotica* **2013**, *31*, 969–980. [CrossRef]
9. Lin, P.; Liu, D.; Yang, D.; Zou, Q.; Du, Y.; Cong, M. Calibration for Odometry of Omnidirectional Mobile Robots Based on Kinematic Correction. In Proceedings of the 14th International Conference on Computer Science & Education (ICCSE), Toronto, ON, Canada, 19–21 August 2019; pp. 139–144. [CrossRef]
10. Maulana, E.; Muslim, M.A.; Hendrayawan, V. Inverse kinematic implementation of four-wheels mecanum drive mobile robot using stepper motors. In Proceedings of the 2015 International Seminar on Intelligent Technology and Its Applications (ISITIA), Surabaya, Indonesia, 20–21 May 2015; pp. 51–56. [CrossRef]
11. Jia, Q.; Wang, M.; Liu, S.; Ge, J.; Gu, C. Research and development of mecanum-wheeled omnidirectional mobile robot implemented by multiple control methods. In Proceedings of the 2016 23rd International Conference on Mechatronics and Machine Vision in Practice (M2VIP), Nanjing, China, 28–30 November 2016; pp. 1–4. [CrossRef]
12. Xu, H.; Yu, D.; Wang, Q.; Qi, P.; Lu, G. Current Research Status of Omnidirectional Mobile Robots with Four Mecanum Wheels Tracking based on Sliding Mode Control. In Proceedings of the 2019 IEEE International Conference on Unmanned Systems and Artificial Intelligence (ICUSAI), Xi'an, China, 22–24 November 2019; pp. 13–18. [CrossRef]
13. Li, Y.; Ge, S.; Dai, S.; Zhao, L.; Yan, X.; Zheng, Y.; Shi, Y. Kinematic Modeling of a Combined System of Multiple Mecanum-Wheeled Robots with Velocity Compensation. *Sensors* **2020**, *20*, 75. [CrossRef]
14. Savaee, E.; Hanzaki, A.R. A New Algorithm for Calibration of an Omni-Directional Wheeled Mobile Robot Based on Effective Kinematic Parameters Estimation. *J. Intell. Robot. Syst.* **2021**, *101*, 28. [CrossRef]
15. Bożek, A. Discovering Stick-Slip-Resistant Servo Control Algorithm Using Genetic Programming. *Sensors* **2022**, *22*, 383. [CrossRef]
16. Prados Sesmero, C.; Buonocore, L.R.; Di Castro, M. Omnidirectional Robotic Platform for Surveillance of Particle Accelerator Environments with Limited Space Areas. *Appl. Sci.* **2021**, *11*, 6631. [CrossRef]
17. Palacín, J.; Rubies, E.; Clotet, E. Systematic Odometry Error Evaluation and Correction in a Human-Sized Three-Wheeled Omnidirectional Mobile Robot Using Flower-Shaped Calibration Trajectories. *Appl. Sci.* **2022**, *12*, 2606. [CrossRef]
18. Sharma, H.P.; Pant, M.; Agarwal, R.; Karatangi, S.V. Self-directed Robot for Car Driving Using Genetic Algorithm. In *Technology Innovation in Mechanical Engineering. Lecture Notes in Mechanical Engineering*; Chaurasiya, P.K., Singh, A., Verma, T.N., Rajak, U., Eds.; Springer: Singapore, 2022. [CrossRef]
19. Tagliani, F.L.; Pellegrini, N.; Aggogeri, F. Machine Learning Sequential Methodology for Robot Inverse Kinematic Modelling. *Appl. Sci.* **2022**, *12*, 9417. [CrossRef]
20. Zhu, Z.; Liu, Y.; He, Y.; Wu, W.; Wang, H.; Huang, C.; Ye, B. Fuzzy PID Control of the Three-Degree-of-Freedom Parallel Mechanism Based on Genetic Algorithm. *Appl. Sci.* **2022**, *12*, 11128. [CrossRef]
21. Cardoza Plata, J.E.; Olguín Carbajal, M.; Herrera Lozada, J.C.; Sandoval Gutierrez, J.; Rivera Zarate, I.; Serrano Talamantes, J.F. Simulation and Implementation of a Mobile Robot Trajectory Planning Solution by Using a Genetic Micro-Algorithm. *Appl. Sci.* **2022**, *12*, 11284. [CrossRef]

22. Rahmaniar, W.; Rakhmania, A. Mobile Robot Path Planning in a Trajectory with Multiple Obstacles Using Genetic Algorithms. *J. Robot. Control.* **2021**, *3*, 1–7. [CrossRef]
23. Batlle, J.A.; Font-Llagunes, J.M.; Barjau, A. Calibration for mobile robots with an invariant Jacobian, Robot. *Auton. Syst.* **2010**, *58*, 10–15. [CrossRef]
24. Clotet, E.; Martínez, D.; Moreno, J.; Tresanchez, M.; Palacín, J. Assistant Personal Robot (APR): Conception and Application of a Tele-Operated Assisted Living Robot. *Sensors* **2016**, *16*, 610. [CrossRef]
25. Palacín, J.; Clotet, E.; Martínez, D.; Martínez, D.; Moreno, J. Extending the Application of an Assistant Personal Robot as a Walk-Helper Tool. *Robotics* **2019**, *8*, 27. [CrossRef]
26. Palacín, J.; Martínez, D.; Clotet, E.; Pallejà, T.; Burgués, J.; Fonollosa, J.; Pardo, A.; Marco, S. Application of an Array of Metal-Oxide Semiconductor Gas Sensors in an Assistant Personal Robot for Early Gas Leak Detection. *Sensors* **2019**, *19*, 1957. [CrossRef]
27. Penteridis, L.; D'Onofrio, G.; Sancarlo, D.; Giuliani, F.; Ricciardi, F.; Cavallo, F.; Greco, A.; Trochidis, I.; Gkiokas, A. Robotic and Sensor Technologies for Mobility in Older People. *Rejuvenation Res.* **2017**, *20*, 401–410. [CrossRef] [PubMed]
28. Palacín, J.; Martínez, D.; Rubies, E.; Clotet, E. Suboptimal Omnidirectional Wheel Design and Implementation. *Sensors* **2021**, *21*, 865. [CrossRef]
29. Bitriá, R.; Palacín, J. Optimal PID Control of a Brushed DC Motor with an Embedded Low-Cost Magnetic Quadrature Encoder for Improved Step Overshoot and Undershoot Responses in a Mobile Robot Application. *Sensors* **2022**, *22*, 7817. [CrossRef]
30. Palacín, J.; Rubies, E.; Clotet, E.; Martínez, D. Evaluation of the Path-Tracking Accuracy of a Three-Wheeled Omnidirectional Mobile Robot Designed as a Personal Assistant. *Sensors* **2021**, *21*, 7216. [CrossRef]
31. Lluvia, I.; Lazkano, E.; Ansuategi, A. Active Mapping and Robot Exploration: A Survey. *Sensors* **2021**, *21*, 2445. [CrossRef] [PubMed]
32. Palacín, J.; Martínez, D.; Rubies, E.; Clotet, E. Mobile Robot Self-Localization with 2D Push-Broom LIDAR in a 2D Map. *Sensors* **2020**, *20*, 2500. [CrossRef] [PubMed]
33. Popovici, A.-T.; Dosoftei, C.-C.; Budaciu, C. Kinematics Calibration and Validation Approach Using Indoor Positioning System for an Omnidirectional Mobile Robot. *Sensors* **2022**, *22*, 8590. [CrossRef]
34. Mora, A.; Prados, A.; Mendez, A.; Barber, R.; Garrido, S. Sensor Fusion for Social Navigation on a Mobile Robot Based on Fast Marching Square and Gaussian Mixture Model. *Sensors* **2022**, *22*, 8728. [CrossRef]
35. Fraser, A. Simulation of genetic systems by automatic digital computers. I. Introduction. *Aust. J. Biol. Sci.* **1957**, *10*, 484–491. [CrossRef]
36. George, P.H. Styan, Hadamard products and multivariate statistical analysis. *Linear Algebra Its Appl.* **1973**, *6*, 217–240. [CrossRef]
37. Boanta, C.; Brișan, C. Estimation of the Kinematics and Workspace of a Robot Using Artificial Neural Networks. *Sensors* **2022**, *22*, 8356. [CrossRef] [PubMed]

applied sciences

MDPI

Article

Three-Dimensional Rendezvous Controls of Multiple Robots with Amplitude-Only Measurements in Cluttered Underwater Environments

Jonghoek Kim

System Engineering Department, Sejong University, Seoul 05006, Republic of Korea; jonghoek@gmail.com

Abstract: This study addresses multi-robot distributed rendezvous controls in cluttered underwater environments with many unknown obstacles. In underwater environments, a Unmanned Underwater Vehicle (UUV) cannot localize itself, since a Global Positioning System (GPS) is not available. Assume that each UUV has multiple signal intensity sensors surrounding it. Multiple intensity sensors on a UUV can only measure the amplitude of signals generated from its neighbor UUVs. We prove that multiple UUVs with bounded speed converge to a designated rendezvous point, while maintaining the connectivity of the communication network. This study further discusses a fault detection method, which detects faulty UUVs based on local sensing measurements. In addition, the proposed rendezvous control is adaptive to communication link failure or invisible UUVs. Note that communication link failure or invisible UUVs can happen due to unknown obstacles in the workspace. As far as we know, our study is novel in developing 3D coordinate-free distributed rendezvous control, considering underwater robots that can only measure the amplitude of signals emitted from neighboring robots. The proposed rendezvous algorithms are provably complete, and the effectiveness of the proposed rendezvous algorithms is demonstrated under MATLAB simulations.

Keywords: distributed rendezvous control; unmanned underwater vehicle; signal amplitude; faulty robot; network connectivity; three dimensional environments; fault tolerant control; underwater robot

Citation: Kim, J. Three-Dimensional Rendezvous Controls of Multiple Robots with Amplitude-Only Measurements in Cluttered Underwater Environments. *Appl. Sci.* **2023**, *13*, 4130. https://doi.org/10.3390/app13074130

Academic Editor: Alessandro Chiolerio

Received: 2 March 2023
Revised: 16 March 2023
Accepted: 23 March 2023
Published: 24 March 2023

1. Introduction

Networked robots have various applications, such as monitoring large environments, rescue missions, and target chasing [1–4]. Multiple coordinated robots are also used to perform many specific tasks, such as rendezvous [5–7], spacecraft docking [8], environmental monitoring [9–11], underwater target chasing [12], and formation control [13–15].

This study addresses multi-robot rendezvous controls in cluttered underwater environments with many unknown obstacles. Unmanned Underwater Vehicles (UUVs) can be divided into two categories based on whether their bodies are streamlined. The UUV's shape is determined by the requirements of the application. For example, a streamlined shape reduces water resistance and is preferable if the UUV is required to move at high speeds. However, if underwater detection or operation tasks are the primary roles of a UUV, a non-streamlined shape is often preferred.

Because of the good water pressure resistance of spherical objects, spherical UUVs can perform rotational motions with a 0 degree turn radius. In the literature, various spherical UUVs have been developed [16–20]. References [16,18–21] addressed a spherical UUV with hybrid propulsion devices including vectored water jet and propeller thrusters. The three Degree-of-Freedom (DoF) motions, including surging, heaving, and yawing, were performed in a swimming pool. References [16,19,21] further demonstrated that by adopting vectored water jets, a spherical UUV can be made to maneuver freely in any direction. Considering a spherical UUV, [22] used fuzzy proportional–integral–derivative (PID) controllers to independently control the robot's movement in all directions. Since spherical UUVs are highly maneuverable, our study considers a spherical UUV [16–20] as our platform.

We address practical application scenarios where multi-robot rendezvous controls are used. Recently, several studies [23–25] handled distributed formation control of multiple robots in environments with obstacles. For instance, a group of robots can be used for tracking a target while measuring the target's signal [26]. During the robots' maneuvering in formation controls, robots need to preserve the network connectivity. In order to make all robots gather close to each other, distributed rendezvous controls can be applied occasionally. In other words, distributed rendezvous algorithms can be used as a multi-robot control module to allow a team of robots to maintain network connectivity while they maneuver. For example, we can run rendezvous controls occasionally, so that all robots get closer to a root robot which leads the formation. Moreover, all robots need to get closer to the root robot, in the case where they move through a narrow tunnel.

Furthermore, rendezvous algorithms can be utilized in multi-robot collection or charging scenarios [27]. Suppose that there is a charging station close to a robot, called the root. All the robots can be charged from the charging station, once they gather close to the root.

The difficulty of the rendezvous problem decreases when a robot is equipped with Global Positioning Systems (GPS). However, GPS is unavailable in underwater environments. In difficult scenarios without GPS, robots cannot move toward a designated rendezvous point directly. Thus, we introduce a distributed rendezvous approach that does not require global localization.

A rendezvous algorithm is considered distributed, when it depends on the local interaction between neighboring robots. A distributed rendezvous algorithm is practical, since a robot cannot communicate with another robot that is too far away. Furthermore, obstacles in cluttered 3D environments can easily block the communication links between robots. In this study, we propose distributed rendezvous algorithms for UUVs in 3D unknown environments with many obstacles. Here, unknown obstacles can lead to communication link failure or invisible robots. The proposed rendezvous control is adaptive to communication link failure or invisible robots.

Considering 3D underwater environments, this study addresses the problem of enabling a network of UUVs to rendezvous at a designated root robot. Any robot in the network can be set as the root robot. Thus, the rendezvous system in this study is referred to as the any-robot rendezvous system.

The rendezvous controls in this study consider a UUV that can only measure the signal intensity emitted from its neighbor UUV. Thus, the proposed rendezvous control is suitable for a cheap UUV, which only has sensors for measuring the signal strength.

Reference [28] addressed a received signal strength (RSS) sensor model for underwater sound propagation. We consider multiple UUVs such that each UUV has multiple signal intensity sensors surrounding it. Each intensity sensor can measure the RSS of sound generated from a neighbor UUV [28].

Multiple intensity sensors on a UUV can only measure the amplitude (intensity) of signals emitted from its neighboring UUVs. To enable a distributed rendezvous based on amplitude-only measurements, we utilize a method of making a UUV approach the source of the signal (neighbor UUV) by measuring the intensity of signal at multiple intensity sensors.

We demonstrate that our distributed rendezvous algorithms are provably complete in achieving multi-robot rendezvous in cluttered 3D environments. The proposed 3D rendezvous controls assure the convergence to a designated UUV, while maintaining the connectivity of the time-varying and position-dependent communication network.

Moreover, the proposed rendezvous controls can handle the case where some UUVs are broken. While controlling the networked system, any UUV, including the root UUV, can become faulty owing to various reasons, such as hardware malfunction. This study, thus, discusses a fault detection method, which detects faulty UUVs based on local sensing measurements. Once a UUV failure is sensed, then we update the network structure, so that healthy UUVs without faults can be controlled effectively.

This study handles scenarios where multiple UUVs are deployed in cluttered 3D environments (e.g., [29]). In obstacle-rich environments, the communication (interaction) link between UUVs may be easily blocked owing to obstacles. Furthermore, invisible UUVs may occur, since line-of-sight can be blocked by unknown obstacles. In such scenarios, it is important to assure that the communication link is preserved, while every UUV maneuvers. The proposed distributed rendezvous control is unique in overcoming communication link failure or invisible UUVs. Note that communication link failure or invisible UUVs may happen due to unknown obstacles in cluttered 3D environments.

We address a distributed rendezvous control for spherical UUVs with bounded speed, such that each UUV can only measure the strength of signals emitted from neighboring UUVs. The contributions of our study are summarized as follows.

1. As far as we know, this study is novel in developing 3D distributed rendezvous controls, considering a UUV that can only measure the strength of signals emitted from neighboring UUVs.
2. The proposed 3D rendezvous controls are provably complete, since we prove the rendezvous to the root UUV, while maintaining the connectivity of the time-varying and position-dependent communication network.
3. To the best of our knowledge, our study is unique in addressing a fault detection method that detects faulty UUVs based on local sensing measurements. In addition, the proposed rendezvous controls are adaptive to communication link failure or invisible UUVs.

The study is organized as follows. Section 2 provides the literature review of this study. Section 3 provides the background information of this study. Section 4 discusses the 3D distributed rendezvous controls introduced in this study. Section 5 provides the simulation results of this study. Section 6 provides the conclusions of this study.

2. Literature Review

Considering 2D environments, there are many studies on distributed controllers to make all robots rendezvous, considering the case where a robot measures the relative position of its nearby robot [7,30–32]. In [33], circumcenter-based consensus algorithms were introduced to achieve distributed rendezvous of multiple robots. However, how to handle faulty robots was not discussed in [33]. Considering 2D environments, [34] introduced rendezvous controllers that are tolerant to faulty robots. However, this method needs a controllable sensing range, which may not be feasible in practice. The authors of [35] considered the optimal consensus problem of asynchronous sampling single-integrator and double-integrator multi-robot systems utilizing distributed model predictive control algorithms.

Event-driven rendezvous strategies [36–38] for multi-agent systems were motivated by the use of embedded microprocessors with limited resources that will gather information and actuate the individual agent controller updates. Considering event-triggered rendezvous controls, [38] proved that if the communication graph is connected, consensus is achieved exponentially. The authors of [37] showed that with appropriate control gains in event-triggering conditions, subsystems employing discrete-time signals from neighbors achieve the state consensus. The authors of [36] studied the rendezvous problem of multi-robot systems by parallel event-triggered connectivity-preserving controls. The event-triggered control laws in [36] can assure the system convergence, and can maintain the connectivity of the time-varying and position-dependent communication network. However, [36–38] considered 2D environments without obstacles. In cluttered underwater environments, obstacles can block the communication link between robots, resulting in the loss of a robot.

Our study handles practical scenarios where multiple underwater robots are deployed in cluttered environments (e.g., [29]). In obstacle-rich environments, the communication (interaction) link between robots may be blocked owing to obstacles. In such scenarios, it is important to assure that the communication link is preserved while every robot maneuvers. As far as we know, the proposed distributed rendezvous control is unique in overcoming communication link failure or invisible robots.

As far as we know, other studies on distributed rendezvous controls [7,30–32,36–38] assumed that a robot can measure the relative position of its neighbor robot using local sensors, such as laser. However, no physical sensors were clearly described for this local interaction. In other words, other rendezvous controls in [7,30–32,36–38] were designed, while the specific sensors that were used for the controls were not considered. However, considering specific sensors is crucial, since robots move based on local sensing measurements.

Considering 2D environments, [39] addressed rendezvous algorithms using received signal strength indicator (RSSI) data from the radio. Reference [40] addressed a simple rendezvous algorithm using RSSI data from the radio. Reference [40] used a small, low-cost, modular robotics platform in 2D environments. Reference [40] explored the potential for using RSSI in platforms equipped with radios in order to rendezvous at a desired location or agent. The authors of [39] addressed multi-robot rendezvous with range-only measurements. However, references [39,40] considered simple 2D environments with no obstacles. In addition, [39,40] did not show that their rendezvous algorithm is provably complete in achieving multi-robot rendezvous. Moreover, [39,40] did not consider how to handle the case where a UUV is broken.

In [41], the authors applied radars with a variable sensing range in order to make all robots rendezvous in 3D environments. However, obstacle environments were not considered in [41]. Unknown obstacles can block the communication link between robots in practice. Moreover, in underwater environments, electromagnetic signals easily dissipate, and thus, radars are not suitable in underwater environments.

In 3D underwater environments, a UUV needs to be equipped with expensive 3D sonar sensors with sensing arrays, in order to measure the relative position of its neighbor UUV. However, the intensity sampling sensors used in our study are not sufficient for measuring the relative position of a neighbor UUV. In other words, intensity sensors used in our study cannot be used for measuring the relative position of a neighbor UUV.

To the best of our knowledge, our study is novel in developing provably complete 3D distributed rendezvous controls, considering a UUV which can only measure the strength of signals emitted from neighboring UUVs. Moreover, our study is unique in introducing a 3D rendezvous control, which is robust to intensity sensor failures or UUV failures.

3. Background Information

3.1. Reference Frames

Two reference frames are used in our study: an inertial reference frame $\{I\}$ and a body−fixed frame $\{B\}$ [42]. The origin of $\{I\}$ is an appropriate position with three axes pointing north, east, and down, respectively. The body−fixed frame $\{B\}$ is fixed to a UUV's body and acts as the moving frame. The origin of $\{B\}$ is fixed at the UUV's center.

Let ϕ, θ, ψ present the euler roll, euler pitch, and euler yaw, respectively. The counter-clockwise (CC) rotation of ψ centered at the z-axis in $\{B\}$ is as follows:

$$\mathbf{R}(\psi) = \begin{pmatrix} \cos(\psi) & -\sin(\psi) & 0 \\ \sin(\psi) & \cos(\psi) & 0 \\ 0 & 0 & 1 \end{pmatrix}. \tag{1}$$

The CC rotation of θ centered at the y-axis in $\{B\}$ is as follows:

$$\mathbf{R}(\theta) = \begin{pmatrix} \cos(\theta) & 0 & \sin(\theta) \\ 0 & 1 & 0 \\ -\sin(\theta) & 0 & \cos(\theta) \end{pmatrix}. \tag{2}$$

The CC rotation of ϕ centered at the x-axis in $\{B\}$ is as follows:

We consider the case where all sensors of a UUV are identical to those of another UUV. Let $P(u_i) \in \mathbb{R}$ denote the average intensity measurement associated with u_i. We have

$$P(u_i) = \frac{1}{6}\sum_{r=1}^{6} P(\mathbf{c}_i^r). \tag{6}$$

It is assumed that P_0, E_p, and σ^P in Equation (5) are known a priori. This is feasible using experiments with UUVs [28]. For realistic simulations of the RSSI for an underwater emitter, we use the model parameters in [28]. In Equation (5), P_0 = 100 dB, $E_p = 2$, $\gamma = 0.05$, $\sigma^P = 1$ are used, according to [28]. In this case, Figure 4 shows the relationship between d_I and $P(u_i)$ using (5) and (6). As d_I increases, $P(u_i)$ decreases.

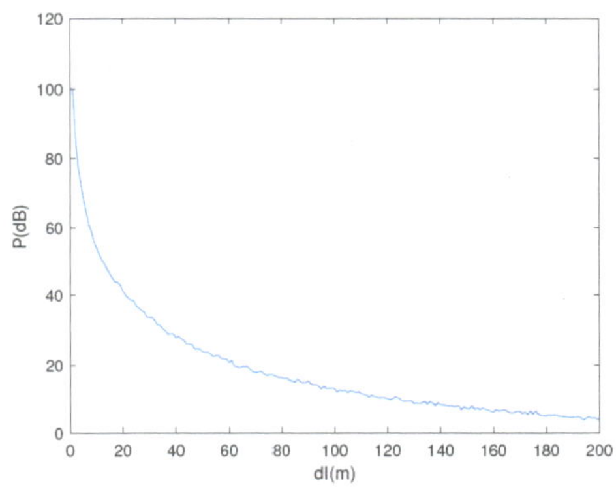

Figure 4. The relationship between d_I and $P(u_i)$ using Equations (5) and (6). As d_I increases, $P(u_i)$ decreases.

It is assumed that if $P(u_i) > P_{thres}$, then SNR is sufficiently large and the intensity sensor can detect the LOS signal generated from the emitter. We say that a UUV u_i is a neighbor to UUV u_j if $P(u_i) > P_{thres}$ and $P(u_j) > P_{thres}$. Here, P_{thres} is a tuning parameter for determining a neighbor.

The neighbor information can be set by mutual communication between u_i and u_j. In the case where u_i and u_j are neighbors, $\|\mathbf{u}_i - \mathbf{u}_j\| < r_s(P_{thres})$ and the LOS between u_i and u_j is not blocked by obstacles. Here, $r_s(P_{thres})$ can be considered as the maximum sensing range for a UUV. Note that $r_s(P_{thres})$ is determined by P_{thres}. As P_{thres} decreases, $r_s(P_{thres})$ increases using (5).

For notation convenience, we use r_s instead of $r_s(P_{thres})$. See Figure 4 for an illustration. If we set $P_{thres} = 52$ dB, then $r_s = 10$ m; if we set $P_{thres} = 20$ dB, then $r_s = 60$ m.

We say that a UUV u_i encounters u_j if $\|\mathbf{u}_i - \mathbf{u}_j\| < \epsilon \approx 0$. Let P^e denote the signal power measured when u_i encounters u_j. Thus, using Equation (5), we obtain the following:

$$P^e = P_0 - 10E_p log_{10}(\epsilon) + \gamma(\epsilon - 1). \tag{7}$$

If both $P(u_i)$ and $P(u_j)$ exceed P^e, then we assume that u_i encountered u_j.

Let $G = (V(G), E(G))$ indicate the graph presenting the networked system. Every node in $V(G)$ indicates a UUV. An edge, e.g., $\{u_i, u_j\} \in E(G)$, indicates that u_i and u_j are neighboring UUVs. Recall that u_i is a neighbor to u_j if $P(u_i) > P_{thres}$ and $P(u_j) > P_{thres}$. Since we check both $P(u_i)$ and $P(u_j)$ for detecting neighbors, G is an undirected graph.

Let $G_0 = (V_0, E_0)$ indicate the initial connectivity network (at sampling step $t = 0$). Without loss of generality, it is assumed that G_0 is connected. As far as we know, this initial connectivity assumption is required for any distributed rendezvous controls in the literature [34,41,51,52]. This is due to the fact that distributed rendezvous controls are based on local interaction between neighboring agents.

Note that the initial connectivity assumption does not assure that the network connectivity is maintained during the maneuver of an agent. As agents move around, the communication link among them may be broken due to obstacles in the environments. Moreover, an agent cannot communicate with another agent if the relative distance between them is larger than r_s. Thus, we need to develop distributed rendezvous controls, which assure that the network connectivity is maintained while all agents rendezvous at u_N.

In the inertial reference frame, let $\mathbf{h}_i(k) \in \mathbb{R}^3$ present the heading direction of u_i at sampling step k. Note that $\mathbf{h}_i(k) \in \mathbb{R}^3$ is a unit vector. Let dt indicate the sampling interval in discrete-time systems. As a UUV's dynamic model in the inertial reference frame, we utilize the following equation:

$$\mathbf{u}_i(k+1) = \mathbf{u}_i(k) + \mathbf{h}_i(k) * dt * s_i(k). \tag{8}$$

This simple motion model is commonly utilized in the literature on multi-agent systems [51,53–60]. In Equation (8), $s_i(k)$ denotes the linear speed of u_i at sampling step k. We assume that $s_i(k) \leq s^{max}$ for all k. This implies that s^{max} is the speed limit of every UUV.

This study considers a spherical UUV. The authors of [16,19,21] showed that by adopting vectored water jets, a spherical UUV can maneuver freely in any direction. The control of a spherical UUV appeared in [22]. The reference [22] proposed a decoupling motion control algorithm based on the robot attitude calculation for an underwater spherical robot. The reference [22] used fuzzy PID controllers to independently control the robot's movement in all directions. Since a spherical UUV is highly maneuverable, the process model in Equation (8) is feasible.

4. Robust Distributed Rendezvous Control

4.1. Distributed Breadth First Search (BFS) Algorithm to Generate a Spanning Tree T

In our study, u_N is set as the root of a tree graph T. In order to generate a tree T rooted at u_N, we use a distributed Breadth First Search (BFS) algorithm in [61]. We acknowledge that [61] addressed a distributed BFS algorithm, so that a node in the network can guide a moving object across the network to the goal. Algorithm 2 in [61] can be applied to make a tree T rooted at u_N. The tree T is rooted at u_N, and T has a unique path from the UUV u_i to u_N.

Algorithm 1 shows a distributed BFS algorithm to generate a tree T containing all UUVs. The goal sensor in Algorithm 2 of [61] represents the root u_N in Algorithm 1. Initially, every UUV u contains $hops_g(u)$, which indicates the hop distance to the root. The root u_N sets $hops_g(u_N) = 0$ initially. For every UUV except for the root, we initially set $hops_g(u) = \infty$, where $u \neq u_N$. Here, $hops_g(u) = \infty$ implies that one has not set the hop distance for $u \neq u_N$.

Algorithm 1 Distributed BFS algorithm to generate a spanning tree T

Every UUV u contains $hops_g(u)$, which indicates the hop distance to the root;
The root u_N sets $hops_g(u_N) = 0$ initially;
We initially set $hops_g(u) = \infty$, where $u \neq u_N$;
Initially, u_N sends its hop distance information $hops_g(u_N)$ to its neighbor UUVs;
repeat
  $u \leftarrow$ every UUV;
  if the UUV u satisfies $hops_g(u) = \infty$, and it receives a hop distance message from its neighbor UUV, e.g., n **then**
    The UUV u updates its hop distance information using (9);
    The UUV u broadcasts $hops_g(u)$ to its neighbors;
  end if
until $hops_g(u) \neq \infty$ for all u;

Initially, u_N sends its hop distance information $hops_g(u_N)$ to its neighbor UUVs. Suppose a UUV u satisfies $hops_g(u) = \infty$, and it receives a hop distance message from its neighbor UUV, e.g., n. Then, u updates its hop distance information using the following equation:

$$hops_g(u) = min(hops_g(u), hops_g(n) + 1). \tag{9}$$

Using Equation (9), $hops_g(u)$ can be updated to $hops_g(n) + 1$. In this case, the parent of u is updated to n. Thereafter, u broadcasts $hops_g(u)$ to its neighbors. In Equation (9), $min(a, b)$ returns a smaller value between a and b.

References [61,62] proved that the number of message broadcasted by every UUV is 1 in this distributed BFS algorithm. This implies that the distributed BFS has the computational complexity $O(1)$.

Utilization of Signal Intensity Sensors to Detect Neighbors

Note that in Algorithm 1, every UUV u_i utilizes its signal intensity sensors to detect its neighbors. Recall that u_i is a neighbor to u_j if $P(u_i) > P_{thres}$ and $P(u_j) > P_{thres}$. Based on the signal intensity measurements, u_i finds its neighbors, e.g., $\mathtt{N}_i$. For instance, if u_i detects two UUVs u_j and u_l, then we set $\mathtt{N}_i = [u_j, u_l]$. By making a UUV u_i stand still while measuring the signals from neighbor UUVs, u_i can determine $\mathtt{N}_i$.

The obstacle environment is not known in advance. In order to build T in unknown obstacle environments, every UUV u_i utilizes its signal intensity sensors to detect its neighbors. As every UUV u_i turns on its signal intensity sensors while standing still, u_i can detect its neighbors.

For the detection of neighbors, we assume that a UUV, e.g., u_j, can identify another UUV, e.g., u_l, by analyzing the signal emitted from u_l. Each UUV emits signal using Binary Phase Shift Keying (BPSK) with a distinct frequency band. Suppose that every UUV shares the frequency band information of all UUVs. Suppose that a UUV, e.g., u_j, receives the signal from another UUV, e.g., u_l. Then, u_j runs a bandpass filter to analyze the frequency of the signal. By running the bandpass filter, u_j can detect that the signal was generated from u_l. In this way, a UUV can distinguish the signal of a UUV from that of another UUV. Moreover, the bandpass filter in a UUV can be used to filter out signal interference, such as a signal generated from unknown transmitters.

We acknowledge that there exists a serious delay of underwater communication. This implies that u_j cannot detect its neighbors instantly. A UUV u_j needs to emit signals and receive signals from its neighbors. However, this delay does not cause problems, since no UUV moves while we generate a spanning tree T using Algorithm 1.

4.2. Distributed Rendezvous Algorithms

While all UUVs stop, Algorithm 1 in Section 4.1 runs to build a spanning tree T in a distributed manner. Based on the tree T, Algorithm 2 runs to achieve rendezvous at u_N. In other words, Algorithm 2 makes every UUV encounter at the root u_N.

We explain Algorithm 2. Initially ($t = 0$), all leaf UUVs begin visiting every UUV along the path to u_N. Each leaf UUV finds a path to u_N using T. Since T is a tree graph, only one path exists from a node to the root in T. Once the path is found, each UUV stores the UUV indexes along the path to u_N. How to make a UUV visit the UUVs along the path to the root u_N is discussed in Section 4.4.

In order to maintain network connectivity, the maneuver of a UUV must not disconnect the network. Therefore, each UUV does not begin moving until it encounters all its descendants in T. This implies that each UUV needs to store the UUV indexes associated with its all descendants.

Let us consider a UUV u' with at least one child. From u' to u_N in T, only one path exists. As soon as u' encounters all its descendants, u' starts visiting every UUV along the path to u_N under Algorithm 2. As time elapses, $p(u')$ encounters all its descendants and $p(u')$ starts visiting every UUV along the path to u_N. This procedure continues until all UUVs encounter at u_N.

Algorithm 2 Distributed rendezvous strategy

While all UUVs stop, Algorithm 1 in Section 4.1 runs to build the tree T rooted at u_N;
repeat
  $u \leftarrow$ every UUV;
  if u is a leaf in T **then**
    The UUV u finds a path to u_N using T;
    The UUV u starts visiting every UUV along the path;
  else if u is not a leaf in T **then**
    if u encounters all its descendants **then**
      The UUV u finds a path to u_N using T;
      The UUV u starts visiting every UUV along the path;
    end if
  end if
  if a UUV is broken or interaction link between two neighboring UUVs in T is broken by moving obstacles **then**
    All UUVs stop moving, and re-build a tree T by re-running Algorithm 1 in Section 4.1;
  end if
  if an invisible UUV blocked by unknown obstacles appears suddenly **then**
    All UUVs stop moving, and re-build a tree T by re-running Algorithm 1 in Section 4.1;
  end if
until every UUV encounters u_N;

Whether a UUV encounters another UUV can be detected utilizing signal power measurements. Recall that if both $P(u_i)$ and $P(u_j)$ exceed P^e, then we assume that u_i encountered u_j.

As long as we consider static obstacles, the connectivity of the path to the root u_N is not broken by obstacles. In other words, as long as obstacles have not moved after the initial tree generation, the maneuver of a UUV along the path to the root is not blocked by obstacles.

Theorem 1 proves that Algorithm 2 is distributed. Furthermore, Theorem 1 proves that network connectivity is preserved while a UUV maneuvers until reaching the root. This implies that under Algorithm 2, every UUV reaches the root, while preserving network connectivity.

Theorem 1. *Algorithm 2 is distributed. Under Algorithm 2, every UUV reaches the root, while preserving network connectivity to the root.*

Proof. Suppose that u has been visiting UUVs along the path to u_N in Algorithm 2. Let $PATH$ indicate the path for convenience. Suppose that $PATH$ consists of a set of UUVs $p_1 \rightarrow p_2 \rightarrow p_3 \ldots \rightarrow p_{end}$ in this order. Here, p_{end} is the root u_N. As u moves along this path, it reaches the root in the end.

We first prove that p_i, where $i \in \{1, 2, \ldots, end-1\}$, starts moving only after u encounters p_i. A UUV, other than a leaf starts moving only after all its descendants in T encounter it. Any UUV on $PATH$ is an ancestor of u. Hence, p_i does not start moving before it encounters u.

In the case where u has just encountered p_i, u can sense p_{i+1} utilizing its local sensors. Since u moves based on local sensing measurements, Algorithm 2 is distributed.

As u moves along $PATH$, it reaches the root in the end. We next prove that u remains connected to the root, during its maneuver along $PATH$ to the root. Consider the case where u has just encountered p_i and starts moving towards p_{i+1}. In this case, u is connected to p_{i+1}. All UUVs in $p_{i+1} \rightarrow p_{i+2} \rightarrow p_{i+3} \ldots \rightarrow p_{end} = u_N$ stand still. Hence, p_{i+1} is connected to u_N. Since u is connected to p_{i+1}, u is also connected to u_N. It is proved that any UUV u remains connected to the root during its maneuver along $PATH$ to the root. □

4.3. Handling of Dynamic Scenarios, including Faulty UUVs

One may have a situation where a UUV is broken or the interaction link between two neighboring UUVs in T is broken by moving obstacles. Moreover, we may have a situation where an invisible UUV blocked by unknown obstacles appears suddenly. In these cases, every UUV stops moving, and we re-build a tree T utilizing Algorithm 1 in Section 4.1, so that network connectivity among UUVs can be re-established. Then, Algorithm 2 re-runs to achieve rendezvous at u_N. In this way, we can handle dynamic scenarios which may happen due to unknown obstacles.

As we re-build a tree T utilizing Algorithm 1 in Section 4.1, we may have a case where a tree T cannot be generated. This implies that a UUV may not be connected to the root, even after running Algorithm 1 in Section 4.1. For handling this case, we need to increase the number of neighbors for each UUV.

For handling the case where a UUV may not be connected to the root, we increase P_0 in Equation (5) for each UUV. As P_0 increases, the number of neighbors can increase, since a UUV u_i is a neighbor to u_j if $P(u_i) > P_{thres}$ and $P(u_j) > P_{thres}$. By increasing the signal emission strength, u_i can detect a UUV that is located far from u_i. This emission power increase was also used in [41], for improving the network connectivity in multi-agent systems. We acknowledge that by increasing the signal emission strength, u_i consumes more power.

4.3.1. Discard Faulty UUVs

Note that a UUV can become faulty owing to various reasons, such as hardware malfunction. We next address how to handle faulty UUVs. Let a healthy UUV denote a UUV having no faults.

Every UUV is encountered by its child under Algorithm 2. Thus, every faulty UUV, including a faulty root, can be detected by its child. Once a healthy UUV meets with a faulty UUV, the healthy UUV sends signal to the faulty UUV. A healthy UUV responds with an acknowledge signal whenever it receives a signal. If a UUV does not respond to the received signal, then the healthy UUV can find that a fault has occurred in the responding UUV.

Algorithm 2 is further designed to cope with a faulty UUV (including a faulty root). We discuss how to handle faulty UUVs from now on.

We first introduce a method of discarding (dropping) faulty UUVs once they are sensed. Once a broken UUV, e.g., u_B, is sensed, then a tree T is updated by applying Algorithm 1 in Section 4.1 without u_B. Thereafter, Algorithm 2 runs utilizing the updated T. The updated T does not contain u_B. This implies that we discarded (dropped) u_B.

If there are too many faulty UUVs, then it may be impossible to build a tree T without them. If we cannot build a tree without faulty UUVs, then we cannot make every healthy UUV encounter at the root UUV.

4.3.2. Using Static Faulty UUVs as Waypoints

There may be a case where a static faulty UUV has a communication ability, and thus, can emit a signal from it. We discuss a method which does not discard a static faulty UUV with communication ability. In this method, every UUV utilizes static faulty UUVs as "waypoints" along the path to the root UUV. Note that we do not discard faulty UUVs under this method.

Note that a healthy UUV, which does not have faults, can still measure the signal from a static faulty UUV. Assume that a healthy UUV, e.g., u, has a static faulty UUV, e.g., f, as its ancestor. In this case, the healthy UUV can still visit UUVs along the path, which contains f, to the root. This implies that u utilizes f as "waypoints" along the path to the root UUV.

Note that if f is healthy, then f waits until it encounters all its descendants. However, since f is faulty and static, f cannot move even after all its descendants encounter f. In this case, $p(f)$, the parent of f, cannot begin moving, since f is faulty and cannot move towards $p(f)$.

In order to resolve this problem, $p(f)$ removes f from its descendants list. In this way, $p(f)$ starts moving after all its descendants, other than f, encounter $p(f)$.

4.4. Visiting UUVs along the Path to the Root, Based on Signal Strength Measurements

At the beginning of Algorithm 2, Algorithm 1 in Section 4.1 runs to build a spanning tree T in a distributed manner. Algorithm 1 has the computational complexity $O(1)$ [61,62]. Once T is generated, the only control applied to each UUV is visiting UUVs along the path, e.g., $PATH$, to the root in T. This control is a high-level control of every UUV.

We introduce a local control to make one UUV visit UUVs along $PATH$ utilizing signal strength measurements. Consider the case where $PATH$ consists of a set of UUVs $p_1 \to p_2 \to p_3 \ldots \to p_{end}$ in this order. Here, p_{end} is the root. In the inertial reference frame, let $\mathbf{p}_j \in \mathbb{R}^3$ indicate the 3D position of p_j for convenience.

Suppose that u_i encountered $\mathbf{p}_{j-1} \in \mathbb{R}^3$ and that the next UUV to encounter is $\mathbf{p}_j \in \mathbb{R}^3$. Since u_i encountered $\mathbf{p}_{j-1}$, u_i can detect the signal emitted from $\mathbf{p}_j$ utilizing its intensity sensors. Each intensity sensor on u_i measures the strength of signals emitted from $\mathbf{p}_j$.

Signal field intensity is maximized at the signal source. See Equation (5) for RSSI. The gradient direction of a field defines the direction which maximizes the increase in the field. We let a UUV u_i move in the gradient direction of the field. Since the gradient is the direction representing the maximum increase of the field, this maneuver makes the UUV move towards the signal source.

The gradient direction is measured in the body−fixed frame of u_i. In the body−fixed frame of u_i, let $\mathbf{gr}_i(P) \in \mathbb{R}^3$ present the gradient of P at the center of u_i. Moving in the direction of $\mathbf{gr}_i(P)$ makes u_i move towards the signal source $\mathbf{p}_j$.

We next discuss how to derive $\mathbf{gr}_i(P)$ utilizing signal intensity measurements. The local coordinates of every intensity sensor are presented in Figure 3.

Assume that $\|\mathbf{c}_i^r\|$ is sufficiently small for all $r \in \{1, \ldots, 6\}$. Through Taylor expansion up to the first-order derivative, one obtains the following:

$$P(\mathbf{c}_i^r) = P(u_i) + \mathbf{c}_i^r * \mathbf{gr}_i(P), \tag{10}$$

which leads to:

$$P(\mathbf{c}_i^r) - P(u_i) = \mathbf{c}_i^r * \mathbf{gr}_i(P). \tag{11}$$

At each sampling step k, one obtains the following:

$$\mathbf{P}_S = [P_c^1, P_c^2, \ldots, P_c^6]. \tag{12}$$

Here, $P_c^r = P(\mathbf{c}_i^r) - P(u_i)$. Furthermore, we utilize the following:

$$\mathbf{C}_S = [\mathbf{c}_i^1; \mathbf{c}_i^2; \ldots \mathbf{c}_i^6]. \tag{13}$$

Utilizing (11)–(13), we derive the following:

$$\mathbf{P}_S^T = \mathbf{C}_S * \mathbf{gr}_i(P). \tag{14}$$

Thereafter, $\mathbf{gr}_i(P) \in \mathbb{R}^3$ is derived using the pseudo-inverse as follows:

$$\mathbf{gr}_i(P) = (\mathbf{C}_S^T * \mathbf{C}_S)^{-1} * \mathbf{C}_S * \mathbf{P}_S^T. \tag{15}$$

where $\mathbf{gr}_i(P) \in \mathbb{R}^3$ defines the gradient of P, measured in the the body−fixed frame of u_i. In order to make u_i head towards the signal source, the heading of u_i is set as the gradient direction. Since the gradient is the direction representing the maximum increase of the field, this maneuver makes the UUV move towards the source.

In Equation (8), the UUV's heading vector $\mathbf{h}_i(k) \in \mathbb{R}^3$ is defined in the inertial reference frame. Since $\mathbf{gr}_i(P)$ is defined in the body−fixed frame of u_i, one changes $\mathbf{gr}_i(P)$ into a gradient vector in the inertial reference frame.

Notice that $\mathbf{R}(\psi(k), \theta(k), \phi(k)) * \frac{\mathbf{gr}_i(P)}{\|\mathbf{gr}_i(P)\|}$ is a gradient vector in the inertial reference frame. In order to set the heading of u_i as the gradient direction, u_i sets its new heading using the following equation:

$$\mathbf{h}_i(k+1) = \mathbf{R}(\psi(k), \theta(k), \phi(k)) * \frac{\mathbf{gr}_i(P)}{\|\mathbf{gr}_i(P)\|}. \tag{16}$$

Note that a UUV does not have to measure its attitude ϕ, θ, ψ for moving towards the signal source. The UUV can move in the gradient direction $\mathbf{gr}_i(P)$ in its body−fixed frame. For instance, suppose that the gradient field $\mathbf{gr}_i(P)$ is estimated as [1,0,0] in the UUV's body−fixed frame. Using (16), the UUV's control command is generated for moving in the direction of [1,0,0] in its body−fixed frame.

We next discuss how u_i can detect the moment when it encounters $\mathbf{p}_j$. If $P(u_i)$ in Equation (6) exceeds P^e in Equation (7), then we assume that u_i encountered $\mathbf{p}_j$. Thus, u_i begins moving towards $\mathbf{p}_{j+1}$ if it exists.

For collision avoidance, we make u_i slow down as it gets closer to p_j. The UUV's linear speed $s_i(k)$ is set as follows. If $P(u_i)$ in Equation (6) is larger than P^e, then u_i begins moving towards $\mathbf{p}_{j+1}$. Otherwise, we set the following:

$$s_i(k) = min(s^{max}, \beta * (P^e - P(u_i))). \tag{17}$$

where $\beta > 0$ is a tuning parameter. In MATLAB simulations, we use $\beta = 10$. In Equation (17), $min(a, b)$ returns a smaller value between a and b.

When a UUV u_i is sufficiently close to $\mathbf{p}_j \in \mathbb{R}^3$, it begins moving towards the next UUV $\mathbf{p}_{j+1}$. In this way, u_i avoids collision with UUVs on its path to the root. In the case where u_i encounters the root, u_i stops moving.

This study considers a spherical UUV. The authors of [16,19–21] showed that by adopting vectored water jets, a spherical UUV can maneuver freely in any direction. The control of a spherical UUV appeared in [22], which used fuzzy PID controllers to independently control the robot's movement in all directions.

While u_i maneuvers, a moving obstacle may abruptly appear in practice. In this case, the UUV u_i can avoid collision using reactive collision avoidance controls. We acknowledge that any reactive control method can be applied for this evasion [63–65].

5. Simulations

The MATLAB R2014a simulator is utilized to demonstrate the effectiveness of the proposed rendezvous controllers. The system environment includes Window 10, Intel Core^{TM} i5-7600K CPU@3.80 GHz. The controllers are implemented in a discrete-time system, and the sampling time interval to discretize the UUV's velocity control (8) is 0.3 s.

We simulate a 3D underwater environment (200 × 200 × 200) with many obstacles. For realistic simulations of the RSSI for an underwater emitter, we use the model parameters used in [28]. In Equation (5), P_0 = 100 dB, $E_p = 2$, $\gamma = 0.05$, and $\sigma^P = 1$ are used, according to [28].

Recall that we said that a UUV u_i is a neighbor to UUV u_j if $P(u_i) > P_{thres}$ and $P(u_j) > P_{thres}$. We set $P_{thres} = 52$ dB. Using Figure 4, the associated maximum sensing range r_s is 10 m. In this way, the relative distance between two neighbor UUVs is less than 10 m.

The maximum speed of every UUV is set as $s^{max} = 2$ (m/s). Recall that if $P(u_i, p_j)$ and $P(u_j, p_i)$ exceeds P^e, then we assume that u_i encountered p_j. We set $P^e = 100$ (in dB), which is identical to P_0. Using (7), $P^e = 100$ (in dB) is associated with $\epsilon = 1$ m.

In this study, a UUV uses six intensity sensors. Figure 3 plots the local coordinates of every intensity sensors positioned on a UUV. dr is the relative distance between the first sensor and any other sensor. dr is a tuning parameter in our control.

If dr is too small, then intensity measurements of six sensors on a UUV are not distinct from each other. In this case, the gradient direction at the UUV position cannot be estimated accurately. If dr is too large, then we cannot apply the Taylor expansion in Equation (10). In the simulations, six intensity sensors are installed such that $dr = 1$ m.

To prove the robustness of the proposed methods, we run 100 Monte Carlo (MC) simulations, such that the initial network satisfies this initial connectivity assumption. As far as we know, this initial connectivity assumption is required for any distributed rendezvous controls in the literature [34,41,51,52].

At the beginning of each MC simulation, 50 UUVs are randomly deployed until the following two conditions are met:

1. No UUV is deployed inside an obstacle boundary.
2. The deployed UUVs satisfy the initial connectivity assumption.

Once these two conditions are satisfied, then 50 UUVs begin to move under Algorithm 2. Otherwise, we keep deploying 50 UUVs until the above two conditions are met.

Once these two conditions are satisfied, then 50 UUVs begin to maneuver under Algorithm 2. This maneuver indicates the beginning of a single MC simulation. In all MC simulations, rendezvous is achieved for every UUV.

Considering one MC simulation, Figure 5 shows the initial position of every UUV. The initial position of every UUV is plotted with a green circle. This figure shows the obstacle boundaries with spheres. We also plot the tree graph T (blue line segments) generated initially.

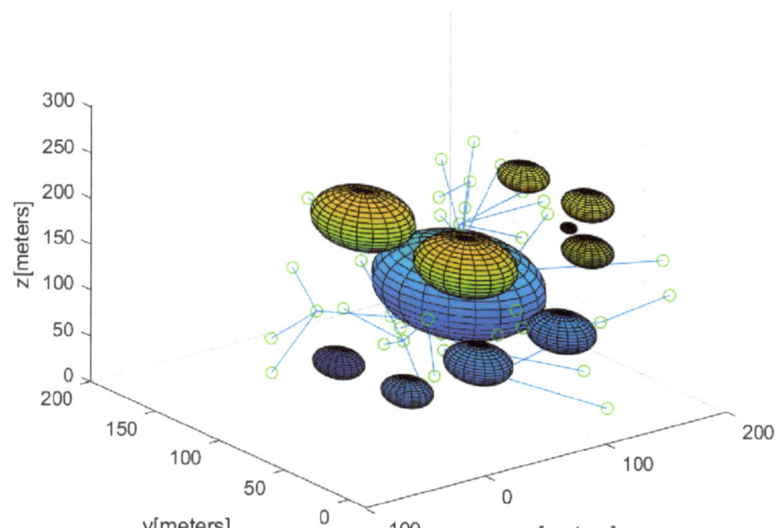

Figure 5. The initial position of every UUV is plotted with a green circle. We also plot the tree graph T (blue line segments) generated at time 0 (one MC simulation). We can see that the initial connectivity assumption is satisfied. Obstacles are plotted with spheres.

Figure 6 shows each UUV's maneuver until $t = 20$ s have passed. We also plot the tree graph T (blue line segments) generated at time 0. Every UUV's maneuver is illustrated as circles with a distinct color. After 148 s have passed, all UUVs encounter at the root UUV while avoiding collision with obstacles. Figure 7 shows each UUV's maneuver once the rendezvous is completed. Figure 8 shows the final position (green circle) of every UUV.

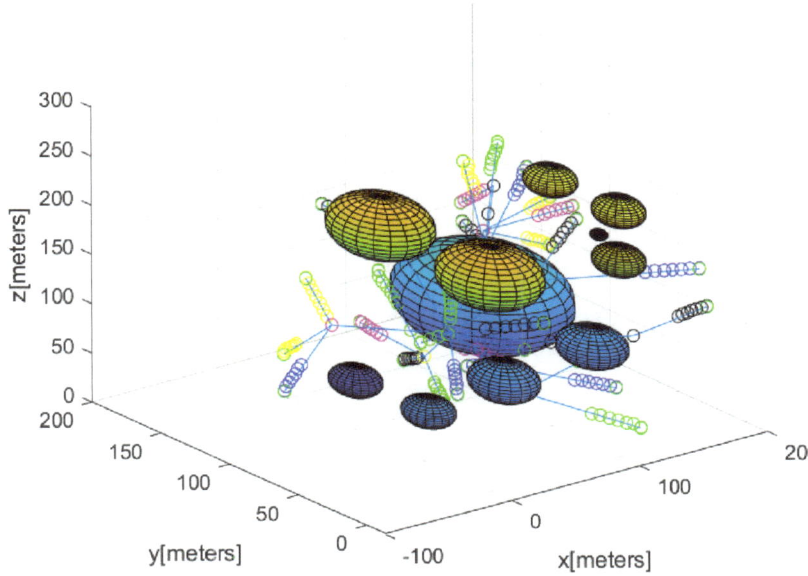

Figure 6. Every UUV's maneuver until $t = 20$ s has passed (one MC simulation). Every UUV's maneuver is illustrated as circles with a distinct color. We also plot the tree graph T (blue line segments) generated at time 0. Obstacles are plotted with spheres.

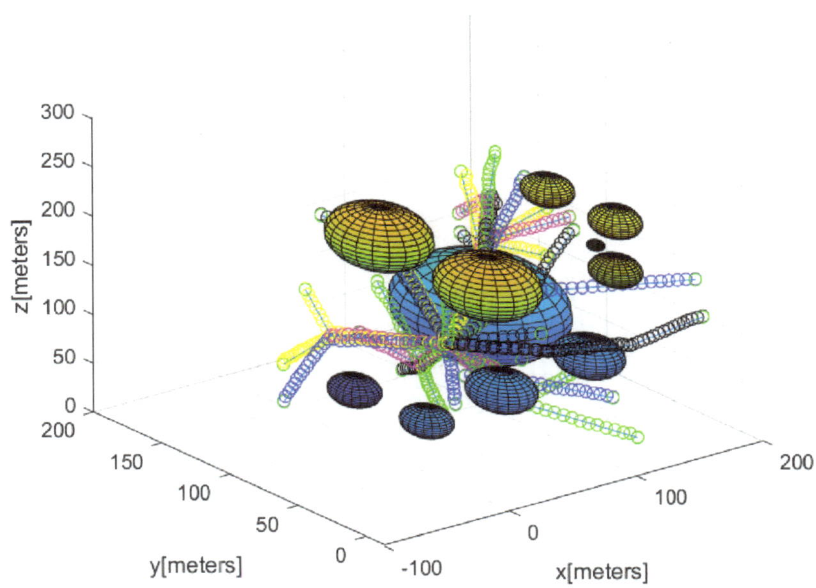

Figure 7. Every UUV's maneuver once the rendezvous is completed (one MC simulation). Every UUV's maneuver is illustrated as circles with a distinct color. Obstacles are plotted with spheres.

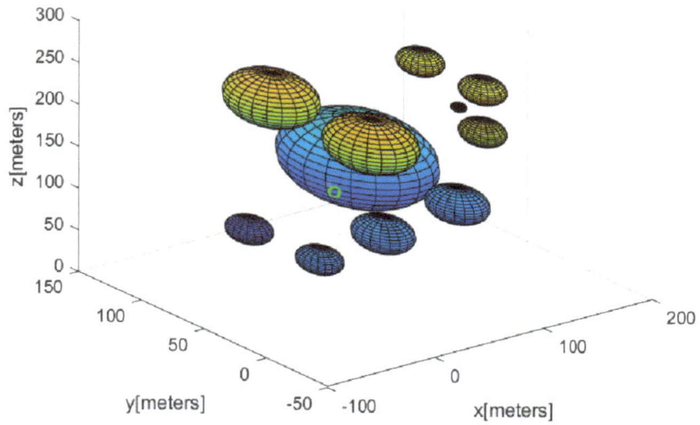

Figure 8. Every UUV's final position (green circle in this figure) after the rendezvous is completed (one MC simulation). Obstacles are plotted with spheres.

5.1. Using the Discard Approach in Section 4.3.1

We utilize the initial position of every UUV as in Figure 5. In this scenario, 25 UUVs among 50 UUVs are broken after 20 s have elapsed. Recall that Figure 6 shows each UUV's maneuver until $t = 20$ s pass.

To handle broken UUVs, we utilize the discard approach in Section 4.3.1. Considering one MC simulation, Figure 9 plots every UUV's maneuver until $t = 20$ s have passed. In this figure, every UUV's maneuver is illustrated as circles with a distinct color. At $t = 0$ s, we build a tree T (blue line segments in Figure 9) containing all UUVs. At $t = 20$ s, 25 broken UUVs are illustrated with red circles in Figure 9. At $t = 20$ s, we re-build a connected tree T without broken UUVs. The re-built tree T is marked with red line segments in Figure 9.

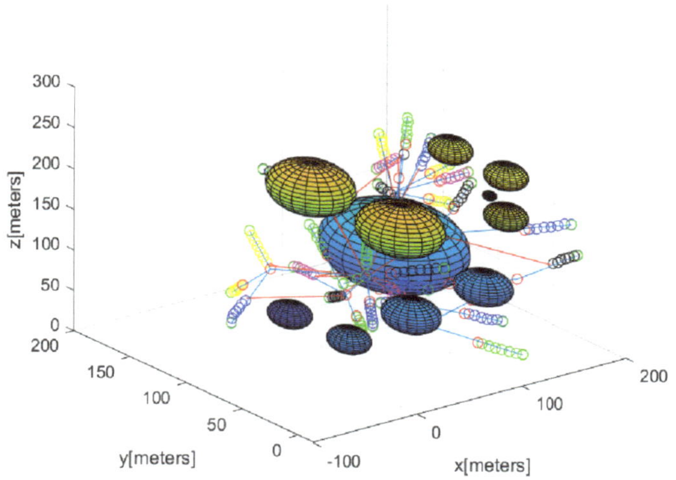

Figure 9. Every UUV's maneuver until $t = 20$ s have passed (one MC simulation). Every UUV's maneuver is illustrated as circles with a distinct color. At $t = 0$ s, we build a tree T (blue line segments) containing all UUVs. At $t = 20$ s, 25 broken UUVs are illustrated with red circles. At $t = 20$ s, we re-build a connected tree T without broken UUVs. The re-built tree T is marked with red line segments. Obstacles are plotted with spheres.

Considering UUVs' faults, Figures 10 and 11 show the movement of every UUV under our 3D distributed rendezvous controllers. In the figures, every UUV's maneuver is illustrated as circles with a distinct color. In total, 25 broken UUVs are illustrated with red circles. To handle broken UUVs, one utilizes the discard approach in Section 4.3.1. In total, 25 healthy UUVs spend 180 s to encounter at the root, while avoiding collision with obstacles. Figure 11 shows every UUV's final position (green circles) after the rendezvous is completed (discard approach is utilized). A total of 25 broken UUVs are illustrated with red circles.

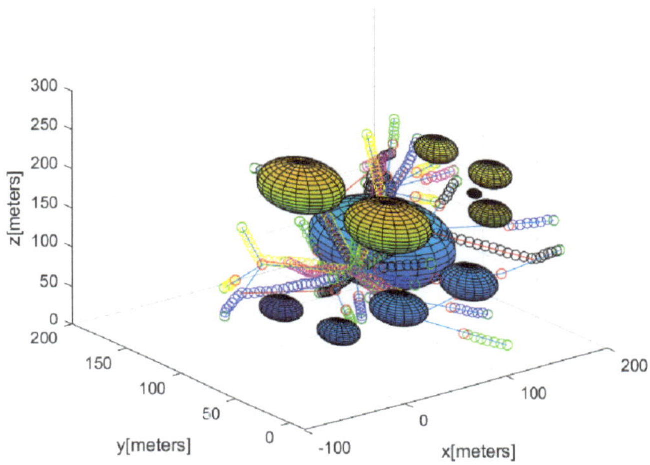

Figure 10. Every UUV's maneuver once the rendezvous is completed (the discard approach is utilized). Every UUV's maneuver is illustrated as circles with a distinct color. A total of 25 broken UUVs are illustrated with red circles. Obstacles are plotted with spheres.

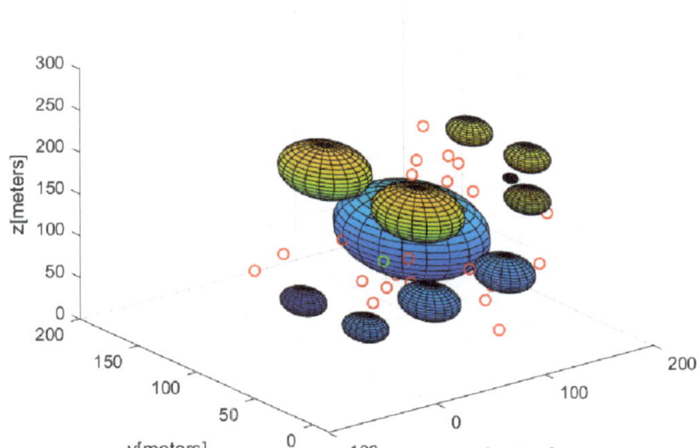

Figure 11. Every UUV's final position (green circles) after the rendezvous is completed (the discard approach is utilized). A total of 25 broken UUVs are illustrated with red circles. Obstacles are plotted with spheres.

5.2. Using the Waypoint Approach in Section 4.3.2

We utilize the initial position of every UUV as in Figure 5. In this scenario, 25 UUVs among 50 UUVs are broken after 20 s have elapsed. Recall that Figure 6 shows each UUV's maneuver until $t = 20$ s pass.

Considering UUVs' faults, Figures 12 and 13 show the movement of every UUV under our 3D distributed rendezvous controllers. In the figures, every UUV's maneuver is illustrated as circles with a distinct color. To handle broken UUVs, we utilize the waypoint approach in Section 4.3.2. At $t = 0$ s, we build a tree T (blue line segments in Figure 12) containing all UUVs. A total of 25 broken UUVs are illustrated with red circles. In Figure 13, all healthy UUVs (green circles) rendezvous at the root. We can see that healthy UUVs use broken UUVs as waypoints for reaching the root. A total of 25 healthy UUVs spend 148 s to encounter at the root UUV, while avoiding collision with obstacles.

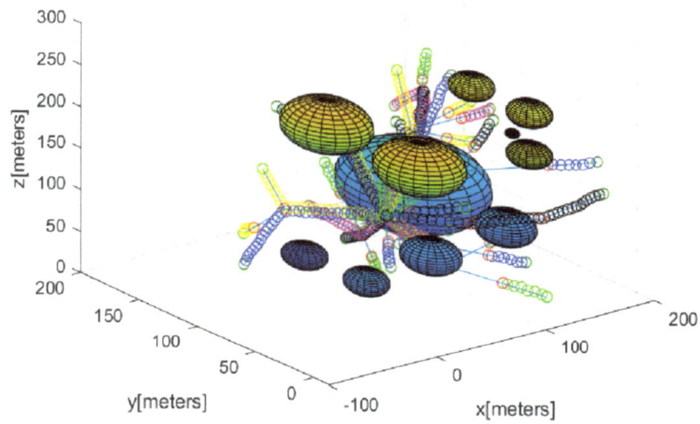

Figure 12. Every UUV's maneuver once the rendezvous is completed (the waypoint approach is utilized). Every UUV's maneuver is illustrated as circles with a distinct color. At $t = 0$ s, we build a tree T (blue line segments) containing all UUVs. A total of 25 broken UUVs are illustrated with red circles. Healthy UUVs use broken UUVs as waypoints for reaching the root. Obstacles are plotted with spheres.

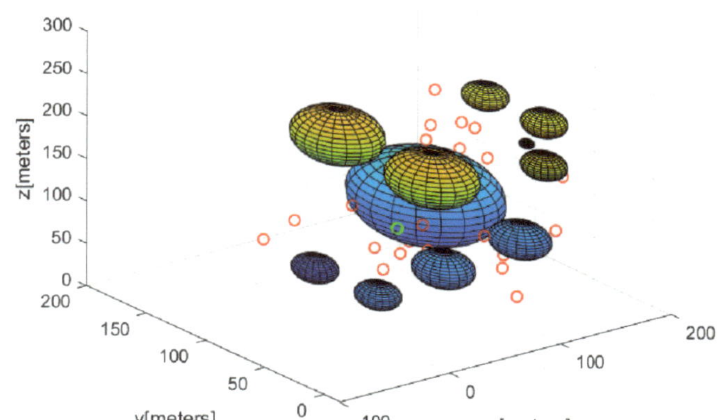

Figure 13. Every UUV's final position after the rendezvous is completed (the waypoint approach is utilized). A total of 25 broken UUVs are illustrated with red circles. All healthy UUVs (green circles in the figure) rendezvous at the root. Obstacles are plotted with spheres.

6. Conclusions

This article proposed distributed rendezvous controllers to solve the multi-robot rendezvous problem in cluttered 3D environments without GPS. The proposed 3D rendezvous controls assure the convergence to the root, while maintaining the connectivity of the time-varying and position-dependent communication network.

This study considers multiple UUVs such that each UUV has multiple signal intensity sensors surrounding it. Multiple intensity sensors on a UUV can measure the strength of signals, and the UUV moves based on the signal strength measurements. Our rendezvous algorithms are robust to faults in the system. This study proves the convergence of the proposed rendezvous algorithms and demonstrates the effectiveness of the proposed algorithms utilizing MATLAB simulations.

Note that the proposed rendezvous algorithms do not rely on the dimension (2D or 3D) of the environment. This implies that our algorithms can be applied for the rendezvous problem of autonomous aerial vehicles, autonomous underwater vehicles, or autonomous ground vehicles. In future studies, we will perform experiments utilizing real UUVs to verify the effectiveness of the proposed rendezvous algorithms.

Funding: This work was supported by the National Research Foundation of Korea (NRF) grant funded by the Korea government (MSIT) (Grant Number: 2022R1A2C1091682). This research was supported by the faculty research fund of Sejong university in 2023.

Institutional Review Board Statement: Not applicable.

Informed Consent Statement: Not applicable.

Data Availability Statement: Not applicable.

Conflicts of Interest: The author declares no conflict of interest.

References

1. Kim, J.; Kim, S. Motion control of multiple autonomous ships to approach a target without being detected. *Int. J. Adv. Robot. Syst.* **2018**, *15*, 1729881418763184. [CrossRef]
2. Kim, J. Cooperative Exploration and Protection of a Workspace Assisted by Information Networks. *Ann. Math. Artif. Intell.* **2014**, *70*, 203–220. [CrossRef]
3. Kim, J. Capturing intruders based on Voronoi diagrams assisted by information networks. *Int. J. Adv. Robot. Syst.* **2017**, *14*. 1–8. [CrossRef]
4. Kim, J. Cooperative Exploration and Networking While Preserving Collision Avoidance. *IEEE Trans. Cybern.* **2017**, *47*, 4038–4048. [CrossRef]
5. Cortés, J.; Martínez, S.; Bullo, F. Robust rendezvous for mobile autonomous agents via proximity graphs in arbitrary dimensions. *IEEE Trans. Autom. Control.* **2006**, *51*, 1289–1298. [CrossRef]
6. Roza, A.; Maggiore, M.; Scardovi, L. A class of rendezvous controllers for underactuated thrust-propelled rigid bodies. In Proceedings of the IEEE International Conference on Decision and Control, Los Angeles, CA, USA, 15–17 December 2014; pp. 1649–1654.
7. Park, H.; Hutchinson, S. A distributed optimal strategy for rendezvous of multi-robots with random node failures. In Proceedings of the IEEE International Conference on Intelligent Robots and Systems, Chicago, IL, USA, 14–18 September 2014; pp. 1155–1160.
8. Muralidharan, V.; Emami, M.R. Concurrent rendezvous control of underactuated spacecraft. *Acta Astronaut.* **2017**, *138*, 28–42. [CrossRef]
9. Yu, X.; Hsieh, M.A.; Wei, C.; Tanner, H.G. Synchronous Rendezvous for Networks of Marine Robots in Large Scale Ocean Monitoring. *Front. Robot. AI* **2019**, *6*, 76. [CrossRef] [PubMed]
10. Wei, C.; Li, C.; Tanner, H.G. Synchronous Rendezvous for Periodically Orbiting Vehicles with Very-Low-Range Interactions. In Proceedings of the 2018 Annual American Control Conference (ACC), Milwaukee, WU, USA, 27–29 June 2018; pp. 1641–1646.
11. Wei, C.; Tanner, H.G.; Yu, X.; Hsieh, M.A. Low-Range Interaction Periodic Rendezvous Along Lagrangian Coherent Structures. In Proceedings of the 2019 American Control Conference (ACC), Philadelphia, PA, USA, 10–12 July 2019; pp. 4012–4017.
12. Wei, W.; Wang, J.; Fang, Z.; Chen, J.; Ren, Y.; Dong, Y. 3U: Joint Design of UAV-USV-UUV Networks for Cooperative Target Hunting. *IEEE Trans. Veh. Technol.* **2023**, *72*, 4085–4090. [CrossRef]
13. Ajorlou, A.; Momeni, A.; Aghdam, A.G. A Class of Bounded Distributed Control Strategies for Connectivity Preservation in Multi-Agent Systems. *IEEE Trans. Autom. Control.* **2010**, *55*, 2828–2833. [CrossRef]
14. Dimarogonas, D.V.; Johansson, K.H. Decentralized connectivity maintenance in mobile networks with bounded inputs. In Proceedings of the IEEE International Conference on Robotics and Automation, Pasadena, CA, USA, 19–23 May 2008; pp. 1507–1512.

15. Chen, C.; Chen, G.; Guo, L. Consensus of flocks under M-nearest-neighbor rules. *J. Syst. Sci. Complex.* **2015**, *28*, 1–15. [CrossRef]
16. Gu, S.; Guo, S.; Zheng, L. A highly stable and efficient spherical underwater robot with hybrid propulsion devices. *Auton. Robot* **2020**, *44*, 759–771. [CrossRef]
17. Prasad, B.; Agrawal, A.; Viswanathan, V.; Chowdhury, A.R.; Kumar, R.; Panda, S.K. A visually guided spherical underwater robot. In Proceedings of the 2015 IEEE Underwater Technology (UT), Chennai, India, 23–25 February 2015; pp. 1–6.
18. He, Y.; Zhu, L.; Sun, G.; Qian, J.; Guo, S. Underwater motion characteristics evaluation of multi amphibious spherical robots. *Microsyst. Technol.* **2019**, *25*, 499–508. [CrossRef]
19. Zheng, L.; Guo, S.; Gu, S. The communication and stability evaluation of amphibious spherical robots. *Microsyst. Technol.* **2019**, *25*, 2625–2639. [CrossRef]
20. He, Y.; Zhu, L.; Sun, G.; Dong, M. Study on formation control system for underwater spherical multi-robot. *Microsyst. Technol.* **2019**, *25*, 1455–1466. [CrossRef]
21. Yue, C.; Guo, S.; Shi, L. Hydrodynamic Analysis of the Spherical Underwater Robot SUR-II. *Int. J. Adv. Robot. Syst.* **2013**, *10*, 247. [CrossRef]
22. Bao, P.; Hu, Y.; Shi, L.; Guo, S.; Li, Z. A decoupling three-dimensional motion control algorithm for spherical underwater robot. *Biomim. Intell. Robot.* **2022**, *2*, 100067. [CrossRef]
23. Lin, J.; Yang, X.; Zheng, P.; Cheng, H. End-to-end Decentralized Multi-robot Navigation in Unknown Complex Environments via Deep Reinforcement Learning. In Proceedings of the 2019 IEEE International Conference on Mechatronics and Automation (ICMA), Tianjin, China, 4–7 August 2019; pp. 2493–2500.
24. Hu, J.; Sun, J.; Zou, Z.; Ji, D.; Xiong, Z. Distributed multi-robot formation control under dynamic obstacle interference. In Proceedings of the 2020 IEEE/ASME International Conference on Advanced Intelligent Mechatronics (AIM), Boston, MA, USA, 6–10 July 2020; pp. 1435–1440.
25. Alonso-Mora, J.; Baker, S.; Rus, D. Multi-robot formation control and object transport in dynamic environments via constrained optimization. *Int. J. Robot. Res.* **2017**, *36*, 1000–1021. [CrossRef]
26. Kim, J. Tracking Controllers to Chase a Target Using Multiple Autonomous Underwater Vehicles Measuring the Sound Emitted From the Target. *IEEE Trans. Syst. Man Cybern. Syst.* **2021**, *51*, 4579–4587. [CrossRef]
27. Mathew, N.; Smith, S.L.; Waslander, S.L. Multirobot Rendezvous Planning for Recharging in Persistent Tasks. *IEEE Trans. Robot.* **2015**, *31*, 128–142. [CrossRef]
28. Zhang, B.; Wang, H.; Xu, T.; Zheng, L.; Yang, Q. Received signal strength-based underwater acoustic localization considering stratification effect. In Proceedings of the OCEANS 2016, Shanghai, China, 10–13 April 2016; pp. 1–8.
29. Giordano, P.R.; Franchi, A.; Seccos, C.; Bulthoff, H.H. A passivity-based decentralized strategy for generalized connectivity maintenance. *Int. J. Robot. Res.* **2013**, *32*, 299–323. [CrossRef]
30. Sabattini, L.; Secchi, C.; Chopra, N.; Gasparri, A. Distributed Control of Multirobot Systems with Global Connectivity Maintenance. *IEEE Trans. Robot.* **2013**, *29*, 1326–1332. [CrossRef]
31. Ajorlou, A.; Momeni, A.; Aghdam, A.G. Connectivity Preservation in Nonholonomic Multi-Agent Systems: A Bounded Distributed Control Strategy. *IEEE Trans. Autom. Control.* **2013**, *58*, 2366–2371. [CrossRef]
32. Gong, C.; Tully, S.; Kantor, G.; Choset, H. Multi-agent deterministic graph mapping via robot rendezvous. In Proceedings of the Robotics and Automation (ICRA), 2012 IEEE International Conference on Robotics and Automation, St. Paul, MN, USA, 14–18 May 2012; pp. 1278–1283.
33. Martinez, S. Practical multiagent rendezvous through modified circumcenter algorithms. *Automatica* **2009**, *45*, 2010–2017. [CrossRef]
34. Park, H.; Hutchinson, S. An efficient algorithm for fault-tolerant rendezvous of multi-robot systems with controllable sensing range. In Proceedings of the Robotics and Automation (ICRA), 2016 IEEE International Conference on Robotics and Automation, Stockholm, Sweden, 16–21 May 2016; pp. 358–365.
35. Wang, Q.; Duan, Z.; Lv, Y.; Wang, Q.; Chen, G. Distributed Model Predictive Control for Linear-Quadratic Performance and Consensus State Optimization of Multiagent Systems. *IEEE Trans. Cybern.* **2021**, *51*, 2905–2915. [CrossRef]
36. Dong, Y.; Xu, S. Rendezvous with Connectivity Preservation Problem of Linear Multiagent Systems via Parallel Event-Triggered Control Strategies. *IEEE Trans. Cybern.* **2022**, *52*, 2725–2734. [CrossRef]
37. Liu, P.; Xiao, F.; Wei, B. Event-Triggered Control for Multi-Agent Systems: Event Mechanisms for Information Transmission and Controller Update. *J. Syst. Sci. Complex.* **2022**, 35, 953–972. [CrossRef]
38. Yi, X.; Liu, K.; Dimarogonas, D.V.; Johansson, K.H. Distributed dynamic event-triggered control for multi-agent systems. In Proceedings of the 2017 IEEE 56th Annual Conference on Decision and Control (CDC), Melbourne, Australia, 12–15 December 2017; pp. 6683–6698.
39. Zheng, R.; Sun, D. Multirobot rendezvous with bearing-only or range-only measurements. *Robot. Biomimetics* **2014**, *1*, 4. [CrossRef]
40. Sabelhaus, A.P.; Mirsky, D.; Hill, L.M.; Martins, N.C.; Bergbreiter, S. TinyTeRP: A Tiny Terrestrial Robotic Platform with modular sensing. In Proceedings of the 2013 IEEE International Conference on Robotics and Automation, Karlsruhe, Germany, 6–10 May 2013; pp. 2600–2605.
41. Cho, C.; Kim, J. Robust Distributed Rendezvous Using Multiple Robots with Variable Range Radars. *Appl. Sci.* **2022**, *12*, 8535. [CrossRef]

42. Fossen, T.I. *Guidance and Control of Ocean Vehicles*; John Wiley and Sons: Hoboken, NJ, USA, 1994.
43. Douglas, B.W. *Introduction to Graph Theory*, 2nd ed.; Prentice Hall: Upper Saddle River, NJ, USA, 2001.
44. Go, S.; Chong, J.W. Improved TOA-Based Localization Method with BS Selection Scheme for Wireless Sensor Networks. *ETRI J.* **2015**, *37*, 707–716. [CrossRef]
45. Guvenc, I.; Chong, C.C. A Survey on TOA Based Wireless Localization and NLOS Mitigation Techniques. *IEEE Commun. Surv. Tutorials* **2009**, *11*, 107–124. [CrossRef]
46. Montminy, M.B. *Passive Geolocation of Low-Power Emitters in Urban Environments Using TDOA*; BiblioScholar: Singapore, 2012.
47. Chen, B.S.; Yang, C.Y.; Liao, F.K.; Liao, J.F. Mobile Location Estimator in a Rough Wireless Environment Using Extended Kalman-Based IMM and Data Fusion. *IEEE Trans. Veh. Technol.* **2008**, *58*, 1157–1169. [CrossRef]
48. Kim, J. Tracking a manoeuvring target while mitigating NLOS errors in TDOA measurements. *IET Radar Sonar Navig.* **2020**, *14*, 495–502. [CrossRef]
49. Liu, D.; Lee, M.C.; Pun, C.M.; Liu, H. Analysis of Wireless Localization in Non-Line-of-Sight Conditions. *IEEE Trans. Veh. Technol.* **2013**, *62*, 1484–1492. [CrossRef]
50. Wann, C.; Chin, H. Hybrid TOA/RSSI Wireless Location with Unconstrained Nonlinear Optimization for Indoor UWB Channels. In Proceedings of the 2007 IEEE Wireless Communications and Networking Conference, Hong Kong, China, 11–15 March 2007; pp. 3940–3945. [CrossRef]
51. Ji, M.; Egerstedt, M. Distributed Coordination Control of Multi-Agent Systems While Preserving Connectedness. *IEEE Trans. Robot.* **2007**, *23*, 693–703. [CrossRef]
52. Kan, Z.; Klotz, J.R.; Shea, J.M.; Doucette, E.A.; Dixon, W.E. Decentralized Rendezvous of Nonholonomic Robots with Sensing and Connectivity Constraints. *J. Dyn. Syst. Meas. Control.* **2016**, *139*. [CrossRef]
53. Wu, W.; Zhang, F. A Speeding-Up and Slowing-Down Strategy for Distributed Source Seeking with Robustness Analysis. *IEEE Trans. Control. Netw. Syst.* **2016**, *3*, 231–240. [CrossRef]
54. Al-Abri, S.; Wu, W.; Zhang, F. A Gradient-Free Three-Dimensional Source Seeking Strategy with Robustness Analysis. *IEEE Trans. Autom. Control.* **2019**, *64*, 3439–3446. [CrossRef]
55. Kim, J. Three-dimensional multi-robot control to chase a target while not being observed. *Int. J. Adv. Robot. Syst.* **2019**, *16*, 1–11. [CrossRef]
56. Garcia de Marina, H.; Cao, M.; Jayawardhana, B. Controlling Rigid Formations of Mobile Agents Under Inconsistent Measurements. *IEEE Trans. Robot.* **2015**, *31*, 31–39. [CrossRef]
57. Krick, L.; Broucke, M.E.; Francis, B.A. Stabilization of infinitesimally rigid formations of multi-robot networks. In Proceedings of the 2008 47th IEEE Conference on Decision and Control, Cancun, Mexico, 9–11 December 2008; pp. 477–482.
58. Paley, D.A.; Zhang, F.; Leonard, N.E. Cooperative Control for Ocean Sampling: The Glider Coordinated Control System. *IEEE Trans. Control. Syst. Technol.* **2008**, *16*, 735–744. [CrossRef]
59. Kim, J. Constructing 3D Underwater Sensor Networks without Sensing Holes Utilizing Heterogeneous Underwater Robots. *Appl. Sci.* **2021**, *11*, 4293. [CrossRef]
60. Luo, S.; Kim, J.; Parasuraman, R.; Bae, J.H.; Matson, E.T.; Min, B.C. Multi-robot rendezvous based on bearing-aided hierarchical tracking of network topology. *Hoc Netw.* **2019**, *86*, 131–143. [CrossRef]
61. Li, Q.; De Rosa, M.; Rus, D. Distributed Algorithms for Guiding Navigation across a Sensor Network. In Proceedings of the Proceedings of the 9th Annual International Conference on Mobile Computing and Networking, MobiCom'03, San Diego, CA, USA, 14–19 September 2003; Association for Computing Machinery: New York, NY, USA, 2003; pp. 313–325.
62. Li, Q.; Aslam, J.; Rus, D. Distributed Energy-conserving Routing Protocols for Sensor Networks. In Proceedings of the IEEE Hawaii International Conference on System Science, Big Island, HI, USA, 6–9 January 2003.
63. Kim, J. Control laws to avoid collision with three dimensional obstacles using sensors. *Ocean. Eng.* **2019**, *172*, 342–349. [CrossRef]
64. Lalish, E.; Morgansen, K. Distributed reactive collision avoidance. *Auton. Robot.* **2012**, *32*, 207–226. [CrossRef]
65. Lavalle, S.M. *Planning Algorithms*; Cambridge University Press: Cambridge, UK, 2006.

applied sciences

MDPI

Article

Hierarchical Sliding Mode Control Combined with Nonlinear Disturbance Observer for Wheeled Inverted Pendulum Robot Trajectory Tracking

Ming Hou *, Xuedong Zhang, Du Chen and Zheng Xu

School of Automation, Beijing Information Science & Technology University, Beijing 100192, China
* Correspondence: houming@bistu.edu.cn

Featured Application: The research presented herein could be used mostly in warehouse logistics transport activities in smart manufacturing.

Abstract: A proposed optimized model for the trajectory tracking control of a wheeled inverted pendulum robot (WIPR) system is presented in this study, which addresses the problem of poor trajectory tracking performance in the presence of unknown disturbances due to the nonlinear and underactuated characteristics of the system. First, a kinematic controller was used to track a reference trajectory and generate a control law that specifies the desired forward and rotation speeds of the system. Next, a nonlinear disturbance observer (NDO) was designed to enhance the system's robustness to external disturbances and improve its tracking performance. Then, the coupled system state variables were decoupled into two subsystems: a forward rotation subsystem and a tilt angle velocity subsystem. An improved hierarchical sliding mode controller was designed to control these subsystems separately. Finally, simulation experiments were conducted to compare the proposed method with a common sliding mode control approach. The simulation results demonstrate that the proposed method achieves better tracking performance in the presence of unknown disturbances.

Keywords: wheeled inverted pendulum robot; underactuated; nonlinear disturbance observer; hierarchical sliding mode control

Citation: Hou, M.; Zhang, X.; Chen, D.; Xu, Z. Hierarchical Sliding Mode Control Combined with Nonlinear Disturbance Observer for Wheeled Inverted Pendulum Robot Trajectory Tracking. *Appl. Sci.* **2023**, *13*, 4350. https://doi.org/10.3390/app13074350

Academic Editor: Jonghoek Kim

Received: 28 February 2023
Revised: 27 March 2023
Accepted: 27 March 2023
Published: 29 March 2023

1. Introduction

With the rapid development of technology, human society has simultaneously achieved increased convenience and comfort [1]. In today's factory warehouses and production lines, a variety of robots add to the possibilities of Industry 4.0. In the warehouses of more advanced companies, transport robots can be found everywhere, replacing traditional manpower and eliminating the need for workers to carry out repetitive lifting and carrying. These "smart" robots can accomplish a task as long as they can follow the requirements of a given transport trajectory. This paper studies the trajectory tracking of a mobile wheeled inverted pendulum on a given reference trajectory to achieve perfect tracking of the ideal motion trajectory of the robots, meeting the requirements of factory transport robots and providing a powerful source of assistance to realizing smart factories [2–5].

Mobile wheeled inverted pendulum models, such as WIPRs, have attracted much attention because of their special advantages, such as compactness, mobility, and human-like functions. WIPRs are widely used to verify the effectiveness of nonlinear underactuated control methods, and compared with the traditional inverted pendulum, WIPRs have more applications than traditional inverted pendulum vehicles, especially in unknown, dynamic, and nonlinear environments, and are commonly used in logistics transportation, commuting, and navigation, as well as in the aforementioned application in the environment of factory transportation. However, a WIPR is classified as a typical model of nonlinear underactuated systems with two input torques driving two wheels and three degrees of freedom (forward, rotation, and tilt angle of the pendulum), and achieving its high-performance motion control is still a challenging task for the control community [6–9].

On the one hand, when a WIPR moves, it is always assumed that the ground can provide enough friction to prevent the robot from side-slipping and wheel-sliding (i.e., the robot is guaranteed to move with purely rolling wheels without skidding phenomena), which is a non-complete constraint at this point. On the other hand, consider that the underdriven inverted pendulum body needs to use the input torque of two driving wheels to control the three degrees of freedom of WIPR forward movement, rotation, and the angle of the inverted pendulum. If we want to use MWIPR to track the trajectory, we need to drive a WIPR in real time to control the three form variables with two input variables, which is a typical underactuated problem. Finally, in the real world, WIPRs operate in factories or other similar environments and always encounter various unknown disturbances that interfere with the system. Therefore, the three problems of incomplete constraints and underactuated and unknown perturbations are the main challenges faced by this particular mobile robot for trajectory tracking control [10–13].

The three issues mentioned above are of importance for the following reasons. First, the incomplete constraint will lead to the WIPR being unable to follow any trajectory movement, especially in the case of high-speed heavy load; if the robot's incomplete constraints are not considered in motion planning, this is likely to lead to untimely obstacle avoidance and unreachable trajectory. Second, underdriven robots often have excellent dynamic performance or price advantages in terms of drive cost, but their biggest problem is the higher requirements in controller design. Finally, unknown disturbances will affect the control accuracy of the system to a certain extent, and more seriously, will affect the stability of the control system [14,15].

Many researchers and practitioners have proposed several control algorithms to overcome the difficulties faced by the problems associated with WIPR systems. One of the widely used methods is fuzzy control, which is an empirical, rule-based control technique that can effectively control nonlinear systems. By establishing a dynamic model of the WIPR, a fuzzy-logic-based controller can be designed to take the position and angle information of the WIPR as the input and output control signals to control its motion state. For instance, Jian Huang, in [16], proposed an Integral Interval Type 2 Fuzzy Logic (IT2FL) method that can maintain the MTWIP equilibrium while obtaining the desired position and orientation to make it work in an uncertain environment. However, the disadvantages of fuzzy control include low control accuracy, strong dependence on control rules, and difficulty in designing control rules.

The second control algorithm type is neural network control. Chenguang Yang [17] decomposed the underdriven WIPR model into two subsystems. The approximation characteristics of the neural network were used for motion control of the fully driven subsystem, and the sub-fully driven system was used to indirectly control the tilt angle motion of the pendulum. However, the method requires a large number of wavelet coefficient vectors, making the neural network computationally intensive.

Finally, sliding mode control, as the most typical robust control method, shows good tracking performance and strong robustness, which support its wide use in linear and nonlinear systems. For underactuated systems, various sliding mode control methods have been proposed by researchers to achieve different control effects, such as integral sliding mode control, terminal sliding mode control, and hierarchical sliding mode control [18–20]. Among them, the application of hierarchical sliding mode control in practical underdriven systems is receiving more and more attention, such as balancing control of a double-inverted pendulum and trajectory tracking control of a wheeled inverted pendulum. Nabanita Adhikary, in [21], proposed an integral inverse-step sliding mode controller for underdriven system control. A feedback control law was designed based on the backpropagation method, and a sliding surface was introduced in the final stage of the algorithm. Jian Huang, in [22], designed two terminal sliding mode controllers to control the speed and braking of a UW-Car based on the dynamic model and the terminal sliding mode control method. He Ping [23] proposed a hierarchical sliding mode controller (HSMC) developed to simultaneously perform speed control and balance control of a two-wheeled self-balancing vehicle (TWSBV).

Hierarchical sliding mode control is a control strategy based on sliding mode control, which divides the sliding surface into two layers. In the first layer, a high-speed sliding surface is introduced, and the control system approaches the desired state quickly. In the second layer, a low-speed sliding surface is introduced, and the control system stabilizes near the desired state. The layered sliding mode control can improve the control accuracy and stability and also has a good effect on the response speed and robustness of the system. Therefore, hierarchical sliding mode control has the same drawback in that it is insensitive to disturbances, which can easily cause the "jitter" phenomenon of the system. To address the shortcomings of sliding mode control, this paper proposes an improved hierarchical sliding mode control method with adaptive exponential convergence law, which can adaptively adjust the control convergence law according to the control state and smooth the sign function, thus effectively improving the problem of the strong jitter of the traditional sliding mode control, and combining the nonlinear disturbance observer (NDO), which is the most powerful method for the control of sliding mode. The NDO can effectively solve the negative impact caused by the unknown disturbance and make the system more robust, and achieve an ideal control effect on the trajectory tracking ability of the WIPR system [24–30].

Overall, this paper includes the following four aspects: the first part constructs the dynamic model of the WIPR system, decouples the multi-coupled state variables, and facilitates the subsequent controller design; the second part establishes the kinematic trajectory tracking controller of the system and solves to obtain the desired speed of the dynamic control system. In the third part, an optimization model of the WIPR system combining nonlinear disturbance observer and hierarchical sliding mode control is designed, and the convergence of the nonlinear disturbance observer and the stability of the improved hierarchical sliding mode controller is demonstrated. The fourth part constructs the simulation model using the MATLAB/Simulink platform and conducts numerical simulation comparison experiments.

The contributions of this paper are as follows:

(1) A wheeled inverted pendulum robot with a transport platform is envisioned for use in warehouses or other application scenarios to move goods.
(2) The convergence law of hierarchical sliding mode control is improved to mitigate the jitter phenomenon of the sliding mode control system, and an adaptive function is introduced to minimize the system jitter.
(3) By combining a nonlinear disturbance observer and hierarchical sliding mode control to estimate unknown external disturbances as input compensation, the system is made to control more accurately.

2. Materials and Methods

2.1. WIPR Model

A WIPR is a wheeled inverted pendulum transport robot with a placement table, as illustrated in Figure 1. Its left and right wheels are independent drive wheels that control the robot's movement speed, rotation direction, and tilt angle of the pendulum using the principle of differential drive to manage the position and posture of WIPR. The generalized world coordinate system is denoted $\Sigma OXYZ$ while (x, y), representing the center coordinate of the robot wheels. The robot's forward velocity and rotational angular velocity are denoted as v and w, respectively. The angle of the robot's direction of motion concerning the X-axis is represented by θ, while α is the tilt angle of the pendulum concerning the Z-axis. M refers to the total weight of the transport platform plus the pendulum, whereas m denotes the weight of each drive wheel. The distance between the two wheels is represented by d, while τ_r, and τ_l are the torque of the right wheel and the left wheel, respectively. The rotational inertia of each driven wheel is denoted by I_w and I_M

represents the rotational inertia of the transport platform and the pendulum together. The length of the pendulum is represented by L. Detailed introduction of robot parameters can be seen in Table 1.

Table 1. Parameter descriptions.

Parameter	Description
m_w	Mass of each wheel
M	The total weight of the transport platform plus the pendulum
I_w	The rotational inertia of each driven wheel
I_M	The rotational inertia of the transport platform and the pendulum
d	The distance between the two wheels
L	The length of the pendulum
τ_l	The torque of the left wheel
τ_r	The torque of the right wheel
v	WIPR forward velocity
w	WIPR rotation velocity
θ	WIPR Yaw angle
α	The tilt angle of the pendulum

Remark 1. *The forward velocity of MWIPR is x_v, and $x_v = \dot{x}\cos\theta + \dot{y}\sin\theta$.*

Assumption 1. *The tires of the MWIPR do not experience any skidding, and there is no potential for lateral deflection during its motion.*

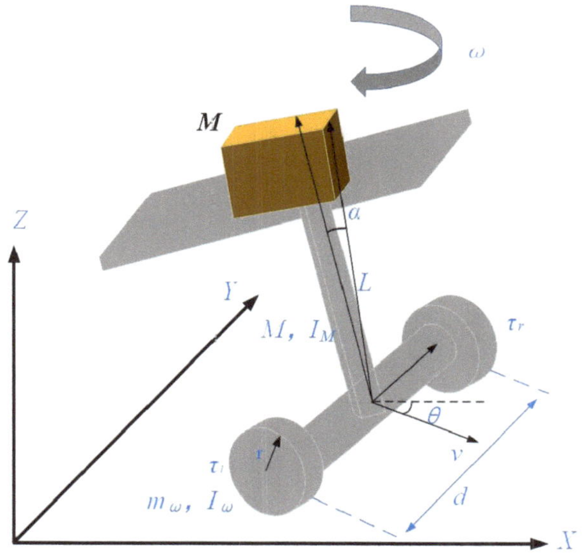

Figure 1. WIPR system.

According to Assumption 1, the incomplete constraint equation of WIPR in Equation (1) can be listed as follows:

$$\dot{x}\sin\theta - \dot{y}\cos\theta = 0, \tag{1}$$

The position and posture of the WIPR in the world coordinate system are represented by $q = [x, y, \theta, \alpha]^T$. As the Lagrangian modeling method does not require the inclusion of internal forces within the system, it is a quick and straightforward method of building a model. This property makes it particularly well-suited for constructing multivariable and nonlinear dynamic models for the WIPR, as demonstrated in this paper. By dividing q

into q_m and α, the position of the robot in the coordinate system is denoted by q_m, while the angle of the pendulum is represented by α. Therefore, the Lagrangian method [31] is employed to establish the dynamic model of the WIPR, and the resulting mathematical model is presented below as Equation (2).

$$\begin{bmatrix} M_m(q) & M_{m\alpha}(q) \\ M_{\alpha m}(q) & M_\alpha(q) \end{bmatrix}\begin{bmatrix} \ddot{q}_m \\ \ddot{\alpha} \end{bmatrix} + \begin{bmatrix} C_m(q) & C_{m\alpha}(q) \\ C_{\alpha m}(q) & C_\alpha(q) \end{bmatrix}\begin{bmatrix} \dot{q}_m \\ \dot{\alpha} \end{bmatrix} + \begin{bmatrix} G_m \\ G_\alpha \end{bmatrix} = \begin{bmatrix} B_m(q_m)\tau_m \\ 0 \end{bmatrix} + \begin{bmatrix} A^T(q_m)\lambda \\ 0 \end{bmatrix} + \begin{bmatrix} d_m \\ d_\alpha \end{bmatrix}, \tag{2}$$

By defining $A(q_m) = [\sin\theta, -\cos\theta, 0]$, the incompleteness constraint of Equation (1) yields the following result:

$$A(q_m)q_m = 0, \tag{3}$$

The WIPR system's incomplete constraint force is $A^T(q_m)\lambda$, where λ is the Lagrange Multiplier. To eliminate the constraint forces in the system, we seek to find a matrix $S(q_m) \in \mathbb{R}^{3\times 2}$ that satisfies $S^T(q_m)A^T(q_m) = 0$.

By defining $\nu = [v, w]^T$, therefore, Equation (4) can be deduced.

$$\dot{q}_m = S(q)\nu, \tag{4}$$

To eliminate the incompetent constraint forces, a new vector $\dot{\zeta} = \left[\dot{\zeta}_1, \dot{\zeta}_2, \dot{\zeta}_3\right]^T = [v, w, \dot{\alpha}]^T$ is defined and used to transform the equation. The transformation involves multiplying both sides of the equation by a scalar $S^T(q_m)$, resulting in Equation (5):

$$M(q)\ddot{\zeta} + C(q,\dot{q})\dot{\zeta} + G(q) = \tau + \tau_d, \tag{5}$$

The dynamics of the system can be described using the following equation, in which $M(q) \in \mathbb{R}^{3\times 3}$ represents the inertia matrix, $C(q,\dot{q})\dot{q} \in \mathbb{R}^{3\times 3}$ is the Coriolis force matrix, $G(q) \in \mathbb{R}^{3\times 1}$ is the gravity matrix, τ is the control input matrix, and τ_d is the total unknown disturbance. The detailed expressions of each vector or matrix are presented below.

$$M(q) = \begin{bmatrix} S^T M_m S & S^T M_{m\alpha} \\ M_{\alpha m} S & M_\alpha \end{bmatrix} = \begin{bmatrix} m_{11} & 0 & m_{13} \\ 0 & m_{22} & 0 \\ m_{31} & 0 & m_{33} \end{bmatrix},\ C(q,\dot{q}) = \begin{bmatrix} S^T C_m S + S^T M_m \dot{S} & C_{m\alpha} \\ C_{\alpha m} S + M_{\alpha m} \dot{S} & C_\alpha \end{bmatrix}$$

$$\begin{bmatrix} 0 & 0 & c_{13} \\ 0 & c_{22} & c_{23} \\ 0 & c_{32} & 0 \end{bmatrix},\ G(q) = \begin{bmatrix} S^T G_m \\ G_\alpha \end{bmatrix} = \begin{bmatrix} 0 \\ 0 \\ g_3 \end{bmatrix} = \begin{bmatrix} 0 \\ 0 \\ -MgL\sin\alpha \end{bmatrix},\ \tau = \begin{bmatrix} S^T B_m(q_m)\tau_m \\ 0 \end{bmatrix} = \begin{bmatrix} \tau_1 \\ \tau_2 \\ 0 \end{bmatrix}$$

$$\tau_d = \begin{bmatrix} S^T d_m \\ d_\alpha \end{bmatrix} = \begin{bmatrix} d_1 \\ d_2 \\ d_3 \end{bmatrix}.$$

The value of each variable in the expression is indicated as $m_{11} = 2m + 2I_M/r^2 + M$, $m_{13} = m_{31} = ML\cos\alpha$, $m_{22} = d^2m/2 + I_M d^2/2r^2 + I_\omega + ML^2\sin^2\alpha$, $m_{33} = ML^2 + I_M$, $c_{22} = (1/2)ML^2\dot{\alpha}\sin^2(2\alpha)$, $c_{23} = -(1/2)\omega ML^2\sin(2\alpha)$, $c_{13} = -ML\dot{\alpha}\sin\alpha$, $c_{32} = -(1/2)\omega ML^2\sin(2\alpha)$.

Due to the coupling of the state variables in the system, Equation (5) is decoupled into Equation (6).

$$\begin{cases} (m_{11}m_{33} - m_{13}m_{31})\ddot{\zeta}_1 - m_{13}c_{32}\dot{\zeta}_2 + m_{33}c_{13}\dot{\zeta}_3 - m_{13}g_3 \\ = m_{33}(\tau_1 + d_1) - m_{13}d_3 \\ m_{22}\ddot{\zeta}_2 + c_{22}\dot{\zeta}_2 + c_{23}\dot{\zeta}_3 = \tau_2 + d_2 \\ (m_{11}m_{33} - m_{13}m_{31})\ddot{\zeta}_3 + m_{11}c_{32}\dot{\zeta}_2 - m_{31}c_{13}\dot{\zeta}_3 + m_{11}g_3 \\ = m_{11}d_3 - m_{31}(\tau_1 + d_1) \end{cases}, \tag{6}$$

By defining $\xi_1 = [\zeta_1, \zeta_2, \zeta_3]^T$, $\xi_2 = \left[\dot{\zeta}_1, \dot{\zeta}_2, \dot{\zeta}_3\right]^T$, Equation (6) is converted to Equation (7).

$$\begin{cases} \dot{\xi}_1 = \xi_2 \\ \dot{\xi}_2 = f(\xi) + g(\xi)\tau + D \end{cases}, \tag{7}$$

where

$$f(\xi) = \begin{bmatrix} f_1 \\ f_2 \\ f_3 \end{bmatrix} = \begin{bmatrix} \frac{1}{\Omega}\left(m_{13}c_{32}\dot{\zeta}_2 - m_{33}c_{13}\dot{\zeta}_3 - m_{13}g_3\right) \\ \frac{1}{m_{22}}\left(-c_{22}\dot{\zeta}_2 - c_{23}\dot{\zeta}_3\right) \\ \frac{1}{\Omega}\left(m_{11}c_{32}\dot{\zeta}_2 - m_{31}c_{13}\dot{\zeta}_3 + m_{11}g_3\right) \end{bmatrix}, g(\xi) = \begin{bmatrix} \frac{m_{33}}{\Omega} \\ \frac{1}{m_{22}} \\ \frac{-m_{31}}{\Omega} \end{bmatrix},$$

$$D = \begin{bmatrix} D_1 \\ D_2 \\ D_3 \end{bmatrix} = \begin{bmatrix} \frac{m_{33}d_1 - m_{13}d_3}{\Omega} \\ \frac{d_2}{m_{22}} \\ \frac{m_{11}d_3 - m_{31}d_1}{\Omega} \end{bmatrix}.$$

where $\Omega = m_{11}m_{33} - m_{13}m_{31}$.

2.2. The Design of the Kinematic Control Law

In kinematic trajectory tracking control for a WIPR, the system can be simplified to a general two-wheeled non-complete mobile robot for trajectory tracking. The process involves utilizing a reference trajectory state vector $q_{mr} = [x_r, y_r, \theta_r]^T$, an actual state vector $q_m = [x, y, \theta]^T$, and a control objective designed to manage linear and angular velocities. The objective is to ensure that the actual robot travel trajectory aligns with the reference trajectory, even if the trajectory error $q_{me} = \left[e_x, e_y, e_\theta\right]^T = [x_r - x, y_r - y, \theta_r - \theta]^T$ approaches zero. To meet the requirement expressed above, the control laws for v_d and w_d can be devised as Equation (8).

$$\lim_{t\to\infty} \|q_{me}\| = 0, \tag{8}$$

The Lyapunov function is selected as Equation (9).

$$V_1 = \frac{1}{2}x_e^2 + \frac{1}{2}y_e^2 + 1 - \cos\theta_e, \tag{9}$$

While the error persists, the value V_1 remains greater than zero, thereby rendering the function positive definite. Equation (10) describes the derivative of V_1.

$$\begin{aligned} \dot{V}_1 &= e_x\dot{e}_x + e_y\dot{e}_y + \dot{e}_\theta \sin e_\theta \\ &= e_x\left(we_y - v + v_r\cos e_\theta\right) + e_y(-we_x + v_r\sin e_\theta) + (w_r - w)\sin e_\theta \\ &= -ve_x - w\sin e_\theta + e_xv_r\cos e_\theta + e_yv_r\sin e_\theta + w_r\sin e_\theta \le 0, \end{aligned} \tag{10}$$

Lyapunov's stability theorem [32] establishes that the system can achieve asymptotic stability if the function is negative definite, i.e., if $\dot{V}_1 \le 0$. Accordingly, the sought control law is as follows (11).

$$\begin{cases} v_d = v_r\cos e_\theta + k_1e_x \\ w_d = w_r + v_re_y + k_2\sin e_\theta \end{cases}, \tag{11}$$

Both k_1 and k_2 are positive constants.

Substituting the control law into $\dot{V}_1$ gives the following equation.

$$\begin{aligned} \dot{V}_1 &= -e_x(v_r\cos e_\theta + k_1e_x) - (w_r + v_re_y + k_2\sin e_\theta)\sin e_\theta + e_xv_r\cos e_\theta + e_yv_r\sin e_\theta + w_r\sin e_\theta \\ &= -e_xv_r\cos e_\theta + e_xv_r\cos e_\theta - k_1e_x^2 - w_r\sin e_\theta + w_r\sin e_\theta - v_re_y\sin e_\theta + e_yv_r\sin e_\theta - k_2\sin^2_{e_\theta} \\ &= -k_1e_x^2 - k_2\sin^2_{e_\theta} \le 0. \end{aligned} \tag{12}$$

Therefore, $\dot{V}_1 \leq 0$ can be proven.

So far, the desired velocity required for the design of the dynamical system is shown in Equation (11), and the velocity tracking problem of the dynamical system and the angle tracking problem of the pendulum will be solved next.

2.3. The Design of NDO

A nonlinear disturbance observer is developed to estimate the actual disturbance in the system for an unknown disturbance D, thereby strengthening the system's robustness. To address practical considerations, it is assumed that any disturbance is bounded as follows [33–35].

Lemma 1. *For initial conditions that are bounded, a Liapunov function is also uniformly bounded $x(t)$ if there exists a continuous positive definite Liapunov function $V(x)$ satisfying the following conditions:*

$$\delta_1(\|x\|) \leq V(x) \leq \delta_2(\|x\|),\ \dot{V}(x) \leq -\kappa V(x) + c, \tag{13}$$

where $\delta_1, \delta_2 : \mathbb{R}^n \to \mathbb{R}$ is the V class function, and κ, c all are positive constants.

Assumption 2. *Since no disturbance can be infinite in the real world, we assume that the perturbations in the WIPR system studied in this paper are all bounded, and their first-order derivatives and second-order derivatives are assumed to be bounded; thus, the following equations can be obtained.*

$$\|\dot{D}\| \leq \varepsilon_1,\ \|\ddot{D}\| \leq \varepsilon_2,\ \varepsilon_1 > 0, \varepsilon_2 > 0$$

$\|\cdot\|$ represents the Euclidean norm of the vector. The NDO is designed as in Equation (14).

$$\begin{aligned} \dot{\hat{D}} &= Z_1 + P_1(\xi_2)\dot{Z}_1 = -L_1(\xi_2)[f(\xi) + g(\xi)\tau + \hat{D}] + \hat{\dot{D}} \\ \dot{\hat{\dot{D}}} &= Z_2 + P_2(\xi_2)\dot{Z}_2 = -L_2(\xi_2)[f(\xi) + g(\xi)\tau + \hat{D}], \end{aligned} \tag{14}$$

$\hat{D}$ and $\hat{\dot{D}}$ represent the estimates of the total perturbation and its derivative, respectively, while Z_1 and Z_2 are intermediate variables in the observer. To meet the requirements of the system, the self-designed nonlinear functions $P_1(\xi_2)$ and $P_2(\xi_2)$ are used and must satisfy the conditions $L_1(\xi_2) = \frac{\partial P_1(\xi_2)}{\partial \xi_2}$ and $L_2(\xi_2) = \frac{\partial P_2(\xi_2)}{\partial \xi_2}$.

Property 1. *The error between the estimated and actual values of the disturbance is represented by $\widetilde{D} = D - \hat{D}$, while$\widetilde{\dot{D}} = \dot{D} - \hat{\dot{D}}$represents the error between the derivative of the actual value of a perturbation and the derivative of the estimated value of the same perturbation.*

The derivation of $\widetilde{D}$ and $\widetilde{\dot{D}}$ substitution of Equations (7) and (14) into the above equation leads to results $\dot{\widetilde{D}}$ and $\dot{\widetilde{\dot{D}}}$, which are the equations of the NDO, (15) and (16), respectively.

$$\begin{aligned} \dot{\widetilde{D}} &= \dot{D} - \dot{\hat{D}} = \dot{D} - \dot{Z}_1 - \frac{\partial P(\xi_2)}{\partial \xi_2} \cdot \frac{d\xi_2}{dt} = \dot{D} - \dot{Z}_1 - L_1(\xi_2)\dot{\xi}_2 \\ &= \dot{D} + L_1(\xi_2)[f(\xi) + g(\xi)\tau + \hat{D}] - \hat{\dot{D}} - L_1(\xi_2)\dot{\xi}_2 \\ &\dot{D} + L_1(\xi_2)[f(\xi) + g(\xi)\tau + \hat{D}] - \hat{\dot{D}} - L_1(\xi_2)[f(\xi) + g(\xi)\tau + D] \\ &= -L_1(\xi_2)\widetilde{D} + \widetilde{\dot{D}}, \end{aligned} \tag{15}$$

$$\dot{\tilde{D}} = \dot{D} - \dot{\hat{D}} = \dot{D} - \dot{Z}_2 - \frac{\partial P(\xi_2)}{\partial \xi_2} \cdot \frac{d\xi_2}{dt} = \dot{D} - \dot{Z}_2 - L_2(\xi_2)\dot{\xi}_2$$
$$= \dot{D} + L_2(\xi_2)[f(\xi) + g(\xi)\tau + \hat{D}] - L_2(\xi_2)[f(\xi) + g(\xi)\tau + D] \quad (16)$$
$$= -L_2(\xi_2)\tilde{D} + \dot{D},$$

By letting $E = \left[\tilde{D}^T, \dot{\tilde{D}}^T\right]^T$ and substituting the appropriate Equations (15)–(17) can be obtained.

$$\dot{E} = LE + \delta\ddot{D}, \quad (17)$$

where $L = \begin{bmatrix} -L_1(\xi_2) & I_3 \\ -L_2(\xi_2) & 0 \end{bmatrix}, \delta = \begin{bmatrix} 0 \\ I_3 \end{bmatrix}$. The observer's stability is examined, and a Lyapunov function is selected to make sure that it can reliably predict the system state despite any nonlinear disturbances. By selecting an appropriate Liapunov function, we rigorously prove the stability of the observer and the precision of its estimation precisely.

Property 2. *L is a skew-symmetric matrix.*

$$V_2 = \frac{1}{2}E^T E, \quad (18)$$

The proof of the derivative of V_2 can be expressed as follows:

$$\dot{V}_2 = E^T\dot{E} = E^T\left(LE + \delta\ddot{D}\right) \leq E^T\left(L + 0.5\|\delta\|^2 I_6\right)E + 0.5\varepsilon_2^2, \quad (19)$$

The above design of an NDO for a WIPR can be summarized in the following theorem.

Theorem 1. *For the existence of an unknown disturbance in a WIPR system, the perturbation estimation error is bounded for the observer designed according to Equation (14).*

Proof. The design parameters $L_1(\xi_2) = \frac{\partial P_1(\xi_2)}{\partial \xi_2}$ and $L_2(\xi_2) = \frac{\partial P_2(\xi_2)}{\partial \xi_2}$ are such that $L + 0.5\|\delta\|^2 I_6$ is a negative definite matrix, according to Lemma 1, then E is bounded. □

2.4. The Design of Improved Slide Mode Control

The sliding mode control algorithm consists of two key elements: (i) the design of the sliding mode surface; and (ii) the design of the convergence rate. The design of the sliding mode surface is mainly based on the system structure as well as the control objective. As for the design of convergence law, there are four different convergence laws: the isokinetic convergence law, exponential convergence law, power convergence law, and general convergence law. In this paper, based on the optimal control objective of WIPPR to cope with nonlinearity and underdrive, as well as unknown disturbances, the traditional exponential convergence law is improved by introducing an adaptive control function, and an improved sliding mode control based on the adaptive exponential convergence law is proposed, which can weaken the system jitter while speeding up the system response, making the sliding mode control more suitable for tracking the reference trajectory of the WIPR system under the action of unknown disturbances [36]. The design for the traditional exponential convergence law is shown in Equation (20).

$$\dot{s} = -\eta \text{sgn}(s) - \mu s, \quad (20)$$

where: s denotes the slip surface function; the parameters η, μ denote the convergence coefficient; and sgn(s) denotes the sign function.

In the traditional exponential convergence law, the isokinetic term is denoted $-\eta\text{sgn}(s)$, and the exponential term is denoted $-\mu s$. When the state of the system is far from the slip surface, the exponential term and the isokinetic term in the convergence law act simultaneously to help the system move toward the slip surface, and the magnitude of the isokinetic term and the exponential term are mainly determined by the reference η, μ. The exponential term is small, and the isokinetic term acts mainly when the system is moving close to the surface.

This paper makes a corresponding improvement based on the traditional exponential convergence law and introduces the adaptive $o(s)$ function to adjust the convergence law in accordance with the control state of the system, as shown in Figure 2, which can accelerate the convergence speed of the sliding mode and weaken the overshoot phenomenon. This allows the sliding mode control to reduce the jitter phenomenon of the system. Following the inclusion of the adaptive function $o(s)$, the new exponential convergence law is as follows:

$$\begin{cases} \dot{s} = -\eta o(s)\mathrm{sgn}(s) - \mu s \\ o(s) = \frac{(|s|+1)[\log_a(|s|+a)]}{|s|+\log_a(|s|+a)+b} \end{cases}, \tag{21}$$

where $a > 0$, $b > 0$, $\eta > 0$, $\mu > 0$.

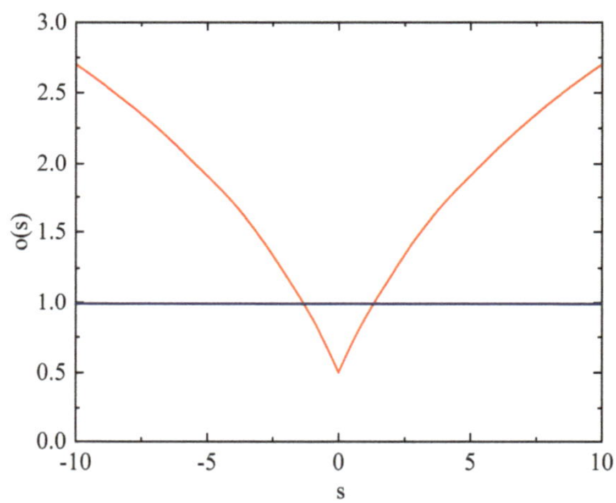

Figure 2. The adaptive function $o(s)$.

Through the analysis, relative to not adding the adaptive function (i.e., o(s) = 1), it can be found that when the system motion point is far away from the sliding surface (namely, when s is far away from the origin 0), the adaptive function $o(s)$ will increase the convergence law, which will speed up the system convergence speed, shorten the system state convergence time to the target state, and reduce the control time; when the system motion point is close to the sliding surface, $|s|$ will converge to 0 and $o(s)$ will be less than 1. The role of $o(s)$ here is to suppress the jitter amplitude and weaken the state variable fluctuation problem after the system is stabilized, and the suppression effect will be more obvious as the parameter b increases. To further weaken the jitter problem of the stabilized system, the smoothing process is carried out for the symbolic function $\mathrm{sgn}(s)$ in this paper, which is known as the traditional symbolic function [37], as shown in the following equation.

$$\mathrm{sgn}(s) = \begin{cases} 1 & s > 0 \\ 0 & s = 0 \\ -1 & s < 0 \end{cases}, \tag{22}$$

The symbolic function after the smoothing process is shown below.

$$\mathrm{sgn}(s) = \frac{s}{|s| + \beta}, \ \beta = 0.01, \tag{23}$$

2.5. The Design of the Forward-Rotation Subsystem

For convenience, the system has been reorganized into the following form (24).

$$\begin{cases} \ddot{\zeta}_1 = \frac{1}{\Omega}\left(m_{13}c_{32}\dot{\zeta}_2 - m_{33}c_{13}\dot{\zeta}_3 - m_{13}g_3\right) + \frac{m_{33}}{\Omega}\tau_1 + D_1 \\ \ddot{\zeta}_2 = \frac{1}{m_{22}}\left(-c_{22}\dot{\zeta}_2 - c_{23}\dot{\zeta}_3\right) + \frac{1}{m_{22}}\tau_2 + D_2 \end{cases}, \tag{24}$$

The state variables in the system described by Equation (24) are highly coupled. To address this issue and to expand the system's asymptotic stability domain, a hierarchical sliding mode controller was designed. The controller's primary objective is to utilize an input control law that can simultaneously control both system variables ζ_1 and ζ_2, thereby, mitigating the problem of system coupling [38].

Having obtained the expected forward velocity (v_d) and angular velocity (w_d) from Equation (11), the error between the actual and expected values can be defined as follows:

$$\begin{cases} e_{x_v} = \zeta_{1d} - \zeta_1 = x_r \cos\theta_r + y_r \sin\theta_r - x_v \\ e_v = \dot{\zeta}_{1d} - \dot{\zeta}_1 = v_d - v \\ e_\theta = \zeta_{2d} - \zeta_2 = \theta_r - \theta \\ e_w = \dot{\zeta}_{2d} - \dot{\zeta}_2 = w_d - w \end{cases}, \tag{25}$$

To design the sliding mode control error tracking scheme for the v-w subsystem, two mutually independent first-layer sliding mode surfaces were initially constructed. The equations used to create these slide surfaces are as follows:

$$s_1 = \gamma_1 e_{x_v} + e_v, \ \gamma_1 > 0, \tag{26}$$

$$s_2 = \gamma_2 e_\theta + e_w, \ \gamma_2 > 0, \tag{27}$$

The results of deriving Equations (26) and (27) are presented below.

$$\dot{s}_1 = \gamma_1 \dot{e}_{x_v} + \dot{e}_v = \gamma_1 \dot{e}_{x_v} + \ddot{\zeta}_{1d} - \ddot{\zeta}_1 = 0 \dot{s}_2 = \gamma_2 \dot{e}_\theta + \dot{e}_w = \gamma_2 \dot{e}_\theta + \ddot{\zeta}_{2d} - \ddot{\zeta}_2 = 0, \tag{28}$$

According to Filippov's equivalent control theory, the equivalent control laws for ζ_1 and ζ_2 are as follows:

$$\begin{cases} \tau_{eq1} = -\frac{\Delta}{m_{33}}\left(\gamma_1 \dot{e}_{x_v} + \ddot{\zeta}_{1d} + \hat{D}_1\right) + \frac{1}{m_{33}}\left(m_{13}c_{32}\dot{\zeta}_2 - m_{33}c_{13}\dot{\zeta}_3 - m_{13}g_3\right) \\ \tau_{eq2} = m_{22}\left(\gamma_2 \dot{e}_\theta + \ddot{\zeta}_{2d}\right) + c_{22}\dot{\zeta}_2 + c_{23}\dot{\zeta}_3 - \hat{D}_2 \end{cases} \tag{29}$$

The second sliding surface can be expressed as a linear combination of the first sliding surface.

$$s_3 = \gamma_3 s_1 + \gamma_4 s_2, \gamma_3, \gamma_4 > 0, \tag{30}$$

To control ζ_1 and ζ_2, the equivalent control law must be included at the same time to control and enter their designed sliding surface, respectively. Therefore, the total control law is shown in the following equation.

$$\tau_u = \tau_{eq1} + \tau_{eq2} + \tau_{sw}, \tag{31}$$

τ_{sw} is the switching law of the converging slide surface phase, and the expressions are as follows.

$$\tau_{sw} = \frac{-\left(\gamma_3 \frac{m_{22}}{\Omega} \tau_{eq2} + \gamma_4 \frac{1}{m_{22}} \tau_{eq1} + \eta_1 o(s_3)\mathrm{sgn}(s_3) + \mu_1 s_3\right)}{\gamma_3 \frac{m_{22}}{\Omega} + \gamma_4 \frac{1}{m_{22}}}, \tag{32}$$

To mitigate the jitter phenomenon of the system, the isokinetic and exponential terms of the sliding mode control are improved, where η_1 and μ_1 are the isokinetic and exponential terms of the previous design convergence law, and both are positive constants; $o(s_3)$ and $\mathrm{sgn}(s_3)$ are the adaptive and symbolic functions designed in the previous paper. To prove that the designed controller is stable, the Lyapunov function is chosen as follows.

$$V_3 = \frac{1}{2} s_3^2 > 0 \tag{33}$$

The derivative of V_3 for time is given by the following expression.

$$\begin{aligned}
\dot{V}_3 &= s_3\dot{s}_3 = s_3(\gamma_3\dot{s}_1 + \gamma_4\dot{s}_2) = s_3[\gamma_3(\gamma_1\dot{e}_{x_v} + \dot{e}_v) + \gamma_4(\gamma_2\dot{e}_\theta + \dot{e}_w)] = s_3[\gamma_3(\gamma_1\dot{e}_{x_v} + \dot{v}_d - \dot{v}) + \gamma_4(\gamma_2\dot{e}_\theta + \dot{w}_d - \dot{w})] \\
&= s_3\left\{\gamma_3\left[\gamma_1\dot{e}_{x_v} + \dot{v}_d - f_1 - \tfrac{m_{33}}{\Omega}(\tau_{eq1} + \tau_{eq2} + \tau_{sw}) - D_1\right] + \gamma_3\left[\gamma_2\dot{e}_\theta + \dot{w}_d - f_2 - \tfrac{1}{m_{22}}(\tau_{eq1} + \tau_{eq2} + \tau_{sw}) - D_2\right]\right\} \\
&= s_3[-\eta_1 o(s_3)\mathrm{sgn}(s_3) - \mu_1 s_3^2 + \gamma_1(\hat{D}_1 - D_1) + \gamma_2(\hat{D}_2 - D_2)] \le -\eta_1 o(s_3)|s_3| - \mu_1 s_3^2 + |s_3|\left(\gamma_1\left|\tilde{D}_1\right| + \gamma_2\left|\tilde{D}_2\right|\right)
\end{aligned} \tag{34}$$

By design $\eta_1 o(s_3) > \gamma_1\left|\tilde{D}_1\right| + \gamma_2\left|\tilde{D}_2\right|$, the result of the following equation can be obtained.

$$\dot{V}_3 \le -\eta_1 s_3^2 \le -\frac{\eta_1}{2} V_3 \tag{35}$$

From Lemma 1 in [38], the following equation can be obtained.

$$V_3(t) \le e^{-\frac{\eta_1}{2}(t-t_0)} V(t_0) \tag{36}$$

It can be seen that the $V_3(t)$ index converges to 0, and the rate of convergence depends on η_1.

As demonstrated by the preceding equation, the error state can attain the slip surface in a finite amount of time. Subsequently, the first layer of slip surfaces s_1 and s_2 can converge asymptotically to zero, leading to the convergence of both the rotational and forward velocities of WIPR to the desired values.

2.6. The Design of the Tilt-Angle Subsystem

The system discussed in the previous section can achieve complete tracking of ζ_1 and ζ_2 within a finite time, which enables us to transform it into the following form:

$$\ddot{\zeta}_3 = \frac{1}{\Omega}\left(m_{11}c_{32}\dot{\zeta}_{2d} - m_{31}c_{13}\dot{\zeta}_3 + m_{11}g_3\right) - \frac{m_{31}}{\Omega}\tau_3 + D_3, \tag{37}$$

As WIPR aims to maintain a vertical and stable direction of the pendulum during its motion, all relevant parameters $(\alpha_d, \dot{\alpha}_d, \ddot{\alpha}_d)$ can be set to zero. As such, the following definitions can be employed:

$$\begin{cases} e_\alpha = \alpha_d - \alpha = -\alpha \\ e_{\dot{\alpha}} = \dot{\alpha}_d - \dot{\alpha} = -\dot{\alpha} \end{cases}, \tag{38}$$

Let the sliding mode surface be defined as Equation (39), with its derivative expressed as Equation (40).

$$s_4 = \gamma_5 e_\alpha + e_{\dot{\alpha}}, \tag{39}$$

$$\dot{s}_4 = \gamma_5\dot{e}_\alpha + \dot{e}_{\dot{\alpha}} = -\gamma_5\dot{\alpha} - \ddot{\alpha} = -\eta_2 o(s_4)\mathrm{sgn}(s_4) - \mu_2 s_4, \tag{40}$$

After substituting Equation (37), the control law for the tilt angle subsystem can be derived as presented in Equation (41).

$$\tau_\alpha = -\frac{\Phi}{m_{31}}\left(\gamma_5\dot{\alpha} + \hat{D}_3\right) - \frac{1}{m_{31}}\left(m_{11}c_{12}\dot{\zeta}_2 - m_{31}c_{13}\dot{\zeta}_3 + m_{11}g_3 - \eta_2 o(s_4)\mathrm{sgn}(s_4)\right) - \mu_2 s_4 \quad (41)$$

To prove the stability of the designed system, the Lyapunov function is chosen as follows.

$$V_4 = \frac{1}{2}s_4^2 \quad (42)$$

The derivative of V_4 for time is given by the following expression.

$$\begin{aligned} \dot{V}_4 &= s_4\dot{s}_4 = s_4(-\gamma_5\dot{\alpha} - \ddot{\alpha}) = s_4\left(-\gamma_5\dot{\alpha} - f_3 - \tfrac{1}{m_{22}}\tau_\alpha - D_3\right) \\ &= s_4\left[-\eta_2 o(s_4)\mathrm{sgn}(s_4) - \mu_2 s_4^2 + \gamma_5\left(\hat{D}_3 - D_3\right)\right] \leq -\eta_2 o(s_4)|s_4| - \mu_2 s_4^2 + \gamma_5\left|\tilde{D}_3\right| \end{aligned} \quad (43)$$

By choosing $\eta_2 o(s_4) > \gamma_5\left|\tilde{D}_3\right|$, it enables $\dot{V}_4 < 0$ to hold, indicating that the system achieves asymptotic stability.

Figure 3 displays the schematic block diagram of the control system.

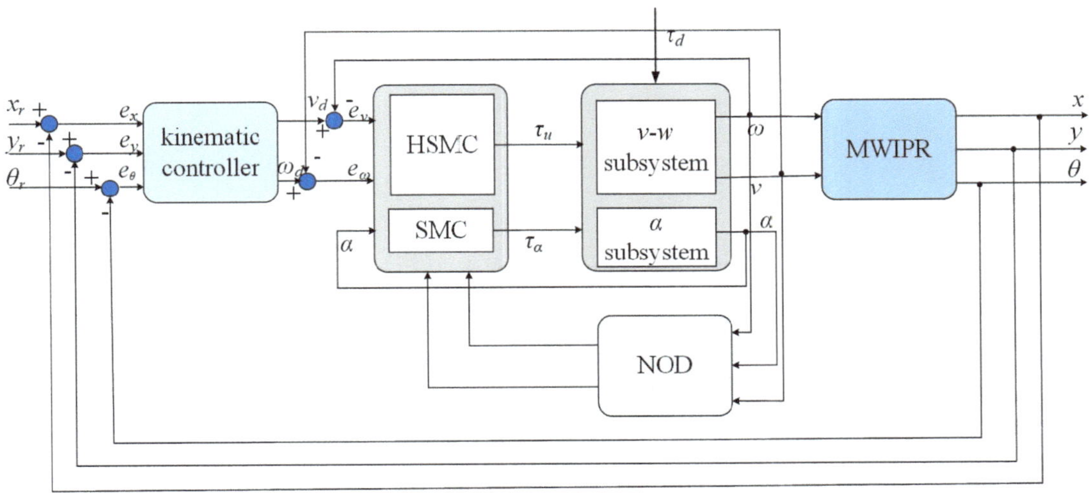

Figure 3. The control system.

3. Simulation

The focus of this section is to discuss the trajectory-tracking effect of the system in a simulation environment, and to verify the feasibility of the proposed control scheme and what the advantages of the proposed method are compared with other control systems in this paper. Next, the simulation results of different control systems in the face of the same disturbance will be compared to verify the control effectiveness of each system. The parameters in the system are shown in the following Table 2.

The simulation experiments in Matlab/Simulink verified the high-precision trajectory tracking capability of the system and the stability of the pendulum in robot motion. During the simulation study, the initial position was set as $q_m = [0,\ 0,\ 0]^T$, and the initial position of the reference trajectory was set as $q_{mr} = [0, 0, \pi/2]^T$, where the desired tracking velocity = 1 m/s and the angular velocity = 1 rad/s. Therefore, the trajectory of the robot should be a circle with a radius of 1 m, and the center of the circle is $(0, 0)$. Therefore, the time function of the reference trajectory was chosen as $\{x = \sin t, y = \cos t\}$.

Table 2. The value of each parameter variable in the system.

Parameter (Unit)	Value
$M(\text{kg})$	8
$m(\text{kg})$	0.5
$I_M(\text{kg}\cdot\text{m}^2)$	5
$I_w(\text{kg}\cdot\text{m}^2)$	0.3
$d(\text{m})$	0.5
$L(\text{m})$	0.5
$r(\text{m})$	0.1

An external perturbation was added, as shown in the following equation.

$$\begin{cases} D_1 = 0.4\sin(0.4t)Nm \\ D_2 = 0.5\cos(0.5t)Nm \\ D_3 = 0.6\sin(0.6t)Nm \end{cases}, \tag{44}$$

To demonstrate the superiority of the proposed method in this paper, three comparative experiments were conducted under the given disturbance conditions: the first experiment involved the simulation results of the unimproved HSMC method, the second experiment involved the simulation results of the improved IHSMC method with adaptive law but without nonlinear disturbance observer, and the third experiment involved the simulation results of the proposed method in this paper (referred to as PC).

First of all, by observing Figures 4–6, it can be concluded that the proposed method in this paper is better than the other two control methods in terms of both the speed of convergence of the error to the steady state and the magnitude of the fluctuation of the error after reaching the steady state when compared with the other two methods. This undoubtedly reflects the effectiveness of the method in this paper, which can track the given reference trajectory very accurately.

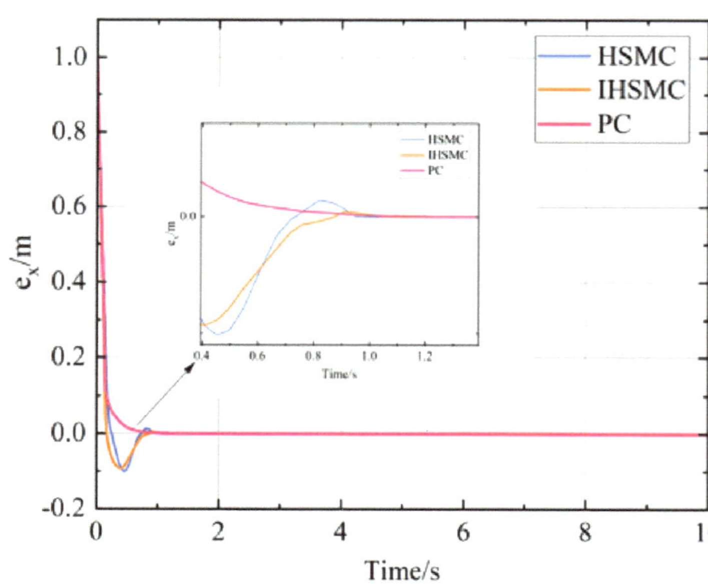

Figure 4. The error of x.

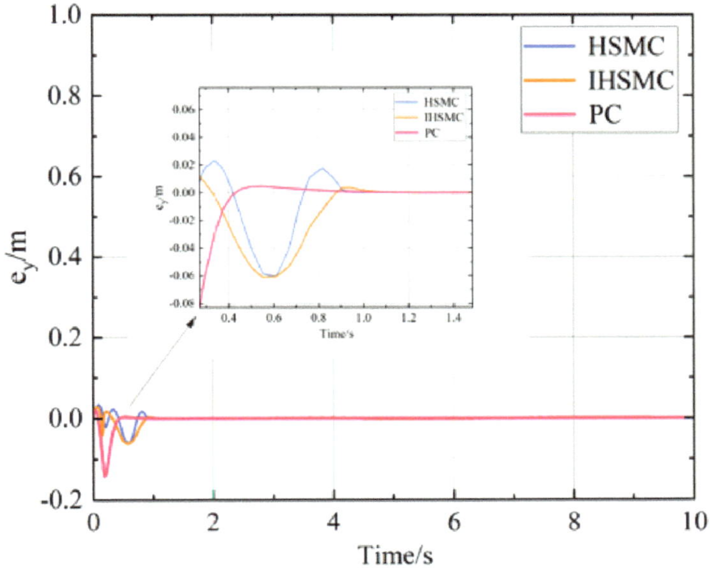

Figure 5. The error of y.

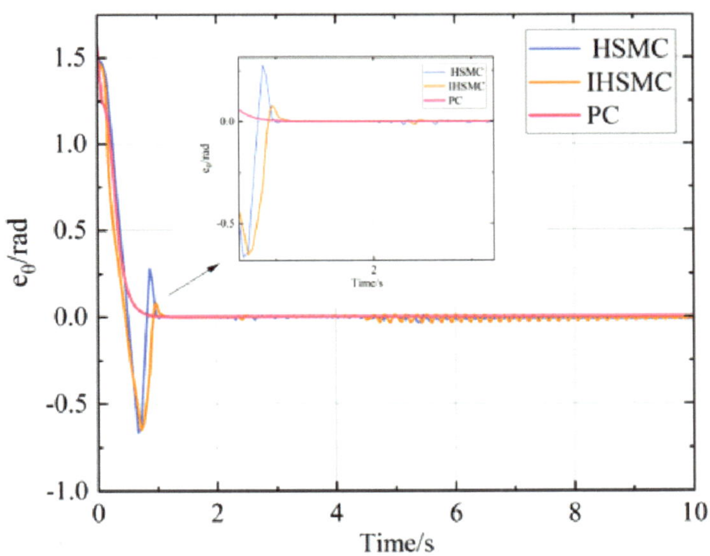

Figure 6. The error of θ.

Figures 7 and 8 give the tracking of the desired speed of the WIPR system under the three control methods. Compared with the other two methods, firstly, the control method in this paper can track the desired speed more rapidly, reaching the effect of tracking the desired speed at 0.7 s, whereas the other two methods track the desired speed in more than 1 s, which is much slower than the method in this paper, and the fluctuation frequency is high, which may affect the stability of the WIPR. As can be seen from Figure 8, the present method exceeds the other two methods in the tracking effect of rotational velocity relative to the forward velocity, for one. The convergence speed is fast, and more importantly, the proposed method is very stable after the velocity tracking reaches the steady state, which can be regarded as showing no fluctuation compared with the other two methods.

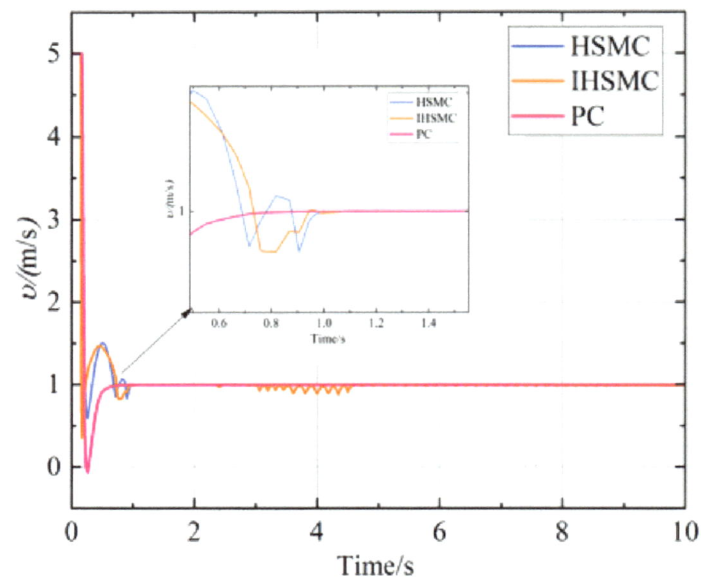

Figure 7. WIPR forward velocity v.

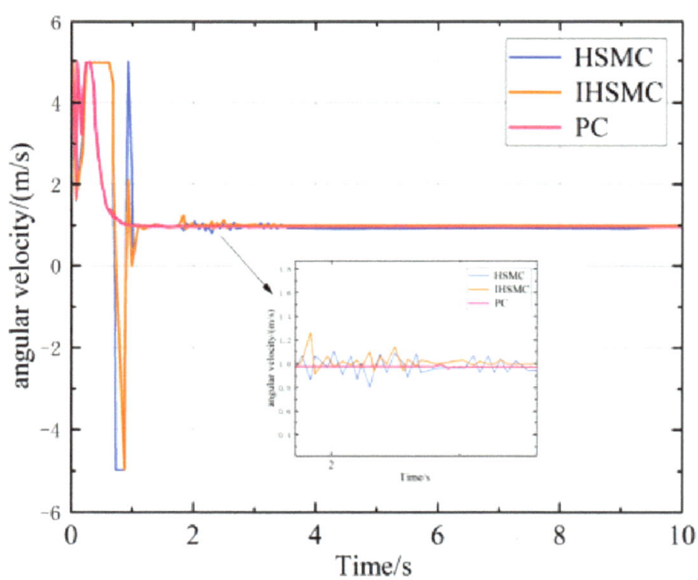

Figure 8. WIPR rotation velocity w.

The HSMC method with general exponential convergence law has more frequent angle oscillations, and the system is more unstable, as can be seen from the angle change of the WIPR pendulum shown in Figure 9, whereas the IHSMC improved convergence law method's pendulum has smoother oscillations after reaching stability, and the control effect is obviously stronger than that of the HSMC with general exponential convergence law. In terms of response time and maximum overshoot, the suggested method outperforms the other two ways, and it can continue to operate smoothly and without oscillations once it has reached the stabilization point.

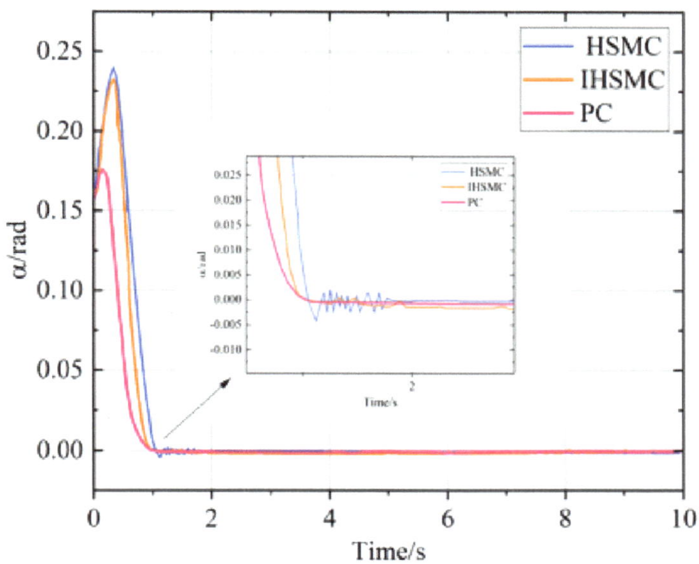

Figure 9. The angle of the WIPR pendulum α.

The variograms of the input torque for the three control methods are presented in Figures 10–12. The results indicate that when the convergence law of HSMC follows the general exponential convergence law, the jitter vibration of the input torque for the left and right wheels of WIPR is evident, which adversely affects the output of the actuator (i.e., affects the output of the drive motors of the left and right wheels). In contrast, Figure 11 illustrates that the improved convergence law significantly reduces the jitter phenomenon, resulting in a more beneficial improvement for the actuator.

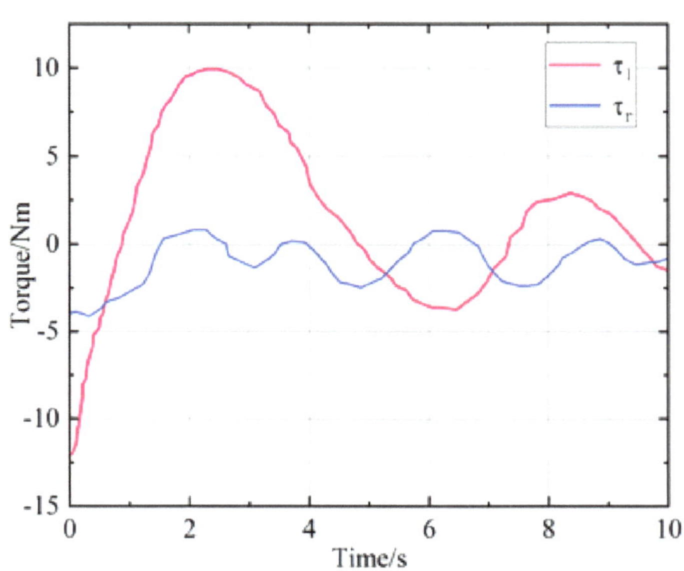

Figure 10. Input torque under HSMC.

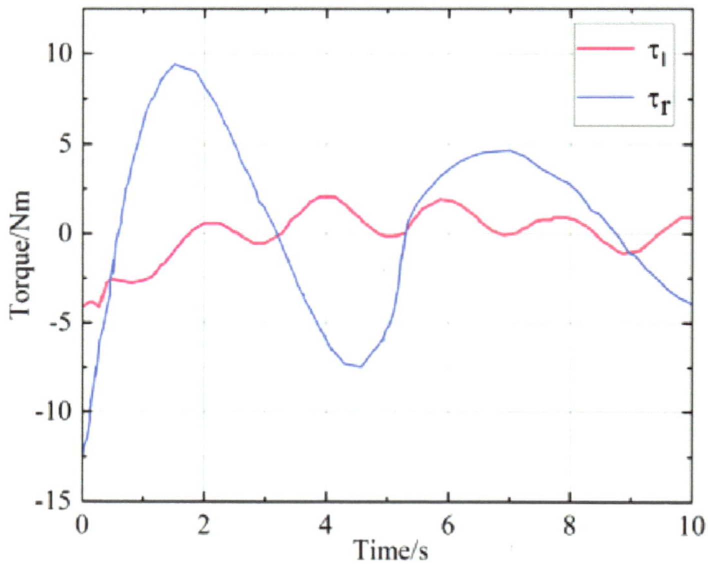

Figure 11. Input torque under IHSMC.

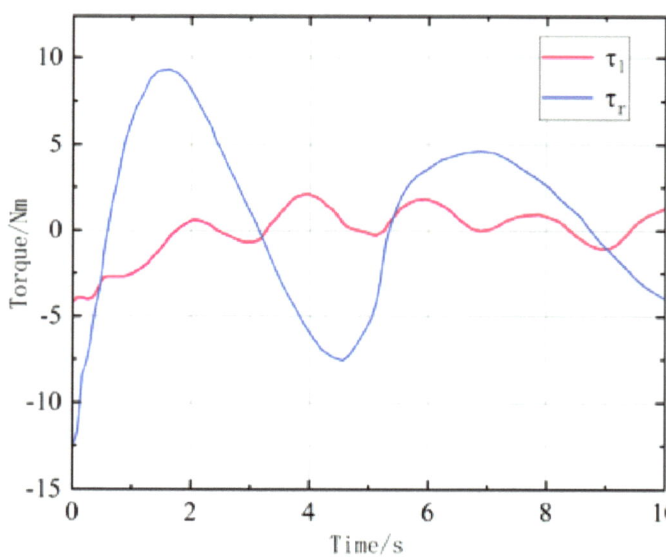

Figure 12. Input torque under PC.

Figure 12 presents the variation of input torque under the proposed control method. It can be observed that the input torque obtained by this method is smoother than IHSMC, and the jitter suppression effect is more satisfactory. This approach achieves a better torque input graph, making it the most effective method among the three for actuator benefits. Therefore, the proposed control method demonstrates superior performance in terms of reducing jitter and enhancing actuator benefits compared with the other two control methods, making it a better solution.

Figure 13 shows the trajectory tracking diagrams of different control systems. From an intuitive point of view, the proposed scheme is also significantly better than the other two schemes. As shown by the simulation comparison experiment, the method proposed in this paper is feasible, and its effect is excellent.

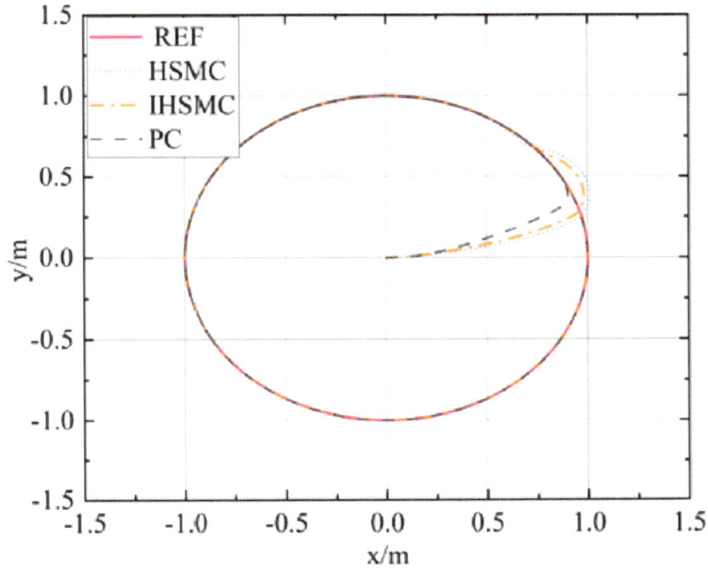

Figure 13. Tracking of circular trajectories.

4. Conclusions

The purpose of this paper was to study the trajectory-tracking problem for WIPRs and propose a hierarchical sliding mode controller with a nonlinear perturbation observer to achieve accurate control of the reference trajectory and maintain the pendulum stability during motion. A nonlinear disturbance observer was designed to make the system more robust to unknown external disturbances. The underdriven coupling of WIPRs was addressed by dividing the system into two subsystems through the decoupling of its control state variables. The hierarchical sliding mode control method with an improved convergence law was then applied to control the system and suppress the "jitter" phenomenon. Finally, the Lyapunov function was chosen to verify the stability of the system mathematically.

The feasibility of the control system was verified using simulation software. However, considering the complexity of the real-world environment and external uncertainty, future work will focus on building a hardware system for the robot to study the real effects of the control method of WIPRs in the real world.

Author Contributions: All of the nominated authors initially made significant contributions to the paper. Conceptualization, M.H. and X.Z.; methodology, X.Z.; software, X.Z.; validation, X.Z., D.C. and M.H.; formal analysis, M.H.; investigation, M.H.; resources, X.Z.; data curation, X.Z., D.C. and Z.X; writing—original draft preparation, M.H.; writing—review and editing, X.Z., D.C. and Z.X.; visualization, D.C.; supervision, M.H.; project administration, M.H.; funding acquisition, M.H. All authors have read and agreed to the published version of the manuscript.

Funding: This research was funded by the National Natural Science Foundation of China (61971048) and the Key Incubation Project of Beijing University of Information Science and Technology (5212110927).

Institutional Review Board Statement: Not applicable.

Informed Consent Statement: Not applicable.

Data Availability Statement: The data presented in this study are available on request from the corresponding author. The data are not publicly available due to internal laboratory confidentiality agreements.

Conflicts of Interest: The authors declare no conflict of interest.

References

1. Kim, S.; Kwon, S. Nonlinear Optimal Control Design for Underactuated Two-Wheeled Inverted Pendulum Mobile Platform. *IEEE/ASME Trans. Mechatron.* **2017**, *22*, 2803–2808. [CrossRef]
2. Chen, M. Robust tracking control for self-balancing mobile robots using disturbance observer. *IEEE/CAA J. Autom. Sin.* **2017**, *4*, 458–465. [CrossRef]
3. Kim, Y.; Kwon, S. Robust Stabilization of Underactuated Two-Wheeled Balancing Vehicles on Uncertain Terrains with Nonlinear-Model-Based Disturbance Compensation. *Actuators* **2022**, *11*, 339. [CrossRef]
4. Liu, J.; Vazquez, S.; Wu, L.; Marquez, A.; Gao, H.; Franquelo, L.G. Extended State Observer-Based Sliding-Mode Control for Three-Phase Power Converters. *IEEE Trans. Ind. Electron.* **2017**, *64*, 22–31. [CrossRef]
5. Yang, C.; Li, Z.; Li, J. Trajectory Planning and Optimized Adaptive Control for a Class of Wheeled Inverted Pendulum Vehicle Models. *IEEE Trans. Cybern.* **2013**, *43*, 24–36. [CrossRef] [PubMed]
6. Mirzaeinejad, H. Optimization-based nonlinear control laws with increased robustness for trajectory tracking of non-holonomic wheeled mobile robots. *Transp. Res. Part C Emerg. Technol.* **2019**, *101*, 1–17. [CrossRef]
7. Wang, F.-C.; Chen, Y.-H.; Wang, Z.-J.; Liu, C.-H.; Lin, P.-C.; Yen, J.-Y. Decoupled Multi-Loop Robust Control for a Walk-Assistance Robot Employing a Two-Wheeled Inverted Pendulum. *Machines* **2021**, *9*, 205. [CrossRef]
8. Yue, M.; Ning, Y.; Yu, S.; Zhang, Y. Composite following control for wheeled inverted pendulum vehicles based on hu-man-robot interaction. *Sci. China Inf. Sci.* **2019**, *62*, 50206.
9. Watson, M.T.; Gladwin, D.T.; Prescott, T.J.; Conran, S.O. Dual-Mode Model Predictive Control of an Omnidirectional Wheeled Inverted Pendulum. *IEEE/ASME Trans. Mechatron.* **2019**, *24*, 2964–2975. [CrossRef]
10. Albert, K.; Phogat, K.S.; Anhalt, F.; Banavar, R.N.; Chatterjee, D.; Lohmann, B. Structure-Preserving Constrained Optimal Trajectory Planning of a Wheeled Inverted Pendulum. *IEEE Trans. Robot.* **2020**, *36*, 910–923. [CrossRef]
11. Ginoya, D.; Shendge, P.D.; Phadke, S.B. Sliding Mode Control for Mismatched Uncertain Systems Using an Extended Disturbance Observer. *IEEE Trans. Ind. Electron.* **2014**, *61*, 1983–1992. [CrossRef]
12. Pathak, K.; Franch, J.; Agrawal, S. Velocity and position control of a wheeled inverted pendulum by partial feedback linearization. *IEEE Trans. Robot.* **2005**, *21*, 505–513. [CrossRef]
13. Takei, T.; Imamura, R.; Yuta, S. Baggage Transportation and Navigation by a Wheeled Inverted Pendulum Mobile Robot. *IEEE Trans. Ind. Electron.* **2009**, *56*, 3985–3994. [CrossRef]
14. Gong, S.; Zhang, A.; She, J.; Zhang, X.; Liu, Y. Trajectory Design and Tracking Control for Nonlinear Underactuated Wheeled Inverted Pendulum. *Math. Probl. Eng.* **2018**, *2018*, e6134764. [CrossRef]
15. Huang, C.-F.; Yeh, T.-J. Anti Slip Balancing Control for Wheeled Inverted Pendulum Vehicles. *IEEE Trans. Control Syst. Technol.* **2019**, *28*, 1042–1049. [CrossRef]
16. Yang, C.; Li, Z.; Cui, R.; Xu, B. Neural Network-Based Motion Control of an Underactuated Wheeled Inverted Pendulum Model. *IEEE Trans. Neural Netw. Learn. Syst.* **2014**, *25*, 2004–2016. [CrossRef] [PubMed]
17. Ren, C.; Li, X.; Yang, X.; Ma, S. Extended State Observer-Based Sliding Mode Control of an Omnidirectional Mobile Robot With Friction Compensation. *IEEE Trans. Ind. Electron.* **2019**, *66*, 9480–9489. [CrossRef]
18. Chen, L.; Wang, H.; Huang, Y.; Ping, Z.; Yu, M.; Zheng, X.; Ye, M.; Hu, Y. Robust hierarchical sliding mode control of a two-wheeled self-balancing vehicle using perturbation estimation. *Mech. Syst. Signal Process.* **2020**, *139*, 106584. [CrossRef]
19. Zhou, Y.; Wang, Z. Robust motion control of a two-wheeled inverted pendulum with an input delay based on optimal integral sliding mode manifold. *Nonlinear Dyn.* **2016**, *85*, 2065–2074. [CrossRef]
20. Adhikary, N.; Mahanta, C. Integral backstepping sliding mode control for underactuated systems: Swing-up and stabilization of the Cart–Pendulum System. *ISA Trans.* **2013**, *52*, 870–880. [CrossRef]
21. Ri, S.; Huang, J.; Wang, Y.; Kim, M.; An, S. Terminal Sliding Mode Control of Mobile Wheeled Inverted Pendulum System with Nonlinear Disturbance Observer. *Math. Probl. Eng.* **2014**, *2014*, e284216. [CrossRef]
22. Ping, H.; Hai, W.; Linfeng, L.; Huifang, K.; Ming, Y.; Canghua, J.; Zhihong, M. A novel hierarchical sliding mode control strategy for a two-wheeled self-balancing vehicle. In Proceedings of the 2017 36th Chinese Control Conference (CCC), Dalian, China, 26–28 July 2017; pp. 3731–3736. [CrossRef]
23. Moness, M.; Mahmoud, D.; Hussein, A. Real-time Mamdani-like fuzzy and fusion-based fuzzy controllers for balancing two-wheeled inverted pendulum. *J. Ambient. Intell. Humaniz. Comput.* **2020**, *13*, 3577–3593. [CrossRef]
24. Sun, W.; Su, S.-F.; Xia, J.; Wu, Y. Adaptive Tracking Control of Wheeled Inverted Pendulums with Periodic Disturbances. *IEEE Trans. Cybern.* **2020**, *50*, 1867–1876. [CrossRef] [PubMed]
25. Palm, R. Sliding mode fuzzy control. In Proceedings of the 1992 IEEE International Conference on Fuzzy Systems, San Diego, CA, USA, 8–12 March 1992; pp. 519–526. [CrossRef]
26. Jmel, I.; Dimassi, H.; Hadj-Said, S.; M'Sahli, F. An adaptive sliding mode observer for inverted pendulum under mass variation and disturbances with experimental validation. *ISA Trans.* **2020**, *102*, 264–279. [CrossRef] [PubMed]
27. Huang, J.; Guan, Z.; Matsuno, T.; Fukuda, T.; Sekiyama, K. Sliding-Mode Velocity Control of Mobile-Wheeled Inverted-Pendulum Systems. *IEEE Trans. Robot.* **2010**, *26*, 750–758. [CrossRef]

28. Fukushima, H.; Muro, K.; Matsuno, F. Sliding-Mode Control for Transformation to an Inverted Pendulum Mode of a Mobile Robot with Wheel-Arms. *IEEE Trans. Ind. Electron.* **2015**, *62*, 4257–4266. [CrossRef]
29. Chen, X.; Komada, S.; Fukuda, T. Design of a nonlinear disturbance observer. *IEEE Trans. Ind. Electron.* **2000**, *47*, 429–437. [CrossRef]
30. Saradagi, A.; Muralidharan, V.; Krishnan, V.; Menta, S.; Mahindrakar, A.D. Formation Control and Trajectory Tracking of Nonholonomic Mobile Robots. *IEEE Trans. Control Syst. Technol.* **2017**, *26*, 2250–2258. [CrossRef]
31. Yoshida, K.; Sekikawa, M.; Hosomi, K. Nonlinear analysis on purely mechanical stabilization of a wheeled inverted pendulum on a slope. *Nonlinear Dyn.* **2016**, *83*, 905–917. [CrossRef]
32. Chen, W.-H.; Yang, J.; Guo, L.; Li, S. Disturbance-Observer-Based Control and Related Methods—An Overview. *IEEE Trans. Ind. Electron.* **2016**, *63*, 1083–1095. [CrossRef]
33. Huang, J.; Ri, S.; Liu, L.; Wang, Y.; Kim, J.; Pak, G. Nonlinear Disturbance Observer-Based Dynamic Surface Control of Mobile Wheeled Inverted Pendulum. *IEEE Trans. Control Syst. Technol.* **2015**, *23*, 2400–2407. [CrossRef]
34. Xie, L.; Yu, X. State Observer Based Robust Backstepping Fault-Tolerant Control of the Free-Floating Flexible-Joint Space Manipulator. *Appl. Sci.* **2023**, *13*, 2634. [CrossRef]
35. Yue, M.; Wei, X.; Li, Z. Adaptive sliding-mode control for two-wheeled inverted pendulum vehicle based on zero-dynamics theory. *Nonlinear Dyn.* **2014**, *76*, 459–471. [CrossRef]
36. Chang, W.-J.; Hsu, F.-L. Sliding mode fuzzy control for Takagi–Sugeno fuzzy systems with bilinear consequent part subject to multiple constraints. *Inf. Sci.* **2016**, *327*, 258–271. [CrossRef]
37. Xu, J.-X.; Guo, Z.-Q.; Lee, T.H. Design and Implementation of Integral Sliding-Mode Control on an Underactuated Two-Wheeled Mobile Robot. *IEEE Trans. Ind. Electron.* **2014**, *61*, 3671–3681. [CrossRef]
38. Liu, Y.; Jing, Y.W.; Liu, X.P.; Li, X.H. Survey on finite-time control for nonlinear systems. *Control Theory Appl.* **2020**, *37*, 1–12.

applied sciences

MDPI

Article

Ship Defense Strategy Using a Planar Grid Formation of Multiple Drones

Jonghoek Kim

System Engineering Department, Sejong University, Seoul 2639, Republic of Korea; jonghoek@gmail.com

Abstract: This article introduces a ship defense strategy using a planar grid formation of multiple drones. We handle a scenario where a high-speed target with variable velocity heads towards the ship. The ship measures the position of the target in real time. Based on the measured target, the drones guidance laws are calculated by the ships on-board computer and are sent to every drone in real time. The drones form a planar grid formation, whose center blocks the Line-Of-Sight (LOS) line connecting the target and the ship. Since the target is guided to hit its goal (ship), the drones can effectively block the target by blocking the LOS line. We enable slow drones to capture a fast target by making the drones stay close to the ship while blocking the LOS at all times. By using a grid formation of drones, we can increase the capture rate, even when there exists error in the prediction of the target's position. To the best of our knowledge, this article is unique in using a formation of multiple drones to intercept a fast target with variable velocity. Through MATLAB simulations, the effectiveness of our multi-agent guidance law is verified by comparing it with other state-of-the-art guidance controls.

Keywords: ship defense; drone; target tracking; multiple drones; multi-agent guidance law; fast target; grid formation; high-speed target

Citation: Kim, J. Ship Defense Strategy Using a Planar Grid Formation of Multiple Drones. *Appl. Sci.* **2023**, *13*, 4397. https://doi.org/10.3390/app13074397

Academic Editor: Vincent A. Cicirello

Received: 7 March 2023
Revised: 23 March 2023
Accepted: 28 March 2023
Published: 30 March 2023

1. Introduction

This article introduces a ship defense strategy using a clustered formation of multiple drones. We handle a scenario where a high-speed target with variable velocity heads towards the ship. The role of the drone team is to protect the ship from the incoming target.

We propose a ship defense approach with multiple drones, such that each drone is not equipped with powerful sensors or an on-board computer. Thus, a drone cannot measure the target, and the target moves based on the commands sent by the ship. In this way, we can decrease the cost of a drone, which can be destroyed once it intersects the target. This enables us to develop rather cheap drones.

Instead, the ship measures the position of the target in real time. Position measurements can be provided by various sensors, such as radar, IR, or laser sensors of the ship. Each drone's guidance law is calculated by the ship's on-board computer and is transmitted to each drone in real time. (This approach relies on the communication between the ship and a drone. Since the signal speed is sufficiently fast (3×10^8 m/s) in the air, we argue that signal delay is negligible in our ship defense scenario.)

Consider a high-speed target whose goal is to hit a ship. The target heads towards its goal (ship) at least in the terminal phase. Otherwise, it is impossible to make a target hit the ship.

Therefore, this paper lets multiple drones form a planar grid formation, whose center lies on the line segment connecting the target and the ship. Moreover, the planar grid formation is generated to be perpendicular to the line segment connecting the target and the formation center. The grid formation can be considered as a "net" structure for capturing the incoming target. The target may perform elusive maneuvers, and there may be measurement noise in measuring the target position. By maximizing the grid formation size, we can increase the capture rate, even when there exists error in the prediction of the target's position. Since the target is guided to hit its goal (ship), the drones can effectively block the target using this grid formation.

As an interceptor, we consider a highly maneuverable drone, such as a quadrotor drone [1–3], which is much slower than the incoming target. This article addresses a high-level path planner, which generates reference position signals for a low-level controller [1–3].

To the best of our knowledge, our paper is novel in developing a ship defense approach using clustered multiple drones. Our paper is novel in addressing a 3D formation of multiple drones for intercepting a high-speed target with variable speeds. We show that in the case where the drones block the LOS line between the target and the ship, the target cannot reach the ship without being captured by the drones. Since the target is guided to hit its goal (ship), the drones can effectively block the target using this strategy. We further let slow drones stay close to the ship in order to protect the ship from the fast target. As far as we know, our paper is novel in showing that slow drones can capture a fast target by staying close to the ship while blocking the LOS at all times.

Our paper is unique in capturing a maneuvering target with variable speeds, which can be faster than the interceptors. In order to estimate the pose of a maneuvering high-speed target, we applied target-tracking filters in [4]. We control the drone formation, based on the prediction of the target's position after one sample-index in the future. This prediction may be erroneous due to the target's elusive maneuvers or measurement noise. This is the motivation for utilizing a grid formation of drones instead of a single drone. Since we use a grid formation, we can increase the capture rate even when the target prediction is erroneous. By maximizing the grid formation size, we can increase the capture rate even when there exists error in the prediction of the target's position.

To the best of our knowledge, our paper is novel in the following aspects:

1. We develop a ship defense approach using clustered multiple drones;
2. We use a 3D formation of multiple drones for intercepting a high-speed target with variable speeds;
3. We let the drones stay close to the ship while blocking the LOS between the ship and the target. Thus, we enable slow drones to capture a fast target.

Through MATLAB simulations, the effectiveness of our multi-agent guidance law is verified by comparing it with other state-of-the-art guidance controls.

We organize this article as follows. Section 2 addresses the literature review of this paper. Section 3 addresses the preliminary information of this paper. Section 4 discusses several definitions and assumptions in this article. Section 5 introduces our multi-agent guidance law. Section 6 shows simulation results to present the effectiveness of the proposed guidance law. Section 7 provides a conclusion.

2. Literature Review

There are many papers on interceptors' guidance laws [5–8]. The authors of [9–16] applied motion camouflage to develop the guidance law of an interceptor. Here, we say that the interceptor is in the motion camouflage state if an interceptor moves in the presence of a target while appearing stationary at a focal point.

References [17,18] developed a motion camouflage guidance law so that the interceptor approaches the target while appearing stationary at a focal point that is infinitely far from the interceptor. The authors of [19] used a neural network architecture to perform motion camouflage in 2D environments. The authors of [12] developed an optimal control approach to derive a 2D motion camouflage position for an interceptor, assuming there is a constant velocity (speed and heading) target. However, assuming a target with constant velocity is not realistic since a maneuvering target could escape from the interceptor. Our paper thus handles a target with variable velocity.

Proportional Navigation Guidance (PNG) laws have been widely applied to let an interceptor hit the target [20–23]. PNG laws are designed considering an interceptor that can measure the bearing of the target by utilizing on-board sensors. PNG laws are based on the fact that two vehicles are on a collision course when their direct Line-Of-Sight (LOS) does not change direction as they get closer to each other. PNG laws are designed so that the interceptor velocity vector rotates at a rate proportional to the rotation rate of the line-of-sight and in the same direction.

Multi-agent systems can be applied for many tasks, such as monitoring environments [24,25], multi-agent herding [26], and sensor deployment [27–30]. References [31,32] controlled multiple mobile sensors to estimate the target position in real time. References [33,34] considered the case where two interceptors, which measure bearings of a target, track the target in two dimensions. The formulation of the homing problem of multiple missiles against a single target, subject to constraints on the impact time, was discussed in [35]. In [36], a fully distributed adaptive method was proposed to solve the simultaneous attack problem with multiple missiles against maneuvering targets. The authors of [37] considered the relative interception angle constraints of multiple interceptors, which is intended to enhance the survivability of multiple interceptors against a defense system with a high value target and also to maximize the collateral target damage. The authors of [38] addressed simultaneous cooperative interception for a scenario where the successful handover cannot be guaranteed by a single interceptor due to the target maneuver and movement information errors at the handover moment.

As far as we know, other guidance laws in the literature make one or more interceptors continue to chase the target. Our paper is unique in making slow interceptors (drones in our paper) stay close to the ship, so they can block a fast target from reaching the ship. This blocking strategy is desirable considering the energy consumption of an interceptor since an interceptor does not have to move far from the ship. Through MATLAB simulations, the effectiveness of this blocking strategy is verified by comparing it with other state-of-the-art guidance controls.

3. Preliminaries

This article utilizes two frames: an inertial reference frame $\{I\}$ and a body-fixed frame $\{B\}$ [39]. We address several definitions in rigid-body dynamics [39].

The origin of $\{I\}$ is a point with three axes pointing North, East, and Down, respectively. We use the *virtual agent* for drone controls. The virtual agent is a virtual drone located at the center of the grid formation. $\{B\}$ is fixed to the virtual agent, such that the origin of $\{B\}$ is at the virtual agent's center.

The virtual agent changes its yaw and pitch while not rotating its body. In rigid-body dynamics [39], θ and ψ define *pitch* and *yaw*, respectively. For convenience, let $c(\eta)$ define $\cos(\eta)$. In addition, let $s(\eta)$ define $\sin(\eta)$. Let $t(\eta)$ define $\tan(\eta)$.

The rotation matrix indicating the counterclockwise (CC) rotation of an angle ψ centered at the z-axis in $\{B\}$ is

$$\mathbf{M}_R(\psi) = \begin{pmatrix} c(\psi) & -s(\psi) & 0 \\ s(\psi) & c(\psi) & 0 \\ 0 & 0 & 1 \end{pmatrix}. \tag{1}$$

The rotation matrix representing the CC rotation of an angle θ centered at the y-axis in $\{B\}$ is

$$\mathbf{M}_R(\theta) = \begin{pmatrix} c(\theta) & 0 & s(\theta) \\ 0 & 1 & 0 \\ -s(\theta) & 0 & c(\theta) \end{pmatrix}. \tag{2}$$

The combined rotation matrix is built by multiplying Equations (1) and (2) to obtain

$$\mathbf{M}_R(\psi,\theta) = \mathbf{M}_R(\psi)\mathbf{M}_R(\theta). \tag{3}$$

4. Assumptions and Definitions

This section discusses assumptions and definitions in our paper. $max(a,b)$ returns a bigger value between two variables (a and b). In addition, $min(a,b)$ returns a smaller value between two variables (a and b). In our paper, bold characters are used to denote vectors and matrices. $\angle(\mathbf{v}_1, \mathbf{v}_2)$ is the angle formed by two vectors ($\mathbf{v}_1$ and $\mathbf{v}_2$). Mathematically, $\angle(\mathbf{v}_1, \mathbf{v}_2) = \arccos(\frac{\mathbf{v}_1 \cdot \mathbf{v}_2}{\|\mathbf{v}_1\| * \|\mathbf{v}_2\|})$. Here, $0 \leq \angle(\mathbf{v}_1, \mathbf{v}_2) \leq \pi$. $l(\mathbf{A}, \mathbf{B})$ is the line segment connecting two locations $\mathbf{A}$ and $\mathbf{B}$. Furthermore, $\|l(\mathbf{A}, \mathbf{B})\|$ indicates the length of $l(\mathbf{A}, \mathbf{B})$.

This article uses the discrete-time system, where T denotes the sample duration. In this article, all drones make a planar grid formation to protect against the incoming target. The grid formation can be considered as a "net" structure for capturing the incoming target.

Let M indicate the total number of drones. M is selected such that

$$M = G^2, \tag{4}$$

where $G \geq 1$ is a positive integer.

In the case where $G = 1$, we use only one drone. In this case, the grid formation cannot be used, and the waypoint of the drone is set as the virtual agent.

In the inertial reference frame, let $\mathbf{r}_{0,k}$ define the 3D Cartesian coordinates of the virtual agent. In the inertial reference frame, let $\mathbf{r}_{i,k}$ ($i \in \{1, 2, \ldots, M\}$) denote the 3D Cartesian coordinates of the i-th drone at sample-index k. Note that the subscript k indicates the sample-index k.

In the inertial reference frame, let $\mathbf{r}^t_k$ denote the target's 3D Cartesian coordinates at sample-index k. In the inertial reference frame, let $\mathbf{r}^s_k$ denote the ship's 3D Cartesian coordinates at sample-index k.

Let v^t_k denote the target's speed at sample-index k. Let $v_{i,k}$ denote the speed of the i-th drone at sample-index k. Let v_{max} indicate the maximum speed of a drone or the virtual agent. Note that v^t_k, $v_{i,k}$, and v_{max} are scalar values.

We say that the target is *captured* when the relative distance between the target and any drone is less than a constant, say Δ. The motion model of the i-th drone ($i \in \{1, 2, \ldots, M\}$) is

$$\mathbf{r}_{i,k+1} = \mathbf{r}_{i,k} + T * v_{i,k} * \mathbf{u}_{i,k}. \tag{5}$$

Here, $\mathbf{u}_{i,k}$ indicates the i-th drone's heading vector at sample-index k. Note that $\mathbf{u}_{i,k}$ is a unit vector presenting the i-th drone's heading direction. The motion model in Equation (5) is commonly used in multi-drone systems [40–46].

In Equation (5), $\mathbf{r}_{i,k+1}$ generates the high-level reference position signal at every sample-index k. For letting the i-th drone move towards $\mathbf{r}_{i,k+1}$ at every sample-index k, one utilizes low-level controls in [1–3].

The motion model of the virtual agent is

$$\mathbf{r}_{0,k+1} = \mathbf{r}_{0,k} + T * v_{0,k} * \mathbf{u}_{0,k}. \tag{6}$$

Recall that the virtual agent is at the center of the grid formation. We say that the virtual agent is *in the lineState* at sample-index k if the line segment $l(\mathbf{r}^t_k, \mathbf{r}^s_k)$ meets the virtual agent position $\mathbf{r}_{0,k}$. At every sample-index k, $v_{0,k}$ and $\mathbf{u}_{0,k}$ are set so that the virtual agent is in the lineState.

Let L_k denote the infinite line crossing both $\mathbf{r}^s_k$ and $\mathbf{r}^t_k$ at sample-index k. Let $\bar{L}_{k+1}$ denote the infinite line crossing both $\mathbf{r}^s_{k+1}$ and $\mathbf{r}^t_{k+1}$. $\mathbf{c}_k$ is the point on $\bar{L}_{k+1}$, which is the closest to $\mathbf{r}_{0,k}$.

Figure 1 depicts the case where the ship moves as the time index changes from k to $k+1$. The ship positions are indicated by crosses. In this figure, L_k, $\bar{L}_{k+1}$, and $\mathbf{c}_k$ are depicted.

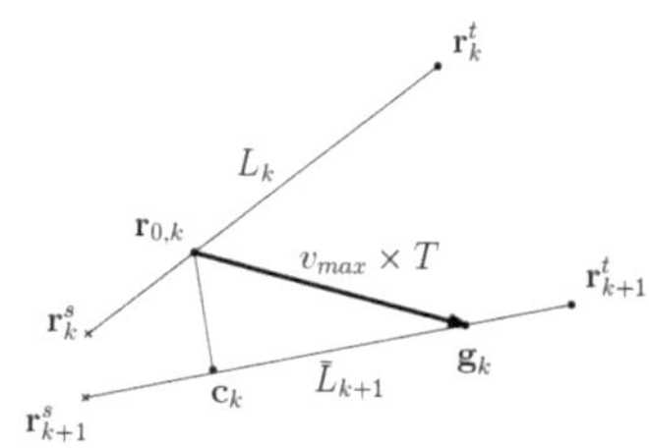

Figure 1. The case where the ship moves as the time index changes from k to $k+1$. The ship positions are indicated by crosses. In this figure, L_k, $\bar{L}_{k+1}$, and $\mathbf{c}_k$ are depicted.

Assumptions

This article assumes that both the ship and a drone's 3D Cartesian coordinates are measured in real time. Global Positioning Systems (GPSs) and Inertial Measurement Units (IMUs) can be used for this localization.

Furthermore, the ship can measure the target's 3D Cartesian coordinates at every sample-index. Position measurements can be provided by various sensors, such as radar sensors or laser sensors. Therefore, the ship at sample-index k can derive L_k.

Furthermore, based on the target's recent trajectory, the ship at sample-index k can predict $\mathbf{r}^t_{k+1}$, the target's 3D Cartesian coordinates, after one sample-index in the future. Section 5.1 shows how to predict the target's 3D Cartesian coordinates after one sample-index in the future.

The ship can also predict $\mathbf{r}^s_{k+1}$, the ship's 3D Cartesian coordinates, after one sample-index in the future. This is feasible because the ship has GPS and IMU. Therefore, the ship can predict $\bar{L}_{k+1}$, which crosses both $\mathbf{r}^t_{k+1}$ and $\mathbf{r}^s_{k+1}$.

It is desirable that as the target is caught, it is sufficiently far from the ship. Otherwise, the ship may be partially caught by the debris of the target. Let $\beta > 0$ define the *safety distance*. The safety distance is set by the operator of the drones. As the target is caught, it is desirable that its distance from the ship is bigger than the safety distance β. This way, we can assure the safety of the ship.

5. Multi-Drone Guidance Law

We consider a high-speed target whose goal is to hit a ship. The target heads towards its goal (ship) at least in the terminal phase. Otherwise, it is impossible to make a target hit the ship.

Therefore, we let multiple drones form a planar grid formation, whose center lies on the line segment connecting the target and the ship. Moreover, the planar grid formation is generated to be perpendicular to the line segment connecting the target and the formation center. The grid formation can be considered as a "net" structure for capturing the incoming target. By maximizing the grid formation size, we can increase the capture rate, even when there exists error in the prediction of the target's position. Since the target is guided to hit its goal (the ship), the drones can effectively block the target using this grid formation.

The proposed multi-drone guidance law is summarized as follows. At every sample-index, the ship measures the 3D Cartesian coordinates of the incoming target. Thereafter, we run the Kalman filter to predict the target's 3D Cartesian coordinates after one sample-index in the future. See Section 5.1 for the prediction of the target's position after one sample-index.

Based on the predicted target pose, the virtual agent is guided to remain in the lineState. See Section 5.2 for the guidance law of the virtual agent. In addition, each drone is guided to generate a grid formation centered at the virtual agent. See Section 5.3 for the guidance law of a drone. Figure 2 shows the block diagram of the proposed multi-drone guidance law.

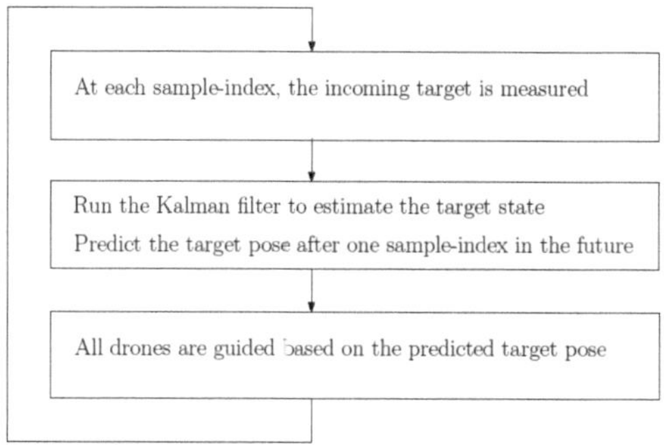

Figure 2. Block diagram of the proposed multi-drone guidance law.

5.1. Prediction of the Target's Position after One Sample-Index

The ship can measure the target's 3D Cartesian coordinates at every sample-index. To track a target with variable velocity, we present how to predict the target one sample-index forward in time. In order to track a maneuvering target, we applied target-tracking filters to [4].

In the inertial reference frame, let $[x_k^t, y_k^t, z_k^t]$ indicate the vector presenting the 3D coordinates of the target at sample-index k. In addition, $[\dot{x}_k^t, \dot{y}_k^t, \dot{z}_k^t]^T$ denotes the vector presenting the target velocity at sample-index k. Furthermore, $[\ddot{x}_k^t, \ddot{y}_k^t, \ddot{z}_k^t]^T$ defines the vector presenting the target acceleration at sample-index k. Let $\mathbf{X}_k = [x_k^t, \dot{x}_k^t, \ddot{x}_k^t, y_k^t, \dot{y}_k^t, \ddot{y}_k^t, z_k^t, \dot{z}_k^t, \ddot{z}_k^t]^T$ define the vector presenting the target state. Based on [4], the target's process model is set as

$$\mathbf{X}_{k+1} = \mathbf{F} * \mathbf{X}_k + \mathbf{w}_k, \tag{7}$$

where $\mathbf{w}_k$ is the process noise with following properties: $\mathbf{w}_k \sim \mathcal{N}(\mathbf{0}, \mathbf{Q})$. Here, $\mathcal{N}(\mathbf{0}, \alpha)$ denotes a Gaussian distribution with a mean of $\mathbf{0}$ and a covariance matrix α. Furthermore, $\mathbf{F}$ in Equation (7) is

$$\mathbf{F} = \begin{pmatrix} \mathbf{M}_F & 0 & 0 \\ 0 & \mathbf{M}_F & 0 \\ 0 & 0 & \mathbf{M}_F \end{pmatrix}, \tag{8}$$

where we use

$$\mathbf{M}_F = \begin{pmatrix} 1, & T, & \frac{a*T-1+e^{-a*T}}{a^2} \\ 0, & 1, & \frac{1+e^{-a*T}}{a^2} \\ 0, & 0, & e^{-a*T} \end{pmatrix}. \tag{9}$$

In $\mathbf{w}_k$, $\mathbf{Q}$ is set as

$$\mathbf{Q} = 2 * a * \sigma_m^2 * \begin{pmatrix} \mathbf{M}_Q & 0 & 0 \\ 0 & \mathbf{M}_Q & 0 \\ 0 & 0 & \mathbf{M}_Q \end{pmatrix}, \tag{10}$$

where

$$\mathbf{M}_Q = \begin{pmatrix} T^5/20, & T^4/8, & T^3/6 \\ T^4/8, & T^3/6, & T^2/2 \\ T^3/6, & T^2/2, & T \end{pmatrix}. \tag{11}$$

a and σ_m in Equation (10) are tuning parameters for tracking a maneuvering target. Detailed derivations of Equation (7) appear in [4].

At every sample-index k, the ship measures the target's 3D position, say $\mathbf{m}_k$. See the first block of Figure 2. The target measurement model is

$$\mathbf{m}_k = \mathbf{H} * \mathbf{X}_k + \mathbf{v}_k \quad (12)$$

where $\mathbf{H}$ is

$$\mathbf{H} = \begin{pmatrix} 1 & 0 & 0 & 0 & 0 & 0 & 0 & 0 & 0 \\ 0 & 0 & 0 & 1 & 0 & 0 & 0 & 0 & 0 \\ 0 & 0 & 0 & 0 & 0 & 0 & 1 & 0 & 0 \end{pmatrix}. \quad (13)$$

Furthermore, $\mathbf{v}_k$ is the measurement noise, such that $\mathbf{v}_k \sim \mathcal{N}(\mathbf{0}, \mathbf{R}_k)$. We assume that $\mathbf{R}_k$ is known a priori.

The Kalman filter (KF) [47] is applied to obtain the estimate vector and its covariance at every sample-index. The KF is composed of the prediction step and the measurement update step. In the KF, the prediction step uses Equation (7), and the measurement update step uses Equation (12).

Let $\hat{\mathbf{X}}_{k|k}$ define the estimation of $\mathbf{X}_k$ derived using all measurements up to sample-index k. Let $\mathbf{P}_{k|k}$ define the error covariance matrix of $\hat{\mathbf{X}}_{k|k}$.

In the prediction step of the KF, we derive the predicted state vector as

$$\hat{\mathbf{X}}_{k+1|k} = \mathbf{F} * \hat{\mathbf{X}}_{k|k}, \quad (14)$$

where Equation (7) is used. Utilizing Equations (7) and (10), the covariance matrix is predicted as

$$\mathbf{P}_{k+1|k} = \mathbf{F} * \mathbf{P}_{k|k} * \mathbf{F}^T + \mathbf{Q}. \quad (15)$$

The measurement update step is

$$\hat{\mathbf{X}}_{k+1|k+1} = \hat{\mathbf{X}}_{k+1|k} + \mathbf{W}_k * (\mathbf{m}_k - \mathbf{H} * \hat{\mathbf{X}}_{k+1|k}), \quad (16)$$

where

$$\mathbf{W}_k = \mathbf{P}_{k+1|k} * \mathbf{H}^T * \mathbf{S}^{-1}. \quad (17)$$

Here, we use

$$\mathbf{S} = \mathbf{H}\mathbf{P}_{k+1|k}\mathbf{H}^T + \mathbf{R}_k. \quad (18)$$

In addition, the covariance matrix is updated using

$$\mathbf{P}_{k+1|k+1} = \mathbf{P}_{k+1|k} - \mathbf{W}_k * \mathbf{S} * \mathbf{W}_k^T. \quad (19)$$

The ship at sample-index k predicts the target state after one sample-index forward in time using Equation (14). Let $\hat{\mathbf{r}}^t_{k+1}$ denote the target position at sample-index $k+1$, which is predicted using all measurements up to sample-index k. Using Equation (14), $\hat{\mathbf{r}}^t_{k+1}$ is predicted as

$$\hat{\mathbf{r}}^t_{k+1} = \mathbf{H} * \hat{\mathbf{X}}_{k+1|k}. \quad (20)$$

Here, recall that $\mathbf{H}$ was defined in Equation (13).

We acknowledge that the prediction of the target's position may not be accurate due to the target's elusive maneuvers or measurement noise. This is the motivation for using a formation of drones instead of a single drone.

In our paper, a formation of drones is used instead of a single drone in order to increase the capture rate. Since we use a drone formation, we can increase the capture rate even when the target prediction is erroneous. Through MATLAB simulations, the effectiveness of our formation-based guidance law is verified by comparing it with other state-of-the-art guidance controls.

5.2. Guidance Law of the Virtual Agent

Using the predicted 3D coordinates of the target, the virtual agent is guided to remain in the lineState. At every sample-index k, the virtual agent is guided to head towards the *guidance point* $\mathbf{g}_k$, which is defined as follows:

1. Suppose that $\|\mathbf{r}^s_k - \mathbf{r}_{0,k}\| < \beta$. This implies that the ship needs to expel the virtual agent away from the ship. Moreover, suppose that

$$\|\mathbf{r}_{0,k} - \mathbf{c}_k\| \leq v_{max} * T \tag{21}$$

holds. Then, we set the guidance point as

$$\mathbf{g}_k = \mathbf{c}_k + \frac{\mathbf{r}^t_{k+1} - \mathbf{c}_k}{\|\mathbf{r}^t_{k+1} - \mathbf{c}_k\|} * \delta_k. \tag{22}$$

Here, $\delta_k > 0$ is defined as

$$\delta_k = \sqrt{(v_{max} * T)^2 - (\|\mathbf{r}_{0,k} - \mathbf{c}_k\|)^2}. \tag{23}$$

This implies that $\|\mathbf{g}_k - \mathbf{r}_{0,k}\|$ is $v_{max} * T$. See Figure 1 for an illustration of this case.

2. Otherwise, we set the guidance point as

$$\mathbf{g}_k = \mathbf{c}_k. \tag{24}$$

At every sample-index k, the virtual agent moves to reach $\mathbf{g}_k$ if possible. Suppose that $\|\mathbf{r}^s_k - \mathbf{r}_{0,k}\| < \beta$. Furthermore, suppose that Equation (21) holds, as depicted in Figure 1. Consider a sphere centered at $\mathbf{r}_{0,k}$, whose radius is $v_{max} * T$. Using Equation (21), L_{k+1} meets this sphere at two points. Between these two points, $\mathbf{g}_k$ is the point that is closer to $\mathbf{r}^t_{k+1}$. In this way, the virtual agent can approach the target while staying on the line segment that connects the target and the ship.

The direction command $\mathbf{u}_{0,k}$ is selected to make the virtual agent move towards $\mathbf{g}_k$ at sample-index $k+1$. At every sample-index k, the direction command is set as follows. *At every sample-index k, the virtual agent sets the new direction command* $\mathbf{u}_{0,k}$ *as*

$$\mathbf{u}_{0,k} = \frac{\mathbf{g}_k - \mathbf{r}_{0,k}}{\|\mathbf{g}_k - \mathbf{r}_{0,k}\|}. \tag{25}$$

Note that the direction command is a unit vector.

In addition, the speed command at sample-index k is set as follows. *At every sample-index k, the virtual agent sets the new speed command* $v_{0,k}$ *as*

$$v_{0,k} = min(\frac{\|\mathbf{g}_k - \mathbf{r}_{0,k}\|}{T}, v_{max}). \tag{26}$$

This implies that the virtual agent moves with the maximum speed v_{max} when it is too far from the guidance point $\mathbf{g}_k$.

Consider the situation in which $\|\mathbf{r}_{0,k} - \mathbf{r}^t_{k+1}\| < v_{max} * T$ holds. In this situation, the virtual agent heads towards $\mathbf{r}^t_{k+1}$ directly while not using the direction command $\mathbf{u}_{0,k} = \frac{\mathbf{g}_k - \mathbf{r}_{0,k}}{\|\mathbf{g}_k - \mathbf{r}_{0,k}\|}$. In this way, the target is caught at sample-index $k+1$.

According to the definitions of the guidance point $\mathbf{g}_k$ (see Equations (22) and (24)), $\mathbf{g}_k$ lies on $l(\mathbf{r}^t_{k+1}, \mathbf{r}^s_{k+1})$.

In the case where $\frac{\|\mathbf{g}_k-\mathbf{r}_{0,k}\|}{T} \leq v_{max}$, the heading command (Equation (25)) and speed command (Equation (26)) lead to

$$\mathbf{r}_{0,k+1} = \mathbf{g}_k. \tag{27}$$

We thus have the following theorem.

Theorem 1. *Suppose that* $\frac{\|\mathbf{c}_k-\mathbf{r}_{0,k}\|}{T} \leq v_{max}$. *The heading command (Equation (25)) and speed command (Equation (26)) make* $\mathbf{r}_{0,k+1}$ *exist on* $l(\mathbf{r}^t_{k+1}, \mathbf{r}^s_{k+1})$.

Theorem 1 implies that in the case where $\frac{\|\mathbf{c}_k-\mathbf{r}_{0,k}\|}{T} \leq v_{max}$, the virtual agent position $\mathbf{r}_{0,k+1}$ is on the line segment $l(\mathbf{r}^t_{k+1}, \mathbf{r}^s_{k+1})$. Since the target's goal is reaching the ship, the target must be hit by the virtual agent eventually.

L_k and $\bar{L}_{k+1}$ Meet at a Point

Next, we consider a special case where L_k and $\bar{L}_{k+1}$ meet at a point. Here, recall that $\bar{L}_{k+1}$ denotes the infinite line crossing both $\mathbf{r}^s_{k+1}$ and $\mathbf{r}^t_{k+1}$. In the inertial reference frame, let $\mathbf{I}_L$ denote the 3D coordinates of the intersection between L_k and $\bar{L}_{k+1}$. For instance, if the ship is static, then L_k and $\bar{L}_{k+1}$ meet at the ship position $\mathbf{r}^s_k = \mathbf{r}^s_{k+1}$.

Suppose that $\mathbf{r}_{0,k}$ lies on $l(\mathbf{r}^t_k, \mathbf{r}^s_k)$. Let $\mathbf{c}^t_k$ denote the point on $\bar{L}_{k+1}$, which is the closest to $\mathbf{r}^t_k$. Let $l_k = \|\mathbf{c}_k - \mathbf{r}_{0,k}\|$ for convenience.

Figure 3 depicts the case where L_k and $\bar{L}_{k+1}$ meet at $\mathbf{I}_L$. Using the geometry in this figure, we have

$$\|\mathbf{c}^t_k - \mathbf{r}^t_k\| = \frac{l_k * \|\mathbf{I}_L - \mathbf{r}^t_k\|}{\|\mathbf{I}_L - \mathbf{r}_{0,k}\|}. \tag{28}$$

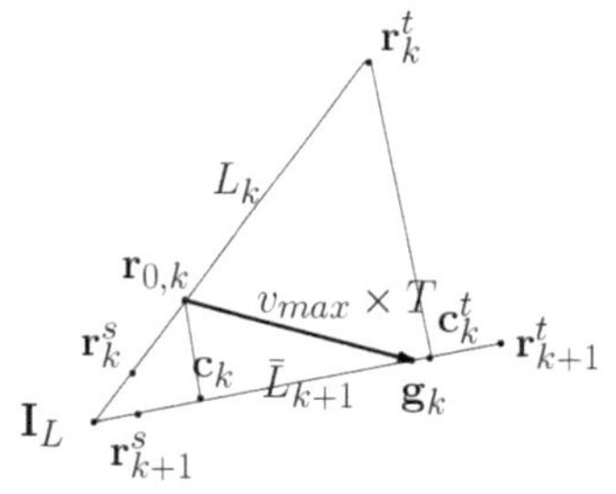

Figure 3. The case where L_k and $\bar{L}_{k+1}$ meet at $\mathbf{I}_L$.

Since the target speed is v^t_k, we have

$$v^t_k * T \geq \|\mathbf{c}^t_k - \mathbf{r}^t_k\|. \tag{29}$$

Using Equations (28) and (29), we have

$$\frac{v^t_k * T * \|\mathbf{I}_L - \mathbf{r}_{0,k}\|}{\|\mathbf{I}_L - \mathbf{r}^t_k\|} \geq l_k. \tag{30}$$

Suppose that the drone's maximum speed v_{max} satisfies

$$v_{max} > \frac{v^t_k * \|\mathbf{I}_L - \mathbf{r}_{0,k}\|}{\|\mathbf{I}_L - \mathbf{r}^t_k\|}. \tag{31}$$

The next theorem addresses the condition for remaining in the lineState at every sample-index.

Theorem 2. *Suppose that L_k and $\bar{L}_{k+1}$ meet at a point, say $\mathbf{I}_L$. Suppose that $\mathbf{r}_{0,k}$ lies on $l(\mathbf{r}_k^t, \mathbf{r}_k^s)$. The target speed is v_k^t. If the drone's maximum speed satisfies Equation (31), then $\mathbf{r}_{0,k+1}$ lies on $l(\mathbf{r}_{k+1}^t, \mathbf{r}_{k+1}^s)$.*

Proof. Suppose the drone's maximum speed satisfies Equation (31). Then, Equations (30) and (31) lead to

$$v_{max} * T > l_k. \tag{32}$$

This implies that Equation (21) is met. In the case where Equation (21) is met, Theorem 1 makes $\mathbf{r}_{0,k+1}$ exist on $l(\mathbf{r}_{k+1}^t, \mathbf{r}_{k+1}^s)$. The proof is complete. □

In practice, the ship moves much slower than the target. Thus, the ship position is close to $\mathbf{I}_L$, as plotted in Figure 3. Consider the case where the virtual agent is sufficiently close to the ship. In this case, $\|\mathbf{I}_L - \mathbf{r}_{0,k}\|$ is small to satisfy Equation (31). Then, using Corollary 2, $\mathbf{r}_{0,k+1}$ lies on $l(\mathbf{r}_{k+1}^t, \mathbf{r}_{k+1}^s)$. Thus, the virtual agent remains in the lineState at sample-index $k+1$.

In the case where the virtual agent stays in the lineState at all times, the target cannot reach the ship without being captured by the virtual agent. In order to stay in the lineState at each sample-index, it is desirable that the virtual agent does not become too far from the ship.

In the case where $\|\mathbf{r}_k^s - \mathbf{r}_{0,k}\| \geq \beta$ is met, the guidance point is set using Equation (24) instead of Equation (22). In this way, drones stay close to the ship while staying in the lineState at all times.

Note that Equation (31) can be satisfied even when $v_{max} < v_k^t$. This implies that even slow drones can capture a fast target when the drones stay close to the ship while staying in the lineState at all times.

5.3. Guidance Law of Every Drone

We let multiple drones form a planar grid formation, such that the formation is perpendicular to the line segment connecting the target and the formation center. The grid formation can be considered a "net" structure for capturing the incoming target. By maximizing the grid formation size, we can increase the capture rate, even when there exists error in the prediction of the target's position. Since the target is guided to hit its goal (ship), the drones can effectively block the target using this grid formation.

We next handle the guidance law of a drone for the generation of the grid formation. At sample-index k, let *grid formation* denote a planar formation composed of $G \times G$ cells, each with *side length* s_k. The grid formation is centered at the virtual agent, and we adjust the grid formation so that it is normal to $l(\mathbf{r}_k^t, \mathbf{r}_{0,k})$ at each sample-index k. We change the pitch and yaw of the virtual agent, but we do not change the roll of the virtual agent. Thus, the grid formation does not roll either.

Let $\mathbf{r}_k^R = \mathbf{r}_k^t - \mathbf{r}_{0,k}$ denote the relative position of the target with respect to the virtual agent at sample-index k. At each sample-index k, the planar grid formation is oriented, such that the formation plane is perpendicular to $\mathbf{r}_k^R$.

For $i \in \{0, 1, 2, \ldots, G-1\}$ and $j \in \{0, 1, \ldots, G-1\}$, let $n[i,j]$ be defined as

$$n[i,j] = 1 + i + G * j. \tag{33}$$

Since $i \in \{0, 1, 2, \ldots, G-1\}$ and $j \in \{0, 1, 2, \ldots, G-1\}$, we have

$$n[i,j] \in \{1, 2, \ldots, M = G^2\}. \tag{34}$$

Here, Equation (4) is used. Using Equations (33) and (34), the drone index $n \in \{1, 2, \ldots, M = G^2\}$ has its associated $n[i, j]$.

For $i \in \{0, 1, 2, \ldots, G-1\}$ and $j \in \{0, 1, \ldots, G-1\}$, let

$$\mathbf{w}^B_{n[i,j],k} = (0, -\frac{s_k * G}{2} + i * s_k, -\frac{s_k * G}{2} + j * s_k)^T \quad (35)$$

denote the n-th drone's waypoint in the body-fixed frame at sample-index k. See that these $G \times G$ waypoints are generated on the yz plane in the body-fixed frame. From now on, n denotes $n[i, j]$ for notation simplicity.

In the case where we set $G = 1$, we use only one drone. In this case, the grid formation cannot be generated. If we use only one drone, then the drone's waypoint in the body-fixed frame is set as

$$\mathbf{w}^B_{1,k} = (0, 0, 0)^T, \quad (36)$$

instead of Equation (35). Equation (36) is used to make the single drone move towards the virtual agent. In other words, the waypoint of the drone is set as the virtual agent.

At sample-index 0, all drones are located inside the grid cell of the ship. At sample-index 0, the planar grid formation's orientation (normal vector) is represented as initial yaw $\psi_0 = 0$ and initial pitch $\theta_0 = \pi/2$, respectively. Under Equation (3), the n-th drone's initial position is located at $\mathbf{w}_{n,0}$, which is given as

$$\mathbf{w}_{n,0} = \mathbf{M}_R(0, \pi/2) * \mathbf{w}^B_{n,0}. \quad (37)$$

Equation (37) implies that all drones are initially located to form the grid formation with side length s_0.

Once the drones are launched from the ship, the planar grid formation at each sample-index k is oriented, such that it becomes perpendicular to $\mathbf{r}^R_k = \mathbf{r}^t_k - \mathbf{r}_{0,k}$. The unit vector associated with $\mathbf{r}^R_k$ is

$$\mathbf{u}^R = (\mathbf{u}^R(1), \mathbf{u}^R(2), \mathbf{u}^R(3))^T, \quad (38)$$

where $\mathbf{u}^R = \frac{\mathbf{r}^R_k}{\|\mathbf{r}^R_k\|}$.

At each sample-index k, the planar grid formation's orientation (normal vector) is represented as ψ_k and θ_k, respectively. Since the virtual agent does not rotate, the formation's orientation does not include rolling motions. Under Equation (3), the n-th drone's waypoint in the inertial frame is

$$\mathbf{w}_{n,k} = \mathbf{r}_{0,k} + \mathbf{M}_R(\psi_k, \theta_k) * \mathbf{w}^B_{n,k}. \quad (39)$$

The axis of the grid formation is oriented towards the target $\mathbf{r}^t_k$ since $\mathbf{u}^R = \frac{\mathbf{r}^R_k}{\|\mathbf{r}^R_k\|}$, where $\mathbf{r}^R_k = \mathbf{r}^t_k - \mathbf{r}_{0,k}$. In other words, the planar grid formation is oriented such that it is perpendicular to $\mathbf{r}^R_k$.

We next calculate the planar grid formation's orientation (normal vector), ψ_k, and θ_k in Equation (39), associated with $\mathbf{u}^R$ in Equation (38). We apply $\mathbf{u}^R(1) = c(\psi_k) * c(\theta_k)$, $\mathbf{u}^R(2) = s(\psi_k) * c(\theta_k)$, and $\mathbf{u}^R(3) = -s(\theta_k)$. Here, $\mathbf{u}^R(j)$ indicates the j-th element of $\mathbf{u}^R$.

Under Equation (38), we obtain

$$\theta_k = atan2(-\mathbf{u}^R(3), \sqrt{\mathbf{u}^R(1)^2 + \mathbf{u}^R(2)^2}). \quad (40)$$

Under Equation (38), we derive ψ_k as follows. If $c(\theta_k) \geq 0$, then we use

$$\psi_k = atan2(\mathbf{u}^R(2), \mathbf{u}^R(1)). \quad (41)$$

If $c(\theta_k) < 0$, then we use

$$\psi_k = atan2(-\mathbf{u}^R(2), -\mathbf{u}^R(1)). \tag{42}$$

Recall that $\mathbf{w}_{n,k}$ defines the waypoint assigned to the n-th drone at sample-index k. The heading command of the n-th drone is set towards $\mathbf{w}_{n,k}$ at every sample-index k. *The heading vector $\mathbf{u}_{n,k}$ from Equation (5) is set as follows.*

$$\mathbf{u}_{n,k} = \frac{\mathbf{w}_{n,k} - \mathbf{r}_{n,k}}{\|\mathbf{w}_{n,k} - \mathbf{r}_{n,k}\|}. \tag{43}$$

At every sample-index k, the n-th drone sets its speed command $v_{n,k}$ from Equation (5) as

$$v_{n,k} = min(\frac{\|\mathbf{w}_{n,k} - \mathbf{r}_{n,k}\|}{T}, v_{max}). \tag{44}$$

The control commands, Equations (43) and (44), are used to satisfy

$$\mathbf{r}_{n,k+1} = \mathbf{w}_{n,k}, \tag{45}$$

if possible. Once Equation (45) is met for all $n \in \{1, 2, \ldots, M\}$, then all drones form a grid formation at sample-index $k+1$.

Consider the case where $\|\mathbf{r}_{n,k} - \mathbf{w}_{n,k}\| < v_{max} * T$ at every sample-index k. In this case, Equation (45) is met at every sample-index k.

At sample-index 0, all drones are located inside the ship. At sample-index 0, all drones form the initial grid formation, as presented in Equation (37). Note that each drone is assigned to a waypoint that is in the body-fixed frame of the virtual agent. See Equation (39) for waypoint assignments.

Moreover, [48] can be used to assign a drone to each waypoint, such that the makespan (time for all robots to reach their waypoints) is minimized while also preventing collisions among drones. The authors of [48] mentioned that their assignment algorithm scales well, such that it can compute the mapping for 1000 robots in less than half a second.

In the worst case, there may be a case where a drone meets another drone while moving toward its assigned waypoint. This case can happen due to localization errors or environment disturbance, such as wind. The drone then uses reactive collision avoidance controls, such as [49–52], to avoid an abrupt collision with another drone. Under reactive collision avoidance controls, the drone can change its speed and heading to avoid a sudden collision. We acknowledge that a drone may not reach its waypoint, as it performs evasive maneuvers to avoid a sudden collision with another drone. For instance, suppose that a drone slows downs to avoid a sudden collision at sample-index k. At the next sample-index $k+1$, the drone needs to speed up to reach its new waypoint at sample-index $k+1$.

Control of the Side Length in the Grid Formation

Considering the uncertainty in the target position, it is desirable to make the formation cover as large of an area as possible. However, in the case where the formation is too sparse, the target can get through the formation without being captured by a drone. Furthermore, in the case where the formation is too dense, a large number of drones must be used to cover a large area.

Recall that s_k denotes the side length at sample-index k. Initially, all drones are stored in the grid cells of the ship, such that $s_0 = 1$ meters in Equation (37).

Let s^u denote the upper bound for the side length. In order to capture a target, the side length is increased gradually using

$$s_k = \eta * s_{k-1} + (1 - \eta) * s^u. \tag{46}$$

Here, $0 < \eta < 1$ is a positive constant. η is a tuning parameter indicating the sensitivity for the radius update. In MATLAB simulations, $\eta = 0.9$ is used. As time elapses, s_k monotonically increases to s^u. This implies that the formation size increases as time elapses.

Recall that a target is captured when the relative distance between the target and a drone is less than Δ. In the simulation section, $\Delta = 10$ m is used. If $s^u \leq \Delta$, then the target cannot get through the grid "net" generated by the drones. Thus, $s^u = \Delta$ is set in our simulations.

6. MATLAB Simulation

This section demonstrates the effectiveness of our multi-agent guidance law through MATLAB simulations. The sample interval is $T = 0.5$ s. The safety distance β is set as 100 m.

In Equation (12), the measurement noise $\mathbf{v}_k$ is generated with $\mathbf{v}_k \sim \mathcal{N}(\mathbf{0}, \mathbf{R}_k)$, where $\mathbf{R}_k$ is the identity matrix such that every diagonal element is 1. This implies that the standard deviation of measurement noise is 1 meter.

Considering the process noise, the motion model of the i-th drone ($i \in \{1, 2, \ldots, M\}$) is

$$\mathbf{r}_{i,k+1} = \mathbf{r}_{i,k} + T * v_{i,k} * \mathbf{u}_{i,k} + \mathbf{n}_i. \tag{47}$$

Here, $\mathbf{n}_i$ indicates the process noise of the i-th drone, such that each term in $\mathbf{n}_i$ has a Gaussian distribution with a mean of 0 and a standard deviation of 1 meter. Because Equation (47) contains the process noise term $\mathbf{n}_i$, Equation (47) is distinct from Equation (5).

At sample-index 0, the target is at (0, 10,000, 5000) in meters, and the virtual agent is at the origin. The ship is located at (0,0,0) at sample-index 0. Furthermore, the maximum speed of a drone is $v_{max} = 20$ meters per second. At every sample-index k, the target's speed, v^t_k, is 200 meters per second. See that the target moves much faster than a drone.

We say that a target is *captured* when the distance between a drone and the target is less than $\Delta = 10$ meters. Moreover, a target is captured at sample-index k when $l(\mathbf{r}^t_k, \mathbf{r}^t_{k-1})$ crosses the grid formation between sample-index $k-1$ and k. Because $s^u = \Delta$, the target is considered to be captured between sample-index $k-1$ and k.

Furthermore, a simulation ends when the distance between the ship and the target is less than Δ. This implies that the ship is hit by the target.

6.1. Monte-Carlo Simulations

Multiple Monte-Carlo (MC) simulations are needed to prove the effectiveness of the proposed method rigorously since the measurements are noisy. We run $mC = 50$ MC simulations while changing generated noise.

At the end of each MC simulation, the target hits the ship or is captured by a drone. We use three metrics ($captureRate$, $endDist$, and $simTime$) to analyze the proposed controls.

Let $captureN$ denote the number of MC simulations where the target is captured by a drone. Considering the analysis of mC MC simulations, $captureRate = \frac{captureN}{mC}$ presents the capture rate (in percents) as we run mC MC simulations. It is desirable that the $captureRate$ is as large as possible.

Let $endDist$ (m) represent the average distance between the ship and all drones' center position when an MC simulation ends. In the case where a single drone is considered, $endDist$ represents the average distance between the ship and the drone when an MC simulation ends. Large $endDist$ implies that at the end of a simulation, a drone is far from the ship.

Let $simTime$ denote the average time (in seconds) spent until each MC simulation ends. Note that an MC simulation ends when the target hits the ship or is captured by a drone.

6.2. 3D PNG Law

The target applies the PNG laws in [20] to approach the ship as time elapses. We briefly introduce the 3D PNG law in [20]. The rotation vector of the line-of-sight at sample-index k is

$$\Omega_{st,k} = \mathbf{R}_{st,k} * \mathbf{V}_{st,k} / (r^2), \tag{48}$$

where $\mathbf{V}_{st,k}$ is the relative velocity of the ship with respect to the target. Furthermore, $\mathbf{R}_{st,k} = \mathbf{r}_k^s - \mathbf{r}_k^t$, and $r = \|\mathbf{R}_{st,k}\|$. The PNG law is set as

$$\mathbf{c}_{png,k} = N_p \mathbf{V}_{st,k} * \Omega_{st,k}, \tag{49}$$

where $N_p = 3$ is a constant. The target's velocity is updated as

$$\mathbf{v}_{k+1}^t = \mathbf{v}_k^t + T * \mathbf{c}_{png,k}. \tag{50}$$

Then, the target's position is updated using

$$\mathbf{r}_{k+1}^t = \mathbf{r}_k^t + T * v_k^t \frac{\mathbf{v}_k^t}{\|\mathbf{v}_k^t\|}. \tag{51}$$

6.3. Scenario 1

In this scenario, the ship moves at a velocity of (−5,0,0) in m/s.

We set $G = 3$ in Equation (4). This implies that we use $M = G^2 = 9$ drones in total. Figure 4 shows the result of one MC simulation using nine drones. In Figure 4, the target's position at every 3 s is marked as blue circles. The position of every drone at every 3 s is depicted as circles with distinct colors. A red asterisk depicts the position of the virtual agent at every 3 s. The position of the ship at every 3 s is indicated by red diamonds.

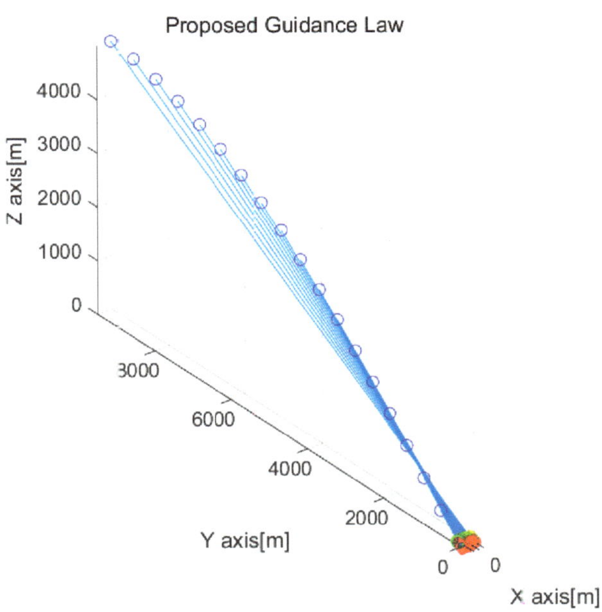

Figure 4. The result of one MC simulation using the proposed multi-agent guidance law. The proposed multi-agent guidance law is applied to hit the target. The position of every drone at every 3 s is depicted as circles with distinct colors. A red asterisk depicts the position of the virtual agent at every 3 s. The target's position at every 3 s is marked as blue circles. The position of the ship at every 3 s is indicated by red diamonds.

Figure 5 is the enlarged figure of Figure 4. The position of every drone at every 3 s is depicted as circles with distinct colors. The drones generate a grid formation for protection against the incoming target.

Considering the scenario in Figure 5, Figure 6 removes the plots for the target's positions in order to see the drones' maneuvers clearly. The position of every drone at every 3 s is depicted as circles with distinct colors. A red asterisk depicts the position of the virtual agent at every 3 s. The position of the ship at every 3 s is indicated by red diamonds. A black asterisk presents the target's position when the target is hit by a drone.

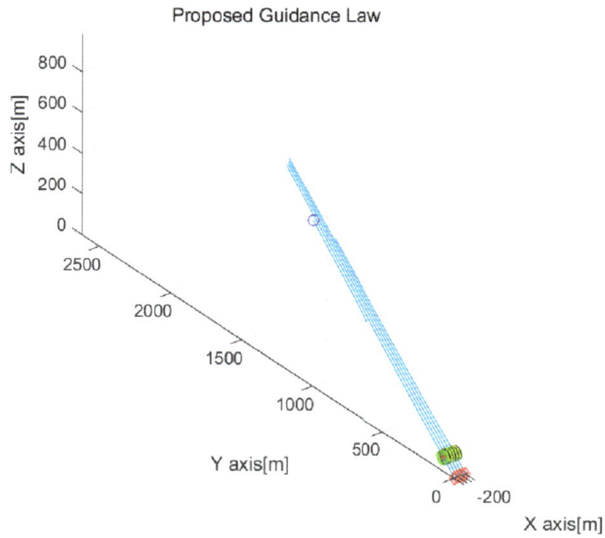

Figure 5. The enlarged figure of Figure 4. The position of every drone at every 3 s is depicted as circles with distinct colors. The drones generate a grid formation for protection against the incoming target.

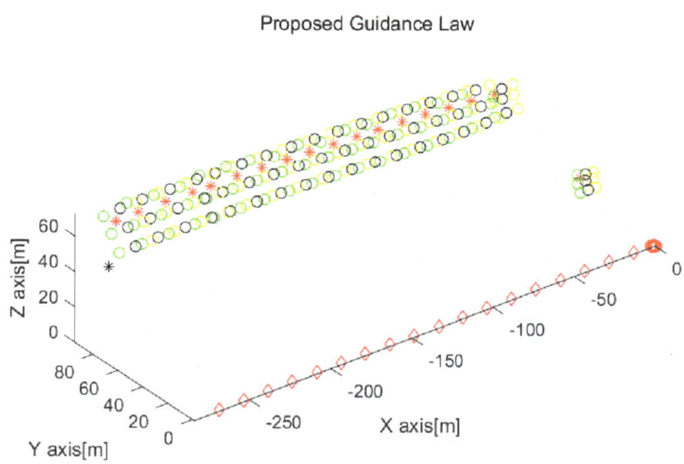

Figure 6. Considering the scenario in Figure 5, we remove the plots for the target's positions in order to see the drones' maneuvers clearly. The position of every drone at every 3 s is depicted as circles with distinct colors. A red asterisk depicts the position of the virtual agent at every 3 s. The position of the ship at every 3 s is indicated by red diamonds. A black asterisk presents the target's position when the target is hit by a drone.

Figure 7a plots the distance between the virtual agent and the target as time elapses. The relative distance keeps decreasing as time elapses. Figure 7b depicts the distance between the virtual agent and the ship as time elapses. You can see that the virtual agent stays close to the safety distance β while protecting the ship. Figure 7c depicts the side length as time elapses. As time elapses, the side length increases to Δ.

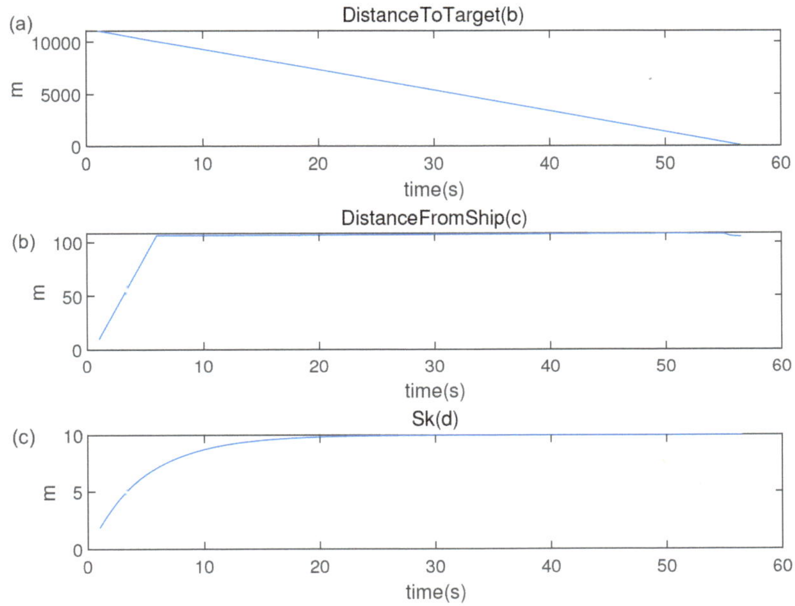

Figure 7. The result of one MC simulation using grid formation. The proposed guidance law is applied to hit the target. (**a**) depicts the distance between the virtual agent and the target as time elapses. The relative distance keeps decreasing as time elapses. (**b**) plots the distance between the virtual agent and the ship as time elapses. See that the virtual agent stays close to the safety distance β while protecting the ship. (**c**) depicts the side length as time elapses.

6.3.1. The Effect of Changing System PARAMETERS (Number of Drones and Noise Strength)

We discuss the effect of the number of drones. We also present the effect of noise on the control performance. Let N_s denote the standard deviation of measurement noise. In Equation (12), the measurement noise $\mathbf{v}_k$ is generated with $\mathbf{v}_k \sim \mathcal{N}(\mathbf{0}, \mathbf{R}_k)$, where $\mathbf{R}_k$ is the diagonal matrix such that every diagonal element is N_s^2. This implies that the standard deviation of measurement noise is N_s in meters.

Table 1 summarizes the MC simulation results representing the effect of system parameters. We apply three metrics ($captureRate(cR)$, $endDist(eD)$, and $simTime(sT)$) to analyze the proposed controls. As the number of drones increases, $captureRate$ increases in general. This is due to the fact that as we deploy more drones, the area covered by the drones increases. Furthermore, as the measurement noise N_s increases to 10 meters, the $captureRate$ decreases. Note that an MC simulation ends when the distance between a drone and the target is less than $\Delta = 10$ m.

Table 1. MC simulation results. The effect of changing system parameters, G and N_s.

G	N_s	cR	eD	sT
3	1	100	95	57
1	1	100	108	57
3	10	100	94	57
1	10	60	105	57

6.3.2. Comparison with Other State-of-the-Art Guidance Laws (Scenario 1)

To the best of our knowledge, this article is novel in developing a ship defense approach using clustered multiple drones. For comparison, we consider the case where only one drone is used, and the virtual agent applies the 3D PNG law in Section 6.2 to capture the target. We also consider the case where only one drone is used, and the drone applies the 3D Motion Camouflage Guidance (MCG) law in [18] to chase the target. We further consider the case where only one drone is used, and the drone applies the 3D Command to Line-Of-Sight (CLOS) guidance law in [53].

Table 2 shows the MC simulation comparison results of Scenario 1. We run mC MC simulations per each control law. Let PRO_q indicate the proposed guidance law using $G = q$ in Equation (4). Note that only one drone is used in PRO_1. Moreover, nine drones are used in PRO_3.

In Table 2, MCG presents the case where the 3D MCG law is used. PNG presents the case where the 3D PNG law is used. $CLOS$ presents the case where the 3D CLOS law is used. See that the proposed control outperforms all other state-of-the-art controls.

Table 2. Comparison with other state-of-the-art controls (Scenario 1).

Control	N_s	cR	eD	sT
PRO_3	1	100	95	57
PRO_1	1	100	108	57
PNG	1	0	280	57
MCG	1	0	280	57
$CLOS$	1	20	276	56

Note that Equation (31) can be satisfied even when $v_{max} < v_k^t$. This implies that even slow drones can capture a fast target when the drones stay close to the ship while staying in the lineState at all times. However, other state-of-the-art guidance laws (MCG, PNG, and $CLOS$) make a drone continue to chase the target. This maneuver makes the drone move away from the ship, which is not desirable for capturing a fast target using a slow interceptor.

Other state-of-the-art guidance laws (MCG, PNG, and $CLOS$) make a single drone chase the target. In our paper, we let multiple drones form a planar grid formation, which can be considered a "net" structure to capture the incoming target. The target may perform elusive maneuvers, and there may be measurement noise in measuring the target position. By maximizing the grid formation size, the capture rate is 100, even when there exists error in the prediction of the target's position. Since the target is guided to hit its goal (ship), the drones can effectively block the target using this grid formation.

6.4. Scenario 2

We introduce Scenario 2 in Figure 8. The distinctions of Figure 8 from Figure 4 are as follows. As the relative distance between the target and the ship is less than 3000 m, the target moves towards the ship directly while increasing its speed to 240 m/s.

Figure 8 depicts the result of one MC simulation. We set $G = 5$ in Equation (4), i.e., we use $G^2 = 25$ drones in total. The position of every drone at every 3 s is depicted as circles with distinct colors. A red asterisk depicts the position of the virtual agent at every 3 s. Figure 8 shows the case where the target is captured by a drone.

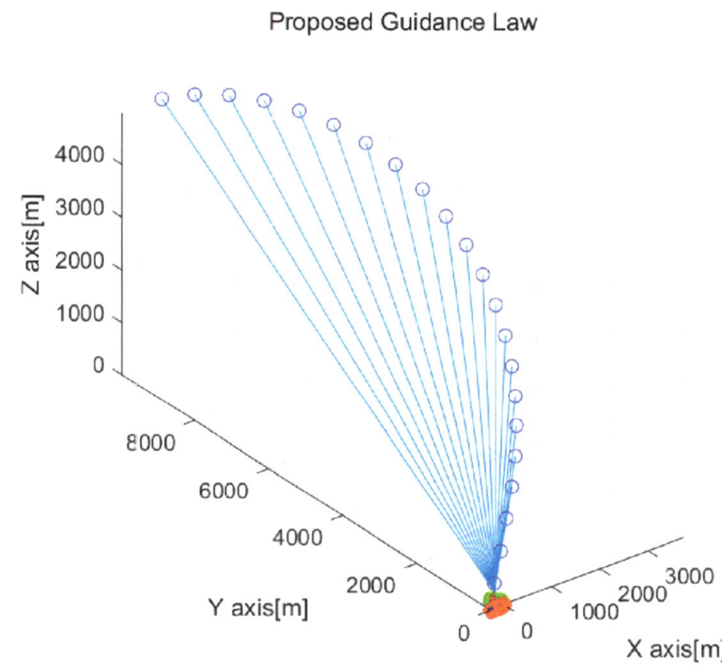

Figure 8. The result of one MC simulation using the proposed grid formation. The position of every drone at every 3 s is depicted as circles with distinct colors. A red asterisk depicts the position of the virtual agent at every 3 s.

Figure 9 is the enlarged figure of Figure 8. The position of every drone at every 3 s is depicted as circles with distinct colors. See that the grid formation is generated to protect the ship from the incoming target.

Considering the scenario in Figure 9, Figure 10 removes the plots for the target's positions in order to see the drones' maneuvers clearly. The position of every drone at every 3 s is depicted as circles with distinct colors. A red asterisk depicts the position of the virtual agent at every 3 s. The position of the ship at every 3 s is indicated by red diamonds. A black asterisk presents the target's position when the target is hit by a drone.

Figure 11a depicts the distance between the virtual agent and the target as time elapses. See that the relative distance continuously decreases over time. Figure 11b depicts the distance between the virtual agent and the ship as time elapses. See that the virtual agent stays close to the safety distance β while protecting the ship. Figure 11c depicts the side length as time elapses. As time elapses, the side length increases to Δ.

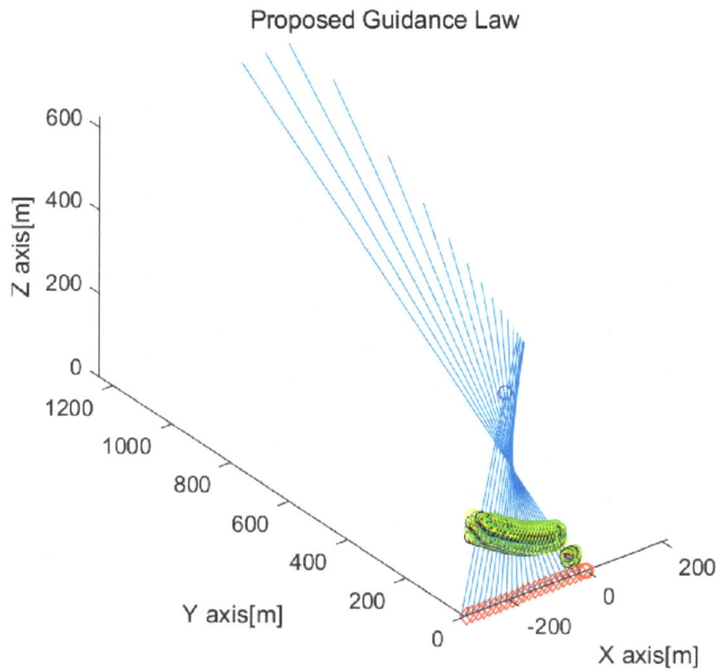

Figure 9. The enlarged figure of Figure 8. The position of every drone at every 3 s is depicted as circles with distinct colors. See that the grid formation is generated to protect the ship from the incoming target.

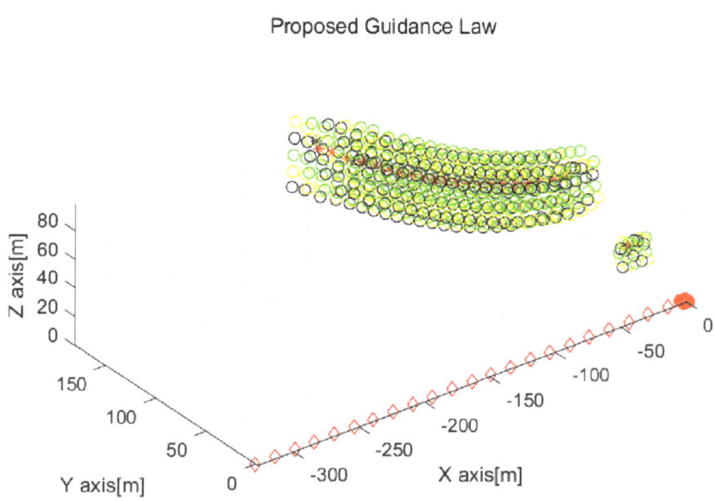

Figure 10. Considering the scenario in Figure 9, we remove the plots for the target's positions in order to see the drones' maneuvers clearly. The position of every drone at every 3 s is depicted as circles with distinct colors. A red asterisk depicts the position of the virtual agent at every 3 s. The position of the ship at every 3 s is indicated by red diamonds. A black asterisk presents the target's position when the target is hit by a drone.

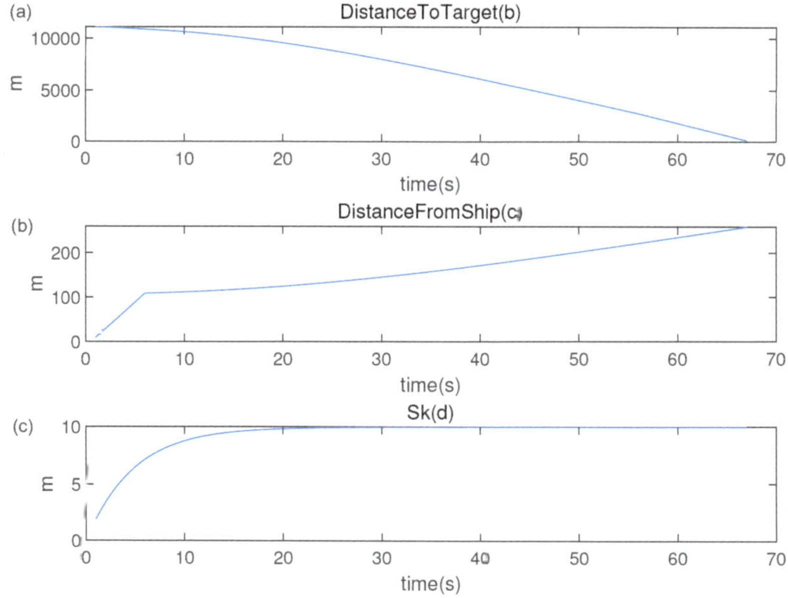

Figure 11. (**a**) depicts the distance between the virtual agent and the target as time elapses. See that the relative distance continuously decreases over time. (**b**) plots the distance between the virtual agent and the ship as time elapses. See that the virtual agent stays close to the safety distance β while protecting the ship. (**c**) depicts the side length as time elapses.

Comparison with Other State-of-the-Art Guidance Laws (Scenario 2)

Table 3 shows the MC comparison results of Scenario 2. In this table, N_s denotes the measurement noise. We run mC MC simulations per each control law.

In Table 3, PRO_q indicates the proposed guidance law using $G = q$ in (4). Note that only one drone is used in PRO_1. Moreover, 25 drones are used in PRO_5.

Table 3 shows that the proposed control outperforms all other controls. *MCG*, *PNG*, and *CLOS* make a drone keep chasing the target. This maneuver makes the drone move away from the ship, which is not desirable for capturing a fast target using a slow interceptor.

Table 3. Comparison with other state-of-the-art controls (Scenario 2).

Control	N_s	cR	eD	sT
PRO_5	10	100	247	67
PRO_1	10	95	260	67
PNG	10	0	339	68
MCG	10	0	339	68
$CLOS$	10	0	339	68

Other state-of-the-art guidance laws (*MCG*, *PNG*, and *CLOS*) make a single drone chase the target. Our strategy is to let multiple drones form a planar grid formation, which can be considered as a "net" for capturing the incoming target. By maximizing the grid formation size, the captureRate is 100 even when there exists error in the prediction of the target's position.

7. Conclusions

This article introduces a multi-agent guidance law so that a formation of drones protects the ship from an incoming high-speed target. The drones generate a planar grid formation, whose center is guided to remain on the line connecting the target and the ship. Moreover, the planar formation is generated to be perpendicular to the line segment connecting the target and the formation center. Since a target heads towards its goal at least in the terminal phase, maintaining a position on this line segment is effective in protecting the ship.

We enable slow drones to capture a fast target by letting the drones stay close to the ship while staying in the lineState at all times. This blocking strategy is desirable considering the energy consumption of the interceptor since an interceptor does not have to move far from the ship.

We control the drone formation based on the prediction of the target's position after one sample-index in the future. Since we use a grid formation of drones, we can increase the capture rate even when the target prediction is erroneous.

As far as we know, this article is novel in developing a ship defense approach using multiple clustered drones. In addition, our paper is novel in addressing the 3D formation control that can handle uncertainty in the target prediction. The effectiveness of our multi-agent guidance law is shown by comparing it with other state-of-the-art guidance laws under MATLAB simulations. We verify that the proposed multi-drone guidance scheme increases the capture probability significantly compared to the case where a single interceptor is used. In the future, we will do experiments to verify our multi-agent guidance law using real drones.

In practice, the presence of wind can affect and modify a drone's path. Many papers handled how to control a drone under the effect of wind [54–58]. The authors of [54] improved the accelerated A-star algorithm with a converted wind vector, and [55] addressed the problem of a drone's path planning operating in complex four-dimensional (time and spatially varying) wind-fields. Additionally, refs. [56–58] handled adaptive path planning in windy conditions. The authors of [58] added a wind model to the existing path planning algorithm and combined it with a drone's control systems. In the future, we will combine the proposed guidance scheme with a wind model so that multiple drones can safely maneuver in time-varying wind-fields.

Note that the proposed guidance scheme can be applied to protect an entity other than a ship, as long as the goal of the target is known a priori. For instance, the proposed multi-agent guidance law can be generalized to protect any vehicles, such as tanks or ground stations.

Funding: This research was supported by the faculty research fund of Sejong university in 2023. This work was supported by the National Research Foundation of Korea (NRF) grant funded by the Korean government (MSIT) (Grant Number: 2022R1A2C1091682).

Institutional Review Board Statement: Not applicable.

Informed Consent Statement: Not applicable.

Data Availability Statement: Not applicable.

Conflicts of Interest: The author declares no conflict of interest.

References

1. Kim, J.; Gadsden, S.A.; Wilkerson, S.A. A Comprehensive Survey of Control Strategies for Autonomous Quadrotors. *Can. J. Electr. Comput. Eng.* **2020**, *43*, 3–16. [CrossRef]
2. Santana, L.V.; Brandao, A.S.; Sarcinelli-Filho, M. Outdoor waypoint navigation with the AR. Drone quadrotor. In Proceedings of the 2015 International Conference on Unmanned Aircraft Systems (ICUAS), Denver, CO, USA, 9–12 June 2015; pp. 303–311.
3. Mellinger, D.; Michael, N.; Kumar, V. Trajectory generation and control for precise aggressive maneuvers with quadrotors. *Int. J. Robot. Res.* **2012**, *31*, 664–674. [CrossRef]
4. Bar-Shalom, Y.; Fortmann, T.E. *Tracking and Data Association*; Academic Press: Orlando, FL, USA, 1988.

5. Minaeian, S.; Liu, J.; Son, Y. Vision-Based Target Detection and Localization via a Team of Cooperative UAV and UGVs. *IEEE Trans. Syst. Man Cybern. Syst.* **2016**, *46*, 1005–1016. [CrossRef]
6. Svec, P.; Thakur, A.; Raboin, E.; Shah, B.C.; Gupta, S.K. Target following with motion prediction for unmanned surface vehicle operating in cluttered environments. *Auton. Robot.* **2014**, *36*, 383–405. [CrossRef]
7. Kim, J.; Kim, S.; Choo, Y. Stealth Path Planning for a High Speed Torpedo-Shaped Autonomous Underwater Vehicle to Approach a Target Ship. *Cyber Phys. Syst.* **2018**, *4*, 1–16. [CrossRef]
8. Kim, J. Target Following and Close Monitoring Using an Unmanned Surface Vehicle. *IEEE Trans. Syst. Man, Cybern. Syst.* **2018**, *50*, 4233–4242. [CrossRef]
9. Xu, Y.; Basset, G. Sequential virtual motion camouflage method for nonlinear constrained optimal trajectory control. *Automatica* **2012**, *48*, 1273–1285. [CrossRef]
10. Srinivasan, M.V.; Davey, M. Strategies for Active Camouflage of Motion. *Proc. R. Soc. B Biol. Sci.* **1995**, *259*, 19–25.
11. Mizutani, A.; Chahl, J.; Srinivasan, M. Insect behaviour: Motion camouflage in dragonflies. *Nature* **2003**, *423*, 604. [CrossRef]
12. Rano, I. Direct collocation for two dimensional motion camouflage with non-holonomic, velocity and acceleration constraints. In Proceedings of the 2013 IEEE International Conference on Robotics and Biomimetics (ROBIO), Shenzhen, China, 13–14 December 2013; pp. 109–114. [CrossRef]
13. Justh, E.; Krishnaprasad, P. Steering laws for motion camouflage. *Proc. R. Soc. A* **2006**, *462*, 3629–3643. [CrossRef]
14. Ghose, K.; Horiuchi, T.K.; Krishnaparasad, P.S.; Moss, C.F. Ecolocating bats use a nearly time-optimal strategy to intercept prey. *PLoS Biol.* **2006**, *4*, 865–873. [CrossRef]
15. Anderson, A.J.; McOwan, P.W. Model of a predatory stealth behaviour camouflaging motion. *Proc. R. Soc. B* **2003**, *270*, 489–495. [CrossRef] [PubMed]
16. Kim, J. Controllers to Chase a High-Speed Evader Using a Pursuer with Variable Speed. *Appl. Sci.* **2018**, *8*, 1976. [CrossRef]
17. Galloway, K.S.; Justh, E.W.; Krishnaprasad, P.S. Motion camouflage in a stochastic setting. In Proceedings of the IEEE International Conference on Decision and Control (CDC), New Orleans, LA, USA, 12–14 December 2007; pp. 1652–1659.
18. Reddy, P.V.; Justh, E.W.; Krishnaprasad, P.S. Motion camouflage in three dimensions. In Proceedings of theDecision and Control, San Diego, CA, USA, 13–15 December 2006; IEEE: Piscataway, NJ, USA, 2006; pp. 3327–3332.
19. Halder, U.; Dey, B. Biomimetic Algorithms for Coordinated Motion: Theory and Implementation. In Proceedings of the International Conference on Robotics and Automation (ICRA), Seattle, DC, USA, 26–30 May 2015; IEEE: Piscataway, NJ, USA, 2015; pp. 5426–5432.
20. Oh, J.; Ha, I. Capturability of the 3-dimensional pure PNG law. *IEEE Trans. Aerosp. Electron. Syst.* **1999**, *35*, 491–503.
21. Song, S.; Ha, I. A lyapunov-like approach to performance analysis of 3-dimensional pure PNG laws. *IEEE Trans. Aerosp. Electron. Syst.* **1994**, *30*, 238–248. [CrossRef]
22. Liu, D.; Lee, M.C.; Pun, C.M.; Liu, H. Analysis of Wireless Localization in Non-Line-of-Sight Conditions. *IEEE Trans. Veh. Technol.* **2013**, *62*, 1484–1492. [CrossRef]
23. Prasanna, H.M.; Ghose, D. Retro-Proportional-Navigation: A new guidance law for interception of high-speed targets. *J. Guid. Control Dyn.* **2012**, *35*, 377–386. [CrossRef]
24. Mischiati, M.; Krishnaprasad, P.S. Mutual motion camouflage in 3D. In Proceedings of the 18th IFAC World Congress, Milan, Italy, 28 August–2 September 2011; pp. 4483–4488.
25. Alonso-Mora, J.; Montijano, E.; N?geli, T.; Hilliges, O.; Schwager, M.; Rus, D. Distributed multi-robot formation control in dynamic environments. *Auton. Robot.* **2019**, *43*, 1079–1100. [CrossRef]
26. Ji, M.; Muhammad, A.; Egerstedt, M. Leader-based multi-agent coordination: Controllability and optimal control. In Proceedings of the American Control Conference, Minneapolis, MN, USA, 14–16 June 2006; pp. 1358–1363.
27. Kim, J. Cooperative Exploration and Networking While Preserving Collision Avoidance. *IEEE Trans. Cybern.* **2017**, *47*, 4038–4048. [CrossRef]
28. Kim, J. Capturing intruders based on Voronoi diagrams assisted by information networks. *Int. J. Adv. Robot. Syst.* **2017**, *14*, 1729881416682693. [CrossRef]
29. Kim, J. Cooperative Exploration and Protection of a Workspace Assisted by Information Networks. *Ann. Math. Artif. Intell.* **2014**, *70*, 203–220. [CrossRef]
30. Parker, L.E.; Kannan, B.; Fu, X.; Tang, Y. Heterogeneous mobile sensor net deployment using robot herding and line-of-sight formations. In Proceedings of the Intelligent Robots and Systems, Las Vegas, NV, USA, 27 October–1 November 2003; Volume 3, pp. 2488–2493.
31. Mirzaei, F.M.; Mourikis, A.I.; Roumeliotis, S.I. On the Performance of Multi-robot Target Tracking. In Proceedings of the 2007 IEEE International Conference on Robotics and Automation, Roma, Italy, 10–14 April 2007; pp. 3482–3489. [CrossRef]
32. Hausman, K.; Muller, J.; Hariharan, A.; Ayanian, N.; Sukhatme, G.S. Cooperative multi-robot control for target tracking with onboard sensing. *Int. J. Robot. Res.* **2015**, *34*, 1660–1677. [CrossRef]
33. Fonod, R.; Shima, T. Wingman-based Estimation and Guidance for a Sensorless PN-Guided Pursuer. *IEEE Trans. Aerosp. Electron. Syst.* **2019**, *56*, 1754–1766. [CrossRef]
34. Fonod, R.; Shima, T. Blinding Guidance Against Missiles Sharing Bearings-Only Measurements. *IEEE Trans. Aerosp. Electron. Syst.* **2018**, *54*, 205–216. [CrossRef]

35. Jeon, I.; Lee, J.; Tahk, M. Homing guidance law for cooperative attack of multiple missiles. *J. Guid. Control Dyn.* **2010**, *33*, 275–280. [CrossRef]
36. Zhang, T.; Yang, J. Guidance law of multiple missiles for cooperative simultaneous attack against maneuvering target. In Proceedings of the 2018 37th Chinese Control Conference (CCC), Wuhan, China, 25–27 July 2018; pp. 4536–4541. [CrossRef]
37. Lee, C.H.; Tsourdos, A. Cooperative Control for Multiple Interceptors to Maximize Collateral Damage. *IFAC-PapersOnLine* **2018**, *51*, 56–61. [CrossRef]
38. Wang, L.; Liu, K.; Yao, Y.; He, F. A Design Approach for Simultaneous Cooperative Interception Based on Area Coverage Optimization. *Drones* **2022**, *6*, 156. [CrossRef]
39. Fossen, T.I. *Guidance and Control of OCEAN Vehicles*; John Wiley and Sons: Hoboken, NJ, USA, 1994.
40. Garcia de Marina, H.; Cao, M.; Jayawardhana, B. Controlling Rigid Formations of Mobile Agents Under Inconsistent Measurements. *IEEE Trans. Robot.* **2015**, *31*, 31–39. [CrossRef]
41. Krick, L.; Broucke, M.E.; Francis, B.A. Stabilization of infinitesimally rigid formations of multi-robot networks. In Proceedings of the 2008 47th IEEE Conference on Decision and Control, Cancun, Mexico, 9–11 December 2008; pp. 477–482.
42. Paley, D.A.; Zhang, F.; Leonard, N.E. Cooperative Control for Ocean Sampling: The Glider Coordinated Control System. *IEEE Trans. Control Syst. Technol.* **2008**, *16*, 735–744. [CrossRef]
43. Ji, M.; Egerstedt, M. Distributed Coordination Control of Multiagent Systems While Preserving Connectedness. *IEEE Trans. Robot.* **2007**, *23*, 693–703. [CrossRef]
44. Kim, J. Constructing 3D Underwater Sensor Networks without Sensing Holes Utilizing Heterogeneous Underwater Robots. *Appl. Sci.* **2021**, *11*, 4293. [CrossRef]
45. Kim, J.; Kim, S. Motion control of multiple autonomous ships to approach a target without being detected. *Int. J. Adv. Robot. Syst.* **2018**, *15*, 1729881418763184. [CrossRef]
46. Luo, S.; Kim, J.; Parasuraman, R.; Bae, J.H.; Matson, E.T.; Min, B.C. Multi-robot rendezvous based on bearing-aided hierarchical tracking of network topology. *Ad Hoc Netw.* **2019**, *86*, 131–143. [CrossRef]
47. Ristic, B.; Arulampalam, S.; Gordon, N. *Beyond the Kalman Filter: Particle Filters for Tracking Applications*; Artech House Radar Library: Boston, MA, USA, 2004.
48. MacAlpine, P.; Price, E.; Stone, P. SCRAM: Scalable Collision-avoiding Role Assignment with Minimal-Makespan for Formational Positioning. In Proceedings of the AAAI Conference on Artificial Intelligence, Austin, TX, USA, 25–30 January 2015; Volume 29.
49. Chakravarthy, A.; Ghose, D. Obstacle avoidance in a dynamic environment: A collision cone approach. *IEEE Trans. Syst. Man Cybern.* **1998**, *28*, 562–574. [CrossRef]
50. Svec, P.; Shah, B.C.; Bertaska, I.R.; Alvarez, J.; Sinisterra, A.J.; von Ellenrieder, K.; Dhanak, M.; Gupta, S.K. Dynamics-aware target following for an automomous surface vehicle pperating under Colregs in civilian traffic. In Proceedings of the IEEE/RSJ International Conference on Intelligent Robots and Systems (IROS), Tokyo, Japan, 3–7 November 2013; pp. 3871–3878.
51. Chang, D.E.; Shadden, S.C.; Marsden, J.E.; Olfati-Saber, R. Collision Avoidance for Multiple Agent Systems. In Proceedings of the IEEE International Conference on Decision and Control, Maui, HI, USA, 9–12 December 2003; pp. 539–543.
52. Lalish, E.; Morgansen, K. Distributed reactive collision avoidance. *Auton. Robot* **2012**, *32*, 207–226. [CrossRef]
53. Zahra, B.F.T.; Shah, S.I.A. Integrated CLOS and PN Guidance for Increased Effectiveness of Surface to Air Missiles. *INCAS Bull.* **2017**, *9*, 141–156.
54. Selecký, M.; Váňa, P.; Rollo, M.; Meiser, T. Wind Corrections in Flight Path Planning. *Int. J. Adv. Robot. Syst.* **2013**, *10*, 248. [CrossRef]
55. Chakrabarty, A.; Langelaan, J. UAV flight path planning in time varying complex wind-fields. In Proceedings of the 2013 American Control Conference, Washington, DC, USA, 17–19 June 2013; pp. 2568–2574.
56. Coombes, M.; Chen, W.H.; Liu, C. Boustrophedon coverage path planning for UAV aerial surveys in wind. In Proceedings of the 2017 International Conference on Unmanned Aircraft Systems (ICUAS), Miami, FL, USA, 13–16 June 2017; pp. 1563–1571.
57. Coombes, M.; Fletcher, T.; Chen, W.H.; Liu, C. Optimal Polygon Decomposition for UAV Survey Coverage Path Planning in Wind. *Sensors* **2018**, *18*, 2132. [CrossRef]
58. McGee, T.; Hedrick, J. Path planning and control for multiple point surveillance by an unmanned aircraft in wind. In Proceedings of the 2006 American Control Conference, Minneapolis, MN, USA, 14–16 June 2006; pp. 4261–4266.

applied sciences

MDPI

Article

RRT*-Fuzzy Dynamic Window Approach (RRT*-FDWA) for Collision-Free Path Planning

Lintao Zhou, Nanpeng Wu, Hu Chen *, Qinge Wu and Yingbo Lu

School of Electrical and Information Engineering, Zhengzhou University of Light Industry, Zhengzhou 450002, China

* Correspondence: chenhu@zzuli.edu.cn

Abstract: Path planning is an important aspect and component in the research of mobile-robot-related technologies. Many path planning algorithms are only applicable to static environments, while in practical tasks, the uncertainty in dynamic environments increases the difficulty of path planning and obstacle avoidance compared with static environments. To address this problem, this paper proposes an RRT*-FDWA algorithm. RRT* first generates a global optimal path, and then, when obstacles exist nearby, an FDWA algorithm fixes the local path in real time. Compared with other path planning algorithms, RRT*-FDWA can avoid local minima, rapidly perform path replanning, generate a smooth optimal route, and improve the robot's maneuvering amplitude. In this paper, the effectiveness of the algorithm is verified through experiments in dynamic environments.

Keywords: RRT*-FDWA; maneuver amplitude; dynamic path planning; local minimum

1. Introduction

In the field of automatic navigation of wheeled mobile robots, path planning is a frequently researched area. The main goal of a path planning algorithm is to plan a collision-free path from the initial state to the target in the constructed space of the robot [1]. Path planning is divided into global path planning and local path planning. Global planning [2–4] is able to accomplish the task in known static environments. However, environments and obstacles change dynamically in most navigation processes. When a mobile robot performs tasks according to global path planning, it must use various devices on board, such as lidar, depth cameras, and IMUs, to sense the surrounding environment and its own state and quickly plan local collision-free paths. For navigation, robots therefore usually use a combination of global path planning and local path planning [5–7].

The Rapid Exploration Randomized Tree [8] (RRT) has been shown to be a very effective global path planning algorithm; RRT can plan paths quickly and simply by generating a tree structure at high-speed increments in the construction space to find a goal. However, RRT algorithms also possess some constraints, such as not having asymptotic optimality [9,10] and an the inability to avoid dynamic obstacles; variants of improved RRT algorithms have, therefore, been proposed. Jin-Gukang [11] et al. proposed a "Post-Triangular Rewiring" method, which reduces the path planning time, improves efficiency, and creates an algorithm close to the global optimum. Moon C [12] et al. proposed a dual-tree fast exploration random tree (DT-RRT) algorithm to decrease the computational complexity of the RRT. In addition, DT-RRT reduces the length of the path and increases the coverage of the sampling points. Nasir J et al. proposed an RRT* [13] algorithm which can make the path asymptotically optimal. Chen L et al. [14] introduced the dual-tree structure into an RRT* algorithm, thereby creating separate extension and optimization processes and improving the convergence speed. In addition, some scholars have attempted to improve global path planning methods such as RRT* to avoid dynamic obstacles. Qi et al. [15] applied the RRT* algorithm to dynamic environments and introduced Pareto

Citation: Zhou, L.; Wu, N.; Chen, H.; Wu, Q.; Lu, Y. RRT*-Fuzzy Dynamic Window Approach (RRT*-FDWA) for Collision-Free Path Planning. *Appl. Sci.* **2023**, *13*, 5234. https://doi.org/10.3390/app13095234

Academic Editor: Jonghoek Kim

Received: 18 March 2023
Revised: 21 April 2023
Accepted: 21 April 2023
Published: 22 April 2023

theory to propose a multi-objective dynamic fast exploration stochastic algorithm (MOD-RRT*), which improved the effective performance of the RRT* algorithm in avoiding unknown obstacles.

However, there are many limitations in global path planning and dynamic environments. Compared with static environments, there are many uncertainties, such as randomly moving obstacles. It is, therefore, very difficult to predict the movement path of obstacles. Thus, local path planning in dynamic environments is currently an area of great research interest.

In recent years, researchers have proposed many local path planning algorithms, such as the dynamic window approach [16] (DWA) and the artificial potential field method [17] (APF). Furthermore, Young-In Choi [18] et al. proposed a collision avoidance algorithm based on the D*Lite algorithm for mobile robots; this algorithm was applied to a logistics delivery scenario to effectively avoid collision and safely reset the robot's optimal movement route when the robot encounters a crossover situation at an intersection. Kumar [19] et al. proposed a combination of the sine cosine algorithm and ant colony algorithm for multiple environments such as static and dynamic, and applied the algorithm in multi-robot formations. Guo [20] et al. constructed a risk region of dynamic obstacles using the Kalman filter state estimation, combining it with a nonlinear model predictive control to achieve safe obstacle avoidance. DWA has been extensively applied because of its dynamic characteristics combined with those of the robot. Later, Zhong [21] et al. proposed an adaptive rolling window method based on the edges of obstacles and the target point, which has good safety and environmental applicability. Chang [22] et al. combined Q learning with the DWA algorithm and proposed two new evaluation functions. This demonstrated a high navigation efficiency and success rates in complex unknown environments, but incurred a higher time cost. Xiang [23] et al. implemented adaptive weight coefficients for the DWA algorithm for complex environments, making the path of the mobile robot smoother when avoiding obstacles. Many new and improved algorithms have been derived for obstacle avoidance in dynamic environments. Wu [24] et al. combined the A* algorithm and the DWA algorithm to produce an algorithm that was closer to the global optimal path but with less smoothness.

Nevertheless, the improved DWA algorithms are still unable to avoid the problem of local optimal solutions and may fail to complete the task due to the lack of global planning. An algorithm may get stuck at local minima during planning and consume more energy. A suitable path planning algorithm should be able to plan a complete collision-free path that satisfies the robot dynamics.

Therefore, to resolve the path planning problem in a dynamic environment, this paper proposes an RRT*-FDWA algorithm. The contribution of the proposed algorithm to the path planning problem is as follows:

- A fuzzy controller is added to the adaptive weight index of the DWA algorithm. This makes the weights adaptive, improves the safety of path planning, and enables timely avoidance of dynamic obstacles.
- By combining the RRT* algorithm with the FDWA algorithm, local planning can avoid hitting local minima. The RRT*-based path re-planning is able to replan the global path after local path planning; this enhances the robot's maneuverability and improves target fit.

The manuscript is divided into the following parts: Section 2 explains the original algorithm, the dynamics required to move the robot, and the related work. Section 3 discusses the FDWA algorithm and the fuzzy control principle. Section 4 discusses the RRT*-FDWA algorithm. Section 5 provides experimental results to demonstrate the effectiveness of the RRT*-FDWA. Section 6 discusses the results and future work.

2. Related Work

Collision-free path planning is usually divided into two stages. The first stage is global path planning—RRT* in this paper. The second stage is local path planning, for which this paper adopts the FDWA algorithm.

2.1. RRT* and DWA Algorithm

The RRT* algorithm is mainly improved by the RRT algorithm. The RRT* algorithm adds *Rewire*, by routing the algorithm to reselect the parent node, thereby achieving the global optimal. The RRT* algorithm, therefore, has probabilistic completeness and asymptotic optimality.

Path planning in the DWA first calculates the current sampled velocity range based on the mobile robot's own characteristics. The sampled acceleration and angular velocity are used to simulate the trajectory of the robot in a certain range, and the trajectory in the range is evaluated using an evaluation function with certain rules; after the trajectory in the range has been obtained, the optimal path is selected to enable the robot to move.

2.2. Kinematic Model

Mobile robots can be divided into omnidirectional and non-omnidirectional robots according to the kinematics of casters, and the robot used in this paper is a non-omnidirectional mobile robot. In Figure 1, XOY represents the global coordinate system. $x_1 P y_1$ is the local coordinate system, x_1 is the linear movement direction of the robot, y_1 is perpendicular to the x_1 axis, P is the center of the robot as the origin of the local coordinate system, and θ is the heading angle. Thus, the robot's global coordinates, position ε, can be obtained, as shown in Equation (1). To map the global coordinate system to the local coordinate system, the rotation matrix R_θ is used. Equation (3) is converted from global coordinates to local coordinates $R\varepsilon$.

$$\varepsilon = \begin{bmatrix} x \\ y \\ \theta \end{bmatrix} \tag{1}$$

$$R_\theta = \begin{bmatrix} \cos\theta & \sin\theta & 0 \\ -\sin\theta & \cos\theta & 0 \\ 0 & 0 & 1 \end{bmatrix} \tag{2}$$

$$R_\varepsilon = R_\theta \times \varepsilon \tag{3}$$

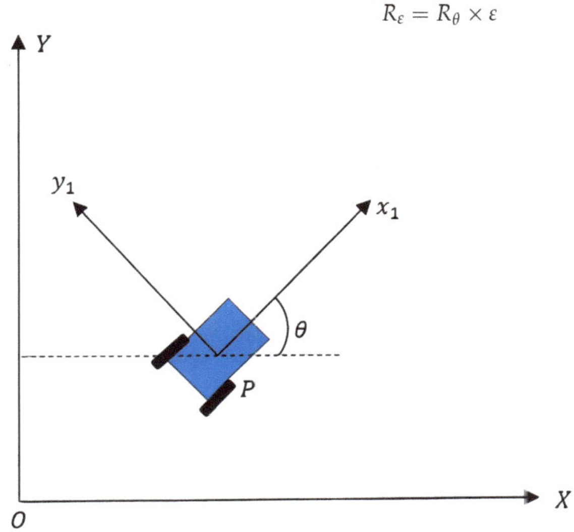

Figure 1. Reference system coordinates.

In a non-omnidirectional moving robot, the robot only has a linear velocity V in the x_1 direction and an angular velocity ω of rotation. Suppose the sampling time is ΔT, the distance the robot travels is ΔS, and the coordinates of the robot are $\Delta X, \Delta Y$.

$$\Delta S = V * \Delta T \tag{4}$$

$$\begin{cases} \Delta X = V * \Delta T \cos \theta_T \\ \Delta Y = V * \Delta T \sin \theta_T \end{cases} \tag{5}$$

According to the coordinates of the robot, $X_{n,} Y_n$, the current sampling time can be obtained by summing up its state and increment.

$$\begin{cases} X_n = X_{n-1} + V * \Delta T \cos \theta_T \\ Y_n = Y_{n-1} + V * \Delta T \sin \theta_T \end{cases} \tag{6}$$

The DWA algorithm [25–27], in Equation (6), has three evaluation metrics. They are the azimuth evaluation function, $heading(v, w)$, which denotes the angular difference between the robot and the target point. $dist(v, w)$, which denotes the distance between the robot and the nearest detected obstacle, is the distance function. $velocity(v, w)$ denotes the relative velocity magnitude of the robot trajectory. α, β, and γ denote the weight coefficients of the function and σ denotes the normalization required to obtain the evaluation function $G(v, w)$.

$$\gamma = 1 - \beta - \alpha \tag{7}$$

$$G(\nu, \omega) = \sigma(\alpha * heading(\nu, \omega) + \beta * dist(\nu, \omega) + \gamma * velocity(\nu, \omega)) \tag{8}$$

3. FDWA Algorithm

The RRT* algorithm is able to perform global path planning in an already established map model environment or with static obstacles. However, in the real environment, there are many uncertainties, such as unknown obstacles. If global path planning alone is used, the robot will easily collide with obstacles in such unknown environments. Therefore, in order to achieve the dynamic obstacle avoidance capability of a mobile robot, this paper incorporates a local path planning algorithm—the FDWA—for local dynamic obstacle avoidance.

3.1. FDWA Fuzzy Distribution

In the DWA algorithm, the percentage of weights has a very strong influence on the results of the path planning algorithm. Therefore, in order to improve the efficiency of the algorithm and be better able to adapt to the complex environment, the fuzzy inference and DWA algorithm are combined, enabling adaption of the evaluation index weights. Fuzzy inference is performed on the azimuth evaluation function weights α and distance function weights β, (β, $\alpha \in [0, 0.5]$), and a dual-input, dual-output fuzzy controller is designed. The angle θd at which the actual trajectory of the robot deviates from the predetermined trajectory and the distance Od to the nearest obstacle are used as dual inputs, and the azimuth evaluation function weight, α, and the distance function weight, β, are used as dual outputs. The membership function of the fuzzy system uses trigonometric functions, which are suitable for describing well-defined ranges and information precisely. The input membership functions θd and θd are shown in Equation (10): ($\mu(u) \in [0, 1]$]). The output affiliation function of weight α is $\mu(o)$, shown in Equation (11): ($\mu(o) \in [0, 1]$). Figure 2a shows the input membership function and Figure 2b shows the output.

$$\mu(u) \in [O_d, \theta_d], \mu(o) \in [\beta, \alpha] \tag{9}$$

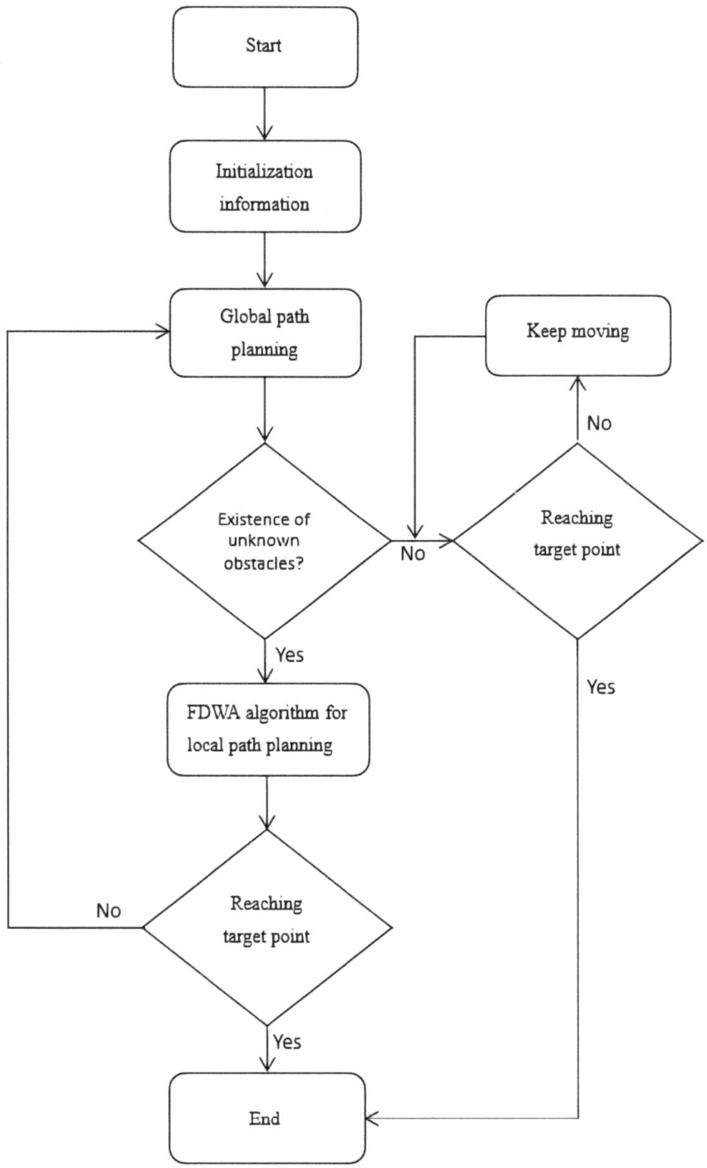

Figure 4. RRT*-FDWA flow chart.

The algorithm flows as follows:

Step 1: Create a map of the known environment and set target points.
Step 2: The RRT* algorithm plans a collision-free global optimal path based on the already established map environment and its own sensors, such as lidar.
Step 3: Determine whether the robot can reach the target point according to the end condition of the algorithm; if so, end the algorithm; if not, continue the execution.
Step 4: Determine whether there is an environment or moving obstacle in the established map; if the lidar determines that the moving obstacle is dangerous, jump to the FDWA algorithm for local path planning to avoid the obstacle.

Step 5: Perform a new global path plan, determine the optimal path, and continue driving along the global path. Judge whether it is possible to reach the target point: if 'No', jump to the Step 3 loop; if 'Yes', end the algorithm. Figure 5 represents the global path planning, encountering moving obstacles in the path replanning process.

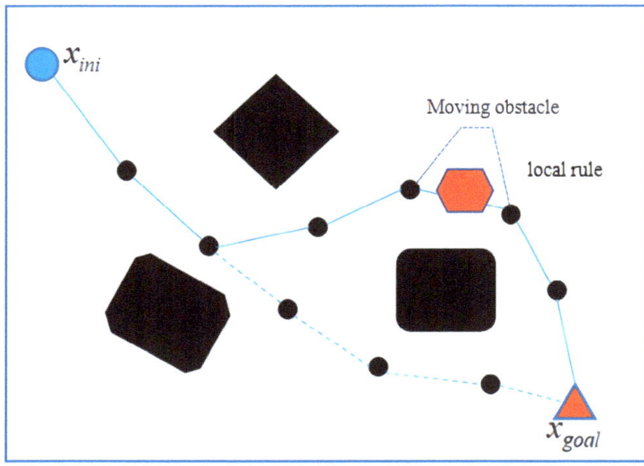

Figure 5. Path planning diagram.

4.2. RRT-FDWA Local Minima and Pseudocode*

In the DWA algorithm, the overall generation value increases as the value of its weight coefficient increases the closer it gets to the obstacle. This will cause the robot to fall into local minima during local path planning. In the RRT*-FDWA algorithm, to avoid local minima, the robot can quickly enter global path planning once it has successfully avoided an obstacle, as shown in Equation (13). K is the reward and punishment function of the DWA algorithm.

$$G'(v,\omega) = K \cdot G(v,\omega) \tag{13}$$

$$\Delta t(v,\omega) = \Delta S/\Delta V \tag{14}$$

$$K = \begin{cases} a & t < 0.5 \\ 1 & t = 0.5 \\ b & 0.5 < t < 1 \\ 0 & t > 1 \end{cases} \tag{15}$$

In Equation (14), $\Delta t(v,\omega)$ is the time function with the obstacle. In Equation (15), a and b represent the reward and punishment function K, where $a < 1 < b$. When the value of t is less than 0.5, it is not caught in the local minimum; when t is greater than 0.5, it is regarded that the robot is caught in the local minimum. Global path planning uses the RRT* [25] algorithm, while local path planning uses the FDWA algorithm. The pseudo code of the RRT*-FDWA algorithm is shown in Algorithm 1; Algorithm 2 is the *Rewire* in Algorithm 1.

Algorithm 1 RRT*-FDWA Algorithm.

```
input: p_ini, p_goal, Map, v, w, K
output: p_ini→p_goal path
WHILE Target_not_reach DO
  FOR i in range(n): p_rand ← Sample(Map
      p_nearest ← Node_list(p_rand, p_ini)
      p_new ← Steer(p_nearest, p_rand)
      Cost(x_new) ←d(p_nerarest, p_new) + Cost(p_nearest)
    IF (Check_collision(p_nearest, p_new)) THEN
       p_near ← findnear_nodes(p_new)
       p_new ← C_parent(p_near, p_new)
       Node_list ← append(p new)
       p_parent ← Rewire(p_new, p_near)
    IF (Check_ Moving obstacle (p_nearest, p_new)) THEN
       DWA(p_near, p_new) ← G'(v, w), K
       G'(v, w), K ← Rule(α, β)
       α, β ← clarification
       IF (Check_collision(p_nearest, p_new)) THEN
       end
    IF(p_new → p_goal)THEN
       IF(Check_conlision(p_new, p_goal)) THEN
       Temppath ← cunrrrent(p_ini → p_goal)
Return path
End while
```

Algorithm 2 Rewire (P_1, P_2).

```
IF Collision_free (p1, p2)THEN
  p_parent2 ← p2[i − 1]
  IFCost(p2) > Cost(p1)) THEN
    IF(Check_collision(p_nearest, p2))THEN
    p_parent ← p2
      Cost ← Cost(p2)
      End if
End if
```

In Algorithms 1 and 2 in this paper, the following terms are used:

Sample (Map): The number of sampling points generated randomly on the map towards the target node is in the range [0, 100].

Node_list (p_1, p_2): The set of Euclidean distances between the sampling point and the parent node, where the coordinates of the random sampling point and the parent node are p_s (x_1, y_2) and p_p (x_2, y_2), respectively.

$p_{nearest}$: Minimum of *Node_list* (p_1, p_2) min (*Node_list*)

Steer (p_1, p_2): The angle between p_1 and the line connecting p_1 and p_2 of its parent node; multiply the steering angle by the step size β to get p_{new}. Let the steering angle be θ; p_{new} is obtained by the following Equations (16) and (17).

$$p_{new,x} = \beta \cos\theta \tag{16}$$

$$p_{new,y} = \beta \sin\theta \tag{17}$$

Check_collision (p_1, p_2): Check if there are obstacles between p_1 and p_2.

Target_not_reach: If or not the target point is reached: the return value is 1 if the target point is reached and 0 if the opposite is true.

Cost(x): Return along the total path length of the target node from node p.

find near_nodes (p_1): Find a new parent node near node p again at random.

C_parent (p_1, p_2): Add a new optional parent node and define its length, angle, and generation value.

D (p_1, p_2): Denotes the Euclidean distance between nodes p_1 and p_2.

Rewire (p_1, p_2): Determine whether p_2 can replace p_1 as the new parent node.

Collision_free (p_1, p_2): Check whether it is feasible and that there are no obstacles between node p_1 and node p_2.

G' (v, ω): Evaluation function of the DWA algorithm with local minimum judgment added.

Rule (α, β): Fuzzy rule table for α and β.

5. Experimental Results

In this paper, the algorithm was simulated in an AMD Ryzen 7 5800H 3.2 GHz and 16 G RAM computer, and a Matlab simulation was used (Matlab version 2019b). A two-wheel differential speed mobile robot with a universal wheel-manipulable wheel in the front was used.

5.1. RRT*-FDWA Global Path Planning

First, RRT* was used to simulate the global planning path. As shown in Figure 6, for the random moving obstacle environment, dashed and solid lines represent the global path. The dashed line shows the process of the RRT* algorithm rewiring and selecting a new node for the first time; the blue circle represents p_{new} and the black objects represent the moving obstacles. A size [10,10] map with a start point of [1.2,1.2] and a target point of [9,9] was used.

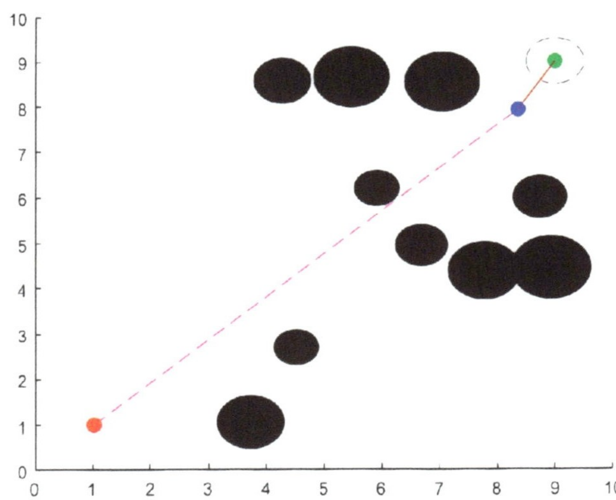

Figure 6. RRT*-FDWA global path planning.

5.2. FDWA Lobal Path Planning

In order to verify its effectiveness, the FDWA algorithm was simulated. First, the initial values of the weight coefficients in the evaluation function were set at $\alpha = 0.2$, $\beta = 0.4$, and $\gamma = 0.4$; the maximum linear and angular velocities were also set to $v = 0.5$ m/s and $\omega = 0.3$ rad/s. a and b represent the reward and punishment function K: $a = 0.5$ and $b = 1.5$. The time for forward simulation was 3 s and dt = 0.1.

Figure 7 shows the FDWA algorithm planning diagram and demonstrates the effectiveness of the algorithm in the presence of unknown obstacles. The solid purple line indicates the path simulated by the FDWA algorithm in the constant forward direction, and the dashed line indicates the final completed trajectory. The blue solid circles indicate stationary obstacles; the blue solid hollow circles indicate the current position of the moving obstacle; and the blue dashed circles indicate the position of the moving obstacle at the previous moment. When the mobile robot simulates the path forward, an optimal local path is found according to the evaluation function and the local minimum problem can be successfully avoided.

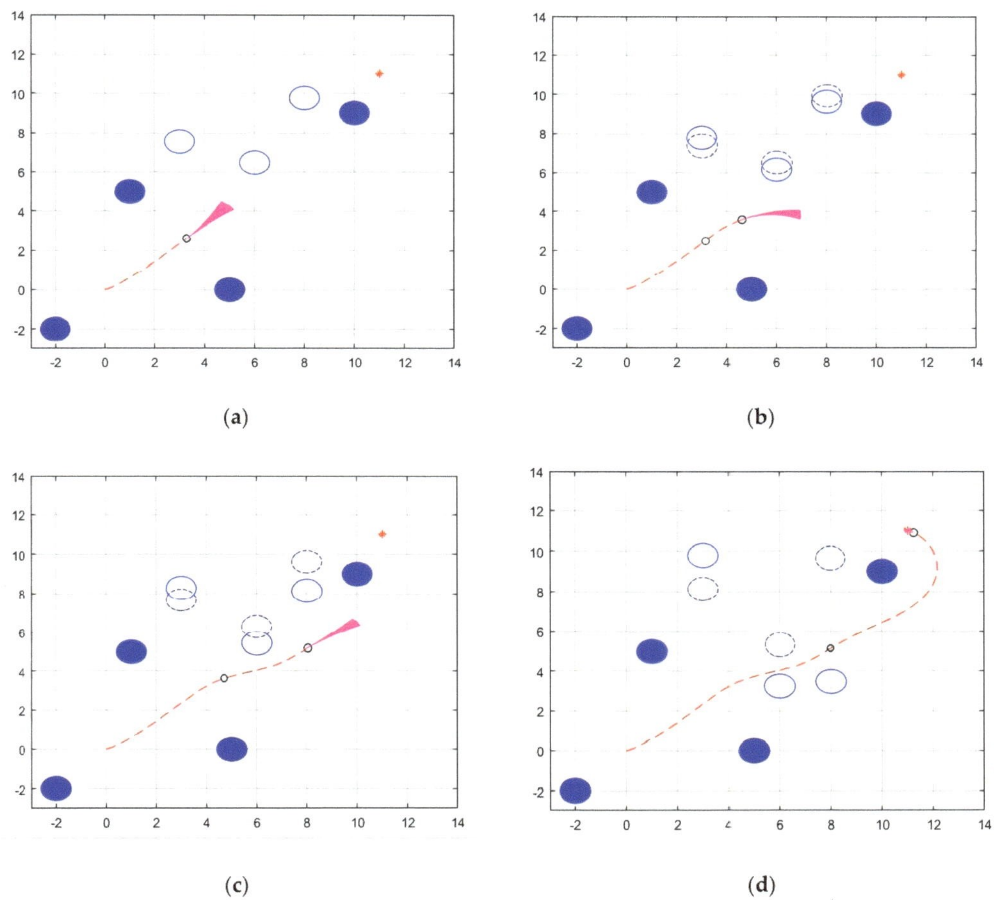

Figure 7. FDWA path planning diagram. (**a**) Position planning diagram at the moment t = 0.5. (**b**) Position planning diagram at the moment t = 1. (**c**) Position planning diagram at the moment t = 3. (**d**) Position planning diagram at the moment t = 5.

To further verify the effectiveness of the FDWA, this paper is validated by a real robot. Figure 8a shows the experimental robot tianbot_mini, a two-wheel differential drive robot with a drive wheel at the rear and a gimbal at the front, and with classic PD control for the drive wheel. It has a ydlidar x2 LIDAR for obstacle detection. The linear velocity is constrained, $v \in (0, 0.5\ \mathrm{m/s})$, and the FDWA algorithm is encapsulated as a local path planner for the ROS navigation package. The experimental environment is the same as the simulation environment, and the white mobile robot in Figure 8b is regarded as a moving obstacle with a handle control and a speed of 0.10 m/s. Figure 8b–f shows the FDWA path planning process, and the experimental results better validate the effectiveness of the algorithm.

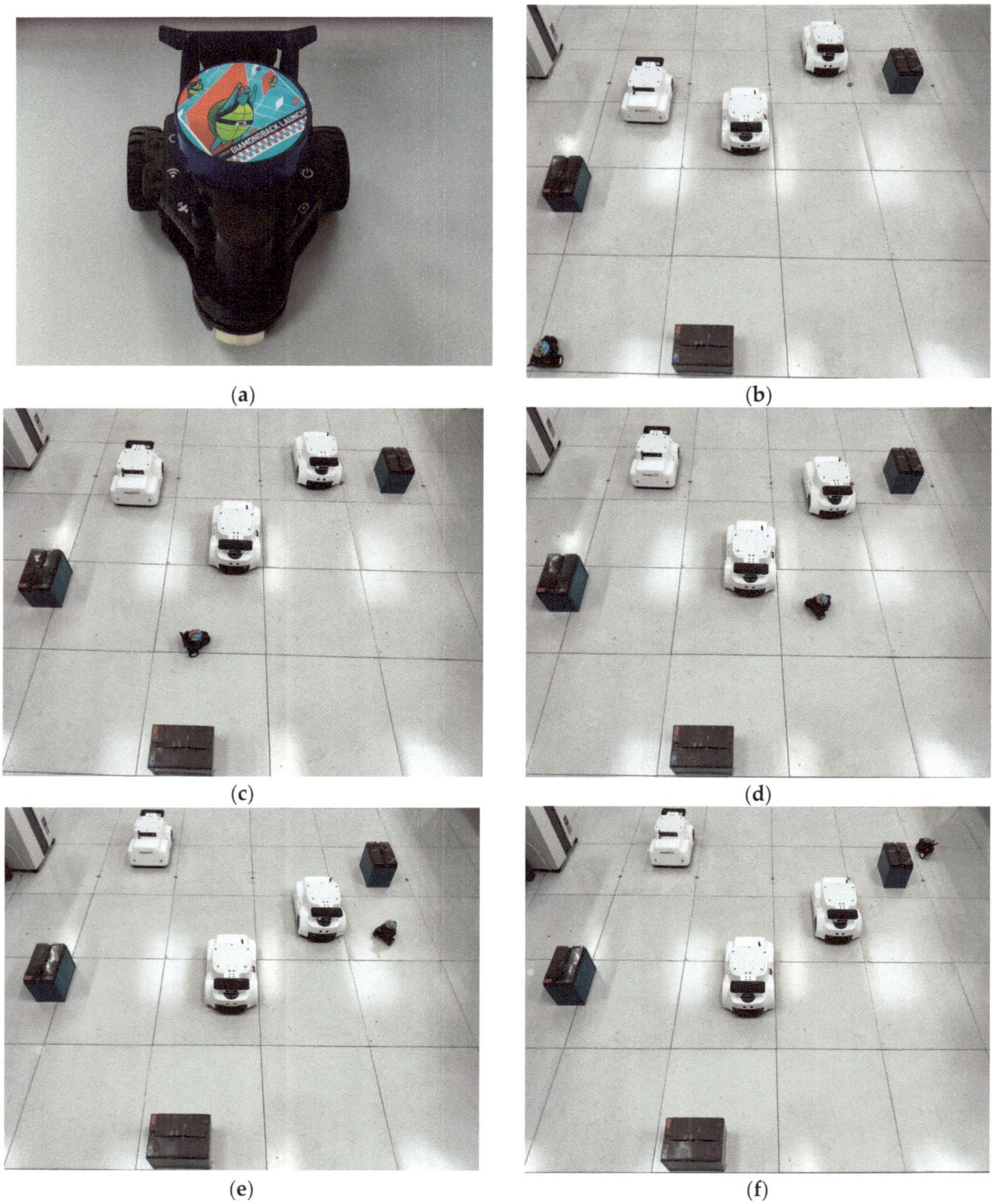

Figure 8. FDWA real environment path planning map. (**a**) the experimental robot (**b**) The robot position map at the moment t = 0. (**c**) The robot position map at the moment t = 3. (**d**) The robot position map at the moment t = 5. (**e**) The robot position map at the moment t = 8. (**f**) The robot position map at the moment t = 10.

5.3. Experimental Comparison

In this paper, we used a two-wheel differential speed robot for the simulation. The maneuverability degree, δ_m, of this mobile robot can, therefore, be obtained as two, and according to Equation (18), $rank[C_\alpha(\beta_\alpha)]$ is the number of maneuverable wheels.

$$\delta_m = 3 - rank[C_a(\beta_a)] \tag{18}$$

In dynamic obstacle avoidance, robot behavior includes deceleration or acceleration to avoid obstacles when it encounters them, but such behaviors may cause damage to safety. In this paper, the behavior of avoiding obstacles is proposed as the standard deviation formula for speed, the smaller value of which represents higher safety, as shown in Equation (19), where v_i indicates the line speed at the current moment. The integrated speed difference formula presents the maneuvering magnitude, which indicates the robot's evasion ability when encountering obstacles, and the higher its value, the better the evasion ability, as shown by Equation (20).

$$\sigma_w^2 = \sqrt{\frac{\sum_{i=1}^{n} (v_i - v)^2}{n}} \tag{19}$$

$$\delta_r = \sigma_w^2 \delta_m / \theta_d \times 100\% \tag{20}$$

Figure 9 shows the RRT*-FDWA path planning process; the solid red circles indicate the current position and solid blue circles indicate p_{new}. Hollow circles indicate the previous moment position and the planned route and dashed hollow circles indicate the moving obstacle at the previous moment.

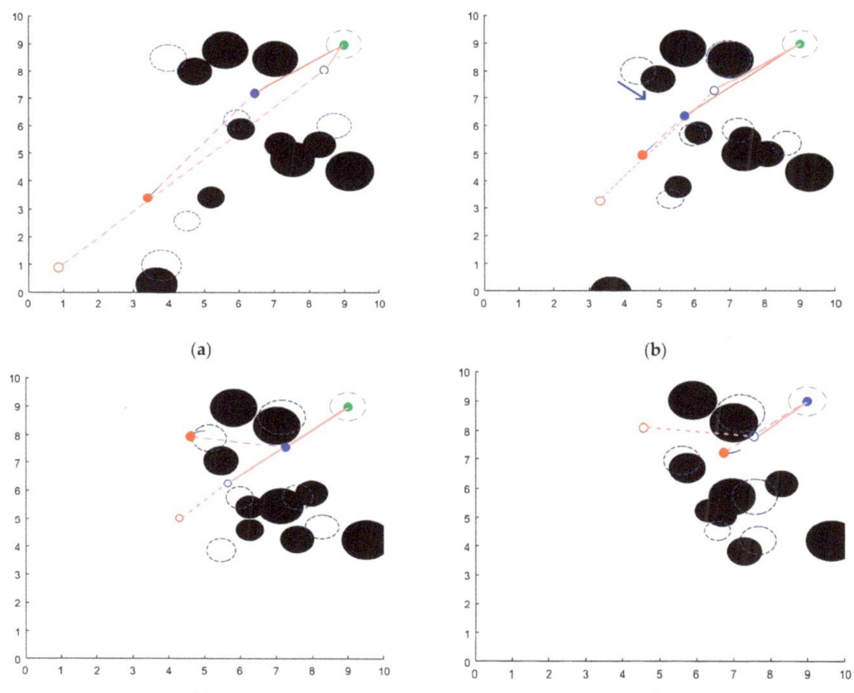

Figure 9. RRT*-FDWA path planning diagram. (**a**) Position planning diagram at the moment $t = 1$. (**b**) Position planning diagram at the moment $t = 2$. (**c**) Position planning diagram at the moment $t = 5$. (**d**) Position planning diagram at the moment $t = 7$.

Figure 10 shows the RRT*-FDWA experimental path planning diagram, and the experimental results show that the algorithm is effective in practical applications. The RRT*-FDWA algorithm was packaged into the path planner in the ROS navigation package before experimental validation. In the figure, the basketball and the white mobile robots are treated as moving obstacles and the experimental environment was the same as the simulation experiment. In the real experimental process, there was an error in the path planning process due to the robot control drive factor.

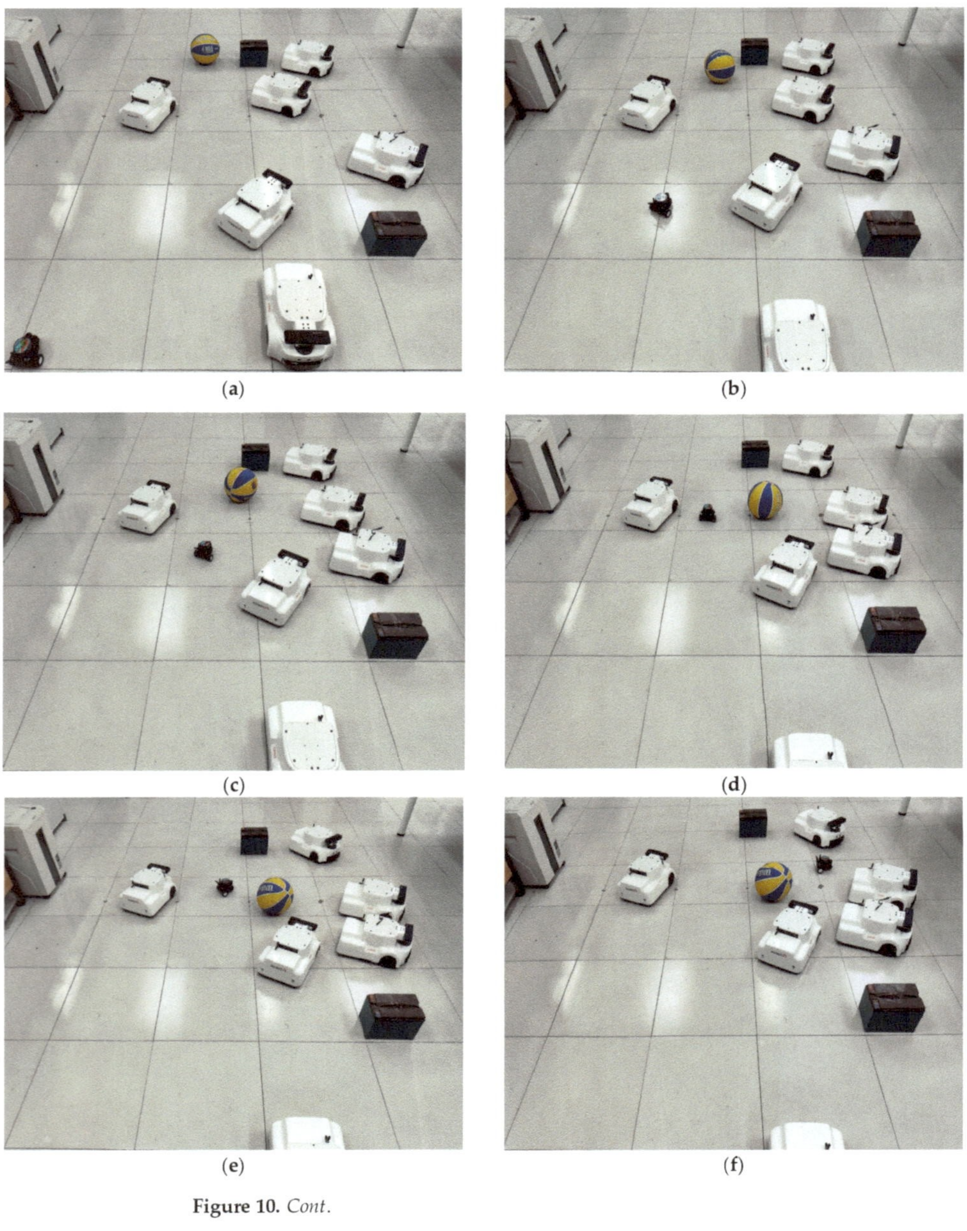

Figure 10. *Cont.*

(g)

(h)

Figure 10. RRT*-FDWA real environment path planning map. (**a**) The robot position map at the moment $t = 0$. (**b**) The robot position map at the moment $t = 3$. (**c**) The robot position map at the moment $t = 5$. (**d**) The robot position map at the moment $t = 7$. (**e**) The robot position map at the moment $t = 9$. (**f**) The robot position map at the moment $t = 13$. (**g**) The robot position map at the moment $t = 16$. (**h**) The robot position map at the moment $t = 18$.

The RRT*-FDWA algorithm proposed in this paper was compared to the hybrid A*—an improved DWA algorithm—the hybrid algorithm being SOTA. Figure 11 shows the final completed path planning graph for both algorithms, the red dot [1,1] is the starting point and the green dot [9,9] is the target point; it can be seen that the A*-DWA algorithm still falls into local minima, while the RRT*-FDWA algorithm can complete the path planning more smoothly.

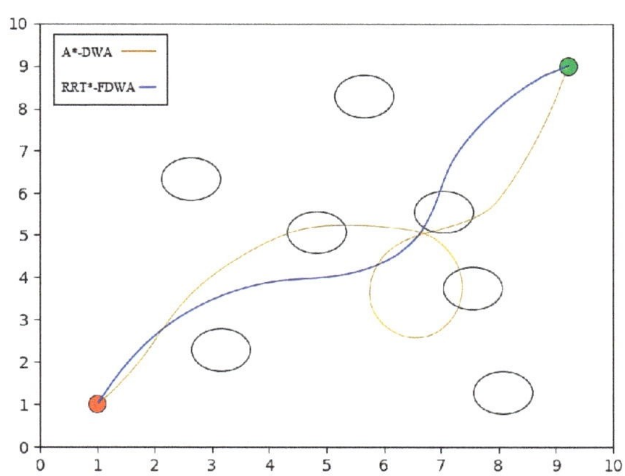

Figure 11. Algorithm comparison procedure.

It can be seen from Table 3 that the σ_w^2 of the RRT*-FDWA algorithm is smaller and δ_r is 12% higher than those of the A*-DWA algorithm, and it takes less time to complete. This indicates that the RRT*-FDWA algorithm is safer, fits better, and plans faster.

Table 3. Algorithm efficiency test comparison.

	σ_w^2	Time (s)	δ_r
A*-DWA	0.548	11.93	79%
RRT*-FDWA	0.424	9.62	91%

The general comparison of the four algorithms is given in Table 4. The RRT*-FDWA algorithm proposed in this paper considers both global and local optimality and can successfully plan a collision-free optimal path in both dynamic and static environments. In Table 4, E means Exist, N means None, L means Low, and H means High.

Table 4. Algorithm comparison.

	Dynamic Obstacle Avoidance	Local Optimality	Global Optimality	Smooth Path
RRT*	N	N	E	L
DWA	N	E	N	L
Fuzzy-DWA	E	E	N	H
RRT*-FDWA	E	E	E	H

6. Conclusions and Future Work

In this paper, mobile robot path planning in a dynamic environment was studied and an RRT*-FDWA algorithm was proposed. First, the RRT* algorithm was used for global path planning to obtain a global optimal route. The uncertainty of moving obstacles increases the difficulty of obstacle avoidance in the path planning process. When obstacles are encountered, the FDWA algorithm is added to improve robot adaptability in the face of a dynamic environment. The combination of the algorithms enables better avoidance of local minima and global replanning according to the new environment after the local planning has been completed. As a result, the robot has good maneuvering magnitude and safety and can complete the task in a shorter time.

The future goal is to apply this algorithm in multi-robot formation path planning to enable multi-robot formations with dynamic obstacle avoidance.

Author Contributions: Conceptualization, L.Z. and N.W.; methodology, N.W.; software, H.C.; validation, L.Z., N.W., and H.C; formal analysis, H.C.; investigation, N.W.; resources, Q.W.; data curation, L.Z.; writing—original draft preparation, N.W.; writing—review and editing, H.C.; visualization, Y.L.; project administration, L.Z.; funding acquisition, Q.W. All authors have read and agreed to the published version of the manuscript.

Funding: This research was funded by the Henan Province Key Science and Technology Program (222102220085); the Henan Province University Key Science and Technology Project (23A413007); the Henan Province Key Science and Technology Project (222102210084); and the Henan Province Postgraduate Education Reform and Quality Improvement Project (YJS2023JD67), respectively.

Institutional Review Board Statement: Not applicable.

Informed Consent Statement: Not applicable.

Data Availability Statement: Data are contained within the article.

Conflicts of Interest: The authors declare no conflict of interest.

References

1. Tan, G.; He, H.; Aaron, S. Global optimal path planning for mobile robot based on improved Dijkstra algorithm and ant system algorithm. *J. Cent. South Univ. Technol.* **2006**, *13*, 80–86. [CrossRef]
2. Chand, P.; Carnegie, D.A. A two-tiered global path planning strategy for limited memory mobile robots. *Robot. Auton. Syst.* **2012**, *60*, 309–321. [CrossRef]
3. Song, X.; Gao, S.; Chen, C.B.; Cao, K.; Huang, J. A new hybrid method in global dynamic path planning of mobile robot. *Int. J. Comput. Commun. Control* **2018**, *13*, 1032–1046. [CrossRef]
4. Persson, S.M.; Sharf, I. Sampling-based A* algorithm for robot path-planning. *Int. J. Robot. Res.* **2014**, *33*, 1683–1708. [CrossRef]
5. Kanayama, Y.J.; Hartman, B.I. Smooth local-path planning for autonomous vehicles1. *Int. J. Robot. Res.* **1997**, *16*, 263–284. [CrossRef]

6. Sedighi, K.H.; Ashenayi, K.; Manikas, T.W.; Wainwright, R.L.; Tai, H.-M. Autonomous local path planning for a mobile robot using a genetic algorithm. In Proceedings of the 2004 Congress on Evolutionary Computation (IEEE Cat. No. 04TH8753), Portland, OR, USA, 19–23 June 2004; IEEE: Piscataway, NJ, USA, 2004; Volume 2, pp. 1338–1345.
7. Henkel, C.; Bubeck, A.; Xu, W. Energy efficient dynamic window approach for local path planning in mobile service robotics. *IFAC-PapersOnLine* **2016**, *49*, 32–37.
8. Karaman, S.; Frazzoli, E. Sampling-based algorithms for optimal motion planning. *Int. J. Robot. Res.* **2011**, *30*, 846–894. [CrossRef]
9. Karaman, S.; Frazzoli, E. Optimal kinodynamic motion planning using incremental sampling-based methods. In Proceedings of the 49th IEEE Conference on Decision and Control (CDC), Atlanta, GA, USA, 15–17 December 2010; pp. 7681–7687.
10. Karaman, S.; Walter, M.R.; Perez, A.; Frazzoli, E.; Teller, S. Anytime motion planning using the rrt*. In Proceedings of the 2011 IEEE International Conference on Robotics and Automation, Shanghai, China, 9–13 May 2011; pp. 1478–1483.
11. Kang, J.-G.; Jung, J.-W. Post Triangular Rewiring Method for Shorter RRT Robot Path Planning. *Int. J. Fuzzy Log. Intell. Syst.* **2021**, *21*, 213–221. [CrossRef]
12. Moon, C.; Chung, W. Kinodynamic planner dual-tree RRT (DT-RRT) for two-wheeled mobile robots using the rapidly exploring random tree. *IEEE Trans. Ind. Electron.* **2014**, *62*, 1080–1090. [CrossRef]
13. Nasir, J.; Islam, F.; Malik, U.; Ayaz, Y.; Hasan, O.; Khan, M.; Muhammad, M.S. RRT*-SMART: A rapid convergence implementation of RRT. *Int. J. Adv. Robot. Syst.* **2013**, *10*, 299. [CrossRef]
14. Chen, L.; Shan, Y.; Tian, W.; Li, B.; Cao, D. A fast and efficient double-tree RRT*-like sampling-based planner applying on mobile robotic systems. *IEEE/ASME Trans. Mechatron.* **2018**, *23*, 2568–2578. [CrossRef]
15. Qi, J.; Yang, H.; Sun, H. MOD-RRT*: A sampling-based algorithm for robot path planning in dynamic environment. *IEEE Trans. Ind. Electron.* **2020**, *68*, 7244–7251. [CrossRef]
16. Molinos, E.J.; Llamazares, A.; Ocaña, M. Dynamic window based approaches for avoiding obstacles in moving. *Robot. Auton. Syst.* **2019**, *118*, 112–130. [CrossRef]
17. Rasekhipour, Y.; Khajepour, A.; Chen, S.K.; Litkouhi, B. A potential field-based model predictive path-planning controller for autonomous road vehicles. *IEEE Trans. Intell. Transp. Syst.* **2016**, *18*, 1255–1267. [CrossRef]
18. Choi, Y.-I.; Cho, J.-H.; Kim, Y.-T. Collision Avoidance Algorithm of Mobile Robots at Grid Map Intersection Point. *Int. J. Fuzzy Log. Intell. Syst.* **2020**, *20*, 96–104. [CrossRef]
19. Kumar, S.; Parhi, D.R.; Muni, M.K.; Pandey, K.K. Optimal path search and control of mobile robot using hybridized sine-cosine algorithm and ant colony optimization technique. *Ind. Robot.* **2020**, *47*, 535–545. [CrossRef]
20. Guo, B.; Guo, N.; Cen, Z. Obstacle avoidance with dynamic avoidance risk region for mobile robots in dynamic environments. *IEEE Robot. Autom. Lett.* **2022**, *7*, 5850–5857. [CrossRef]
21. Zhong, X.Y.; Peng, X.F.; Zhou, J.H.; Huang, X.C. Mobile robot path planning based on local environment modeling and adaptive window. *Appl. Mech. Mater.* **2011**, *48*, 679–683. [CrossRef]
22. Chang, L.; Shan, L.; Jiang, C.; Dai, Y. Reinforcement based mobile robot path planning with improved dynamic window approach in unknown environment. *Auton. Robot.* **2021**, *45*, 51–76. [CrossRef]
23. Xiang, L.; Li, X.; Liu, H.; Li, P. Parameter Fuzzy Self-Adaptive Dynamic Window Approach for Local Path Planning of Wheeled Robot. *IEEE Open J. Intell. Transp. Syst.* **2021**, *3*, 1–6. [CrossRef]
24. Wu, B.; Chi, X.; Zhao, C.; Zhang, W.; Lu, Y.; Jiang, D. Dynamic Path Planning for Forklift AGV Based on Smoothing A* and Improved DWA Hybrid Algorithm. *Sensors* **2022**, *22*, 7079. [CrossRef] [PubMed]
25. Trinh, L.A.; Ekström, M.; Cürüklü, B. Petri net based navigation planning with dipole field and dynamic window approach for collision avoidance. In Proceedings of the 2019 6th International Conference on Control, Decision and Information Technologies (CoDIT), Paris, France, 23–26 April 2019; pp. 1013–1018.
26. Zeng, T.; Si, B. Mobile robot exploration based on rapidly-exploring random trees and dynamic window approach. In Proceedings of the 5th International Conference on Control, Automation and Robotics (ICCAR), Beijing, China, 19–22 April 2019; pp. 51–57.
27. Fox, D.; Burgard, W.; Thrun, S. The Dynamic Window Approach to Collision Avoidance. *IEEE Robot. Autom. Mag.* **1997**, *4*, 23–33. [CrossRef]

applied sciences

MDPI

Article

Research on Real-Time Obstacle Avoidance Motion Planning of Industrial Robotic Arm Based on Artificial Potential Field Method in Joint Space

Yu Chen *, Liping Chen, Jianwan Ding and Yanbing Liu

School of Mechanical Science and Engineering, Huazhong University of Science and Technology, Wuhan 430074, China; chenlp@hust.edu.cn (L.C.); dingjw@hust.edu.cn (J.D.); d202180319@hust.edu.cn (Y.L.)
* Correspondence: d202180318@hust.edu.cn

Abstract: The artificial potential field method is a highly popular obstacle avoidance algorithm which is widely used in the field of industrial robotics due to its high efficiency. However, the traditional artificial potential field method has poor real-time performance, making it less suitable for modern factory work patterns, and it is difficult to handle situations when the robotic arm encounters singular configurations. In this paper, we propose an improved artificial potential field method in joint space, which effectively improves the real-time performance of the algorithm, and still performs well when the robotic arm falls into a singular configuration. This method solves the gradient of the repulsive potential field in advance by defining the shortest distance from each joint of the robotic arm to the obstacle, and only needs to calculate potential field function once per cycle, which significantly reduces the calculation time. In addition, when a robotic arm falls into a local minimum position in potential field, the algorithm adds a virtual obstacle to make it leave the position, while this virtual obstacle does not require additional input information. Experimental results show that the algorithm obtains short movement paths and requires very little computing time in the face of different obstacles.

Keywords: robotic arm; real-time obstacle avoidance; artificial potential field method; joint space

Citation: Chen, Y.; Chen, L.; Ding, J.; Liu, Y. Research on Real-Time Obstacle Avoidance Motion Planning of Industrial Robotic Arm Based on Artificial Potential Field Method in Joint Space. *Appl. Sci.* **2023**, *13*, 6973. https://doi.org/10.3390/app13126973

Academic Editor: Alberto Doria

Received: 7 May 2023
Revised: 4 June 2023
Accepted: 7 June 2023
Published: 9 June 2023

1. Introduction

The utilization of robotic manipulators enhances the efficiency and automation of industrial manufacturing processes. In the 3c industry (computer, communication, consumer electronics), rapidly updated products and complex production patterns require highly flexible robotic arms to complete tasks. In the process of performing the task, the robotic arm may collide with obstacles, so a real-time obstacle avoidance algorithm is needed to control the robotic arm [1–6].

The obstacle avoidance algorithm is widely studied in the field of robot motion planning, and it is a computational method to find the path from the starting point to the target point without colliding with obstacles when the robotic manipulator is working in an environment with obstacles. At present, researchers have proposed various obstacle-avoidance algorithms, such as genetic algorithm, rapidly exploring random tree (RRT), and probabilistic roadmap (PRM). However, they all have limitations in practical applications [7].

The RRT algorithm begins by selecting the starting point as the root node and proceeds to generate a randomly expanded tree structure by iteratively sampling and adding leaf nodes, and the shortest path from the starting point along the leaf node to the target position is the obstacle avoidance path. Nevertheless, the calculation speed of RRT is slow, and the smoothness of the trajectory obtained by this algorithm is unsatisfactory, which makes the manipulator shake easily [8]. The genetic algorithm finds an intermediate point with position and velocity information, and connects the starting position to the intermediate point and the intermediate point to the target position through two polynomial paths. The two paths are concatenated to obtain an obstacle avoidance path. Although the genetic algorithm can obtain a smooth obstacle avoidance trajectory, the computational efficiency is very low [9,10]. The PRM algorithm establishes a complete undirected graph by sampling and collision detection, and then obtains a feasible obstacle avoidance path through a graph search algorithm (such as the A* algorithm). However, the process of establishing the map using the PRM algorithm is complex and time-consuming. Moreover, the connection between two nodes in the undirected graph does not take into account the robot's motion model [11,12]. Although the above algorithms can obtain obstacle avoidance trajectories, their real-time performance is weaker than the artificial potential field method. This paper only discusses the artificial potential field method.

The artificial potential field (APF) method is a widely recognized motion planning strategy in the field of robotics. In 1986, Oussama Khatib first proposed the artificial potential field method (APF) for the motion planning of robots [13]. The APF method extensively borrows the concept of potential fields from physics and transforms the path planning problem of robots or moving objects into a problem of searching for the optimal path in a virtual potential field. The core idea is to consider the target location as an attractive source and obstacles or infeasible areas as repulsive sources. High potential fields are assigned around obstacles, while low potential fields are assigned around the target location. By calculating the forces and directions, the APF guides the robot or object to move along the direction of decreasing potential energy, thereby achieving the goal of avoiding obstacles and reaching the target location.

The APF method is characterized by its simplicity, intuitiveness, ease of implementation, and real-time capabilities. As a result, it has been widely applied in the industrial field. For example, in automated warehousing systems, robots utilize the APF method to navigate around obstacles and find the shortest path based on the layout and obstacle positions within the warehouse. This potential field guidance enables them to efficiently and swiftly accomplish tasks such as cargo handling [14,15]. In the field of autonomous driving, the APF method establishes a potential field within the road network. It considers other vehicles, pedestrians and obstacles as repulsive sources while regarding the target destination as an attractive source. This enables vehicles to navigate to their destination safely and smoothly [16,17]. In multi-agent cooperation, the APF method guides robots to avoid collisions and collaboratively accomplish complex production tasks. It achieves this by considering robots and target positions as attractive sources, and other robots as repulsive sources [18,19]. These application cases highlight the significant role of the APF method in achieving efficient, safe, and flexible industrial operations.

The coordinate systems of the APF method can be divided into two categories: Cartesian space and joint space. If attraction and repulsion forces are calculated in Cartesian space, the coordinate system for this method is the Cartesian space. If attraction and repulsion forces are calculated in joint space, the coordinate system for this method is the joint space.

In studies based on Cartesian space, Hui Zhang et al. proposed an obstacle avoidance strategy for the dual-arm robot by improving the potential field into a velocity field [20]. Sun-Oh Park et al. proposed a numerical method based on a Jacobian matrix to solve the obstacle avoidance problem of a redundant robotic arm [21]. All the above obstacle avoidance strategies in Cartesian space need to be mapped to joint space, and the inverse of the Jacobian matrix cannot be solved when the robotic arm falls into a singular configuration. In studies based on joint space, Lufeng Luo et al. uses a sampling method to generate anti-collision path points [22]. Anoush Sepehri et al. used a numerical method to solve the gradient of potential field [23]. All of the above obstacle avoidance strategies in joint space require the multiple calculations of the potential field function, so the real-time performance is low. In Section 3, we introduced these above algorithms and compared them with the algorithm in this paper.

In an intelligent workshop, the robotic arm is required to grip the workpieces on the conveyor belt and move them to the target position without collisions. Due to the uncertain initial position of the workpieces, a real-time obstacle avoidance algorithm is needed. This paper proposes an improved artificial potential field method specifically designed for the aforementioned industrial scenario. The algorithm generates a suitable trajectory for the robotic arm, so that it can start from any specified starting point and reach the target point without colliding with obstacles in the process. We propose a way to calculate the shortest distance from the obstacle to the joint of the robotic arm, whereby an analytic formula for the gradient of repulsive potential field can be solved in advance. In addition, we propose a way to make the robotic arm jump out of the local minimum.

The contributions of this study are novel and unique in various aspects. (1) Compared with the APF method in Cartesian space, this algorithm does not need to calculate the inverse of the Jacobian matrix. (2) Compared with others' studies on the APF method in joint space, this algorithm only needs to calculate the potential field function once during each cycle. Therefore, the algorithm has a very high real-time performance. (3) In addition, the algorithm provides an effective way to avoid the robotic arm falling into the local optimum by introducing a virtual obstacle, which does not need additional information.

The subsequent sections of this paper are structured as follows: in Section 2, the D-H parameters and kinematic model of the robot are presented. Section 3 describes the principle and flow of the algorithm, which is the focus of this paper. In Section 4, the simulations and experiments are presented. Section 5 discusses the advantages and disadvantages of the algorithm proposed in this paper compared to other algorithms. Section 6 summarizes the conclusions.

2. Kinematics Model of n-DOF Robot

It is firstly necessary to establish the kinematics model of the robotic manipulator to prevent collision.

To describe the structure and coordinate system of robot, we used 1-n to represent each joint of the manipulator, and x1y1z1-xnynzn to represent the coordinate systems of each link. We used the modified D-H parameter method to establish the kinematics model of the robotic manipulator. Equation (1) describes the relationship between any two adjacent joints.

$$
{}_{i-1}^{\ i}T = Rot_{x_{i-1}}(\alpha_{i-1}) \times Trans_{x_{i-1}}(a_{i-1}) \times Rot_{z_i}(\theta_i) \times Trans_{z_i}(d_i) = \begin{bmatrix} C\theta_i & -S\theta_i & 0 & a_{i-1} \\ S\theta_i C\alpha_{i-1} & C\theta_i C\alpha_{i-1} & -S\alpha_{i-1} & -S\alpha_{i-1} d_i \\ S\theta_i S\alpha_{i-1} & C\theta_i S\alpha_{i-1} & C\alpha_{i-1} & C\alpha_{i-1} d_i \\ 0 & 0 & 0 & 1 \end{bmatrix} \tag{1}
$$

where θ_i represents the angle of rotation of the i-th joint, $S\theta_i = sin\theta_i, C\theta_i = cos\theta_i$, and ${}_{i-1}^{\ i}T,\ i = 1, 2, \cdots, n$ is the homogeneous transformation matrix built from D-H parameters. If the D-H parameters of the n-DOF robot are known, the homogeneous transformation matrix of each link can be obtained by substituting these parameters into Equation (1), so that the coordinate system transformation from the base to the end of the manipulator can be expressed as Equation (2):

$$
{}_n^0T = {}_1^0T \times {}_2^1T \times {}_3^2T \times \cdots \times {}^{n-1}_{\ \ n}T \tag{2}
$$

${}_n^0T$ represents the position and orientation of the end-effector in Cartesian space, and it can be expressed as (3):

$$
{}_n^0T = \begin{bmatrix} n_x & o_x & a_x & p_x \\ n_y & o_y & a_y & p_y \\ n_z & o_z & a_z & p_z \\ 0 & 0 & 0 & 1 \end{bmatrix} \tag{3}
$$

where n, o, a denote the orientation and p denotes the position. Our industrial background involves a robotic arm gripping objects on a conveyor belt in workshops and moving them to the end position. In order to maintain a stable grip and prevent object dropping, the robotic arm's end-effector should predominantly move along the z axis of Cartesian space when approaching and placing the objects [24]. According to the requirements of the industrial task, the positions and orientations of both the starting point and the endpoint of the task are known. The Cartesian coordinates of the task starting point and goal point can be mapped to joint space by inverse kinematics.

As evidenced from (1) and (2), the position of joint i depends on a series of variables related to joint angles θ_i, $i = 1, \cdots, n$, with these variables ranging from θ_1 to θ_{i-1}, so the position of joint i can be expressed as $\{x(\theta), y(\theta), z(\theta)\}$.

3. Improved Artificial Potential Field Method Based on Joint Space

3.1. Traditional Artificial Potential Field Algorithm

In APF, the robotic arm is subjected to two forces when moving to the target point. One force is the attractive force that pulls the robot towards the target position, while the other force is the repulsive force exerted by obstacles on the robot. Under the combined force of these two forces, the robotic arm can quickly reach the target position without colliding with obstacles in this process. The force analysis of the robotic arm in the potential field is shown in Figure 1. The force on each joint and link of the robotic arm is shown in Figure 2.

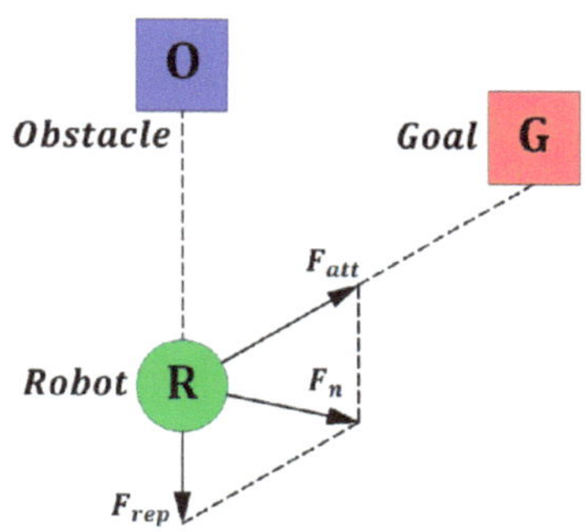

Figure 1. The force of the robotic arm in the APF.

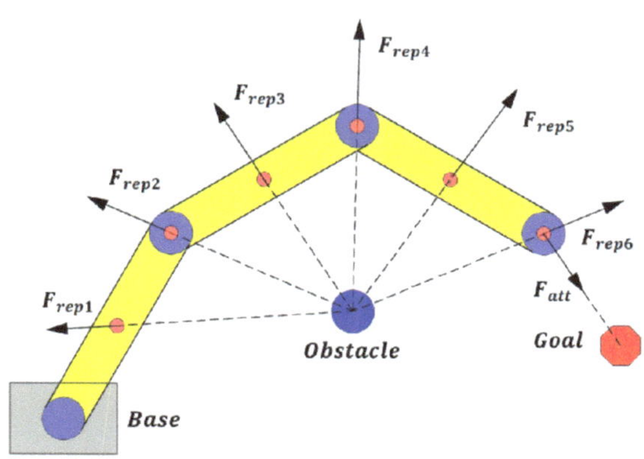

Figure 2. Force on each joint and link of the robotic arm.

F_{rep} represents the repulsive force; F_{att} represents the attractive force; and F_n is the combined force of the two forces, which makes the manipulator reach the target position.

The attraction field produced by F_{att} can be obtained by Equation (4):

$$U_{att}(\theta_1, \cdots, \theta_n) = \frac{1}{2}\mu d^2\left(q, q_{goal}\right) \tag{4}$$

where $U_{att}(\theta_1, \cdots, \theta_n)$ is the attraction field, μ is the correction coefficient of the attractive potential field, q represents the current position configuration of the robot, q_{goal} represents the position configuration of the robot when reaching the target position, and $d\left(q, q_{goal}\right)$ is the distance between the robot and the target position.

The repulsive potential field produced by F_{rep} can be obtained by Equation (5):

$$U_{rep}(\theta_1, \cdots, \theta_n) = \begin{cases} \frac{1}{2}k\left(\frac{1}{d(q,q_{obs})} - \frac{1}{d_0}\right)^2, d(q, q_{obs}) \leq d_0 \\ 0, d(q, q_{obs}) > d_0 \end{cases} \tag{5}$$

where $U_{rep}(\theta_1, \cdots, \theta_n)$ is the repulsive potential field, k is the correction coefficient of the repulsion potential field, q_{obs} represents the position configuration of the robot when colliding with an obstacle, $d(q, q_{obs})$ is the real-time distance between the robotic arm and the obstacle, and d_0 represents the influence radius of the obstacle. When $d(q, q_{obs}) > d_0$, the repulsion field will not affect the movement of the robot.

The total potential energy function can be expressed as Equation (6):

$$U(\theta_1, \cdots, \theta_n) = U_{att}(\theta_1, \cdots, \theta_n) + U_{rep}(\theta_1, \cdots, \theta_n) \tag{6}$$

3.2. The Gradient of the Potential Field Function

The potential energy function is only dependent on joint angles $(\theta_1, \cdots, \theta_n)$ for a robotic arm. We use the gradient descent method to minimize the potential energy function and determine a set of angles so that the robotic arm can reach the target position in the direction of the fastest potential energy descent without a collision with obstacles.

The attraction F_{att} is the negative of the gradient of the attraction potential field U_{att}, and it can be expressed as (7):

$$F_{att} = -\nabla U_{att}(\theta_1, \cdots, \theta_n) = -\mu d\left(q, q_{goal}\right)\nabla d(q, q_{goal}) \tag{7}$$

q is the current position configuration of the robot, and q_{goal} is the position configuration of the robot when reaching the goal position. If the manipulator is entirely composed of revolute joints, q can be assumed to be $[\theta_1, \cdots, \theta_n]$, and q_{goal} can be assumed to be $[\theta_1^*, \cdots, \theta_n^*]$. Then, $d\left(q, q_{goal}\right)$ can be expressed as (8):

$$d\left(q, q_{goal}\right) = \sqrt{\sum_{i=1}^{n}(\theta_i - \theta_i^*)^2} \tag{8}$$

If the manipulator comprises both prismatic joints (from joint 1 to joint j) and revolute joints (from joint $j+1$ to joint n), in this case, q can be assumed to be $\left[d_1, \cdots, d_j, \theta_{j+1}, \cdots, \theta_n\right]$, q_{goal} can be assumed to be $\left[d_1^*, \cdots, d_j^*, \theta_{j+1}^*, \cdots, \theta_n^*\right]$, and we assign a weight ω_1 to the prismatic joint and a weight ω_2 to the revolute joint to eliminate the units of nonhomogeneous form. Then, $d\left(q, q_{goal}\right)$ can be expressed as (9):

$$d\left(q, q_{goal}\right) = \omega_1 \times \sqrt{\sum_{i=1}^{j}(d_i - d_i^*)^2} + \omega_2 \times \sqrt{\sum_{i=j+1}^{n}(\theta_i - \theta_i^*)^2} \tag{9}$$

The repulsive force F_{rep} is the negative of the gradient of the repulsive potential field U_{rep} can be expressed as (10):

$$F_{rep} = -\nabla U_{rep}(\theta_1, \cdots, \theta_n) = \begin{cases} k\left(\frac{1}{d(q,q_{obs})} - \frac{1}{d_0}\right)\frac{1}{d^2(q,q_{obs})}\nabla d(q, q_{obs}), d(q, q_{obs}) \leq d_0 \\ 0, d(q, q_{obs}) > d_0 \end{cases} \tag{10}$$

We use the gradient descent to minimize the potential energy function in Equation (6). Equation (11) presents the generalized formula utilized for the sth iteration of gradient descent. and Equation (12) shows how the learning rate λ is calculated. q_s represents the position configuration of the robot arm at the sth cycle.

$$q_s = q_{s-1} - \lambda \times \nabla U(\theta_1, \cdots, \theta_n) \quad (11)$$

$$\lambda = \epsilon 1 + \epsilon 2 \times d\left(q, q_{goal}\right) + \epsilon 3 \times d(q, q_{obs}) \quad (12)$$

The learning rate λ is dynamic, and it depends on the distance between the end-effector and goal point and the distance between the joints of robot and obstacle. $\epsilon 1, \epsilon 2, \epsilon 3$ are positive constants. $\epsilon 1$ specifies the basic learning rate and $\epsilon 2, \epsilon 3$ specify the sensitivity to the position of manipulator. As the end-effector approaches the goal point and the joint of the robot approaches the obstacle, the learning rate decreases. This is performed to avoid collision with obstacles and reach the goal point more accurately. If faster algorithm execution is desired, the values of $\epsilon 1$, $\epsilon 2$, and $\epsilon 3$ can be increased accordingly. In this paper, $\epsilon 1$, $\epsilon 2$, and $\epsilon 3$ are adjusted within the range of 0.5–1.

The gradient of the attractive potential field is easy to calculate, while the gradient of the repulsive potential field is difficult to calculate, because the shortest distance $d(q, q_{obs})$ between the robotic arm and the obstacle is difficult to calculate. This section focuses on the calculation of the shortest distance $d(q, q_{obs})$ when facing different obstacles.

There are two types of obstacles on the factory assembly line usually, which are obstacles with complex shapes and obstacles with regular shapes. The obstacle can be simulated using the ball envelope method [25], as shown in Figures 3 and 4.

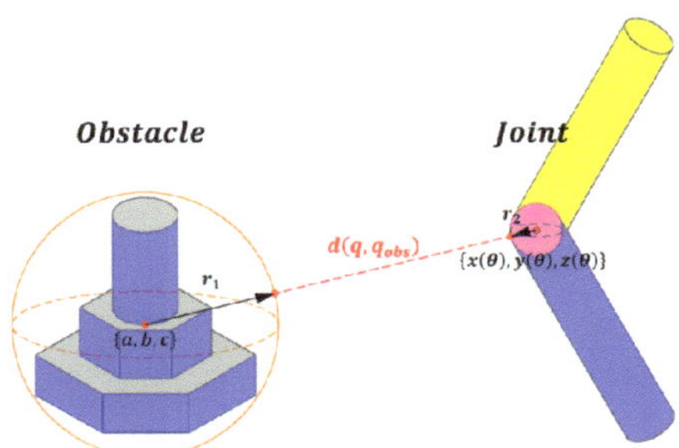

Figure 3. The shortest distance $d(q, q_{obs})$ between the point on the joint and an obstacle with a complex shape.

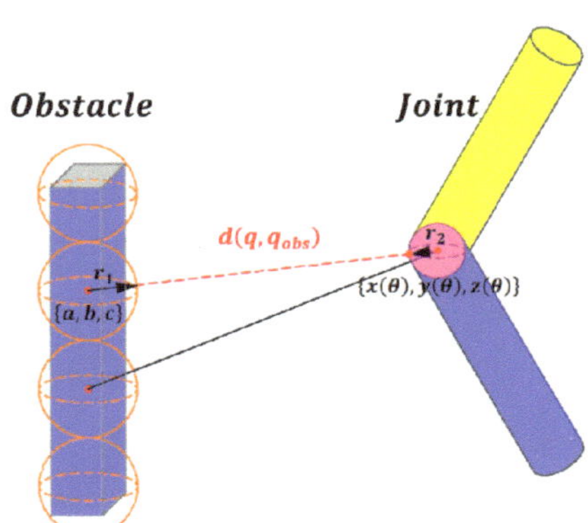

Figure 4. The shortest distance $d(q, q_{obs})$ between the point on the joint and an obstacle with a regular shape.

The parameters describing the obstacle are $\{a, b, c, r\}$, where a, b, c represent the spatial location coordinates of the obstacle and r represents the radius of the ball envelope obstacle.

For an obstacle with a complex shape, it can be enveloped by a single sphere. According to the conclusion of Section 2, the position configuration coordinates of q are determined by a series variable of the joint angle, and can be solved by kinematics. Therefore, the coordinates of q can be expressed as $q : \{x(\theta), y(\theta), z(\theta)\}$. Then, the shortest distance between the joint of the robot and the obstacle can be expressed as (13):

$$d(q, q_{obs}) = \sqrt{(x(\theta) - a)^2 + (y(\theta) - b)^2 + (z(\theta) - c)^2} - r_1 - r_2 \tag{13}$$

where r_1 denotes the radius of the enveloping obstacle sphere and r_2 denotes the thickness of the joint.

For an obstacle with a regular shape, it can be enveloped by a series of spheres whose coordinate parameters are denoted by $\{a_i, b_i, c_i, r_i\}$. The distance between the joint and the nearest sphere is the shortest distance $d(q, q_{obs})$ between the joint and the obstacle with a regular shape. In order to make the motion of the robotic arm smooth, the manipulator receives repulsive forces from the two sources, one is the nearest ball, and the other is the ball adjacent to the nearest ball.

On the left side of Figure 5, ball 2 is closest to the joint, and the robotic arm is subject to the repulsive force of ball 2 and the adjacent ball 1. When the robotic arm moves to the next state, as shown in the right side of Figure 5, ball 3 is closest to the joint and the robot is subject to the repulsive force of ball 3 and the adjacent ball 2.

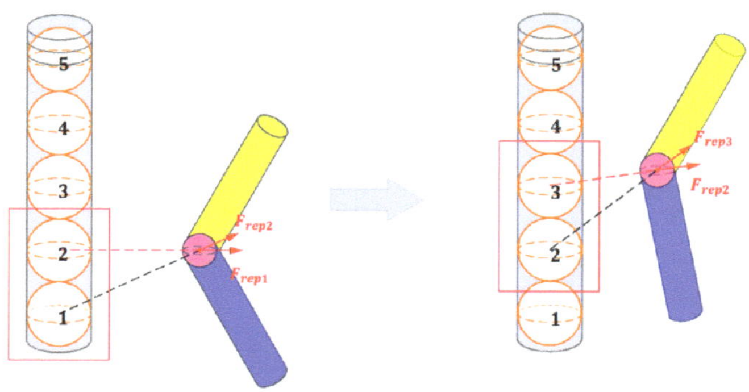

Figure 5. The repulsive force on the robotic arm during movement.

The derivative of Equation (13) can lead to Equation (14):

$$\nabla d(q, q_{obs}) = \frac{(x-a)\frac{\partial x(\theta)}{\partial \theta} + (y-b)\frac{\partial y(\theta)}{\partial \theta} + (z-c)\frac{\partial z(\theta)}{\partial \theta}}{\sqrt{(x-a)^2 + (y-b)^2 + (z-c)^2}} \tag{14}$$

The analytic formula of the gradient of the repulsive potential field ∇U_{rep} can be obtained by substituting Equations (13) and (14) into Equation (10), and it is determined by the series variable of the joint angle. This algorithm has a very high real-time performance because it only needs to calculate the potential field function once per cycle.

3.3. The Virtual Obstacle

Sometimes, the robotic arm has not reached the target, but the resultant force is zero. At this time, the robotic arm falls into the local minimum state, and it can be expressed as (15):

$$-\nabla U(\theta_1, \cdots, \theta_n) = F_{att} + F_{rep} = 0 \tag{15}$$

We make the robotic arm jump out of the local minimum state by adding a virtual obstacle. In Figure 6, we treat the center of the ball which envelops the obstacle as a virtual obstacle, and q_1 is the position configuration of the robot when colliding with the center of the ball. The position coordinates of this virtual obstacle come from the obstacle, so there is no need to introduce additional information. F_{rep_vir} is the repulsive force generated by q_1 when the robot is trapped in a local minimum state, and it is calculated as shown in (16) and (17).

$$F_{rep\_vir} = k\left(\frac{1}{d(q,q_1)} - \frac{1}{d_0}\right)\frac{1}{d^2(q,q_1)}\nabla d(q,q_1) \tag{16}$$

$$d(q,q_1) = \sqrt{(x(\theta)-a)^2 + (y(\theta)-b)^2 + (z(\theta)-c)^2} \tag{17}$$

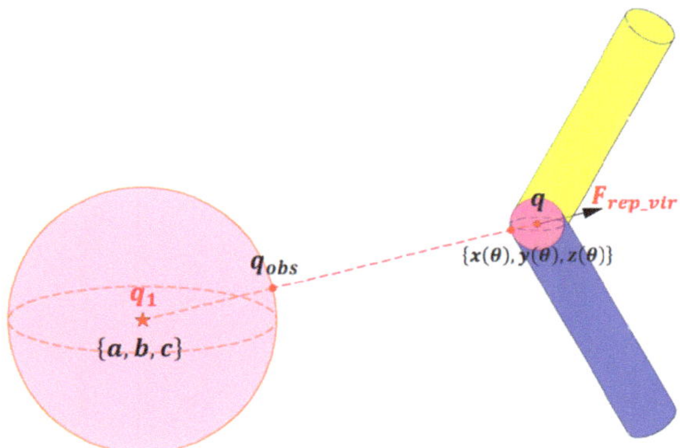

Figure 6. The repulsive force of a virtual obstacle to joint.

To ensure successful obstacle avoidance, the repulsion coefficient k of the virtual obstacle is set to 1–10^3 times the general obstacle, and the influence radius of the virtual obstacle is set as 1.5–3 times that of general obstacles [26].

3.4. Joint Angle Constraints

Typically, each joint of a robotic arm has angle limitations. To ensure that the joint angles of the robotic arm do not exceed their respective limits during obstacle avoidance, the algorithm proposed in this paper must satisfy the following requirements.

Taking an n-degree-of-freedom robotic arm as an example, we assume that the joint angle limits for each joint of the robotic arm are in the range of $[\theta_{i\_min}, \theta_{i\_max}]$.

Solution 1: In the first step of the algorithm, the initial and target positions of the robotic arm can be transformed from Cartesian space to joint space using inverse kinematics. q_{start} can be assumed to be $[\theta_1, \cdots, \theta_n]$ and q_{goal} can be assumed to be $[\theta_1^*, \cdots, \theta_n^*]$. Due to the existence of multiple solutions in inverse kinematics, it is necessary to select an appropriate solution in this step, ensuring that $\theta_{i\_min} < \theta_i < \theta_{i\_max}$ and $\theta_{i\_min} < \theta_i^* < \theta_{i\_max}$.

Solution 2: The algorithm imposes limitations on the magnitude of attraction *Fatt* and repulsion forces *Frep* during the computation (such as $F_{att} \leq 0.1, F_{rep} \leq 0.1$). This serves to prevent the joint angles from exceeding the constraint range.

Solution 3: During obstacle avoidance, it is possible that the strong repulsive force from an obstacle may cause one of the joint angles of the robotic arm to exceed the specified limit. In the case depicted in Figure 7, we can appropriately reduce the repulsive force by adjusting the repulsion coefficient k. This allows the robotic arm to complete the obstacle avoidance task while ensuring that the joint angles remain within the allowed range.

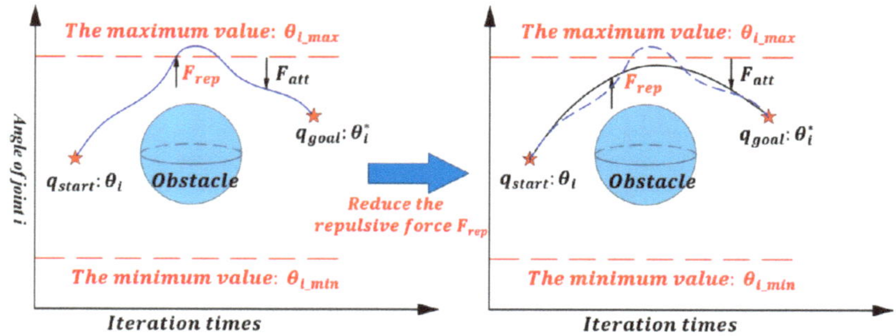

Figure 7. Reducing F_{rep} to prevent the angle of joint i from exceeding the limit range.

3.5. Algorithm Comparison

The flowchart of the artificial potential field algorithm in joint space is shown in Figure 8, and the pseudocode is shown in Algorithm 1. The values of ε1, ε2, and ε3 can be adjusted appropriately according to the requirements. It can be seen that each iteration only requires the calculation of the potential field function once from the algorithm flowchart. The configuration at the starting position and the target position are obtained by inverse kinematics. In order to obtain a smooth obstacle avoidance trajectory, the values of F_{rep} and F_{rep_vir} are limited to less than ε3.

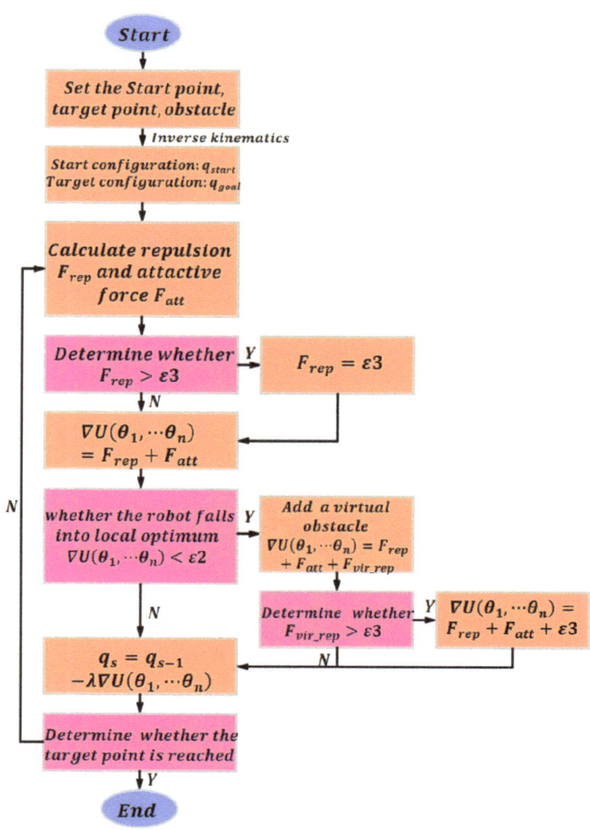

Figure 8. The flowchart of the algorithm.

Algorithm 1 The pseudocode of the algorithm.

Algorithm 1: Improved artificial potential field

Set *Start position:* q_{start} and *Goal position:* q_{goal}
$q_{start} = [\theta_1, \cdots, \theta_n]_{start}, q_{goal} = [\theta_1, \cdots, \theta_n]_{goal}$ by inverse kinematics
While $d(q, q_{goal}) > \varepsilon 1$ *%Determine whether the target point has been reached*
For $i = 1 : n$ *%Calculate the attraction and repulsive forces on each joint*
$F_{att}(i) = -\mu d\left(q, q_{goal}\right); d(q, q_{obs}) = \sqrt{(x(\theta) - a)^2 + (y(\theta) - b)^2 + (z(\theta) - c)^2} - r1 - r2$
$F_{rep}(i) = \begin{cases} k\left(\frac{1}{d(q,q_{obs})} - \frac{1}{d_0}\right)\frac{1}{d^2(q,q_{obs})}\nabla d(q, q_{obs}), d(q, q_{obs}) \leq d_0 \\ 0, d(q, q_{obs}) > d_0 \end{cases}$
if $F_{rep} > \varepsilon 3$ $F_{rep} = \varepsilon 3$ end *%Ensure smooth trajectory*
$F_{att} = \text{sum}(F_{att}(i)), F_{rep} = \text{sum}(F_{rep}(i))$ end
$\nabla U(\theta_1, \cdots, \theta_n) = F_{att} + F_{rep}$
if $\nabla U(\theta_1, \cdots, \theta_n) < \varepsilon 2$ *%Determine whether the robot falls into local optimum*
$d(q, q_1) = \sqrt{(x(\theta) - a)^2 + (y(\theta) - b)^2 + (z(\theta) - c)^2}$
$F_{rep\_vir} = k\left(\frac{1}{d(q,q_1)} - \frac{1}{d_0}\right)\frac{1}{d^2(q,q_1)}\nabla d(q, q_1)$ *%adding a virtual obstacle*
$\nabla U(\theta_1, \cdots, \theta_6) = F_{att} + F_{rep} + F_{vir\_rep}$ end
$\lambda = \epsilon 1 + \epsilon 2 \times d\left(q, q_{goal}\right) + \epsilon 3 \times d(q, q_{obs})$ *%Determine iteration step size*
$q_s = q_{s-1} - \lambda \times \nabla U(\theta_1, \cdots, \theta_n)$ *%Update joint angles using gradient descent method*
end

3.5.1. Comparison with APF Algorithm in Cartesian Space

This type of method [20] calculates repulsion and attraction in Cartesian space, and maps the resultant force from Cartesian space to joint space using the Jacobian matrix. The process of this type of method is as Algorithm 2 follows:

Algorithm 2: APF in Cartesian space

STEP1: Calculate the coordinates of each joint using kinematics
STEP2: Calculate attraction Fa and repulsive forces Fr in the Cartesian space, and convert them into the velocity $V = [v_x, v_y, v_z]$ *of the end-effector*
STEP3: Calculate the Jacobian matrix $J(q)$*and its inverse* $J^{-1}(q)$
STEP4: Converting velocity V into joint space using Jacobian matrix $\dot{q} = J^{-1}(q)V$
STEP5: $q_{s+1} = q_s + h \times \dot{q}$ *%h is the iteration step size*

The coordinate system of this method is constantly transformed between Cartesian space and joint space, and this method cannot solve the inverse matrix of the Jacobian matrix when the robotic arm falls into a singular configuration ($J(\theta) = 0$). Because the algorithm in this paper is directly in joint space so that it does not require *STEP3* and *STEP4*. It has much higher real-time performance and efficiency than the APF algorithm in Cartesian space.

3.5.2. Comparison with APF Algorithm Using a Sampling Method

This type of methods [22] uses a sampling method to calculate the fastest decreasing direction of the potential field function. Its process is as Algorithm 3 follows:

Algorithm 3: APF using a sampling method

STEP1: Each joint angle θ_i *has three variations in the next moment*
For example: $\theta_1' \leftarrow \{\theta_1 - h, \theta_1, \theta_1 + h\}$ *%h is the iteration step size*
STEP2: Calculate all cases of $U(\theta')$ *and find the lowest value of* $U(\theta')$
%$\theta' = [\theta_1', \theta_2', \theta_3', \cdots, \theta_n']$. *The joint angle* θ *changes to* θ' *along the negative gradient direction of the potential field function.*
STEP3: Update joint angle from θ *to* θ'

This type of method needs to calculate the potential field function 3^n times in each cycle, while our algorithm only needs to calculate the potential field function once in each cycle. Thus, the algorithm in this paper has a higher real-time performance.

3.5.3. Comparison with APF Algorithm Using a Numerical Method

This type of method [23] uses the Euler method to calculate the gradient of the potential field function, as shown in Equations (18) and (19):

$$\nabla U(\theta_1, \cdots \theta_n) = (\frac{\partial U}{\partial \theta_1}, \cdots, \frac{\partial U}{\partial \theta_n}) \tag{18}$$

$$\frac{\partial U}{\partial \theta_i} = \frac{U(\theta_i + h \cdots, \theta_n) - U(\theta_i - h \cdots, \theta_n)}{2h} \tag{19}$$

h is the iteration step size. This method requires calculating the potential field function twice to obtain the gradient direction of a joint angle, and each cycle requires calculating the gradient of six joint angles. Therefore, the method requires 2n calculations of the potential field function per cycle, while our algorithm only needs to calculate the potential field function once in each cycle. Thus, the algorithm in this paper has a higher real-time performance.

4. Experiment and Result

In this paper, the ROCR6 is our research object. ROCR6 is a six-axis robotic arm developed by a Chinese company, Si Valley. Figure 9 shows the structure and coordinate system of the robot, and Table 1 shows the modified D-H parameters of the 6-DOF robot.

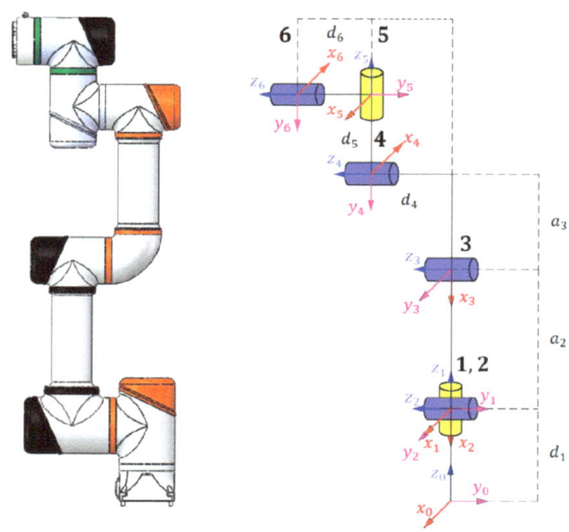

Figure 9. The structure and coordinate system of the robot.

Table 1. Modified D-H parameters of the 6-DOF robot.

Joint	θ_i	α_{i-1}	a_{i-1} (mm)	d_i (mm)
1	θ_1	0	0	122.3
2	$\theta_2 - \pi/2$	$\pi/2$	0	0
3	θ_3	0	−270	0
4	$\theta_4 - \pi/2$	0	−253	123.3
5	θ_5	$\pi/2$	0	207.1
6	θ_6	$-\pi/2$	0	99.1

We performed kinematic simulations in Simscape in Matlab, and these simulations were divided into two cases, namely obstacle avoidance for a sphere and obstacle avoidance for a rectangular beam. To demonstrate the real-time capability of our algorithm in this paper, we compared it with the algorithms discussed in Section 3.5. Additionally, we compared it with the genetic algorithm to showcase the performance of the results obtained by our algorithm.

4.1. The Obstacle Avoidance for a Sphere

Table 2 shows the task information and the parameter details of the spherical obstacle. The motion process of the robotic arm using the APF algorithm is shown in Figure 10, where the red trajectory represents the running trajectory of the middle joint and end-effector. The change in each joint angle using the improved APF algorithm is shown in Figure 11, and the change in each joint angle using the genetic algorithm is shown in Figure 12, and their operation indexes are shown in Table 3. Table 4 compares the running time of our algorithm with each of the algorithms in Section 3.5.

Table 2. The information of the spherical obstacle and the configuration of the robot.

Spherical Obstacle and Robot	Information
Starting position	[590,−131,112] (mm)
The configuration at the starting position	$\begin{bmatrix} -0.0130, -1.0585, -0.6853 \\ 0.1730, 1.5780, 1.5578 \end{bmatrix}$ (rad)
Target position	[104,−600,120] (mm)
The configuration at the target position	$\begin{bmatrix} -1.1953, -1.0796, -0.6122 \\ 0.1210, 1.5780, 0.3755 \end{bmatrix}$ (rad)
Obstacle coordinates	[250,−300,112] (mm)
Obstacle radius	75 mm

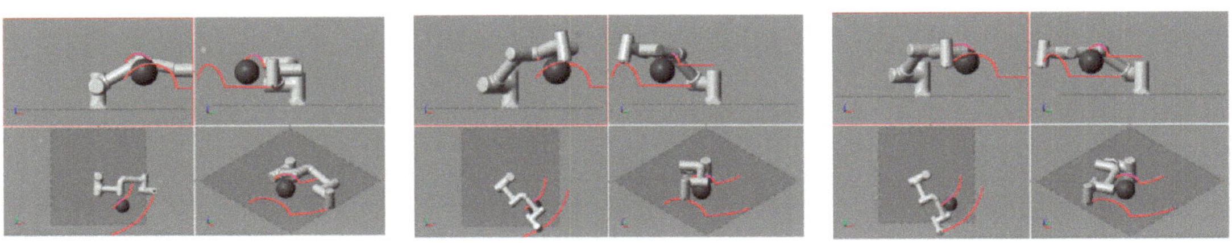

Figure 10. The motion process of the robotic arm using APF.

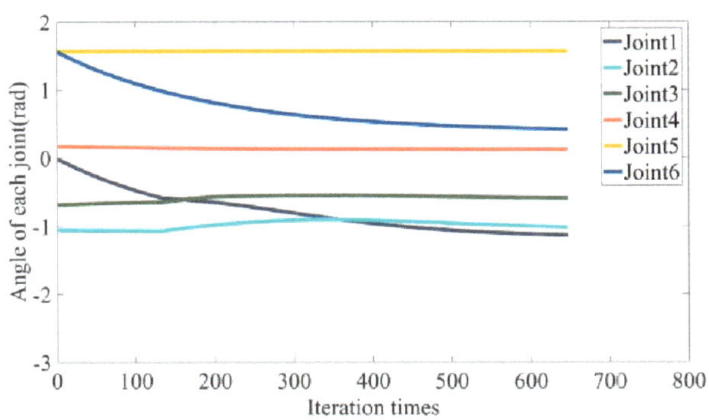

Figure 11. Change of the joint angle using APF.

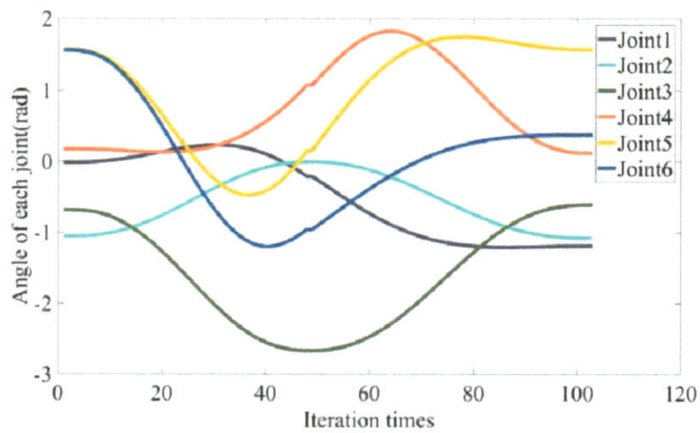

Figure 12. Change of the joint angle using genetic algorithm.

Table 3. The performance of algorithms result.

Algorithm	Improved APF	Genetic
Joint angle increment	2.9510 (rad)	10.8894 (rad)
Path length	880.6657 (mm)	895.9892 (mm)

Table 4. The run time of algorithms.

Algorithm	Run Time
Improved APF	0.3558 (s)
APF in Cartesian space	8.8696 (s)
APF using a sampling method	260.3108 (s)
APF using a numerical method	6.2696 (s)

4.2. *The Obstacle Avoidance for a Rectangular Beam*

Table 5 shows the parameter details of the rectangular beam. The motion process of the robotic arm using the APF algorithm is shown in Figure 13, where magenta trajectory represents the running trajectory of the middle joint and end-effector. The change of each joint angle using the improved APF algorithm is shown in Figure 14, the change of each joint angle using genetic algorithm is shown in Figure 15, and their operation indexes are shown in Table 6. Table 7 shows the running time of each algorithm.

Table 5. The information of the rectangular beam and the configuration of the robot.

Rectangular Beam Obstacle and Robot	Information
Starting position	[590, −131, 110] (mm)
The configuration at the starting position	$\begin{bmatrix} -0.0130, -1.0716, -0.7080 \\ 0.2088, 1.5780, 1.5578 \end{bmatrix}$ (rad)
Target position	[104, −600, 160] (mm)
The configuration at the target position	$\begin{bmatrix} -1.1953, -1.0659, -0.4804 \\ 0.0245, 1.5780, 0.3755 \end{bmatrix}$ (rad)
Obstacle coordinates	[250, −300, 0] (mm)
Obstacle bottom edge	60 × 60 (mm)
Obstacle height	300 mm

Table 6. The performance of the algorithms.

Algorithm	Improved APF	Genetic
Joint angle increment	4.1690 (rad)	10.1739 (rad)
Path length	1246.6496 (mm)	838.3892 (mm)

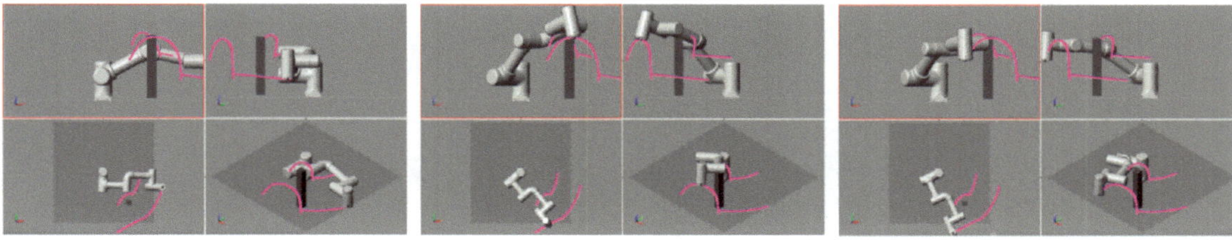

Figure 13. The motion process of the robotic arm using APF.

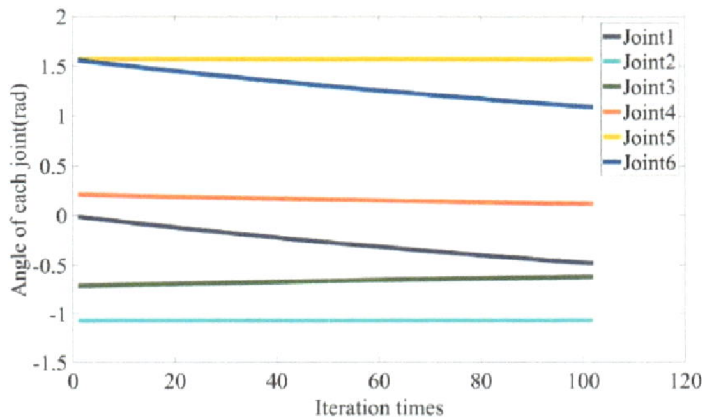

Figure 14. Change of the joint angle using APF.

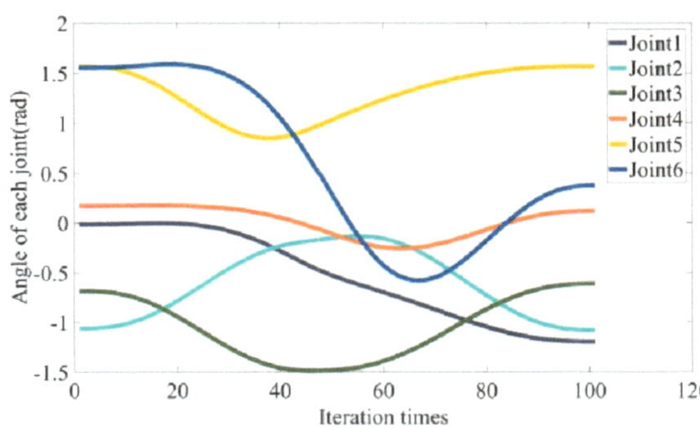

Figure 15. Change of the joint angle using genetic algorithm.

Table 7. The run time of algorithms.

Algorithm	Run Time
Improved APF	0.3912 (s)
APF in Cartesian space	9.7521 (s)
APF using a sampling method	286.2102 (s)
APF using a numerical method	6.8694 (s)

4.3. Experimentation in the Real World

The improved APF algorithm has now been tested in the real world for the rectangular beam obstacle in Section 4.2. Figure 16 shows each joint of the real robotic arm and the obstacle. Figure 17 shows the experimental result in the real world, and the robot avoids the obstacle and reaches the target position.

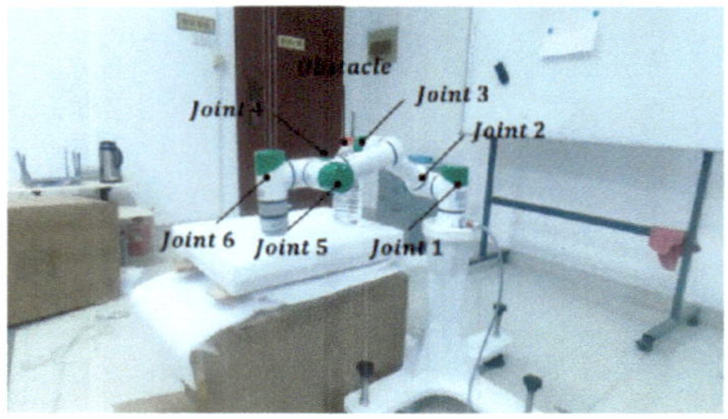

Figure 16. The experiment detail of the real robotic arm.

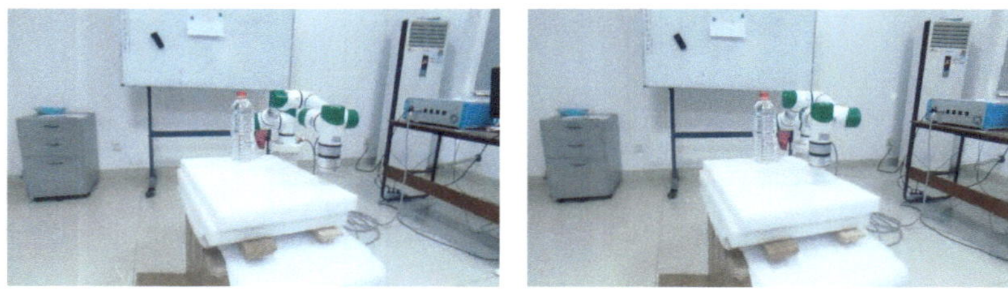

(**a**)

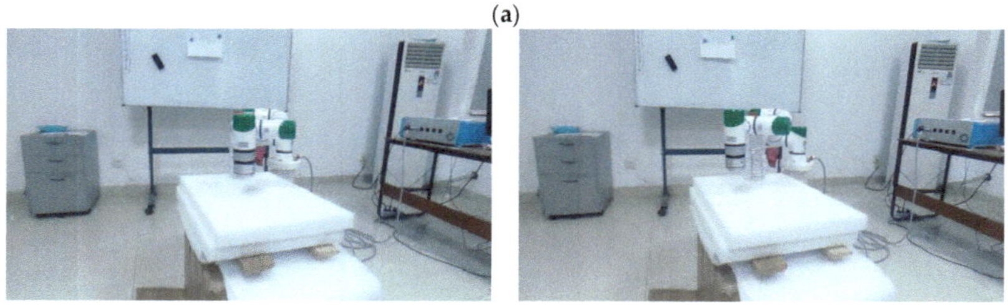

(**b**)

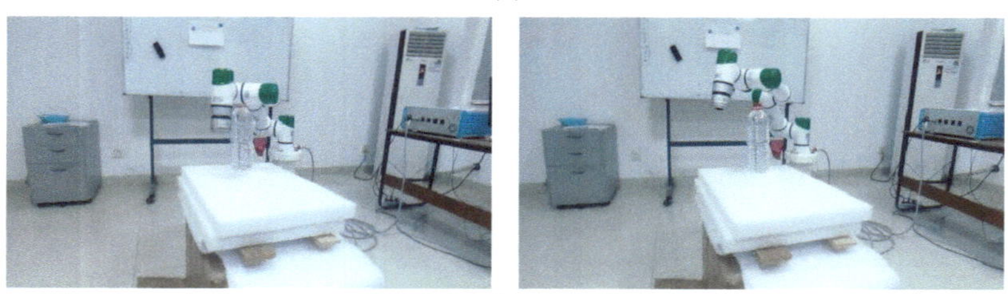

(**c**)

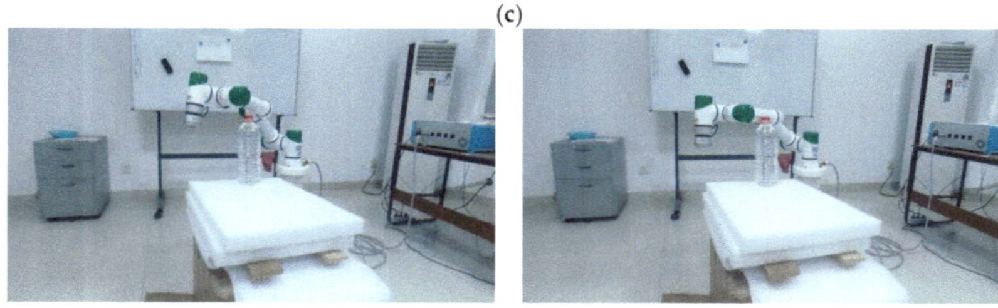

(**d**)

Figure 17. The motion process of the real robotic arm. (**a**) depicts the motion of the robotic arm from 0 to 1 s. (**a**) depicts the motion of the robotic arm from 0 to 1 s. (**b**) depicts the motion of the robotic arm from 1 to 2 s. (**c**) depicts the motion of the robotic arm from 2 to 3 s. (**d**) depicts the motion of the robotic arm from 3 to 4 s.

5. Discussions

In smart factories, trajectory planning algorithms require high real-time capability to adapt to complex and dynamic industrial tasks. The algorithm proposed in this paper meets this requirement by demonstrating an excellent real-time performance. As evident in Tables 4 and 7, the improved artificial potential field method presented in this paper outperforms other APF methods in terms of real-time performance, as it can complete trajectory planning tasks within a mere 0.4 s.

The algorithm should strive to compute excellent trajectories, as excellent trajectories can save the energy consumed by the robotic arm, reduce the impacts during motion, and enhance the operational efficiency. From Tables 4 and 7, it can be observed that the algorithm presented in this paper exhibits smaller joint angle variations compared with the genetic algorithm. This indicates that using the algorithm proposed in this paper can reduce the energy consumption during the operation of the robotic arm. However, the algorithm proposed in this paper still has some drawbacks. In comparison to the genetic algorithm, this algorithm may cause the robotic arm's end effector to move slightly further, thereby reducing the efficiency of the robotic arm's operation. Additionally, the smoothness of the paths obtained by this algorithm is lower than that of the genetic algorithm, which may result in a greater impact on the robotic arm. Lastly, taking the rectangular beam obstacle in Section 4.2 as a case study, we conducted real-world testing using the algorithm presented in this paper. It can be observed that the robotic arm can effectively avoid obstacles.

6. Conclusions

Robotic arms are widely used in factory assembly lines. A real-time obstacle avoidance algorithm that can adapt to changes in the starting position and goal position at any time is needed. In this paper, we propose an improved artificial potential field method in joint space, which can realize the real-time obstacle avoidance of the robotic arm. The experimental results show that the algorithm can achieve good performance under different obstacles. The algorithm completes the computation very quickly, and the performance of the obstacle avoidance trajectory obtained by this algorithm is higher than the trajectory obtained by genetic algorithm. Its main advantages are as follows:

1. The algorithm prevents the transformation of coordinate system from Cartesian space into joint space, and the robotic arm still performs well when it is caught in a singular configuration.
2. The algorithm can solve the gradient expression of repulsive potential field in advance. The potential field function only needs to be calculated once during each iteration, and the calculation efficiency is very fast.
3. The algorithm proposes an effective method to prevent the robotic arm falling into a local minimum state without introducing additional information.

In future work, we will further improve this algorithm to make the motion trajectory smoother. In addition, the algorithm can obtain the angle, velocity, and acceleration information of each joint during operation. We intend to combine this algorithm with dynamics to control robotic arms in the future.

Author Contributions: Y.C.: Conceptualization, Writing—original draft, Data curation, Formal analysis, Validation. L.C.: Writing—Review and editing, Conceptualization, Data curation, Formal analysis, Validation. J.D.: Writing—review and editing, Conceptualization, Data curation, Formal analysis, Validation. Y.L.: Writing—review and editing. All authors have read and agreed to the published version of the manuscript.

Funding: This research was funded by the National Key R&D Program of China, grant number (2019YFB1706501).

Institutional Review Board Statement: Not applicable.

Informed Consent Statement: Not applicable.

Data Availability Statement: Not applicable.

Conflicts of Interest: The authors declare no conflict of interest.

References

1. Xu, F. Research and Development of Control Technology of Visual Guided Grasping Robot. Master's Thesis, Jiangnan University, Jiangsu, China, 2014.
2. Ghadge, K.; More, S.; Gaikwad, P.; Chillal, S. Robotic ARM for pick and place application. *Int. J. Mech. Eng. Technol.* **2018**, *9*, 125–133.
3. Pal; Raj, A.A.B.; Shinde, S.R. Performance study of different robotic manipulator systems designed for space debris management. In Proceedings of the 2020 5th International Conference on Communication and Electronics Systems (ICCES), Coimbatore, India, 10–12 June 2020; pp. 148–153.
4. Yenorkar, R.; Chaskar, U.M. GUI based pick and place robotic arm for multipurpose industrial applications. In Proceedings of the 2018 Second International Conference on Intelligent Computing and Control Systems (ICICCS), Madurai, India, 14–15 June 2018; pp. 200–203.
5. Grunwald, G.; Schreiber, G.; Albu-Schaffer, A.; Hirzinger, G. Touch: The direct type of human interaction with a redundant service robot. In Proceedings of the 10th IEEE International Workshop on Robot and Human Interactive Communication ROMAN, Paris, France, 18–21 September 2001; pp. 347–352.
6. Nuchkrua, T.; Chen, S.-L. Precision contouring control of five degree of freedom robot manipulators with uncertainty. *Int. J. Adv. Robot. Syst.* **2017**, *14*, 1729881416682703. [CrossRef]
7. Tang, J.; Liu, G.; Pan, Q. A Review on Representative Swarm Intelligence Algorithms for Solving Optimization Problems: Applications and Trends. *IEEE/CAA J. Autom. Sin.* **2021**, *8*, 1627–1643. [CrossRef]
8. Wang, X.; Li, X.; Guan, Y.; Song, J.; Wang, R. Bidirectional Potential Guided RRT* for Motion Planning. *IEEE Access* **2019**, *7*, 95046–95057.
9. Siar, M.V.S.; Fakharian, A. Energy effificiency in the robot arm using genetic algorithm. In Proceedings of the 2018 8th Conference of AI & Robotics and 10th RoboCup Iranopen International Symposium (IRANOPEN), Qazvin, Iran, 10 April 2018; pp. 14–20.
10. Kalra, P.; Mahapatra, P.B.; Aggarwal, D.K. On the solution of multimodal robot inverse kinematic functions using real-coded genetic algorithms. In Proceedings of the IEEE International Conference on Systems, Man and Cybernetics. Conference Theme-System Security and Assurance, Washington, DC, USA, 8 October 2003; Volume 2, pp. 1840–1845.
11. Noohi, E.; Moradi, H.; Parastegari, S.; Ahmadabadi, M.N. Object Manipulation Using Unlimited Rolling Contacts: 2-D Kinematic Modeling and Motion Planning. *IEEE Trans. Robot.* **2015**, *31*, 790–797. [CrossRef]
12. Liu, C.; Xie, S.; Sui, X.; Huang, Y.; Ma, X.; Guo, N.; Yang, F. PRM-D* Method for Mobile Robot Path Planning. *Sensors* **2023**, *23*, 3512. [CrossRef] [PubMed]
13. Khatib, O. Real-time obstacle avoidance for manipulators and mobile robots. *Int. J. Robot. Res.* **1986**, *5*, 90–98. [CrossRef]
14. Yang, X.; Yang, W.; Zhang, H.; Chang, H.; Chen, C.-Y.; Zhang, S. A new method for robot path planning based artifificial potential field. In Proceedings of the IEEE 11th Conference on Industrial Electronics and Applications (ICIEA), Hefei, China, 5–7 June 2016; pp. 1294–1299.
15. Zhu, Q.; Yan, Y.; Xing, Z. Robot path planning based on artifificial potential field approach with simulated annealing. In Proceedings of the 6th International Conference on Intelligent Systems Design and Applications, Jinan, China, 16–18 October 2006; pp. 622–627.
16. Song, J.; Hao, C.; Su, J. Path planning for unmanned surface vehicle based on predictive artifificial potential field. *Int. J. Adv. Robot. Syst.* **2020**, *17*, 1729881420918461. [CrossRef]
17. Chiang, H.-T.; Malone, N.; Lesser, K.; Oishi, M.; Tapia, L. Pathguided artifificial potential fifields with stochastic reachable sets for motion planning in highly dynamic environments. In Proceedings of the IEEE International Conference on Robotics and Automatio, Seattle, WA, USA, 26–30 May 2015; pp. 2347–2354.
18. Zhong, X.; Tian, J.; Hu, H.; Peng, X. Hybrid Path Planning Based on Safe A* Algorithm and Adaptive Window Approach for Mobile Robot in Large-Scale Dynamic Environment. *J. Intell. Robot. Syst.* **2020**, *99*, 65–77. [CrossRef]
19. Chen, X.; Huang, Z.; Sun, Y.; Zhong, Y.; Gu, R.; Bai, L. Online on-Road Motion Planning Based on Hybrid Potential Field Model for Car-Like Robot. *J. Intell. Robot. Syst.* **2022**, *105*, 7. [CrossRef] [PubMed]
20. Zhang, H.; Zhu, Y.; Liu, X.; Xu, X. Analysis of Obstacle Avoidance Strategy for Dual-Arm Robot Based on Speed Field with Improved Artificial Potential Field Algorithm. *Electronics* **2021**, *10*, 1850. [CrossRef]
21. Park, S.O.; Lee, M.C.; Kim, J. Trajectory Planning with Collision Avoidance for Redundant Robots Using Jacobian and Artificial Potential Field-based Real-time Inverse Kinematics. *Int. J. Control Autom. Syst.* **2020**, *18*, 2095–2107. [CrossRef]
22. Luo, L.; Wen, H.; Lu, Q.; Huang, H.; Chen, W.; Zou, X.; Wang, C. Collision-Free Path-Planning for Six-DOF Serial Harvesting Robot Based on Energy Optimal and Artificial Potential Field. *Complexity* **2018**, *2018*, 3563846. [CrossRef]
23. Sepehri, A.; Moghaddam, A.M. A Motion Planning Algorithm for Redundant Manipulators Using Rapidly Exploring Randomized Trees and Artificial Potential Fields. *IEEE Access* **2021**, *9*, 26059–26070. [CrossRef]
24. Chen, S. Research on the Technology of Zero Moment Control and Collision Detection of Collaborative Robot. Ph.D. Dissertation, University of Science and Technology of China, Hefei, China, 2018.

25. Cui, S. The Application of Bounding Volume Algorithm in Collision Detection of Medical VR. Master's Thesis, Qingdao University, Qingdao, China, 2004.
26. Wang, J.; Zhang, G.; Yang, F.; Jing, B. Improved artificial potential field method for robotic arm obstacle avoidance path planning. *Comput. Eng. Appl.* **2013**, *49*, 266–270.

applied sciences

MDPI

Article

Machine Learning-Based Shoveling Trajectory Optimization of Wheel Loader for Fuel Consumption Reduction

Yanhui Chen [1,2,*], Gang Shi [2], Cheng Tan [1] and Zhiwen Wang [3]

[1] Department of Mechanical and Electrical Engineering, Guangxi Vocational College of Water Resources and Electric Power, Nanning 530023, China
[2] School of Mechanical and Automotive Engineering, Guangxi University of Science and Technology, Liuzhou 545006, China
[3] School of Computer Science and Telecommunication Engineering, Guangxi University of Science and Technology, Liuzhou 545006, China
* Correspondence: 100001535@gxust.edu.cn

Abstract: The difference in fuel consumption of wheel loaders can be more than 30% according to different shoveling trajectories for shoveling operations, and the optimization of shoveling trajectories is an important way to reduce the fuel consumption of shoveling operations. The existing shoveling trajectory optimization method is mainly through theoretical calculation and simulation analysis, which cannot fully consider the high randomness and complexity of the shoveling process. It is difficult to achieve the desired optimization effect. Therefore, this paper takes the actual shoveling operation data as the basis. The factors that have a high impact on the fuel consumption of shoveling are screened out through Kernel Principal Component Analysis. Moreover, the mathematical model of fuel consumption of shoveling operation is established by Support Vector Machine and combined with the Improved Particle Swarm Optimization algorithm to optimize the shoveling trajectory. To demonstrate the generalization ability of the model, two materials, gravel, and sand, are selected. Meanwhile, the influence of different engine speeds on the shoveling operation is considered. We optimize the shoveling trajectories for three different engine speeds. The optimized trajectories are verified and compared with the sample data and manually controlled shoveling data. The results show that the optimized trajectory can reduce the fuel consumption of shoveling operation by 27.66% and 24.34% compared with the manually controlled shoveling of gravel and sand, respectively. This study provides guidance for the energy-efficient operation of wheel loaders.

Keywords: trajectory optimization; machine learning; wheel loader; fuel consumption

Citation: Chen, Y.; Shi, G.; Tan, C.; Wang, Z. Machine Learning-Based Shoveling Trajectory Optimization of Wheel Loader for Fuel Consumption Reduction. *Appl. Sci.* **2023**, *13*, 7659. https://doi.org/10.3390/app13137659

Academic Editor: Jonghoek Kim

Received: 19 April 2023
Revised: 24 June 2023
Accepted: 27 June 2023
Published: 28 June 2023

1. Introduction

The wheel loader is a relatively versatile construction vehicle widely used in many civil engineering and mining projects. It usually performs a variety of tasks, including shoveling, transporting, and dumping [1]. Among them, the most important form of operation is shoveling. The fuel consumption in the process of shoveling is one of the critical indicators for evaluating the operational efficiency of wheel loaders [2]. Reducing the fuel consumption of wheel loaders in shoveling operations has become an urgent problem. Numerous studies have shown that the fuel consumption of wheel loader shoveling operation is closely related to the shoveling trajectory. The shoveling trajectory is the track of the wheel loader bucket moving in the pile. The fuel consumption of wheel loader shoveling operations with different trajectories has obvious differences [3,4]. Thus, optimizing the shoveling trajectory is an effective way to reduce the fuel consumption of shoveling operations.

The wheel loader is usually divided into three phases, namely the insertion phase, scooping phase, and lifting phase, when carrying out the shoveling operation. Among them, the optimization of the shoveling trajectory in the insertion phase and the shoveling trajectory in the scooping phase is the key to reducing the fuel consumption of the wheel loader shoveling operation. The main point for the shoveling trajectory in the insertion phase is to determine the insertion depth. Sarata et al. [5] concluded that if the wheel loader is inserted too deep into the pile, the bucket resistance may exceed the maximum power of the wheel loader, and the bucket may get stuck in the pile. Hong et al. [6] planned three shoveling trajectories based on different insertion depths. The different insertion depths resulted in large differences in fuel consumption, and the best insertion depth was obtained in the three trajectories. Xu et al. [7] analyzed the relationship between insertion depth and resistance of the wheel loader and established a mathematical model for the optimization of insertion depth. The results showed that there exists an optimal bucket depth when the wheel loader is shoveling the pile. For the scooping phase, which is the phase with the highest fuel consumption in the whole shoveling process, it is important to determine the shoveling trajectory of the scooping phase to reduce the resistance and fuel consumption. Gong et al. [8] compared the shoveling operation with linear shoveling trajectory and curved shoveling trajectory by ADAMS software and concluded that the energy consumption of linear shoveling trajectory is smaller. Zhang et al. [9] proposed the concept of parallel trajectory, which means the trajectory of the scooping phase should be parallel to the pile plane, and with this trajectory, the resistance of the bucket will be the smallest. The corresponding fuel consumption will be lower. Filla et al. [10] generated 800 sets of trajectories based on four shoveling methods and compared them. The shoveling trajectories with parallel trajectories have obvious advantages. The above studies only start from the part of the shoveling process and ignore the continuity of the shoveling process, which has a limited effect on the optimization of the shoveling trajectory. Meng et al. [11] established the resistance model of the bucket based on the Coulomb theory and obtained the optimized trajectory based on the principle of minimum energy consumption. Yu et al. [12] established the agent model of shoveling trajectory and shoveling efficiency by the Kriging method and obtained the optimized shoveling trajectory under different working conditions by joint RecurDyn-EDEM simulation. Osumi et al. [13] established the bucket resistance model and optimized the shoveling trajectory with the objectives of resistance reduction and energy consumption reduction. Some other scholars optimized the shoveling trajectory by establishing a mathematical model of the shoveling process. Zhang et al. [14] established the trajectory motion model of the electric shovel by kinetic, combined with the pseudospectral method to convert the kinetic model into the algebraic format. They obtained the optimal excavation trajectory by solving the nonlinear programming. Shen et al. [15] modeled the whole excavation process and obtained the excavation trajectory by a rule-based planning method, which significantly reduced fuel consumption. Frank et al. [16] used a dynamic planning method to generate the shoveling trajectory. They conducted tests in a gravel pile, and the results showed a 15% improvement in fuel economy with this method. Yao et al. [17] established a mathematical model of the shoveling process to optimize the shoveling process to reduce fuel consumption, which can be reduced by 30% through simulation analysis. However, in these studies, the optimization of shoveling trajectory is mainly carried out by theoretical analysis and simulation analysis. It ignores the high randomness and complexity of the shoveling operation process, and there is a specific error between the dynamics or mathematical model and the actual shoveling operation [18–20]. As a result, it is difficult to achieve the desired effect in the actual shoveling operation.

Research has always focused on modeling the fuel consumption of shoveling operations. At present, the common fuel consumption modeling method is the theoretical calculation method [21–23]. This method is mainly based on theoretical analysis of energy consumption during the shoveling operation of wheel loaders and then calculates the overall fuel consumption. This method usually cannot consider the large randomness in the process of wheel loader shoveling operation, and there is a large error between it and the actual shoveling operation. In recent years, machine learning methods have shown excellent properties in solving real engineering problems. There have been several successful applications of using machine learning to achieve modeling of operational fuel consumption of engineering vehicles. Dindarloo et al. [24] analyzed the fuel consumption of mining dump trucks at each stage of operation with the objective of fuel consumption of a single operating cycle and constructed a fuel consumption model using machine learning methods. Siami-Irdemoosa et al. [25] established an Artificial Neural Network fuel consumption model to achieve fuel consumption prediction of mining trucks in each cycle with a model error of 10%. Alamdari et al. [26] used machine learning methods such as multiple linear regression, Random Forest, Artificial Neural Network, Support Vector Machine, and Kernel nearest neighbor to predict the fuel consumption of transport trucks and compared the accuracy of each model. Gong et al. [27] used a binary logistic regression model to analyze 21 fuel consumption influencing factors of heavy trucks and obtained 8 influencing factors that have a significant effect on fuel consumption. Meanwhile, the accuracy of the fuel consumption model was compared using Decision Tree, Artificial Neural Network, and Random Forest. Shen et al. [28] constructed an excavator data collection system and proposed an improved PSO-BP model for excavator energy consumption prediction, which achieved good prediction accuracy.

However, there are fewer applications for modeling the fuel consumption of wheel loaders using machine learning. Liu et al. [29] constructed a BP neural network model with wheel loader driver, pile type, and bucket capacity as input features, although the accuracy of the model was high. However, in this study, there are problems of fewer data samples and unreasonable selection of fuel consumption influencing factors, which are difficult to be extended to other application scenarios. Since the shoveling process itself is a high randomness process. There are many factors that can have a great impact on the fuel consumption of shoveling operations. Therefore, determining the reasonable influence factors on the fuel consumption of shoveling is the key task to establishing the fuel consumption model by machine learning. Therefore, this study selects the Kernel Principal Component Analysis (KPCA) method to filter the factors that have a strong influence on the fuel consumption of shoveling operations. Compared with the traditional Principal Component Analysis (PCA) method, KPCA has better nonlinear data processing capability [30]. After obtaining the factors with a strong influence on the fuel consumption of the shoveling operation, a model of the fuel consumption of the shoveling operation is constructed using Support Vector Machine (SVM), a typical machine learning method. It replaces the empirical risk in traditional neural networks with the structured risk minimization principle. It solves the drawback that it is difficult to overcome the local extremes when the sample size is small [31]. After establishing the fuel consumption model, the Improved Particle Swarm Optimization (IPSO) algorithm is selected in this study to optimize the factors with a strong influence on fuel consumption. Compared with the Particle Swarm Optimization (PSO) algorithm, IPSO greatly improves the ability of local optimization search. The optimization algorithm has been widely used to improve operational efficiency by optimizing the operational trajectory. For example, Yuasa et al. [32] optimized the excavating trajectory of an excavator by a Genetic Algorithm based on a parametric mathematical model of the excavating trajectory, loading, and linkage mechanism, which effectively improved the operational efficiency of the excavator. Bi et al. [33] obtained the optimal shoveling trajectory of a cable shovel by a Multi-Objective Genetic Algorithm with the operating time and energy consumption per payload as the optimization objectives.

In response to the above issues and discoveries, this study proposed a method to construct a fuel consumption model by SVM and optimize the shoveling trajectory by IPSO. Firstly, the wheel loader was shoveled several times with three different engine speeds on the gravel pile and sand pile to obtain samples. After that, the factors with a strong influence on the fuel consumption of shoveling operation were selected by KPCA and combined with a Grey relation analysis (GRA) to validate, and the fuel consumption model of shoveling operation was established by SVM. Finally, the factors with a strong influence on the fuel consumption of shoveling operation were optimized by the IPSO algorithm. The lowest energy consumption trajectory was obtained at different engine speeds. The optimized shoveling trajectory was used for validation experiments and compared with the sample data and manually controlled shoveling data to verify the optimization effect.

2. Shoveling Experiments and Analysis

2.1. Shoveling Principle and Experiment Design

The shoveling method is characterized by the insertion of the wheel loader into the pile to a certain depth. Then, the bucket is turned and lifted by the action of the lift arm, tilt level, and connecting rod to complete the shoveling. Finally, the bucket is lifted. Among them, the shoveling trajectory of the scooping phase is parallel to the surface of the pile, as shown in Figure 1. Thus, the wheel loader can be regarded as a system with three degrees of freedom when carrying out the shoveling operation, which is wheel loader advancing, bucket lifting, and bucket turning. Among them, the bucket lifting and turning are controlled by the lift cylinder and the tilt cylinder, respectively, as shown in Figure 2. The wheel loader displacement, lift cylinder displacement, and tilt cylinder displacement are not the same in different shoveling trajectories, which is the reason for the large difference in fuel consumption during the shoveling operation of wheel loaders. Meanwhile, the different engine speeds provide different power to the wheel loader. Even if the same shoveling trajectory is used, however, due to the difference in engine speed, the fuel consumption of the shoveling operation will also appear as a large gap. Therefore, optimizing the shoveling trajectory at different engine speeds is a critical way to reduce the fuel consumption of the shoveling operation.

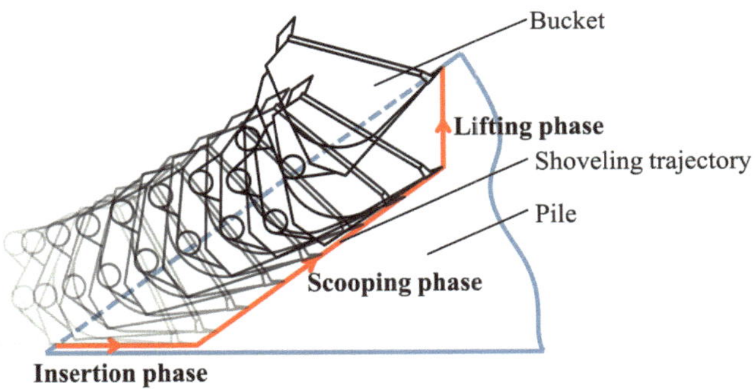

Figure 1. Shoveling trajectory.

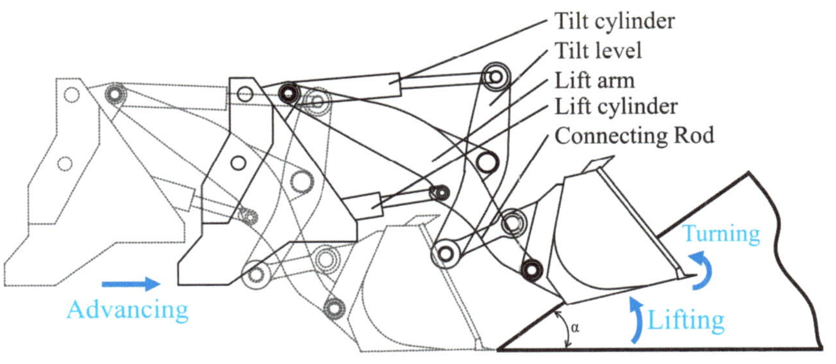

Figure 2. The process of shoveling operation.

To execute the shoveling trajectory better and avoid the errors caused by manual operation. And to obtain more accurate data about the shoveling process, this study has developed an automatic shoveling experiment platform for a wheel loader. The platform will carry out automatic shoveling by preplanning the shoveling trajectory and setting the relevant shoveling parameters. The relevant parameters during the shoveling operation are recorded, including the wheel loader displacement, cylinder displacement, fuel consumption, engine speed, operation time, and shoveling weight. The recording frequency is 500 Hz, which is recorded once every 0.002 s. The automatic shoveling platform and the main sensors are shown in Figure 3.

Figure 3. Automatic shoveling platform and sensors. As shown above, 1 fifth wheel gauge, 2 flow sensor, 3 vehicle PC system, 4 fifth wheel gauge control module, 5 signal receiving system, 6 all data acquisition system, 7 lift cylinder displacement sensor, 8 shoveling weight acquisition system, 9 tilt cylinder displacement sensor.

2.2. Trajectory Planning

Due to the different insertion depths, the shoveling trajectories are also different. The fuel consumption of shoveling operations with different shoveling trajectories has obvious differences. In this study, a variety of insertion depths are randomly selected, ranging from 600 mm to 1000 mm. Two types of materials are selected, including gravel and sand. Gravel is characterized by non-uniform, high-density, and large particles. At the same time, sand is characterized as uniform, low-density, and with small particles. Both above materials have high representativeness, and the relevant parameters of the two materials are shown in Table 1. Due to the different angles of repose of the two piles, their shoveling trajectories are also different, as shown in Figure 4.

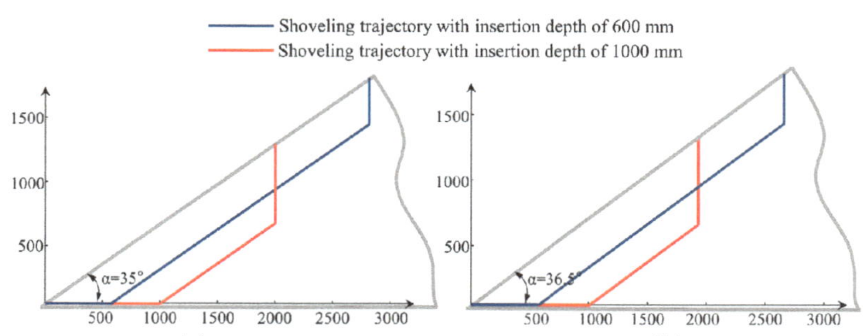

Figure 4. Range of shoveling depths. (**a**) Gravel pile; (**b**) Sand pile.

Table 1. Pile parameters.

Type of Pile	Density (kg/m^3)	Angle of Repose (°)	Particle Diameter (mm)
Gravel	2672	35.0	20–60
Sand	2392	36.5	<1

The shoveling area S consisting of the shoveling trajectory and the surface of the pile is shown in Figure 5. According to the geometric relationship in Figure 5, the shoveling area can be expressed as shown in Equation (1). However, the shoveling trajectory is not the same. However, the shoveling area composed of the shoveling trajectory and surface of the pile is always constant and equal to the bucket cross-section. When the length of l_{AB} is determined, both l_{BC} and l_{CD} can be expressed as a function of l_{AB} only. This is shown in Equation (2). When the shoveling area is held constant, different shoveling trajectories can be obtained by adjusting the length of l_{AB}.

$$S = \frac{V}{L} = \frac{1}{2} l_{AB} l_{BQ} + l_{BC} l_{BP} \tag{1}$$

where V is the bucket volume, and L is the bucket length.

$$\begin{cases} l_{BP} = l_{AB} \sin(\alpha) \\ l_{CD} = l_{BQ} = l_{AB} \tan(\alpha) \\ l_{BC} = \frac{S}{l_{BP}} - \frac{l_{BP}}{\sin(2\alpha)} \\ l_{BC'} = l_{BC} \cos(\alpha) \end{cases} \tag{2}$$

where l_{AB} is the insertion depth of the insertion phase, l_{BC} is the shoveling length of the scooping phase, l_{CD} is the shoveling length of the lifting phase, l_{BP} is the horizontal distance between the trajectory of the scooping phase and the surface of the pile, and $l_{BC'}$ is the wheel loader displacement of the scooping phase.

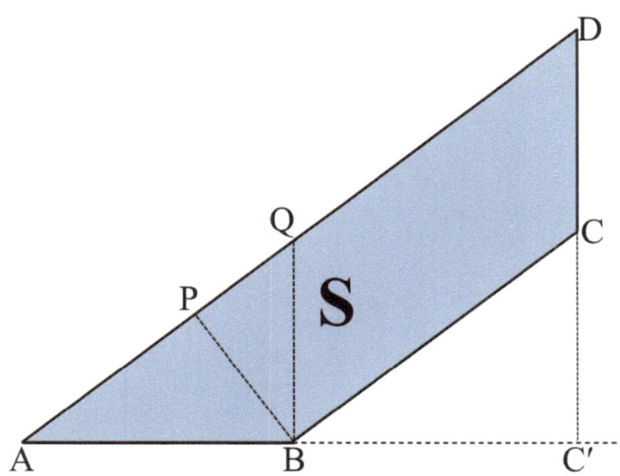

Figure 5. Shoveling area.

2.3. Engine Speed Analysis

To obtain the optimal trajectory with different engine speeds. Three different engine speeds are selected, 900 RPM, 1200 RPM, and 1500 RPM, and 40 shoveling operations are performed on the gravel pile and the sand pile with each engine speed. The relevant parameters are recorded during the shoveling process. The pile is recovered after each shoveling operation. The process of shoveling gravel and the process of shoveling sand are shown in Figure 6. The experimental environment is an open flat area. The average temperature is 22 °C, and the average relative humidity is 83%.

Figure 6. Shoveling gravel and shoveling sand.

When the engine speed is 1500 RPM, the engine speed for shoveling gravel and shoveling sand is shown in Figure 7. The shoveling operation process can be divided into four phases with the change of displacement of the lift cylinder and the tilt cylinder, which are the insertion phase, scooping phase, stopping phase, and lifting phase, corresponding to S1, S2, S3, and S4 in Figure 7, respectively.

1. Insertion phase: There is no change in the displacement of the lift cylinder and the displacement of the tilt cylinder. The engine speed is maintained at 1500 RPM, but a smaller oscillation will occur. This oscillation is more violent at the beginning of the insertion phase. However, as the insertion depth increases, it gradually calms down and stays at 1500 RPM.
2. Scooping phase: The displacement of the tilt cylinder increases rapidly, and the displacement of the lift cylinder steps up. With the increasing displacement of the cylinder, there is a large change in engine speed. When the displacement of the cylinder starts to increase, the engine speed decreases at first, then increases rapidly and is higher than 1500 RPM. When the displacement of the cylinder is stable, the engine speed also falls back to 1500 RPM rapidly.
3. Stopping phase: The displacement of the tilt cylinder and lift cylinder has no change, and the engine speed decreases to a lower state in a short time.
4. Lifting phase: The displacement of the lift cylinder increases rapidly, and the displacement of the tilt cylinder is almost unchanged. The engine speed decreases to the lowest point in the whole shoveling process in a short period and then increases.

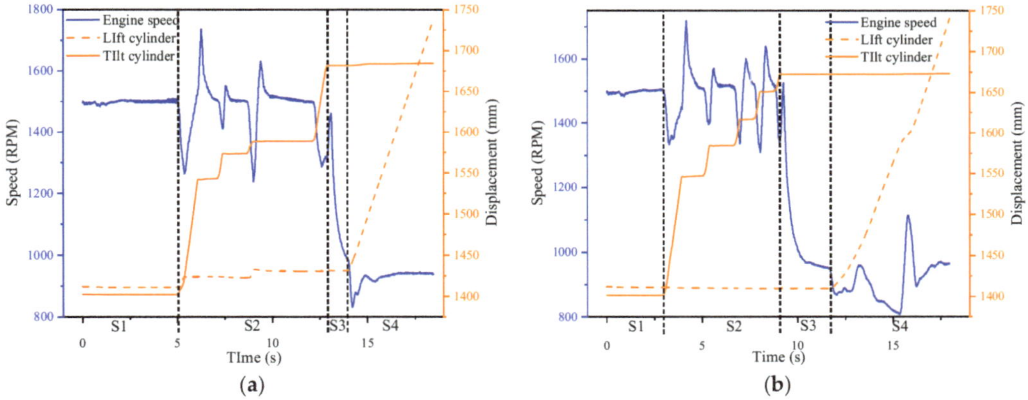

Figure 7. Engine speed during the shoveling process. (a) Shoveling gravel; (b) Shoveling sand.

2.4. Fuel Consumption Analysis

The wheel loader fuel consumption per unit time, wheel loader displacement, lift cylinder displacement, and tilt cylinder displacement during the wheel loader shoveling process are shown in Figure 8.

1. Insertion phase: The wheel loader is inserted into the pile at a certain distance with traction. In the early insertion phase, the insertion speed of the wheel loader is fast. However, as the insertion depth increases, the resistance to the bucket also increases rapidly, and the wheel loader insertion speed gradually decreases. Meanwhile, the fuel consumption per unit of time rises rapidly after the insertion of the pile and then remains in a relatively stable state.
2. Scooping phase: The wheel loader continues to keep moving forward. In the early scooping phase, the wheel loader can move forward rapidly because the material is scooped up by the bucket to provide space for the wheel loader to move forward. In the later scooping phase, the wheel loader advances at a slower speed due to resistance. It is obvious that when the lift cylinder and tilt cylinder remain unchanged, the fuel consumption per unit time remains at a relatively stable level. The fuel consumption per unit time increases rapidly when the displacement of the lift cylinder and tilt cylinder increases and then is in a stable state.
3. Stopping phase: The resistance of the wheel loader increases sharply, and the wheel loader is forced to stop moving forward. However, to prevent the wheel loader from moving backward, the wheel loader will continue to keep moving ahead and move forward for a short distance. The fuel consumption per unit time is then reduced to the lowest state during the entire shoveling operation.
4. Lifting phase: The wheel loader is subjected to inertia in the process of lifting the bucket and continues to keep moving forward. And the fuel consumption per unit time increases with the displacement of the lift cylinder and then remains stable.

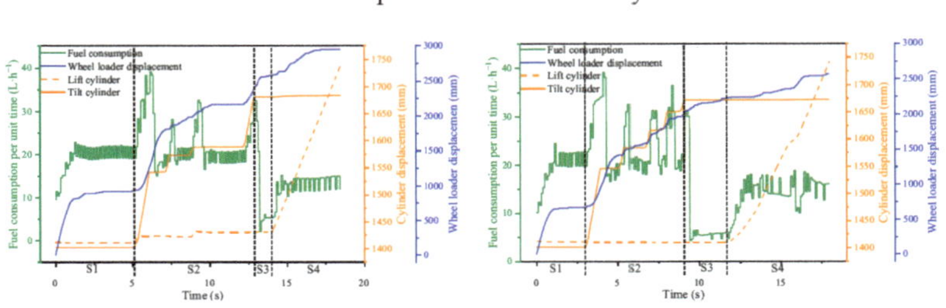

Figure 8. Fuel consumption during the shoveling process. (a) Shoveling gravel; (b) Shoveling sand.

The sensor on the wheel loader is a flow sensor. The total fuel consumption needs to be obtained by the calculation formula, which is shown in Equation (3). As an example, in Figure 8, the total fuel consumption, the fuel consumption of each phase, and the percentage of the total fuel consumption by each phase are shown in Table 2.

$$C = \frac{\sum_{i=1}^{n} c_i}{f \cdot 3600} \cdot 1000 \tag{3}$$

where C is the total fuel consumption, mL, c is the fuel consumption per unit time at a certain moment, $L \cdot h^{-1}$, f is the frequency, Hz.

Table 2. Fuel consumption.

Type of Pile	Parameter	Insertion Phase	Scooping Phase	Stopping Phase	Lifting Phase	Total
Gravel	Fuel consumption (mL)	27.12	49.27	4.10	15.61	96.10
	Percentage (%)	28.22	51.27	4.27	16.24	100.00
Sand	Fuel consumption (mL)	15.97	41.15	6.45	24.68	88.26
	Percentage (%)	20.51	52.83	8.29	31.68	100.00

According to Figure 8 and Table 2. The fuel consumption in the scooping phase is the highest, and the fuel consumption in the stopping phase is the lowest. In shoveling gravel, the fuel consumption in the insertion phase is higher than that in the lifting phase. While shoveling sand, the fuel consumption in the insertion phase is less than that in the lifting phase. This is related to the material properties, as gravel is a denser and less loose material. The wheel loader needs more operation time and energy consumption when inserting the gravel pile. Sand, on the other hand, is less dense and relatively looser. The wheel loader is easier to insert from the sand pile and therefore generates less fuel consumption when inserting into the sand pile. In short, the fuel consumption of shoveling gravel is higher than that of shoveling sand with the same engine speed, which is determined by the material characteristics. From Figure 8, the lift cylinder displacement and tilt cylinder displacement within different phases all affect fuel consumption, which increases with the expansion of cylinder displacement. Meanwhile, the different engine speed of the wheel loader leads to the different initial velocity of the wheel loader when inserted into the pile, which will directly affect the time of wheel loader shoveling operation. Moreover, the engine speed is related to the wheel loader power, which will have a direct impact on the fuel consumption of the shoveling operation. In addition, the final shoveling weight will also have an impact on the fuel consumption of the shoveling operation. It is mainly reflected in the lifting phase; if the shoveling weight is too high, it will increase the extra fuel consumption.

3. Research Methodology

3.1. KPCA and GRA

KPCA is an improved processing method of PCA. As for PCA, it only applies to linear data, and there is a significant error when dealing with nonlinear data. In contrast, KPCA can largely preserve the local structure of nonlinear data by introducing kernel functions. Meanwhile, PCA will map data points into a low-dimensional space when dealing with high-dimensional data, facing dimensional disaster and leading to the loss of distance information between data points. Therefore, KPCA is more advantageous than PCA [34]. The basic idea is to project the samples that are linearly indistinguishable in the low-dimensional space to the high-dimensional space and make them linearly distinguishable by the kernel function [35]. For a set of samples $X = [x_1, x_2, \ldots, x_n] \in R^m$, the mapping function is

$$\sum_{i=1}^{n} \varphi(x_i) = 0(\varphi : R^m \rightarrow R^k(k \gg m)) \tag{4}$$

where $\varphi(x_i)$ is the mapping function, R^m is an m-dimensional vector, and R^k is a k-dimensional vector.

Then the covariance matrix can be expressed as

$$C = \frac{1}{n}\sum_{i=1}^{n} \varphi(x_i)\varphi(x_i)^T \tag{5}$$

Suppose the eigenvalue of C is λ, and its corresponding eigenvector is v, the result is

$$\lambda v = Cv = (\frac{1}{n}\sum_{i=1}^{n} \varphi(x_i)\varphi(x_i)^T)v = \frac{1}{n}\sum_{i=1}^{n} (\varphi(x_i)^T v)\varphi(x_i) \tag{6}$$

For $\lambda v = Cv$, by multiplying each left and right by $\varphi(x_i)$, we can obtain $\lambda\varphi(x_i)v = Cv\varphi(x_i)$. The introduced kernel function matrix can be expressed as $K_{ij} = K(x_i, x_j) = \varphi(x_i)\varphi(x_j)$. Then, we can obtain $n\lambda a = Ka$, where a is the eigenvector of K. Then, the projection of x into the higher dimensional space can be expressed as

$$v\varphi(x) = \sum_{i=1}^{n} a_i\varphi(x_i)\varphi(x) = \sum_{i=1}^{n} a_i K(x_i, x) \tag{7}$$

In this study, the Gaussian kernel function is chosen as the kernel function, which is defined as follows:

$$K(x, q) = \frac{\exp(-\|x-q\|^2)}{\sigma^2} \tag{8}$$

where σ is the function parameter, and q is the center of the function.

GRA is an analytical method based on grey system theory, mainly used to study the strength of association between influencing factors. It converts the grey information between multiple factors into black-and-white information, then uses the magnitude of the grey relation coefficient value as the basis for evaluating the strength of the correlation. The specific steps are as follows.

1. Select the parent sequence Y and the subsequence X.
2. To eliminate the influence of the magnitude, the parent sequence and the subsequence are normalized, and the normalized parent sequence is recorded as Y' and the subsequence is recorded as X'.
3. The formula is shown in Equation (9) to calculate the grey relation coefficient.

$$\xi(i) = \frac{\min_i(\min|Y' - X'(i)|) + \rho\max_i(\max|Y' - X'(i)|)}{|Y' - X'(i)| + \rho\max_i(\max|Y' - X'(i)|)}, \quad i = 1\cdots n \tag{9}$$

where ρ is the discriminant coefficient, which takes values from 0 to 1 and is taken as 0.5 in this paper.

3.2. SVM

SVM is to use the kernel function to map the sample data eigenvalues in the low-dimensional space to the corresponding high-dimensional space. And the optimal hyperplane is determined in the high-dimensional space based on the structural risk minimization principle [36]. The expression of the classification hyperplane is shown in Equation (9), the objective function is shown in Equation (10), and the constraints are shown in Equation (11).

$$f(x) = w\cdot x + b \tag{10}$$

$$\min\frac{1}{2}\|w\|^2 + g\sum_{i=1}^{m} (\xi_i + \xi^*) \tag{11}$$

$$\text{s.t.}\begin{cases} y_i - (w \cdot \varphi(x_i) + b) \leq \varepsilon + \xi_i \\ (w \cdot \varphi(x_i) + b) - y_i \leq \varepsilon + \xi_i^* \\ \xi_i \geq 0, \xi_i^* \geq 0 \ (i = 1, 2, \cdots, m) \end{cases} \tag{12}$$

where w is the normal vector of the hyperplane, x is the input vector, b is the hyperplane translation distance, ξ is the relaxation factor, g is the penalty factor, m is the number of samples, ε is the insensitivity coefficient factor, and φ is the mapping function.

The output function of the SVM can be expressed as

$$g(x) = \sum_{i=1}^{m} a_i K(x_i, x_j) + b \tag{13}$$

where a_i is the Lagrange multiplier, and K is the kernel function, and the radial basis kernel is chosen as the kernel function in this paper, as shown in Equation (13):

$$K(x_i, x_j) = \exp\left(\frac{-\|x_i - x_j\|^2}{2c^2}\right) \tag{14}$$

where c is the kernel function parameter.

For machine learning models, optimizing hyperparameters is decisive for the model performance [37]. A reasonable choice of hyperparameters is beneficial to improve the performance of the model. Studies have shown that the optimization of penalty factor g and kernel function parameter c in SVM is an effective way to improve the accuracy of SVM [38]. To obtain reasonable g and c and avoid overfitting of the model. In this study, we choose the method of K-fold Cross Validation (CV) to obtain reasonable parameters, and the pseudo-code of K-fold CV optimize g and c is shown in Algorithm 1.

Algorithm 1 K-fold Cross Validation

```
begin
bestaccuracy; c_best; g_best;
  for c = c_min^2:c_max^2
    for g = g_min^2:g_max^2
    divide the dataset equally into K groups;
      for = 1:K
        divide the dataset equally into K groups;
        train(K) as the test set, the rest as the training set;
        record the accuracy of the test set ace(K);
      end;
      cv = (ace(1) + ace(2) + ··· + ace(K))/K;
        if cv > bestaccuracy
          bestaccuracy = cv; c_best = c; g_best = g;
        end;
    end;
  end;
end.
```

3.3. IPSO

PSO is inspired by the foraging behavior of bird populations in nature, and the basic units in PSO are particles. Each particle is described by three features: fitness value, position, and velocity. Each particle in the algorithm is measured by the fitness function to measure the degree of merit of this particle. The velocity of the particle is influenced by itself and the population and can be adjusted during each iteration. And the position describes the current location of the particle and is updated by the velocity. The updated formulas for velocity and position are shown in Equations (14) and (15), respectively [39].

$$v_i = wv_i + c_1r_1(p_i - x_i) + c_2r_2(g_{best} - x_i) \tag{15}$$

$$x_i = x_i + v_i \tag{16}$$

where w is the inertia coefficient, c is the acceleration factor, and r is a random number of [0, 1].

The choice of the inertia coefficient w directly affects the convergence of the PSO. In traditional PSO, w is taken as 1 by default. w tends to be the global search for larger PSO and the local search for smaller PSO. To avoid premature maturation of the algorithm and oscillation of particles near the global optimal solution in the late stage of the algorithm, the value of w is usually taken to decrease linearly with the increase of the number of iterations. However, it requires repeated trials to determine the maximum, minimum, and number of iterations, and the optimal value may not be found. Thus, an Improved Particle Swarm algorithm is applied in this study for implementing a nonlinear dynamic adaptive adjustment of the inertia coefficient w such that the inertia coefficient w can follow the change of the fitness value, calculated as shown in Equation (16) [40].

$$w = \begin{cases} w_{\min} + \frac{(w_{\max} - w_{\min})(f - f_{\min})}{f_{avg} - f_{\min}} \ (f \le f_{avg}) \\ w_{\max}(f > f_{avg}) \end{cases} \tag{17}$$

4. SVM Fuel Consumption Model and IPSO Optimization

4.1. KPCA Processing and GRA Verification

From the previous analysis, it can be seen that the wheel loader displacement, lift cylinder displacement, and tilt cylinder displacement in different phases will influence fuel consumption. Meanwhile, the initial velocity, operation time, and shoveling weight will also influence fuel consumption. Since the displacement of the lift cylinder and the tilt cylinder in some phases do not change. The displacement of the wheel loader in the four phases, the displacement of the lift cylinder and the tilt cylinder in the scooping phase, the displacement of the lift cylinder in the lifting phase, the initial velocity, the operation time, and the shoveling weight, a total of 10 factors, are taken as the initial fuel consumption influencing factors. The partial data of the gravel sample and the sand sample are shown in Table 3.

Table 3. Part of the sample data.

Type of Pile	Engine Speed (RPM)	Number	Initial Velocity (m·s^{-1})	Displacement of Wheel Loader				Displacement of Tilt Cylinder in Scooping Phase (mm)	Displacement of Lift Cylinder		Operation Time (s)	Shoveling Weight (kg)	Fuel Consumption (mL)
				Insertion Phase (mm)	Scooping Phase (mm)	Stopping Phase (mm)	Lifting Phase (mm)		Scooping Phase (mm)	Lifting Phase (mm)			
Gravel	900	1	0.48	802	1431	259	182	27	301	281	13.62	4771	73.63
		2	0.49	789	1536	105	39	20	307	265	13.31	5116	74.34
		3	0.35	995	1061	134	125	21	304	277	19.04	4966	141.51
		...	...	...	...	...	...	...	...	...	...	...	...
		40	0.42	789	1563	134	202	26	303	279	18.58	4927	142.70
	1200	1	0.81	749	1508	462	103	20	308	267	13.03	4609	62.54
		2	0.83	718	1738	105	125	28	300	282	12.64	4934	63.88
		3	0.75	745	1699	77	163	25	302	270	14.99	5103	110.32
		...	...	...	...	...	...	...	...	...	...	...	...
		40	0.79	779	1431	249	144	32	296	251	15.79	4284	111.48
	1500	1	1.17	806	1334	164	355	28	299	278	14.62	3894	73.49
		2	1.09	819	1392	240	67	20	309	258	11.73	4914	75.12
		3	1.18	730	1724	202	422	28	299	278	15.96	3796	76.96
		...	...	...	...	...	...	...	...	...	...	...	...
		40	1.16	729	1669	173	345	28	298	279	16.57	3770	102.71
Sand	900	1	0.43	787	1521	307	173	71	263	270	11.22	5409	55.49
		2	0.50	739	1651	183	230	93	241	260	13.55	5474	68.65
		3	0.48	682	1787	39	537	89	244	225	17.72	6722	81.59
		...	...	...	...	...	...	...	...	...	...	...	...
		40	0.37	748	1632	173	250	90	243	271	15.87	6449	83.71
	1200	1	0.86	707	1755	192	67	94	237	270	13.08	5864	62.71
		2	0.83	768	1579	221	134	73	261	276	12.11	5844	63.72
		3	0.84	960	1110	96	105	23	328	275	16.27	5441	70.59
		...	...	...	...	...	...	...	...	...	...	...	...
		40	0.81	729	1684	240	211	68	265	275	14.31	6878	74.78
	1500	1	1.16	682	1841	192	259	99	233	263	14.16	5909	61.06
		2	1.17	670	1899	96	250	94	237	265	12.78	6117	61.33
		3	1.17	826	1418	144	96	94	239	270	11.77	6241	63.60
		...	...	...	...	...	...	...	...	...	...	...	...
		40	1.20	889	1234	173	364	91	237	271	13.36	6052	64.94

Due to the different magnitudes among the initial fuel consumption influence factors, they need to be normalized. After the normalization process and then the KPCA operation, the contribution rate and cumulative contribution rate are calculated as shown in Equation (17). The contribution rate and cumulative contribution rate of each factor are shown in Table 4. The factors that have the greatest influence on the fuel consumption of shoveling operation are initial velocity and insertion phase displacement of the wheel loader. The initial velocity is related to the engine speed, which is responsible for providing power for the whole shoveling operation process. While the wheel loader displacement in the insertion phase is related to the insertion resistance, and the wheel loader reaches the peak of resistance in the insertion phase, which is the key factor that influences fuel consumption. In the sand sample, the contribution of shoveling weight is greater than the operation time, while in the gravel sample, the contribution of operation time is slightly greater than the shoveling weight. This is related to the experimental environment, which is affected by the climate, where the humidity in the air increases and the sand appears to stick, so the shoveling weight is higher when shoveling sand. The scooping phase is the most important phase in the shoveling operation process, so the displacement of the wheel loader in this phase also has a greater influence on the fuel consumption of the shoveling operation. From Figure 8, the cylinder displacement will influence the fuel consumption of the shoveling operation, but compared with other factors, the influence of cylinder displacement on the fuel consumption of the shoveling operation is not great. According to the principle of KPCA preference, the requirement of replacing all factors can be achieved for factors with a cumulative contribution of 90% [41]. In the gravel samples and sand samples, the initial velocity, wheel loader displacement in the insertion phase, shoveling weight, operation time, and wheel loader displacement in the scooping phase, the cumulative contribution rate is greater than 90%. Thus, these five factors are selected as the key factors for the influence of fuel consumption of shoveling operation, and further research is conducted.

$$\begin{cases} \gamma = \dfrac{\lambda_j}{\sum\limits_{i=1}^{n} \lambda_i} \\ \gamma\prime = \dfrac{\sum\limits_{i=1}^{m} \lambda_j}{\sum\limits_{i=1}^{n} \lambda_i} \end{cases} \quad (j = 1,2,\cdots,n) \tag{18}$$

where λ is the eigenvalue, γ is the contribution rate, and γ' is the cumulative contribution rate.

Table 4. Contribution rate and the cumulative contribution rate of each factor.

Gravel			Sand		
Parameters	γ (%)	γ' (%)	**Parameters**	γ (%)	γ' (%)
Initial velocity	31.64	31.64	Initial velocity	31.00	31.00
Displacement of wheel loader in insertion phase	27.04	58.68	Displacement of wheel loader in insertion phase	24.42	55.42
Operation time	15.36	74.03	Shoveling weight	13.99	69.41
Shoveling weight	15.35	89.39	Operation time	13.79	83.20
Displacement of wheel loader in scooping phase	6.56	95.95	Displacement of wheel loader in scooping phase	10.70	93.89
Displacement of wheel loader in stopping phase	2.65	98.60	Displacement of wheel loader in stopping phase	5.64	99.53
Displacement of wheel loader in lifting phase	1.13	99.73	Displacement of wheel loader in lifting phase	0.29	99.82
Displacement of lift cylinder in scooping phase	0.24	98.97	Displacement of lift cylinder in scooping phase	0.16	98.99
Displacement of lift cylinder in lifting phase	0.02	99.98	Displacement of lift cylinder in lifting phase	0.01	99.99
Displacement of tilt cylinder in scooping phase	0.01	100.00	Displacement of tilt cylinder in scooping phase	0.01	100.00

In order to verify the correlation between the key factors of fuel consumption influence and fuel consumption, the GRA is selected to test the correlation. The fuel consumption is chosen as the parent sequence, and the key factors of fuel consumption influence the subsequence. After calculation, the grey relation coefficients of gravel samples and sand samples are shown in Table 5. The more the grey relation coefficient value tends to 1, the higher the correlation between the subsequence and the parent sequence. The grey relation coefficient values are greater than 0.75 in the gravel samples and sand samples. There is a greater correlation between fuel consumption and key factors of fuel consumption influence. It proves that the key factors of fuel consumption influence screened by KPCA are effective.

Table 5. Grey relation coefficients.

Type of Pile	Initial Velocity	Displacement of Wheel Loader in Insertion Phase	Shoveling Weight	Operation Time	Displacement of Wheel Loader in Scooping Phase
Gravel	0.792	0.782	0.834	0.789	0.779
Sand	0.762	0.795	0.894	0.846	0.822

4.2. SVM Model

According to the results of KPCA, the initial velocity, the displacement of the wheel loader in the insertion phase, the shoveling weight, the operation time, and the displacement of the wheel loader in the scooping phase are selected as the input layer, and the fuel consumption as the output layer. Among 120 sets of gravel samples and sand samples, respectively, 100 sets of samples are randomly selected as the training set, and the other 20 sets are used as the test set. Meanwhile, the penalty factor g and the kernel function parameter c are optimized by combining them with the CV method. The optimal c and g for the gravel fuel consumption model are 3.03 and 143.58, respectively, and for the sand fuel consumption model, they are 4.32 and 123.52, respectively. The accuracy effect of the fuel consumption model is shown in Figure 9. Meanwhile, root mean square error (RMSE), average relative error (ARE), and goodness of fit R^2 are used in this study as indicators to evaluate the performance of the fuel consumption model. The smaller the RMSE and ARE, the smaller the deviation between the predicted value and the true value, and the higher the accuracy of the model. The value of R^2 is closer to 1, which indicates a higher degree of fit and higher accuracy of the model, and its calculation formula is as follows. The evaluation indicators of the fuel consumption model are shown in Table 6.

$$RMSE = \sqrt{\frac{1}{n}\sum_{i=1}^{n} (y_i{}^* - y_i)^2} \tag{19}$$

$$ARE = \frac{1}{n}\sum_{i=1}^{n}\left|\frac{y_i - y_i{}^*}{y_i}\right| \tag{20}$$

$$R^2 = 1 - \frac{\sum_{i=1}^{n} (y_i - y_i{}^*)^2}{\sum_{i=1}^{n} (y_i - \overline{y}_i)^2} \tag{21}$$

where y_i is the true value, $y_i{}^*$ is the predicted value, $\overline{y}_i$ is the mean of the true value, and n is the number of samples.

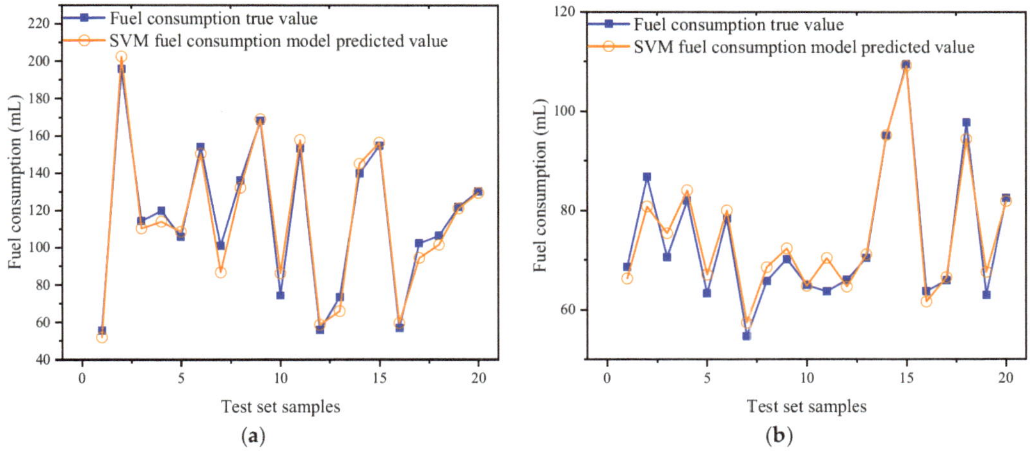

Figure 9. Prediction performance of fuel consumption model. (**a**) Gravel fuel consumption model; (**b**) Sand fuel consumption model.

Table 6. Fuel consumption model evaluation indicators.

Type of Pile	ARE (%)	RMSE	R^2
Gravel	4.91	5.91	0.9764
Sand	3.49	3.08	0.9494

In the gravel fuel consumption prediction model, the ARE is 4.91%, the RMSE is 5.91, and the R^2 is 0.9764. While in the sand fuel consumption prediction model, the ARE is 3.49%, the RMSE is 3.08, and the R^2 is 0.9494. In terms of prediction accuracy, the prediction accuracy of the gravel fuel consumption prediction model is lower than that of the sand fuel consumption prediction model. And in terms of fitting ability, the gravel fuel consumption prediction model has a better fitting ability. Overall, the SVM fuel consumption prediction model showed better prediction ability in both the gravel samples and the sand samples. The deviation between the predicted and true values of the model is slight, and the model has high stability and generalization ability. To further test the rationality of KPCA. Different dimensions of fuel consumption influence factors are selected as input layers in the SVM fuel consumption model. The same training set and test set are selected, and ARE is used as the evaluation indicator. The prediction performance of the gravel model and sand model is shown in Figure 10. When the dimension increases, the value of ARE decreases rapidly, and when the dimension is 5, the ARE of the gravel model and sand model reaches the lowest point. When the dimension continues to increase, the values of ARE both have a small increase.

4.3. IPSO Optimization Process

To obtain the optimal shoveling trajectory with different engine speeds. IPSO is used to optimize the SVM fuel consumption model. The indicators of optimization are the input features of the SVM model, which are the initial velocity, the displacement of the wheel loader in the insertion phase, the shoveling weight, the operation time, and the displacement of the wheel loader in the scooping phase. Among them, the initial velocity exists in different intervals depending on the engine speed. The displacement of the wheel loader in the scooping phase is not involved in the optimization search due to the limitation of trajectory planning, and the results are obtained by the calculation of wheel loader displacement in the insertion phase. As for the operation time and shoveling weight, which is a posteriori data, as long as they correspond to the upper and lower limits of the sample data, the optimization intervals of each parameter are shown in Table 7.

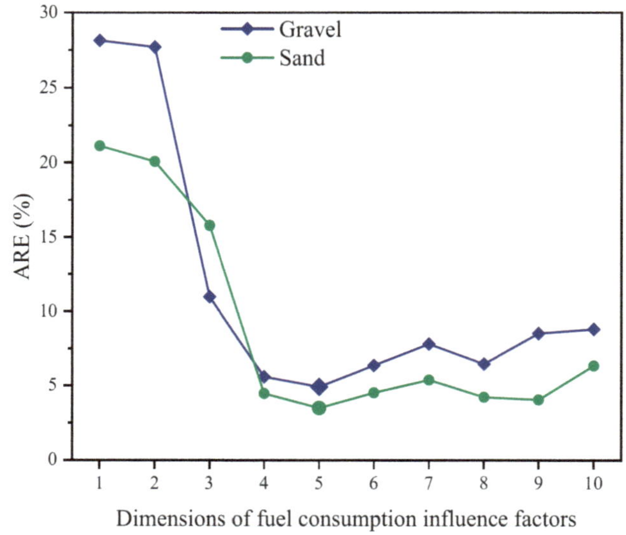

Figure 10. Dimension of fuel consumption influence factors and model accuracy.

Table 7. Optimization intervals for each parameter.

Parameters	Interval	Gravel Model			Sand Model		
		900 RPM	1200 RPM	1500 RPM	900 RPM	1200 RPM	1500 RPM
Initial velocity ($m \cdot s^{-1}$)	Upper limit	0.6	0.95	1.3	0.6	0.95	1.3
	Lower limit	0.3	0.65	1.05	0.3	0.65	1.05
Displacement of wheel loader in insertion phase (mm)	Upper limit	1000					
	Lower limit	600					
Shoveling weight (kg)	Upper limit	5526	5545	5558	7294	7441	7520
	Lower limit	4459	4609	3913	5400	5773	5519
Operation time (s)	Upper limit	25	23	21	25	26	23
	Lower limit	12	12	10	11	12	10
Displacement of wheel loader in scooping phase (mm)		Calculated according to Equation (2)					

In the IPSO algorithm, c_1 and c_2 are set to 1.5. The number of iterations is set to 200. The population size is set to 30. The optimal trajectories of the gravel model and the sand model are optimized for three different engine speeds, respectively. The optimization process is shown in Figure 11, and the optimization results are shown in Table 8.

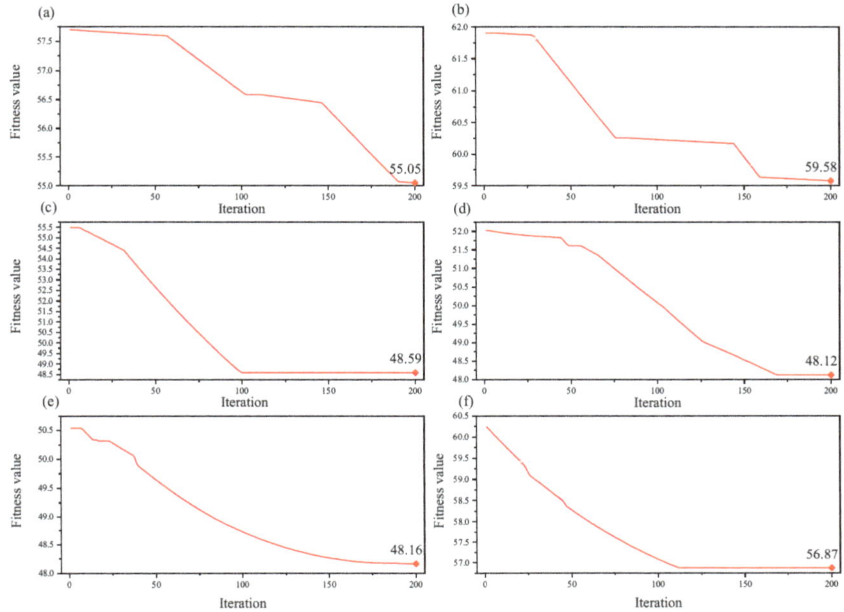

Figure 11. IPSO optimization iterative process. (**a**) Gravel-900 RPM; (**b**) Gravel-1200 RPM; (**c**) Gravel-1500 RPM; (**d**) Sand-900 RPM; (**e**) Sand-1200 RPM; (**f**) Sand-1500 RPM.

Table 8. Optimization results.

Type of Pile	Engine Speed (RPM)	Initial Velocity ($m{\cdot}s^{-1}$)	Displacement of Wheel Loader in Insertion Phase (mm)	Displacement of Wheel Loader in Scooping Phase (mm)	Fuel Consumption (mL)
Gravel	900	0.59	704	1778	55.05
	1200	0.86	686	1843	59.58
	1500	1.16	698	1799	48.59
Sand	900	0.48	839	1272	48.12
	1200	0.79	749	1520	48.16
	1500	1.25	733	1569	56.87

5. Validation Experiment and Discussion

5.1. Validation Experiment

To verify the performance of the optimized shoveling trajectory, the corresponding trajectories were planned at different engine speeds according to the data in Table 8, and the research line of thought is shown in Figure 12. Three experiments at each of three different engine speeds with optimized trajectories in a gravel pile and sand pile and the results of the shoveling experiments with the shoveling operation parameters are shown in Table 9. Due to the large randomness of the shoveling operation itself and the accuracy problem in the data collection process, which led to certain errors in the testing process, it was difficult to ensure that the trajectory remained the same as the theoretical trajectory even by using the automatic shoveling platform. Therefore, in the insertion phase, when the error between the actual insertion distance of the wheel loader and the theoretical distance is less than 20 mm, it can be considered as basically conforming to the theoretical trajectory.

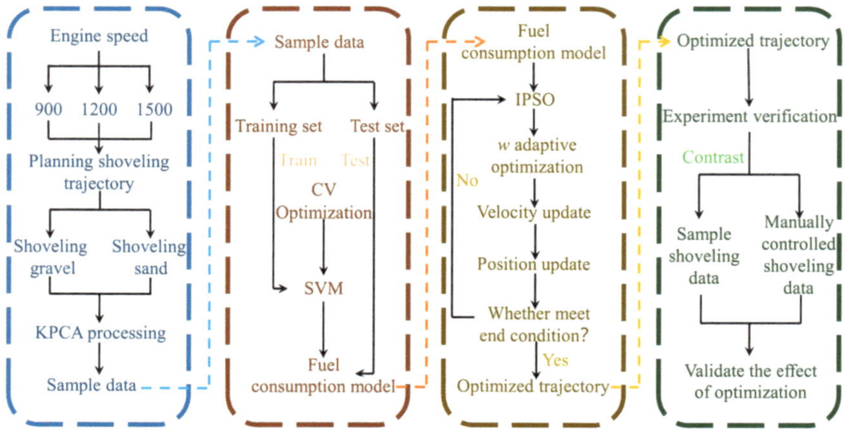

Figure 12. The research line of thought.

Table 9. Validation experiment shoveling operation parameters and fuel consumption.

Type of Pile	Engine Speed (RPM)	Number	Initial Velocity ($m \cdot s^{-1}$)	Displacement of Wheel Loader in Insertion Phase (mm)	Displacement of Wheel loader In Scooping Phase (mm)	Operation Time (s)	Shoveling Weight (kg)	Fuel Consumption (mL)	Average Fuel Consumption (mL)
Gravel	900	1	0.56	719	1690	10.99	4492	53.50	54.50
		2	0.53	713	1710	11.79	4602	51.91	
		3	0.49	722	1703	11.20	4570	58.09	
	1200	1	0.83	702	1804	12.02	4641	67.64	61.97
		2	0.81	704	1785	12.49	4713	58.68	
		3	0.81	694	1748	12.55	4596	59.60	
	1500	1	1.14	726	1737	8.01	4388	45.09	44.59
		2	1.12	704	1844	10.56	4414	47.30	
		3	1.09	725	1745	7.33	3770	41.38	
Sand	900	1	0.52	826	1432	10.52	4232	49.24	49.98
		2	0.47	855	1342	9.85	5357	50.23	
		3	0.49	819	1436	9.90	5207	50.48	
	1200	1	0.81	747	1668	9.31	4966	49.78	50.00
		2	0.77	758	1601	10.10	4550	47.54	
		3	0.87	730	1643	10.39	4843	52.69	
	1500	1	1.14	728	1594	10.65	6514	56.50	57.42
		2	1.14	739	1603	11.12	6442	64.64	
		3	1.15	749	1488	11.37	4758	51.11	

The average fuel consumption of the validation experiment, the theoretical fuel consumption of IPSO optimization, the minimum fuel consumption of the sample, the maximum fuel consumption of the sample, and the average fuel consumption of the sample are shown in Figure 13. The deviation between the average fuel consumption of the validation experiment and the theoretical fuel consumption of IPSO optimization is small. The maximum deviation value is 8.23%, and the accuracy of the fuel consumption model is high. The deviation from the actual shoveling operation is small. In Gravel-1200 RPM, the experimental average fuel consumption is slightly higher than the minimum fuel consumption of the sample, and the fuel consumption is improved by 0.62%. However, in the other models, the average fuel consumption of the validation experiment is reduced compared with the minimum fuel consumption of the sample. Among them, the maximum reduction in fuel consumption is 22.63% in Gravel-900 RPM. The average fuel consumption of the validation experiment is substantially reduced compared to the sample maximum fuel consumption, with a maximum reduction of 73.65% and a minimum reduction of 55.70%. In the gravel model with three different engine speeds, the average fuel consumption of the validation experiment is reduced by 60.93%, 45.79%, and 49.10%, respectively, compared to the sample average fuel consumption. In the sand model with three different engine speeds, the average fuel consumption of the validation experiment is reduced by 30.93%, 35.56%, and 28.59%, respectively, compared with the sample average fuel consumption. The optimized shoveling trajectory effectively reduced the fuel consumption of the shoveling operation.

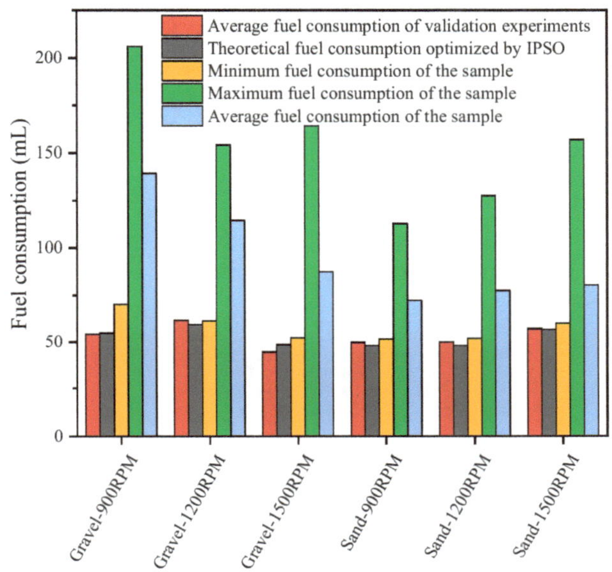

Figure 13. Fuel consumption comparison.

To further compare the performance of the optimized shoveling trajectory, a comparison with the shoveling operation of a manually operated wheel loader is selected. In manually controlled shoveling, there are no restrictions on the parameters such as the engine speed of the wheel loader, displacement of the wheel loader in the insertion phase, and displacement of the wheel loader in the scooping phase. Wheel loader drivers will conduct random shoveling according to their driving habits and operating experience. To increase the comparability, two drivers are selected to carry out 10 shoveling operations on the gravel pile and sand pile, respectively, in the same experimental conditions and record the parameters during the shoveling operations. The parameters of shoveling operations are shown in Figure 14, and the fuel consumption of shoveling operations is shown in Table 10.

Table 10. Fuel consumption by manually controlled shoveling.

Driver	Shoveling Operation	Fuel Consumption										Average Fuel Consumption
Driver A	Shoveling gravel	71.67	64.06	92.31	72.56	75.38	71.89	63.60	71.72	76.12	66.70	72.60
	Shoveling sand	72.65	62.36	72.07	64.77	68.66	92.63	76.10	72.38	65.51	66.17	71.33
Driver B	Shoveling gravel	84.85	65.72	74.56	77.76	72.16	74.00	65.73	68.77	96.22	78.53	75.83
	Shoveling sand	61.17	57.76	67.28	61.05	74.34	66.57	69.00	74.26	68.23	74.09	67.38

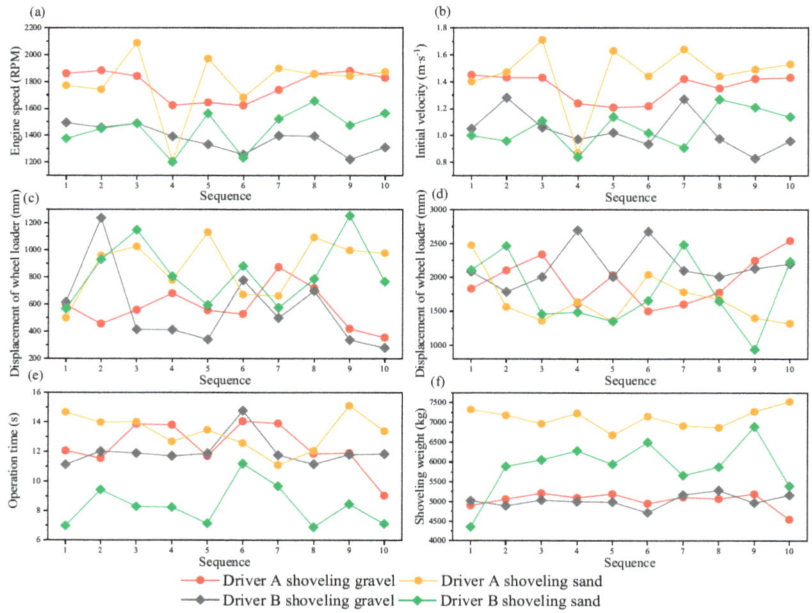

Figure 14. Parameters of the shoveling process for driver A and driver B. (**a**) Engine speed; (**b**) Initial velocity; (**c**) Displacement of wheel loader in insertion phase; (**d**) Displacement of wheel loader in scooping phase; (**e**) Operation time; (**f**) Shoveling weight.

The average fuel consumption for the validation experiments of the three optimized trajectories decreased by 29.54%, 19.84%, and 42.33%, respectively, compared to driver A's control of shoveling gravel and by 30.39%, 21.42%, and 43.46%, respectively, compared to driver B. The average fuel consumption of the validation experiments for the three optimized trajectories decreased by 33.63%, 33.60%, and 23.76%, respectively, compared to driver A shoveling sand, and by 29.99%, 29.96%, and 19.58%, respectively, compared to driver B. The average fuel consumption for shoveling gravel with optimized trajectory with three engine speeds is 53.69 mL, which is 26.05% and 29.20% lower compared to Driver A and Driver B, respectively. And the average fuel consumption of shoveling sand with optimized trajectory with three engine speeds is 52.47 mL; compared with Driver A and Driver B, the average fuel consumption is reduced by 26.44% and 22.12%, respectively. The average fuel consumption for manually controlled shoveling of gravel is 74.22 mL, and the average fuel consumption for manually controlled shoveling of sand is 69.35 mL. The fuel consumption can be, on average, reduced by 27.66% when shoveling gravel with an optimized trajectory and by 24.34% when shoveling sand with an optimized trajectory. In conclusion, compared with the manually controlled shoveling operation, the shoveling operation, according to the optimized trajectory, can effectively reduce fuel consumption of shoveling operation.

In order to test the significance of the reduction in fuel consumption for the optimized shoveling trajectory, a *t*-test is chosen to test it in this paper. The samples are selected from the fuel consumption of the optimized shoveling trajectory and the fuel consumption of drivers A and B shoveling gravel and sand, respectively. The *t*-test allows us to obtain statistical conclusions about the significance of the difference between the means of the two groups of samples and to determine their level of statistical significance. First, the hypothesis is established. The null hypothesis H_0 is that there is no significant difference between the fuel consumption data of the optimized shoveling trajectory and the fuel consumption data of the manually controlled shoveling. The alternative hypothesis H_1 is that the fuel consumption data from the optimized shoveling trajectory performs significantly better than the fuel consumption data from the manually controlled shoveling, as shown in Equation (22). After that, the significance level needs to be determined. In this paper, the significance level α is chosen as 0.05. Finally, the T-value is calculated and compared with the critical value of the two-sided test, and the formula for calculating the T-value is shown in Equation (23). The fuel consumption of shoveling gravel with optimized shoveling trajectory is 5.02 and 5.50 compared to driver A and B, respectively, while the fuel consumption of shoveling sand with optimized shoveling trajectory is 5.80 and 5.83 compared to driver A and B, respectively, with all T-value greater than the critical value $t_{0.05/2}(17) = 2.11$. Therefore, the null hypothesis can be rejected, and the alternative hypothesis is selected. Thus, we can conclude that the fuel consumption with an optimized shoveling trajectory performs significantly better than the fuel consumption with manually controlled shoveling. At the restricted significance level, fuel consumption is significantly reduced with the optimized shoveling trajectory.

$$\begin{cases} H_o : \mu_1 = \mu_2 \\ H_1 : \mu_1 > \mu_2 \end{cases} \tag{22}$$

where μ_1 is the overall mean value of fuel consumption for manually controlled shoveling, and μ_2 is the overall mean value of fuel consumption for optimized shoveling trajectory.

$$T = \frac{\overline{x}_1 - \overline{x}_2}{\sqrt{\frac{s_1^2}{n_1} + \frac{s_2^2}{n_2}}} \tag{23}$$

where $\overline{x}_1$ is the sample mean value of fuel consumption of manually controlled shoveling, $\overline{x}_2$ is the sample mean value of fuel consumption of optimized shoveling trajectory, s_1 is the sample standard deviation of fuel consumption of manually controlled shoveling, s_2 is the sample standard deviation of fuel consumption of optimized shoveling trajectory, n_1 is the sample number of fuel consumption of manually controlled shoveling, n_2 is the sample number of fuel consumption of optimized shoveling trajectory.

5.2. Discussion

This paper is based on the constructed automatic operation platform of the wheel loader. The mathematical model of fuel consumption is established by SVM. And the optimized shoveling trajectory with three different engine speeds in the gravel pile and the sand pile is obtained by combining IPSO. By comparing with the fuel consumption of manually controlled shoveling, the fuel consumption with the optimized shoveling trajectory has been significantly reduced. In this study, only the effect of different engine speeds and different shoveling trajectories on fuel consumption has been considered. There are other influencing factors for fuel consumption. The main factors include the driver's operating ability and the age of the wheel loader. The driver's operating ability will directly affect the fuel consumption, varying from two to three times for different drivers [42]. The influence of manual operation on fuel consumption mainly includes the selection of insertion depth [43], the bucket attitude control during shoveling [44], and the cooperation between the gas pedal and the brake pedal [45]. The power possessed by the wheel loader is different at different engine speeds. When the engine speed is low, if the insertion is too deep, the bucket will get stuck in the pile and cause the tires to slip, increasing fuel consumption. Moreover, if the bucket is turned too much during the shoveling process, the resistance to the bucket will surge, which requires additional fuel consumption. It is worth noting that the shoveling process is not a uniform speed process, which needs a moderate increase and decrease, that is, better cooperation between the gas pedal and the brake pedal. For the age of the wheel loader, it is sure that with the increase in use time, there will be different degrees of wear between the mechanisms, and fuel efficiency will be reduced. In particular, the bucket tip of the bucket, as the part that contacts the pile, suffers the most wear during the long-term shoveling process [46]. Timely replacement or repair of bucket tips is an effective way to reduce fuel consumption. Automatic shoveling of wheel loaders has undoubtedly become the mainstream development trend in the future [47,48], and automatic shoveling means a more standard operation mode, higher fuel economy, and a corresponding reduction in wear for wheel loaders. At the same time, human safety is ensured to a large extent. Most importantly, automatic shoveling is an effective way to reduce fuel consumption. Azulay et al. [49] developed a controller based on deep reinforcement learning. They conducted experiments on a shoveling robot with three degrees of freedom, which included lift, turn, and speed, and the shoveling efficiency of the robot with this controller was significantly higher than that of the robot with manual control. Dadhich et al. [50] proposed a time-lag neural network-based shoveling operation method for wheel loaders was conducted, and several experiments were successful, with only 26% longer operation time and improved fuel economy for the same shoveling weight. Huang et al. [51] established an automatic shoveling model for wheel loaders by Q-learning, which could reduce fuel consumption by 8.0% and 10.6%, respectively, compared to manual operation by two drivers. Our study promotes the development of automatic shoveling by planning a reasonable shoveling trajectory for it. Moreover, it demonstrates excellent advantages and potential compared to manually controlled shoveling.

In this study, we consider only the fuel consumption with an optimized shoveling trajectory. However, the shoveling weight is the critical indicator for evaluating the efficiency of wheel loaders. The importance of shoveling weight and fuel consumption varies for different shoveling environments. It is essential to recognize that higher shoveling weights are not better. If the wheel loader is overloaded for a long time, it may lead to greater wear on the wheel loader and reduce its service life. In conclusion, the weighting between shoveling weight and fuel consumption must be determined in actual operation. In the future, we will optimize the shoveling trajectory for the actual shoveling environment to balance fuel consumption and shoveling weight. In the meantime, we have only tested in gravel piles and sand piles. Although these two materials are highly representative, we plan to extend them to other applications, such as large rocks. As the material size and irregularities increase, the bucket will shake significantly during shoveling and may even deviate from the planned shoveling trajectory. We must ensure that this deviation is within an acceptable range. Of course, the control of the shoveling process will also become a new challenge.

6. Conclusions

In this study, the shoveling trajectories of gravel and sand are optimized at different engine speeds. The corresponding optimized trajectories were obtained with the optimization objective of reducing the fuel consumption of shoveling operations. The results show that the optimized shoveling trajectory effectively reduces the fuel consumption of shoveling operation, and the main conclusions are as follows.

1. In this study, the factors with a strong influence on the fuel consumption of shoveling operations are screened by experimental analysis and KPCA. The factors with a strong influence are initial velocity, the wheel loader displacement in the insertion phase, the shoveling weight, the operation time, and the wheel loader displacement in the scooping phase. KPCA results show that the cumulative contribution rate of the above five factors can reach 90%. Meanwhile, the grey relation coefficients between key factors of fuel consumption influence and fuel consumption are all greater than 0.75. It proves that the screening of key factors of fuel consumption influenced by KPCA is effective.
2. In this study, the fuel consumption model is established by SVM, and the penalty factor and kernel function parameters in SVM are optimized by CV. The SVM fuel consumption model has high stability and generalization ability, and the deviation between the predicted and true values of the model is small. In the gravel samples, its ARE is 4.91%, RMSE is 5.91, and R^2 is 0.9764. While in the sand samples, the ARE is

3.49%, RMSE is 3.08, and R^2 is 0.9494. The deviation between the true values and the predicted values from the SVM fuel consumption model is small.
3. In this study, the optimized shoveling trajectories are obtained by IPSO for shoveling gravel and shoveling sand at three engine speeds. Validation experiments are conducted based on the optimized trajectory and compared with the sample data fuel consumption and manually controlled shoveling fuel consumption. The results show that the optimized trajectory can significantly reduce the fuel consumption of shoveling operations. For shoveling gravel, the average fuel consumption of the validation experiment is reduced by 60.93%, 45.79%, and 49.10%, respectively, compared with the sample average fuel consumption with three engine speeds. For shoveling sand, the average fuel consumption in the validation experiment is reduced by 30.93%, 35.56%, and 28.59%, respectively, compared with the sample average fuel consumption with three engine speeds. Compared with manually controlled shoveling gravel and sand, fuel consumption can be reduced by 27.66% and 24.34%, respectively.

Author Contributions: Conceptualization, Y.C. and G.S.; methodology, Y.C.; software, G.S.; validation, G.S.; formal analysis, Y.C.; investigation, C.T.; resources, G.S.; data curation, G.S.; writing—original draft preparation, G.S.; writing—review and editing, Y.C.; visualization, C.T.; supervision, Y.C.; project administration, Z.W.; funding acquisition, Y.C. and Z.W. All authors have read and agreed to the published version of the manuscript.

Funding: This research was funded by the National Natural Science Foundation of China (61962007), the Key Projects of Guangxi Natural Science Foundation (2018GXNSFDA294001), and the S&T Fund of Guangxi Province (1598021-2).

Institutional Review Board Statement: Not applicable.

Informed Consent Statement: Not applicable.

Data Availability Statement: The data in this study are available from the corresponding authors upon request.

Conflicts of Interest: The authors declare no conflict of interest.

References

1. Zhang, G.J.; Xiao, C.Y. Dynamic simulation analysis on loader's working device. *Aust. J. Mech. Eng.* **2018**, *16*, 2–8.
2. Xu, J.; Yoon, H.S. A review on mechanical and hydraulic system modeling of excavator manipulator system. *J. Constr. Eng.* **2016**, *2016*, 9409370. [CrossRef]
3. Koizumi, T.; Tsujiuchi, N.; Yoshida, T.; Andou, H. Evaluation process of digging performance for hydraulic excavator by bucket tip trace. *J. Syst. Des. Dyn.* **2011**, *5*, 1005–1016. [CrossRef]
4. Wang, X.; Sun, W.; Li, E.; Song, X. Energy-minimum optimization of the intelligent excavating process for large cable shovel through trajectory planning. *Struct. Multidisc. Optim.* **2018**, *58*, 2219–2237. [CrossRef]
5. Sarata, S.; Koyachi, N.; Sugawara, K. Field test of autonomous loading operation by wheel loader. In Proceedings of the 2008 IEEE/RSJ International Conference on Intelligent Robots and Systems, Nice, France, 22–28 September 2008.
6. Hong, Q.G.; Chen, Y.H.; Huai, Z.P. Experimental Study on Automatic Shovel Loading of Loader. *Coal Mine. Mach.* **2021**, *42*, 195–197.
7. Xu, L.C.; Ge, R.H. Optimization design of shovel depth when loader shovelling original raw soil. *TELKOMNIKA Indones. J. Electr. Eng.* **2013**, *11*, 4639–4645.
8. Gong, J.; Cui, Y.X. Track planning for a wheel Loader in a digging. *J. Mech. Eng.* **2009**, *45*, 29–34. [CrossRef]
9. Zhang, H.; Cao, S.J.; Kazuhiko, S. Test Investigation on Autonomous Scooping Control Strategy of Wheel Loader. *Appl. Mech. Mater.* **2010**, *42*, 156–159. [CrossRef]
10. Filla, R.; Frank, B. Towards finding the optimal bucket filling strategy through simulation. In Proceedings of the 15th Scandinavian International Conference on Fluid Power, Linköping, Sweden, 7–9 June 2017.
11. Meng, Y.; Fang, H.Z.; Liang, G.D.; Gu, Q.; Liu, L. Bucket trajectory optimization under the automatic scooping of LHD. *Energies* **2019**, *12*, 3919. [CrossRef]
12. Yu, X.J.; Huai, Y.H.; Li, X.F.; Wang, D.W.; Yu, A. Shoveling trajectory planning method for wheel loader based on kriging and particle swarm optimization. *J. Jilin Univ. (Eng. Technol. Ed.)* **2020**, *50*, 437–444.
13. Osumi, H.; Uehara, T.; Okada, N. Efficient scooping of rocks by autonomous controlled wheel loader. *J. Rob. Mechatron.* **2012**, *24*, 924–932. [CrossRef]

14. Zhang, T.; Fu, T.; Song, X. Multi-objective excavation trajectory optimization for unmanned electric shovels based on pseudospectral method. *Autom. Constr.* **2022**, *136*, 104176. [CrossRef]
15. Shen, W.; Jiang, J.; Su, X.; Reza Karimi, H. Control strategy analysis of the hydraulic hybrid excavator. *J. Frankl. Inst. Eng. Appl. Math.* **2015**, *352*, 541–561. [CrossRef]
16. Frank, B.; Kleinert, J.; Filla, R. Optimal control of wheel loader actuators in gravel applications. *Autom. Construct.* **2018**, *91*, 1–14. [CrossRef]
17. Yao, J.; Edson, C.P.; Yu, S. Bucket Loading Trajectory Optimization for the Automated Wheel Loader. *IEEE Trans. Veh. Technol.* **2023**, *72*, 6948–6958. [CrossRef]
18. Liang, G.; Liu, L.; Meng, Y. Dynamic Modeling and Analysis of Loader Working Mechanism Considering Cooperative Motion with the Vehicle Body. *Machines* **2022**, *11*, 9. [CrossRef]
19. Brinkschulte, L.; Hafner, J.; Geimer, M. Real-time load determination of wheel loader components. *ATZheavy Duty Worldw.* **2019**, *12*, 62–68. [CrossRef]
20. Wang, S.; Yin, Y.; Wu, Y.; Hou, L. Modeling and Verification of an Acquisition Strategy for Wheel Loader's Working Trajectories and Resistance. *Sensors* **2022**, *22*, 5993. [CrossRef] [PubMed]
21. Trani, M.L.; Bossi, B.; Gangolells, M.; Casals, M. Predicting fuel energy consumption during earthworks. *J. Clean. Prod.* **2016**, *112*, 3798–3809. [CrossRef]
22. Ma, W.X.; Zhang, Y.B.; Liu, C.B.; Wang, S.L. Prediction method of the fuel consumption of wheel loaders in the V-type loading cycle. *Math. Prob. Eng.* **2015**, *2015*, 538176. [CrossRef]
23. Chen, Y.; Jiang, H.; Shi, G.; Zheng, T. Research on the Trajectory and Operational Performance of Wheel Loader Automatic Shoveling. *Appl. Sci.* **2022**, *12*, 12919. [CrossRef]
24. Dindarloo, S.R.; Siami-Irdemoosa, E. Determinants of fuel consumption in mining trucks. *Energy* **2016**, *112*, 232–240. [CrossRef]
25. Siami-Irdemoosa, E.; Dindarloo, S.R. Prediction of fuel consumption of mining dump trucks: A neural networks approach. *Appl. Energy* **2015**, *151*, 77–84. [CrossRef]
26. Alamdari, S.; Basiri, M.H.; Mousavi, A.; Soofastaei, A. Application of machine learning techniques to predict haul truck fuel consumption in open-pit mines. *J. Min. Environ.* **2022**, *13*, 69–85.
27. Gong, J.; Shang, J.; Li, L.; Zhang, C.; He, J.; Ma, J. A Comparative Study on Fuel Consumption Prediction Methods of Heavy-Duty Diesel Trucks Considering 21 Influencing Factors. *Energies* **2021**, *14*, 8106. [CrossRef]
28. Shen, Y.C.; Song, H.; Li, G.Q.; Cherouat, A. Analysis and Prediction of Excavator Energy Consumption Based on Improved PSO-BP Neural Network. In Proceedings of the 12th International Workshop of Advanced Manufacturing and Automation, Xiamen, China, 11–12 October 2012.
29. Liu, Y.H.; Zhang, Z.D.; Cai, Z.G. Research on the prediction of working fuel consumption for wheel loader based on BP neural network. *Constr. Mach.* **2018**, *503*, 81–83.
30. Zhao, Y.; Zhou, M.; Wang, L. Bearing Fault Diagnosis of Single-Channel Data by a 3D DCN with Bilinear LBP and Modified KPCA. *J. Electr. Eng. Technol.* **2023**. [CrossRef]
31. Wang, Y.C.; Yang, S.J.; Guo, Q.P.; Yin, Y. Prediction of Blasting Muckpile Morphology in Throw Blasting of Coal Mine Based on MPA-SVM. *Explos. Mater.* **2023**, *52*, 58–64.
32. Yuasa, T.; Ishikawa, M.; Ogawa, S. An Optimal Design Methodology for the Trajectory of Hydraulic Excavators Based on Genetic Algorithm. *J. Rob. Mechatron.* **2021**, *33*, 1248–1254. [CrossRef]
33. Bi, Q.; Wang, G.; Wang, Y.; Yao, Z.; Hall, R. Digging Trajectory Optimization for Cable Shovel Robotic Excavation Based on a Multi-Objective Genetic Algorithm. *Energies* **2020**, *13*, 3118. [CrossRef]
34. Zhao, X.; Jia, M. A new local-global deep neural network and its application in rotating machinery fault diagnosis. *Neurocomputing* **2019**, *366*, 215–233. [CrossRef]
35. Lu, X.; Yang, C.; Wu, Q.; Wang, J.; Lu, Z.; Sun, S.; Liu, K.; Shao, D. Research on Analog Circuit Soft Fault Diagnosis Method Based on Mathematical Morphology Fractal Dimension. *Electronics* **2023**, *12*, 184. [CrossRef]
36. Zhou, T.; Lu, H.L.; Wang, W.W.; Yong, X. GA-SVM based feature selection and parameter optimization in hospitalization expense modeling. *Appl. Soft Comput.* **2019**, *75*, 323–332.
37. Patange, A.D.; Pardeshi, S.S.; Jegadeeshwaran, R. Augmentation of Decision Tree Model through Hyper-Parameters Tuning for Monitoring of Cutting Tool Faults Based on Vibration Signatures. *J. Vib. Eng. Technol.* **2022**, *10*, 1–19. [CrossRef]
38. Bajaj, N.S.; Patange, A.D.; Jegadeeshwaran, R. Application of metaheuristic optimization based support vector machine for milling cutter health monitoring. *Intell. Syst. Appl.* **2023**, *18*, 200196. [CrossRef]
39. Jain, M.; Saihjpal, V.; Singh, N.; Singh, S.B. An Overview of Variants and Advancements of PSO Algorithm. *Appl. Sci.* **2022**, *12*, 8392. [CrossRef]
40. Han, D.; Yang, X.; Li, G. Highway traffic speed prediction in rainy environment based on APSO-GRU. *J. Adv. Transp.* **2021**, *2021*, 4060740. [CrossRef]
41. Lin, J.; Sheng, G.; Yan, Y.; Dai, J.; Jiang, X. Prediction of Dissolved Gas Concentrations in Transformer Oil Based on the KPCA-FFOA-GRNN Model. *Energies* **2018**, *11*, 225. [CrossRef]
42. Ahluwalia, R.K.; Wang, X.; Star, A.G. Performance and cost of fuel cells for off-road heavy-duty vehicles. *Int. J. Hydrogen Energy* **2022**, *47*, 10990–11006. [CrossRef]

43. Cao, B.; Liu, X.; Chen, W. Intelligent energy-saving operation of wheel loader based on identifiable materials. *J. Mech. Sci. Technol.* **2020**, *34*, 1081–1090. [CrossRef]
44. Li, J.; Chen, C.; Li, Y. Difficulty assessment of shoveling stacked materials based on the fusion of neural network and radar chart information. *Autom. Constr.* **2021**, *132*, 103966. [CrossRef]
45. Fei, X.; Han, Y.; Wong, S.V. An Overview of and Prospects for Research on Energy Savings in Wheel Loaders. *Automot. Exper.* **2023**, *6*, 133–148. [CrossRef]
46. Dong, Z.; Jiang, F.; Tan, Y.; Wang, F.; Ma, R.; Liu, J. Review of the Modeling Methods of Bucket Tooth Wear for Construction Machinery. *Lubricants* **2023**, *11*, 253. [CrossRef]
47. Xiao, W.; Liu, M.; Chen, X. Research Status and Development Trend of Underground Intelligent Load-Haul-Dump Vehicle—A Comprehensive Review. *Appl. Sci.* **2022**, *12*, 9290. [CrossRef]
48. Dadhich, S.; Bodin, U.; Andersson, U. Key Challenges in Automation of Earth-Moving Machines. *Autom. Constr.* **2016**, *68*, 212–222. [CrossRef]
49. Azulay, O.; Shapiro, A. Wheel loader scooping controller using deep reinforcement learning. *IEEE Access* **2021**, *9*, 24145–24154. [CrossRef]
50. Dadhich, S.; Sandin, F.; Bodin, U. Field test of neural-network based automatic bucket-filling algorithm for wheel-loaders. *Autom. Constr.* **2019**, *97*, 1–12. [CrossRef]
51. Huang, J.; Kong, D.; Gao, G.; Cheng, X.; Chen, J. Data-Driven Reinforcement-Learning-Based Automatic Bucket-Filling for Wheel Loaders. *Appl. Sci.* **2021**, *11*, 9191. [CrossRef]

applied sciences

MDPI

Article

Area Division Using Affinity Propagation for Multi-Robot Coverage Path Planning

Nikolaos Baras * and Minas Dasygenis

Department of Electrical and Computer Engineering, University of Western Macedonia, 50100 Kozani, Greece
* Correspondence: nbaras@uowm.gr

Abstract: In the wake of advancing technology, autonomous vehicles and robotic systems have burgeoned in popularity across a spectrum of applications ranging from mapping and agriculture to reconnaissance missions. These practical implementations have brought to light an array of scientific challenges, a crucial one among them being Coverage Path Planning (CPP). CPP, the strategic planning of a path that ensures comprehensive coverage of a defined area, while being widely examined in the context of a single-robot system, has found its complexity magnified in the multi-robot scenario. A prime hurdle in multi-robot CPP is the division and allocation of the operation area among the robots. Traditional methods, largely reliant on the number of robots and their initial positions to segment the space, often culminate in suboptimal area division. This deficiency can occasionally render the problem unsolvable due to the sensitivity of most area division algorithms to the robots' starting points. Addressing this predicament, our research introduced an innovative methodology that employs Affinity Propagation (AP) for area allocation in multi-robot CPP. In our approach, the area is partitioned into 'n' clusters through AP, with each cluster subsequently assigned to a robot. Although the model operates under the assumption of an unlimited robot count, it offers flexibility during execution, allowing the user to modify the AP algorithm's similarity function factor to regulate the number of generated clusters. Serving as a significant progression in multi-robot CPP, the proposed model provides an innovative approach to area division and path optimization, thereby setting a strong foundation for future exploration and practical enhancements in this field.

Keywords: affinity propagation; area allocation; coverage path planning

Citation: Baras, N.; Dasygenis, M. Area Division Using Affinity Propagation for Multi-Robot Coverage Path Planning. *Appl. Sci.* **2023**, *13*, 8207. https://doi.org/10.3390/app13148207

Academic Editor: Jonghoek Kim

Received: 3 June 2023
Revised: 4 July 2023
Accepted: 10 July 2023
Published: 14 July 2023

1. Introduction

Over the past few decades, we have witnessed an exponential advancement in technology that has fundamentally reshaped the global landscape [1]. With a broad gamut ranging from telecommunications and computing to artificial intelligence and robotics, these technological breakthroughs have drastically altered the ways in which humans operate and perform tasks [1–3]. Once labor-intensive or monotonous tasks are now executed with unparalleled speed, precision, and efficiency, largely due to automation and the advent of intelligent systems.

A predominant player in this transformational journey has been the field of robotics. Robots, once confined to the realms of science fiction, now pervade numerous sectors, such as manufacturing [4], healthcare [5], agriculture [6–8], and logistics [9], serving as invaluable tools for enhancing productivity and improving the quality of services. In disaster management, robots have been instrumental in executing search-and-rescue missions in environments too hazardous for human intervention [10,11]. In agriculture, robotic systems are leveraged for tasks ranging from seeding to harvesting, contributing significantly to precision farming.

However, as with any emergent technology, the rapid rise of robotics has posed a unique set of challenges, requiring innovative solutions. A prominent challenge in the domain of robotics is Coverage Path Planning (CPP). In essence, CPP is the process of developing a route or path that allows a robot to cover an entire operational area in an efficient manner [12,13]. A vital aspect of numerous applications, CPP is integral to tasks such as field inspection in agriculture [14–16], where a robotic system needs to cover a farm field to assess crop health.

The complexity of CPP intensifies when we move from single-robot systems to multi-robot systems. Multi-robot CPP involves devising paths for multiple robots to ensure comprehensive and efficient coverage of a larger or more complex area [17–26]. A pressing issue in multi-robot CPP is the division of the operational area among the robots, a problem known as the 'area division problem'. For example, consider a team of drones deployed for large-scale environmental monitoring or a fleet of autonomous vehicles performing a search-and-rescue mission in a disaster-stricken area. The optimal allocation of specific regions to individual robots can drastically enhance the coverage efficiency and operational coordination, reducing redundancies and saving valuable time. Addressing the area division problem effectively is paramount for the successful deployment of multi-robot systems across a range of real-world applications. As such, it remains a vibrant area of research, inviting novel solutions and methodologies that can meet the evolving demands of modern robotic systems.

Several researchers have proposed models to resolve the multi-robot CPP problem, yet they often encounter significant limitations. A prevalent limitation of multi-robot CPP algorithms lies in the arbitrary selection of the number of robots and their starting locations. Both these factors wield substantial influence over the efficacy of the algorithm. For example, an inadequate number of robots may yield incomplete coverage, resulting in overlooked regions. Conversely, deploying an excessive number of robots can render the task inefficient and unnecessarily complicated, thus elevating the cost and computational burden. Likewise, the robots' initial placement crucially impacts the effectiveness of the algorithm, dictating the distribution of robots and the thoroughness of area coverage. Another major constraint is the limited scalability of the algorithms, referring to the capability of an algorithm to perform optimally as the number of inputs or variables increases. In the context of multi-robot CPP, scalability pertains to the algorithm's ability to manage expansive environments and multiple robots cooperating to execute a task.

This paper presents a significant contribution to the field of multi-robot CPP by introducing a novel model for area division predicated on the Affinity Propagation (AP) algorithm. The fundamental premise of the proposed approach is to deconstruct the multi-robot CPP problem into a collection of single-robot CPP tasks, which may or may not be interdependent. While numerous models in the literature adopt a similar divide-and-conquer strategy, our model distinguishes itself in three vital respects:

(i) The proposed approach assigns territory to robots without arbitrarily defining the number of robots (and hence the number of areas). The AP clustering algorithm leverages a weighted similarity index (SI) function to discern similar cells and ascertain the optimal number of robots for the multi-robot CPP task. This effectively liberates the user from the responsibility of determining the appropriate number of robots for the task. To our knowledge, we are the first to incorporate an AP algorithm to address the multi-robot CPP problem.

(ii) The algorithm proposed in this paper presents a novel perspective on multi-robot area division, where the number of robots and their respective areas are not predesignated arbitrarily by the user. A common drawback of many existing algorithms is their inability to handle scenarios where the robots' initial positions are in close proximity, leading to inefficient area division and path planning. Unlike these approaches, our proposed algorithm demonstrates exceptional resilience to such situations. It intelligently circumvents the constraints of predefined robot counts and positions, thereby offering a more flexible and efficient solution for area division in multi-robot coverage path planning.

(iii) The similarity function of the AP algorithm takes into consideration several factors, such as layer type, the spatial connectivity of cells, and their "normalized distance", before incorporating them into a cluster. This results in reduced area blending rates, meaning that the cells within the final clusters created by this method are less likely to be enveloped by cells from other clusters.

The structure of the remaining manuscript is as follows:

Section 2 of the document introduces similar works from the literature and highlights their advantages and disadvantages. Following that, in Section 3, the formal problem definition is provided. The subsequent section, Section 4, elaborates on the algorithm we propose. Section 5 presents the experimental data and techniques employed to assess the effectiveness of our algorithm. The algorithm's limitations, possible performance improvements, and the usage of parallel computing are discussed in Section 6. Finally, Section 7 contains concluding remarks.

2. Literature Review

This section will delve into research endeavors in the literature that confront the multi-robot coverage path planning problem, particularly emphasizing area division and allocation. The objective of this section is to furnish a comprehensive review of extant algorithms and methodologies employed to mitigate the problem of multi-robot coverage. An analysis, comparison, and critical evaluation of related works were conducted to highlight their respective merits and demerits, thereby establishing a foundation for assessing the scientific contribution and novelty of the model proposed in this paper and identifying lacunae in the prevailing state-of-the-art.

A widely recognized offline multi-robot CPP algorithm is the one proposed by Tang et al. [21] using the MSTC* framework. This algorithm primarily aims to generate coverage paths for multiple robots while considering realistic physical constraints such as obstacles and communication paths among robots. The MST technique, employed to divide the target environment into smaller sub-regions, forms the backbone of the algorithm. Robots are then allocated to these sub-areas according to their capabilities and workload requirements. This algorithm possesses the advantage of addressing physical constraints, a pivotal aspect in real-world scenarios where robots need to maneuver through complex and dynamic environments. However, the algorithm does not consider environmental uncertainties, leading to sub-optimal performance in volatile and dynamic situations. Further, it predetermines the number of robots, thereby deciding the number of sub-areas. It is notable that Tang et al.'s problem formulation aligns with the problem formulated in this paper in terms of area division and allocation, although the specific problem details the authors address are distinct.

Another noteworthy offline multi-robot CPP algorithm is the one developed by Rahman et al. [24] for autonomous radiation mapping using a mobile robot. The primary objective of this algorithm is to create coverage paths that a single robot can traverse to conduct radiation mapping in a designated area. The method draws from a genetic algorithm, utilized to generate a multitude of potential coverage paths, and, subsequently, the optimal path is selected based on criteria such as coverage efficiency and path length. The flexibility of this algorithm lies in its adaptability; it can be modified to cater to varying radiation mapping scenarios. Moreover, the algorithm's computational efficiency makes it suitable for extensive application. However, it uses the K-means clustering technique to partition the overall space into smaller sub-spaces, implying that the algorithm necessitates a completely arbitrary selection of the number of robots.

The DARP algorithm [18] represents a notable approach in the domain of multi-robot coverage path planning. It offers a systematic solution by dividing the total environment into distinct sub-areas, each allocated to a specific robot. The primary objective of DARP is to minimize the total coverage time, accomplished by intelligently dividing the environment based on its characteristics and the robotic fleet's capabilities. However, the algorithm makes predetermined assumptions about the number of robots and their initial positions, leading to potential limitations in more complex environments.

An interesting approach was proposed by Idir et al. [19], who suggested a multi-robot CPP algorithm based on the DARP algorithm. The authors sought to tackle the problem of weight allocation for multi-robot coverage using a weighted approach. The algorithm employs a grid-based representation of the environment and segregates the environment into a collection of cells necessitating coverage. The robots are assigned weights and the DARP algorithm is utilized to distribute these weights among the robots and determine their coverage paths. The strengths of this approach encompass handling the weight allocation problem, thereby enhancing coverage performance, and the ability to manage large-scale scenarios. Despite its improvements over the original DARP algorithm, it retains the limitation of being unable to find a solution when the initial positions of the robots are proximate.

Another DARP-based algorithm was proposed in [25]. The authors presented an A*-modified DARP algorithm. This modified version of the DARP algorithm assigns tasks to the appropriate robot and based on an Up-First approach the Spanning Trees are constructed in order to ensued full coverage of the initial area. The authors claim that, compared to the original DARP algorithm, their modifications yield higher efficiency and a higher coverage rate.

A different approach for multi-robot CPP was proposed in [26] that uses Ant Colony Optimization. The authors introduced an improved Ant Colony Optimization (ACO) algorithm for single-robot CPP, which optimizes the energy and time consumption by building the best possible Spanning Tree (ST) and, consequently, an optimal path. For multi-robot scenarios, the study employed the DARP algorithm [18], dividing the total area into smaller, equally sized sections, thus simplifying the complex computation. Each subarea then constructs a spanning tree using the improved ACO. In the final stage, the end nodes are shared among subareas to develop ideally-shaped spanning trees that minimize the number of turns in the coverage path. The algorithms are proven to be near polynomial, and simulation results highlight benefits including complete coverage, no backtracks, minimum path length, zero preparation time, and the least number of turns. However, the implementation of this approach still suffers from the drawbacks of the DARP algorithm, meaning the arbitrary choice of robot's initial positions.

Given the literature review, it is evident that multi-robot CPP has garnered considerable attention from researchers. However, a direct evaluation and comparison of these research works is challenging as each paper addresses a subtly different problem. A commonality that most offline multi-robot CPP algorithms share, including those mentioned above, is their reliance on arbitrarily defining the number of robots and sensitivity to the initial environmental conditions. Further research is imperative to devise new multi-robot CPP algorithms that are devoid of these limitations, are more efficient, and tailored for real-world applications. The Affinity Propagation algorithm proposed in this paper demonstrates the potential to address these issues, and thus presents a significant advancement in this field.

3. Problem Definition

The effective division of a given environment into distinct sub-areas for the deployment of multiple robots presents a challenging problem, particularly in the context of ensuring contiguous access within each sub-region. To lay a solid foundation for our investigation, we embarked on a mathematical formulation of this problem, succinctly capturing the essential aspects and constraints involved. This mathematical model serves as an unambiguous representation of the underlying problem, illuminating its core elements and thus guiding the development of algorithmic solutions.

Our environment is described as a binary matrix representation, defining accessible areas and obstacles. We considered robots that are assigned to different sections of this environment, each section described as a 'sub-area'. We also incorporated the distinct nature of the terrain, assigning different types to each cell in the environment. The fundamental goal was to find a valid division of the environment into sub-areas, each assigned to a unique robot, with the sub-areas satisfying a set of constraints related to accessibility and continuity. The mathematical model elucidated below presents a precise description of the problem, facilitating further discussion and solution design.

Given:

- An environment A of size $X \times Y$ represented as a binary matrix, $A = [a_{i,j}]$, where $a_{i,j} \in \{0,1\}$ for all $1 \leq i \leq X$ and $1 \leq j \leq Y$. In the environment matrix A, $a_{i,j} = 0$ signifies an obstacle, whereas $a_{i,j} = 1$ represents an accessible cell. By definition, robots can only traverse accessible cells.
- A set of robots $R = \{r_1, r_2, \ldots, r_n\}$, where n is the number of robots.
- A matrix $T = [t_{i,j}]$, where, that describes the type of each cell $a_{i,j}$.
- A matrix $E = [e_{i,j}]$, where, that describes the elevation level of each cell $a_{i,j}$.
- An elevation weight value E_W which represents the elevation weight importance factor, and a floor type F_W which represents the floor type elevation weight importance factor.
- A matrix $T_w = [w_1, w_2, \ldots, w_n]$ where w_i represents the weight assigned to the i th floor type. The number of values in T_w is equal to the number of different floor types. Based on this information, we seek:
- A set of sub-areas $S = \{s_1, s_2, \ldots, s_n\}$, where each sub-area s_i is a contiguous partition of A assigned to robot r_i. Formally, we define the following constraints:
- Each sub-area s_i maintains 4-neighbor continuity for all its accessible cells (Figure 1), i.e., for each pair of accessible cells c_{m_1,n_1} and c_{m_2,n_2} in s_i, there exists a sequence of accessible cells $c_{m_k,n_k}, k = 1, \ldots, p$, such that $c_{m_1,n_1} = c_{m_1,n_1}, c_{m_p,n_p} = c_{m_2,n_2}$, and c_{m_k,n_k} is a 4-neighbor of $c_{m_{k+1},n_{k+1}}$ for all $k = 1, \ldots, p-1$. The 4-neighbor criterion specifies that connectivity between cells must be either horizontal or vertical—not diagonal. Thus, in essence, every cell in the sequence is horizontally or vertically adjacent to its successor, for all values of k ranging from 1 to $p-1$.
- This condition ensures a continuity or chain of 4-neighbor connections between any two accessible cells within a given sub-area, thereby preserving the rule of 4-neighbor connectivity throughout the entire grid.
- The environment A is the union of all sub-areas S, i.e., $A = \cup_{i=1}^{n} s_i$.
- The intersection of any two distinct sub-areas s_i and s_j for $i \neq j$ is empty, i.e., $s_i \cap s_j = \varnothing$ for all $i \neq j$.
- The problem is to find a bijective function $f: R \rightarrow S$ such that $f(r_i) = s_i$ for all $1 \leq i \leq n$, fulfilling the constraints mentioned above.

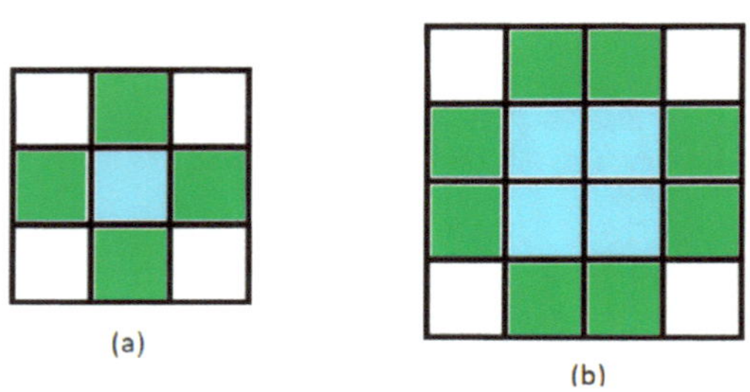

Figure 1. Image (**a**) depicts a cluster that consists of a single cell (blue cell). Green cells represent the cells that have 4-neighbor connectivity with the blue cell. Image (**b**) depicts a larger environment along with its 4-neighbors. Essentially, 4-neighbor connectivity between cells prevents a robot from moving diagonally within the environment.

4. The Proposed Algorithm

4.1. Data Initialization

To develop a strong solution for the multi-robot CPP problem, we introduced a new algorithm based on AP [27] (Figure 2). AP is a powerful clustering algorithm commonly used for dividing and assigning areas. In our approach, we rely on AP to group data points into meaningful clusters. The algorithm operates by exchanging messages containing real values among the data points. These messages help the algorithm determine the best number of clusters and how to assign the data points to those clusters. The algorithm continues this message exchange process until a consensus is reached on the optimal cluster configuration. This methodology allows us to bypass the necessity to predetermine the number of robots or the dimensions of the respective sub-areas arbitrarily. As opposed to conventional applications, we employed a meticulous transformation process to adapt our grid-like environment, A, into a dataset suitable for AP (Figure 3).

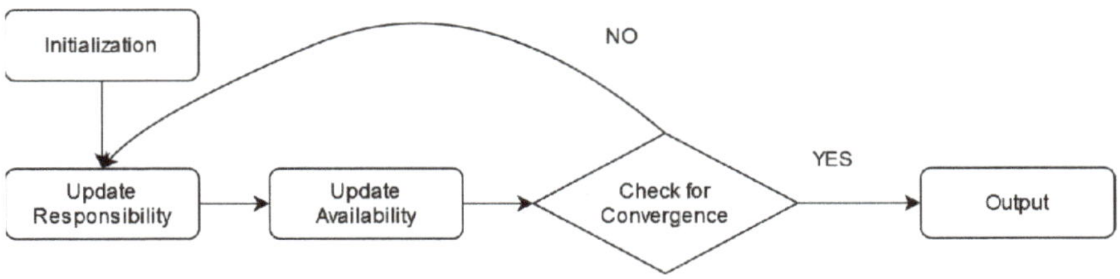

Figure 2. The initialization step prepares the data, converts the grid to a set of data points, and calculates the similarity matrix S. Then, the responsibility matrix R is updated. The responsibility matrix reflects how well suited a data point is to serve as the exemplar for the other data points. Next, the Availability matrix is updated. This matrix reflects how suitable an exemplar is for each data point to serve as its exemplar. Finally, the algorithm iterates until the maximum number of iterations set, or until there are no changes from the last iteration.

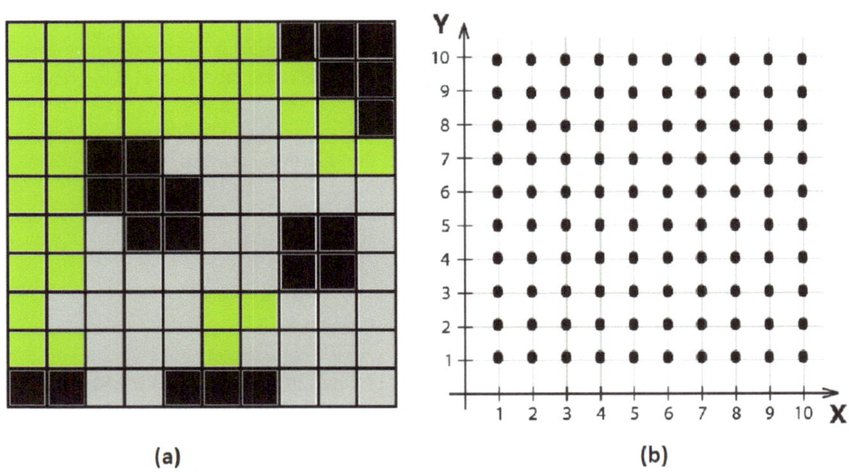

Figure 3. The initial environment (**a**) consists of obstacles (black cells), cells of type grass (green cells), and cells of type asphalt (gray cells). In order to divide the environment into multiple sub-areas, we first have to convert the environment into a set of data points. (**b**) depicts the data points in a X, Y Cartesian coordinate system.

4.2. Calculating Normalized Distance

In this process, each cell within the initial environment is mapped onto a data point in a two-dimensional coordinate system. To discern the relationship between data points, we define a similarity function. In many standard applications of AP, similarity is gauged by the negative of Euclidean distance. This inversely proportional relationship implies a smaller similarity for larger absolute distances between data points.

However, the application of Euclidean distance to our context of CPP within a two-dimensional grid, where we aim to preserve each cluster's continuity and its cell's 4-neighbor relationship, is not fitting. This metric fails to account for obstacles that may disrupt the path between two locations. This discrepancy between Euclidean distances and actual traversable distances can lead to significant errors, with the disparity being exacerbated by the presence of numerous obstacles.

Understanding the inherent constraints of the traditional approach, we adapted our similarity function to include a more tailored and nuanced model (Algorithm 1).

Algorithm 1: Calculating the normalized 4-neighbor distance between two points

Input: Binary matrix A with dimensions X by Y, starting point (x_1, y_1), and target point (x_2, y_2)
Output: minimum distance D between the starting and target points
Function 4_neighbor_actual_distance $(A,\ x_1, y_1,\ x_2,\ y_2)$:
    Step 1: Initialize a distance matrix D with dimensions X by Y, set all elements to infinity
    Step 2: Initialize a queue Q
    Step 3: Set $D\,[x_1, y_1] = 0$ and add (x_1, y_1) to Q
    Step 4: While Q is not empty:
        Step 4.1: Dequeue a point (x, y) from Q
        Step 4.2: Loop through each of its four neighbors $(x_n,\ y_n)$ in the environment A:
            Step 4.2.1: If $(x_n,\ y_n)$ is an obstacle $(A[x_n,\ y_n] = 0)$, skip this neighbor
            Step 4.2.2: If $D[x_n,\ y_n] > \mathrm{D}[x,\ y] + 1$:
            Step 4.2.2.1: Update $D[x_n,\ y_n] = D[x,\ y] + 1$
            Step 4.2.2.2: Add $(x_n,\ y_n)$ to Q
    Step 5: Return $D[x_2,\ y_2]$ as the minimum distance D between the starting and target points
End Function

Our novel algorithm not only calculates the minimum 4-neighbor distance between cells, similar to a BFS approach, but also takes into account the type of floor and elevation of each cell, thereby capturing crucial information about the landscape's unique characteristics. It thereby ensures a more accurate and practical representation of the environment's traversability. Incorporating these parameters directly into the similarity function provides a more realistic framework for area division in 2D grid environments. This ultimately leads to enhanced efficiency and effectiveness in our CPP solutions, as it ensures more prudent and strategic allocation of sub-areas to robots, factoring in complex grid conditions that could impact their performance.

4.3. Calculating the Similarity Matrix

In the similarity matrix, each element holds a singular metric quantifying its resemblance to the other elements in the adjacent vicinity (Algorithm 2). The degree of similarity between any two elements augments in direct proportion with the increase in the similarity value. It warrants highlighting that, due to the intrinsic characteristics of the AP method, the similarity function $S(p_1, p_2)$ may not necessarily be identical to $S(p_2, p_1)$. Although this directional feature is not incorporated in the current implementation of our proposed methodology (Figure 4), potential future adaptations of the algorithm may consider its integration to facilitate directional clustering of cells and other specific elements. Utilizing the AP methodology, we classified data points into distinct sub-regions for each robot, following an assessment of the similarity quotient between each pair of data points. The AP algorithm operates through the transmission of messages that denote a data point's proclivity towards a particular cluster. Subsequent to the resolution on the number of clusters and the allocation of data points to respective clusters, these messages are subject to iterative refinement.

Algorithm 2: The procedure that calculates the weighted Similarity Matrix S

```
Input:
    Environment matrices: binary A, weight W, elevation E, floor F
    Significance multipliers: t, q, r
Output: similarity matrix S
Function similarity_calculation(A, X, Y, W):
    Step 1: Initialize an X by Y matrix S
    Step 2: Loop through each pair of data points (x1, y1) and (x2, y2) in the environment A:
        Step 2.1: Calculate the distance between the data points:
            d = 4_neighbor_actual_distance (x1, y1, x2, y2)
        Step 2.2: Calculate the elevation difference
            e = |E[x1, y1] − E[x2, y2]|
        Step 2.3: Calculate the floor discrepancy
            f = |F[x1, y1] − F[x2, y2]|
        Step 2.2: Multiply each metric by its weight factor:
            d = t × d
            e = q × e
            f = r × f
        Step 2.3: Store the similarity between the data points in the similarity matrix S:
            S[x1, y1, x2, y2] = −(d + e + f)
    Step 3: Return the similarity matrix S
End Function
```

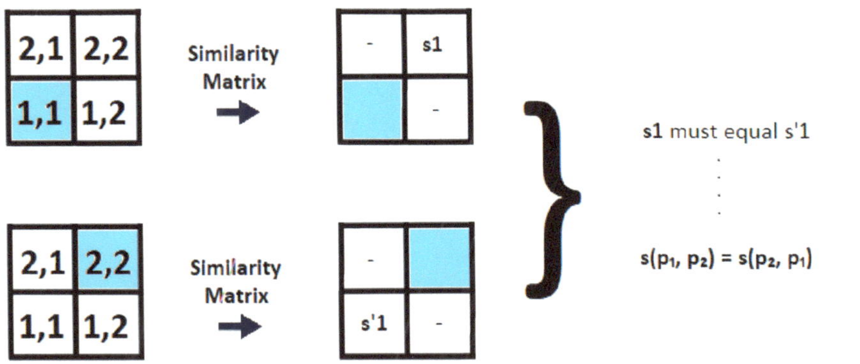

Figure 4. Two points (denoted with blue color) will always have the same similarity matrix with each other. In practice, this means that going from point p_1 to point p_2 has the same cost as going from point p_2 to p_1.

4.4. Generation of Clusters

After calculating the similarity between all pairs of data points, the AP algorithm was used to cluster the data points into sub-areas for each robot. The AP algorithm works by passing messages between data points, which indicate their preference for a particular cluster. These messages are updated iteratively until a consensus is reached on the number of clusters and which points belong to which cluster.

Upon the conclusion of the iterative message-passing phase, the AP algorithm proceeds towards the establishment of sub-areas (Algorithm 3). This crucial stage determines the most suitable exemplar for each cluster—a data point that accrues the maximum preference value when both availability and responsibility are taken into account. Consequently, each data point (cell) is assigned to the exemplar of its respective cluster.

Algorithm 3: Generating the sub-areas using AP

1. Input: grid-like area A with dimensions X by Y
2. Output: sub-areas for each robot
3. Step 1: Convert the grid-like area A into a set of data points
4. Represent each grid cell as a data point in a two-dimensional coordinate system
5. Step 2: Calculate the similarity between every pair of data points
6. Calculate the similarity based on Algorithms 1 and 2 (taking into consideration the weight factor)
7. Store the similarity in a matrix S
8. Step 3: Initialize messages between data points
9. Initialize two matrices, R and A, to store the messages between data points
10. Initialize the self-similarity matrix, S, to store the similarity between a data point and itself
11. Step 4: Iterate until convergence
12. Update the responsibility matrix, R
13. Update the availability matrix, A
14. Step 5: Identify exemplars
15. Identify the data points with the highest responsibility and availability values as exemplars
16. Step 6: Assign sub-areas to each robot
17. Assign each non-exemplar data point to the closest exemplar (only if they are spatially connected using the 4-neighbor scheme)
18. Group the data points assigned to each exemplar into a sub-area
19. Step 8: Return the sub-areas for each robot

A key attribute of the proposed algorithm lies in its capacity to autonomously ascertain the optimal number of clusters (sub-areas), without any need for user-defined inputs. This ability to self-regulate cluster formation ensures flexibility and adaptability, which is particularly beneficial in complex real-world applications.

The final output of the algorithm is a list of cluster labels for each data point in the initial environment, indicating the respective robot assignment for each cell (Figure 5). This clustered set of data points serves as the foundation for subsequent stages of the proposed model, leading to the final path planning for the robots.

Figure 5. The initial environment as depicted in (**a**) contains two layer types (denoted by green and gray cells). The output of the algorithm in (**b**) shows the two clusters (blue and yellow), as they were generated by the algorithm. It is worth mentioning that different values of the importance value T_w may result in slightly different clusters.

5. Experimental Results

In this section, we present the findings obtained from our in-depth analysis of the proposed algorithm. The experimental results validate the efficiency and adaptability of our algorithm in effectively decomposing the environment into suitable sub-areas. The evaluation metrics employed primarily centered on the criteria of computational time and the quality of the generated clusters.

In terms of comparison with existing algorithms, it is important to highlight that our proposed AP-based algorithm addresses a unique problem in the field of multi-robot CPP. In the literature, many algorithms attempt to solve the issue of dividing an initial area into multiple sub-areas for each robot. However, the identical problem that our algorithm solves, without arbitrarily pre-determining the number of robots and their initial positions, is lacking in the current literature. Nevertheless, we embarked on an indirect comparative analysis, contrasting our algorithm's performance with traditional algorithms for area division in CPP, which rely on pre-specified robot counts and locations. The comparative study primarily emphasized the improvements in efficiency, flexibility, and adaptability introduced by the proposed algorithm. Despite the inherent differences in the problem contexts, the comparative analysis provided a clear demonstration of the strides made by the AP-based approach, particularly in situations where the initial robot positions are in close proximity.

The conducted simulations predominantly operated in two meticulously constructed environments of distinct sizes, each featuring distinct characteristics. Both environments were generated using a pseudo-random process, offering a unique combination of accessible areas and obstacles along with varying environmental types, ensuring the robustness of the simulated scenarios.

The smaller 24 × 24 environment served as an essential proving ground for our proposed algorithm, presenting a grid with a variety of unique parameters and diverse characteristics. The total area consisted of 576 cells, with approximately 80% of the grid being accessible and approximately 20% assigned as obstacles. These obstacles were evenly scattered throughout the grid to simulate potential hindrances that robots might encounter in real-world scenarios. Beyond mere accessibility, cells in the environment were characterized by two distinct terrain types—grass and asphalt. The grass cells accounted for roughly 60% of the accessible area, while the asphalt cells constituted the remaining 40%. This bifurcation of terrain types aimed to emulate real-world environments where robots might encounter varied terrains requiring different path planning strategies and navigation capabilities. To augment the realism of the simulation, cells were assigned varying elevation levels. The elevations ranged from 0 to 10 units, with a standard deviation of three units to ensure a substantial variation in elevation across the grid. An elevation weight factor of 0.1 was applied to indicate the importance of elevation. Similarly, the type of terrain also had an associated weight factor to indicate the importance of terrain type.

Subsequently, a second, larger, and more complex environment was introduced for a more in-depth simulation. The dimensions of this environment are 100 × 100. The rest of the parameters were the same as those of the previous smaller environment. Due to its substantial size, detailed visualization was rendered impractical. Nevertheless, we present empirical data and statistics to illustrate the algorithm's performance. This larger grid serves to emulate more complex real-world scenarios, thereby demonstrating the scalability and adaptability of the proposed AP algorithm in diverse, challenging situations.

The computational experiments were conducted on a dedicated testbed configured to ensure accurate and consistent results. The hardware setup encompassed a high-performance workstation equipped with an Intel Core i7-9700K 8-core processor clocked at 3.60 GHz, bolstered with 32 GB of DDR4 RAM and enabling efficient data handling and manipulation, which were particularly critical given the size of the datasets and complexity of operations involved in Affinity Propagation.

Table 1 shows the experimental results for each setup. For each algorithm, we conducted 20 experiments using the aforementioned parameters. The table shows the average values for the number of generated clusters and the cluster quality.

Table 1. Experimental results for two environments using three clustering algorithms for CPP.

Environment Size	Algorithm	Generated Clusters	Cluster Quality
24 × 24	[18] n = 2	2	0.66
24 × 24	[19] n = 2	2	0.84
24 × 24	[25] n = 2	2	0.89
24 × 24	Proposed	2.19	0.91
100 × 100	[18] n = 2	2	0.39 (often failed to find solution)
100 × 100	[19] n = 2	2	0.85
100 × 100	[25] n = 2	2	0.92
100 × 100	Proposed	6.52	0.94

It is worth mentioning that it is difficult to evaluate the proposed algorithm by directly comparing it with others found in the literature, since these algorithms are not configurable to identify the different cell types, elevation, and importance factor. Therefore, we could only compare the results of these algorithms by taking into account the quality of the clustering. The quality of clustering was calculated using the Silhouette Coefficient (SC) [28].

The SC, also known as Silhouette Score, is a well-established and popular metric for evaluating the quality of a clustering algorithm. The essence of this method lies in its dual-faceted measurement approach, quantifying both cohesion and separation simultaneously for each individual cell. It operates by comparing the average distance of a cell to all other points within its own cluster (cohesion) against the average distance to points in the nearest cluster (separation). The coefficient thus provides an aggregate measure of how similar a given data point is to its own cluster relative to other clusters. Higher values of the SC suggest that the cell is well-clustered and lower values imply that the specific cell might have been better assigned to a neighboring cluster. It is widely regarded well due to its intuitive interpretation, its capability to work with any distance metric (in our example the normalized distance as presented in Algorithm 1), and its agnosticism towards the specific clustering algorithm used. To properly evaluate the proposed algorithm, we calculated the SC not only for the average distances but also for the similarity with regard to elevation and floor type. The total SC calculated was weighted based on the respective weights of floor type and elevation.

A more nuanced evaluation of the proposed algorithm was achieved by conducting multiple runs in the same environment but with varied importance factors attached to floor type and elevation. This exploratory approach aimed to investigate the algorithm's capability to adapt and respond to shifts in preference and the importance of environmental features. For this experiment, the main focus was on gauging the homogeneity of the resulting clusters. Homogeneity here was defined as the proportion of a cluster that exhibits uniformity in terms of either floor type or elevation. When the importance factor assigned to floor type was modified, we anticipated observing clusters that largely contain the same floor type. Consequently, the algorithm's sensitivity to floor type would be reflected by the degree of homogeneity of the resulting clusters. Analogously, when the emphasis is shifted to elevation, the homogeneity of the clusters, in terms of their elevation, becomes the pivotal measure of algorithm performance. The experimental results are presented in Table 2.

The experimental results indicate that the importance factors affect the final clusters and their cells. It is important, however, to fine tune the exact values for each multi-robot CPP task, based on the capabilities of the available robots. A visual representation of a 10 × 10 environment and the output of the algorithm with different importance factor T_W are depicted in Figure 6. A larger 24 × 24 environment is depicted in Figure 7.

Table 2. Experimental results from the execution of the proposed algorithm in the same environment using different elevation E_w and floor type F_W weight factors. These values range from 0 to 1 and describe how important the preservation of elevation or floor type within a generated sub-area is. Increasing or decreasing both values at the same time reduces the importance of both variables within the clustering process.

Environment	Importance Factors (E_w, F_W)	Height Homogeneity	Floor Homogeneity
100 × 100	(0, 0)	0.61	0.49
100 × 100	(0.8, 0)	0.75	0.45
100 × 100	(0.8, 0.8)	0.59	0.53
100 × 100	(0, 0.8)	0.53	0.78

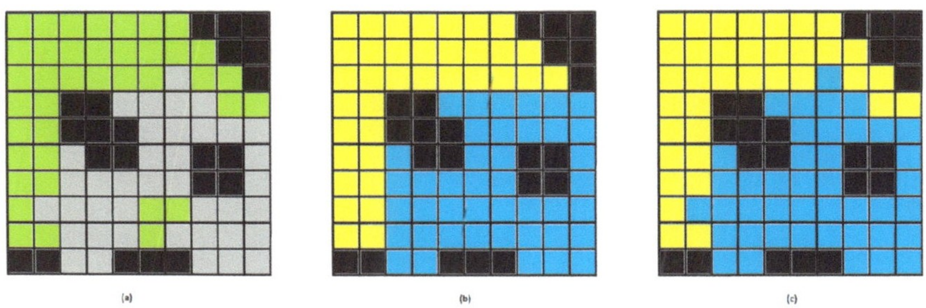

Figure 6. The initial environment as depicted in (**a**) contains two layer types (green and gray). The second image (**b**) shows the output clusters of the algorithm (blue and yellow) with an importance factor $T_W = 0$. The third image (**c**) shows the output of the algorithm for the same input environment where the importance factor T_W is equal to 0.8. Increasing the importance factor of floor type increases the likelihood that the algorithm will consider two cells of the same type to be more similar.

Figure 7. An example environment with dimensions 24 × 24. The initial environment (**a**) contains two layer types (green and gray). The second image (**b**) shows the output clusters of the algorithm (blue and yellow) with an importance factor $T_W = 0.3$.

6. Discussion

6.1. Limitations

While this paper proposed an innovative application of AP in the field of multi-robot CPP, it is not without limitations. A clear understanding of these potential constraints is essential for refining the algorithm, enhancing its applicability, and identifying areas for future research.

The primary limitation arises from the AP's inherent computational complexity. With a time complexity of $O(N^{2T})$, where N represents the number of data points and T denotes the number of iterations, AP can be computationally expensive for large-scale environments. This computational cost is primarily due to the calculation of the similarity matrix and the iterative exchange of "responsibility" and "availability" messages. This renders the algorithm less practical for real-time operations, particularly for larger environments, as it may lead to increased latency in area division and subsequent path planning.

Another important limitation is the static nature of the AP algorithm, which is very hard if not impossible to adjust for dynamic environments. The AP algorithm, as applied in this context, assumes a static environment where the position of obstacles and the type of each cell are known beforehand. In scenarios where the environment changes over time, the algorithm would need to be rerun, potentially leading to delays and inefficiencies. The ability to adapt to dynamic environments remains a significant challenge in the field of multi-robot CPP and represents a key area for future research.

6.2. Performance Improvement

The performance of the AP algorithm, as is the case with most computational procedures, is pivotal in real-world applications. The urgency for efficiency and execution speed improvements is even more pronounced when dealing with multi-robot coverage path planning, given the scale of the task and the inherent complexity associated with environment mapping and path planning. In light of this, several strategies can be considered to enhance the execution speed and overall efficiency of the AP algorithm.

One crucial step of the AP algorithm is the calculation of the similarity matrix. Given that this phase accounts for a significant portion of the computations (approximately 40%), improving its efficiency is imperative. Given its intrinsic parallelizable nature, where the similarity between each pair of data points can be calculated independently, we could potentially harness the power of parallel computing. By distributing the calculation of similarity measures across multiple cores or nodes in a parallel computing environment, we can expedite this process markedly, thereby increasing the overall efficiency of the AP algorithm.

On the other hand, the message passing phase of the AP algorithm, which is pivotal for its iterative structure, is more challenging to parallelize. Although in theory, each "responsibility" and "availability" message update could be computed in parallel, the iterative nature of the AP algorithm necessitates the results of each preceding iteration. Nonetheless, we could explore certain forms of "soft" parallelization, such as utilizing vectorized operations or parallel map functions provided by high-level languages and libraries. Even though this approach would not offer true parallelization due to each iteration still needing to await the completion of all message updates, it could still provide substantial speed improvements.

Besides parallel computing, other potential strategies for improving the efficiency of the AP algorithm could include optimization of the algorithm's parameters or the application of hardware accelerators such as Graphics Processing Units (GPUs) [29]. Fine-tuning parameters like the damping factor or the preference value could potentially reduce the number of iterations required for convergence, thereby accelerating the execution speed. Similarly, using GPUs, which are particularly suited for parallelizable tasks, could lead to substantial reductions in computation time. However, such strategies would require careful evaluation to balance efficiency gains against the potential impact on the quality of the results.

7. Conclusions

The presented research introduces a paradigm shift in the domain of multi-robot CPP by utilizing AP for optimally dividing the operational area among the robots. Instead of using traditional methods, which largely rely on the number of robots and their initial positions, this innovative methodology partitions the area into 'n' clusters using AP and subsequently assigns each cluster to a robot. This model, while functioning under the assumption of an unlimited number of robots, provides a unique flexibility by allowing the modification of the AP algorithm's similarity function factor to control the number of generated clusters.

One field where the proposed algorithm can find profound applications is precision agriculture. This industry, already substantially automated, requires precision farming, which involves the distribution of multiple tasks, such as seeding, fertilizing, and harvesting, across a fleet of robotic entities. Identifying the optimal number of sub-areas becomes paramount to prevent overlap and redundancy in operations. The proposed algorithm, by facilitating automatic partitioning of farmland into sub-areas based on factors such as crop type and topography paves the way for improved resource management. It ensures optimal task distribution amongst autonomous agricultural machines, enhancing their overall operational efficiency, thereby contributing to a significant reduction in the time and cost associated with agricultural practices.

Additionally, this algorithm can significantly revolutionize urban search and rescue operations. Typically, these operations are time-sensitive, requiring the division of large, affected areas into smaller manageable sub-areas to enable quick and efficient search strategies. The conventional method of dividing areas based on available rescuers may not be effective, especially in scenarios where the rescuers are robotic entities. By employing this algorithm, we could efficiently partition the search area into the appropriate number of sub-areas (clusters) regardless of the number of robots, optimizing the search strategy and increasing the likelihood of successful rescue operations. Moreover, with the AP algorithm's adaptable similarity function factor, the rescue team has flexibility in regulating the number of generated clusters, facilitating a more efficient and coordinated search operation.

As a significant progression in multi-robot CPP, this methodology paves the way for novel research directions and practical enhancements in this field. The capability to deliver effective area division and path optimization, without burdening the user with the arbitrary decision of the number of robots or their initial positions, sets a new benchmark for multi-robot CPP implementations. Moreover, our work provides a strong foundation for the development of enhanced strategies that can address the existing complexities of multi-robot CPP and further expedite the deployment of autonomous systems in diverse fields ranging from agriculture to reconnaissance missions. Future work will focus on refining the proposed model, incorporating more complex environmental factors, and exploring the potential of integrating this method with different path planning algorithms for better performance.

Author Contributions: Conceptualization, N.B. and M.D.; methodology, N.B.; writing—original draft preparation, N.B.; writing—review and editing, N.B.; visualization, N.B.; supervision, M.D.; funding acquisition, N.B. All authors have read and agreed to the published version of the manuscript.

Funding: This research was carried out as part of the project «Smart Safe Navigation for Electric Bicycles and Skateboards» (Project code:KMP6-0292520) under the framework of the Action «Investment Plans of Innovation» of the Operational Program «Central Macedonia 2014 2020», that is co-funded by the European Regional Development Fund and Greece.

Institutional Review Board Statement: Not applicable.

Informed Consent Statement: Not applicable.

Data Availability Statement: Not applicable.

Conflicts of Interest: The authors declare no conflict of interest.

References

1. Mourtzis, D.; Angelopoulos, J.; Panopoulos, N. A Literature Review of the Challenges and Opportunities of the Transition from Industry 4.0 to Society 5.0. *Energies* **2022**, *15*, 6276. [CrossRef]
2. Kagermann, H.; Wahlster, W. Ten Years of Industrie 4.0. *Science* **2022**, *4*, 26. [CrossRef]
3. Demir, K.A.; Döven, G.; Sezen, B. Industry 5.0 and human-robot co-working. *Procedia Comput. Sci.* **2019**, *158*, 688–695. [CrossRef]

4. Wang, E.Z.; Lee, C.C.; Li, Y. Assessing the impact of industrial robots on manufacturing energy intensity in 38 countries. *Energy Econ.* **2022**, *105*, 105748. [CrossRef]
5. Kyrarini, M.; Lygerakis, F.; Rajavenkatanarayanan, A.; Sevastopoulos, C.; Nambiappan, H.R.; Chaitanya, K.K.; Makedon, F. A survey of robots in healthcare. *Technologies* **2021**, *9*, 8. [CrossRef]
6. Oliveira, L.F.; Moreira, A.P.; Silva, M.F. Advances in agriculture robotics: A state-of-the-art review and challenges ahead. *Robotics* **2021**, *10*, 52. [CrossRef]
7. Aivazidou, E.; Tsolakis, N. Transitioning towards human–robot synergy in agriculture: A systems thinking perspective. *Syst. Res. Behav. Sci.* **2022**, *40*, 536–551. [CrossRef]
8. Marinoudi, V.; Sørensen, C.G.; Pearson, S.; Bochtis, D. Robotics and labour in agriculture. A context consideration. *Biosyst. Eng.* **2019**, *184*, 111–121. [CrossRef]
9. Shamout, M.; Ben-Abdallah, R.; Alshurideh, M.; Alzoubi, H.; Kurdi, B.A.; Hamadneh, S. A conceptual model for the adoption of autonomous robots in supply chain and logistics industry. *Uncertain Supply Chain. Manag.* **2022**, *10*, 577–592. [CrossRef]
10. Jorge, V.A.M.; Granada, R.; Maidana, R.G.; Jurak, D.A.; Heck, G.; Negreiros, A.P.F.; dos Santos, D.H.; Gonçalves, L.M.G.; Amory, A.M. A Survey on Unmanned Surface Vehicles for Disaster Robotics: Main Challenges and Directions. *Sensors* **2019**, *19*, 702. [CrossRef]
11. Angelopoulos, G.; Baras, N.; Dasygenis, M. Secure autonomous cloud brained humanoid robot assisting rescuers in hazardous environments. *Electronics* **2021**, *10*, 124. [CrossRef]
12. Dogru, S.; Marques, L. ECO-CPP: Energy constrained online coverage path planning. *Robot. Auton. Syst.* **2022**, *157*, 104242. [CrossRef]
13. Patle, B.K.; Pandey, A.; Parhi, D.R.K.; Jagadeesh, A.J.D.T. A review: On path planning strategies for navigation of mobile robot. *Def. Technol.* **2019**, *15*, 582–606. [CrossRef]
14. Moysiadis, V.; Sarigiannidis, P.; Vitsas, V.; Khelifi, A. Smart farming in Europe. *Comput. Sci. Rev.* **2021**, *39*, 100345. [CrossRef]
15. Utamima, A.; Reiners, T.; Ansaripoor, A.H. Optimisation of agricultural routing planning in field logistics with Evolutionary Hybrid Neighbourhood Search. *Biosyst. Eng.* **2019**, *184*, 166–180. [CrossRef]
16. Moysiadis, V.; Tsolakis, N.; Katikaridis, D.; Sørensen, C.G.; Pearson, S.; Bochtis, D. Mobile Robotics in Agricultural Opera-tions: A Narrative Review on Planning Aspects. *Appl. Sci.* **2020**, *10*, 3453. [CrossRef]
17. Almadhoun, R.; Taha, T.; Seneviratne, L.; Zweiri, Y. A survey on multi-robot coverage path planning for model reconstruction and mapping. *SN Appl. Sci.* **2019**, *1*, 847. [CrossRef]
18. Kapoutsis, A.C.; Chatzichristofis, S.A.; Kosmatopoulos, E.B. DARP: Divide areas algorithm for optimal multi-robot coverage path planning. *J. Intell. Robot. Syst.* **2017**, *86*, 663–680. [CrossRef]
19. Idir, O.; Renzaglia, A. Multi-Robot Weighted Coverage Path Planning: A Solution based on the DARP Algorithm. In Proceedings of the 2022 17th International Conference on Control, Automation, Robotics and Vision (ICARCV), Singapore, 11–13 December 2022; pp. 98–104.
20. Huang, X.; Sun, M.; Zhou, H.; Liu, S. A multi-robot coverage path planning algorithm for the environment with multiple land cover types. *IEEE Access* **2020**, *8*, 198101–198117. [CrossRef]
21. Tang, J.; Sun, C.; Zhang, X. MSTC: Multi-robot Coverage Path Planning under Physical Constrain. In Proceedings of the 2021 IEEE International Conference on Robotics and Automation (ICRA), Xi'an, China, 30 May–5 June 2021; pp. 2518–2524.
22. Collins, L.; Ghassemi, P.; Esfahani, E.T.; Doermann, D.; Dantu, K.; Chowdhury, S. Scalable coverage path planning of multi-robot teams for monitoring non-convex areas. In Proceedings of the 2021 IEEE International Conference on Robotics and Automation (ICRA), Xi'an, China, 30 May–5 June 2021; pp. 7393–7399.
23. Qin, Y.; Fu, L.; He, D.; Liu, Z. Improved Optimization Strategy Based on Region Division for Collaborative Multi-Agent Coverage Path Planning. *Sensors* **2023**, *23*, 3596. [CrossRef]
24. Abd Rahman, N.A.; Sahari KS, M.; Hamid, N.A.; Hou, Y.C. A coverage path planning approach for autonomous radiation mapping with a mobile robot. *Int. J. Adv. Robot. Syst.* **2022**, *19*, 17298806221116483. [CrossRef]
25. Huang, Y.; Li, M.; Zhao, T. A Multi-robot Coverage Path Planning Algorithm Based on Improved DARP Algorithm. *arXiv* **2023**, arXiv:2304.09741.
26. Gao, C.; Kou, Y.; Li, Z.; Xu, A.; Li, Y.; Chang, Y. Optimal multirobot coverage path planning: Ideal-shaped spanning tree. *Math. Probl. Eng.* **2018**, *2018*, 3436429. [CrossRef]
27. Frey, B.J.; Dueck, D. Clustering by passing messages between data points. *Science* **2007**, *315*, 972–976. [CrossRef] [PubMed]
28. Dinh, D.T.; Fujinami, T.; Huynh, V.N. Estimating the optimal number of clusters in categorical data clustering by silhouette coefficient. In *Knowledge and Systems Sciences, Proceedings of the 20th International Symposium 2019, KSS 2019, Da Nang, Vietnam, 29 November–1 December 2019*; Springer: Singapore, 2019.
29. Hijma, P.; Heldens, S.; Sclocco, A.; Van Werkhoven, B.; Bal, H.E. Optimization Techniques for GPU Programming. *ACM Comput. Surv.* **2023**, *239*, 81. [CrossRef]

applied sciences

MDPI

Article

CERRT: A Mobile Robot Path Planning Algorithm Based on RRT in Complex Environments

Kun Hao, Yang Yang, Zhisheng Li *, Yonglei Liu and Xiaofang Zhao

School of Computer and Information Engineering, Tianjin Chengjian University, Tianjin 300384, China; kunhao@tcu.edu.cn (K.H.); yhpl01@163.com (Y.Y.); sanxiong_1@163.com (Y.L.); zhaoxftju@tju.edu.cn (X.Z.)

* Correspondence: lzs@tcu.edu.cn

Abstract: In complex environments, path planning for mobile robots faces challenges such as insensitivity to the environment, low efficiency, and poor path quality with the rapidly-exploring random tree (RRT) algorithm. We propose a novel algorithm, the complex environments rapidly-exploring random tree (CERRT), to address these issues. The CERRT algorithm builds upon the RRT approach and incorporates two key components: a pre-allocated extension node method and a vertex death mechanism. These enhancements aim to improve vertex utilization and overcome the problem of becoming trapped in concave regions, a limitation of traditional algorithms. Additionally, the CERRT algorithm integrates environment awareness at collision points, enabling rapid identification and navigation through narrow passages using local simple sampling techniques. We also introduce the bidirectional shrinking optimization strategy (BSOS) based on the pruning optimization strategy (POS) to further enhance the quality of path solutions. Extensive simulations demonstrate that the CERRT algorithm outperforms the RRT and RRV algorithms in various complex environments, such as mazes and narrow passages. It exhibits shorter running times and generates higher-quality paths, making it a promising approach for mobile robot path planning in challenging environments.

Keywords: path planning; RRT; path optimization; complex environments

Citation: Hao, K.; Yang, Y.; Li, Z.; Liu, Y.; Zhao, X. CERRT: A Mobile Robot Path Planning Algorithm Based on RRT in Complex Environments. *Appl. Sci.* **2023**, *13*, 9666. https://doi.org/10.3390/app13179666

Academic Editor: Jonghoek Kim

Received: 4 August 2023
Revised: 25 August 2023
Accepted: 25 August 2023
Published: 26 August 2023

1. Introduction

Path planning is a crucial research field in the robotics industry. Its purpose is to find a safe, collision-free path for a mobile robot to traverse from its starting position to its destination in a specified area that contains obstacles [1]. It has widespread applications in complex environments such as urban roads, factory production lines, and outdoor exploration. Currently, path planning mainly uses algorithms based on search, heuristics, and sampling. Among them, sampling-based path planning algorithms have become a research hotspot due to their wide applicability, ease of implementation, and lack of need to construct complex structures [2].

In the class of sampling-based algorithms, the rapidly-exploring random tree (RRT) algorithm, which is widely used, can avoid complex space constructions by implementing a collision check module, making it suitable for solving high-dimensional or multi-constraint planning problems [3]. However, the efficiency of the RRT algorithm is typically affected when it faces complex environments such as multiple obstacles, mazes, narrow passages, and concave traps. In recent years many researchers have proposed improved RRT algorithms to address these issues. For instance, Kuffner et al. propose a straightforward and efficient bidirectional random tree algorithm, denoted as RRT-Connect [4], alternately expands two trees to improve the algorithm's efficiency. However, its performance still suffers in complex environments. Tahirovic et al. introduced a rapid exploration algorithm named Rapid Random Vine (RRV) [5] for efficient exploration. It determines the local environment type using principal component analysis (PCA), a dimensionality reduction technique that captures the most significant variations in the data. By analyzing the relationships among environmental features, RRV selects an appropriate direction in which to expand, thereby solving the narrow passage problem well. However, its performance is poor in complex environments without passages. Hsu et al. [6] augmented the Rapidly-Exploring Random Tree (RRT) algorithm with a bridge-testing technique, which increased the sampling probability in narrow passages and decreased the sampling probability in non-interest regions, thereby addressing narrow passage issues. However, the unevenness of the sampling probability may cause the algorithm to ignore some feasible paths. Wu et al. proposed the Fast-RRT [7] algorithm, which detects narrow passages by re-randomizing the expansion direction at collision points, but the algorithm's stability is poor. Cai et al. combined RRV with bridge testing [8], enabling efficient identification and expansion in complex environments without the need for additional collision detection, greatly reducing computational intensity, but the algorithm can generate a large number of useless vertices in open areas. Building upon RRV and RRT-Connect, Li et al. proposed an adaptive random tree algorithm called ARRT-Connect [9], which effectively improves the algorithm's performance, but the algorithm may fall into concave traps. Chi et al. [10] introduce a heuristic path-planning algorithm based on the Generalized Voronoi Diagram (GVD), which significantly improves the algorithm's performance in maze environments but requires preprocessing of the map and does not consider the narrow passage problem. Taheri et al. proposed a Fuzzy Greedy Rapidly Exploring Random Tree (FG-RRT) [11] algorithm, which significantly reduces computation time in maze, narrow passage, and convex obstacle environments, but the algorithm requires the setting of nine fuzzy rules, and the parameter settings are complex.

To address the problem of low-quality generated paths, the RRT* algorithm was introduced by Karaman et al. [12], which introduced the ChooseParent and Rewire processes when adding new nodes to the tree, making the algorithm asymptotically optimal. As the number of iterations tends to infinity, the probability of finding the optimal solution approaches 100%. RRT* is a milestone in the development of RRT. To improve the convergence speed of the RRT* algorithm, numerous scholars have conducted extensive research, mainly optimizing the sampling, ChooseParent, and Rewire processes of RRT*. Islam et al. put forth an intelligent sampling tree named RRT*-Smart [13] to expedite the convergence rate of the algorithm, but the quality of the generated path depends largely on the initial solution. Inspired by node exclusion, Gammell et al. [14] employ a direct sampling method within the hyperellipsoid to enhance algorithm performance, but the algorithm is no longer applicable when the ellipsoid is larger than the planning domain. P-RRT* [15] combines APF and RRT* to provide feasible directions for sampling exploration, which speeds up the convergence speed. Jeong et al. improved the ChooseParent and Rewire procedures using the triangle inequality to propose the Quick-RRT* [16], which generates better initial paths and faster convergence. Inspired by Quick-RRT*, F-RRT* [17] creates a parent node near the obstacle for each sampled point, obtaining better initial solutions and faster convergence speed than Quick-RRT* and RRT* under the same conditions. Although algorithms based on the RRT* framework can find the optimal or approximate optimal solution, they all require a large number of samples to gradually search for the optimal path. Therefore, when the important parameter for an algorithm is its running speed, optimizing the path directly generated by RRT is necessary. To optimize the initial path generated by RRT, Qian et al. [18] proposed a method to optimize the initial path generated by RRT by merging trees based on the initial path to form a closed-loop path and then performing optimization to obtain the relatively optimal path. Chen et al. [19] introduced a bidirectional pruning optimization strategy that prunes redundant nodes from both the starting and ending points of the path and selects the shortest optimized path, effectively improving the quality of the path.

In conclusion, extensive research has been conducted on path planning utilizing the RRT algorithm in complex environments. However, no algorithm currently exists that effectively and simply addresses the dual issues of subpar performance in complex environments and inferior path quality. To remedy this, the present paper proposes the complex environments rapidly-exploring random tree (CERRT) path planning algorithm inspired by RRV, which greatly improves the efficiency of the algorithm in complex environments and optimizes the generated initial feasible paths.

The main contributions of this paper are as follows:

(1) We have designed a new process for environmental perception. This process determines the type of environment by sampling the local area, eliminating the need for principal component analysis and significantly reducing computational complexity.
(2) We propose a pre-allocated vertex expansion method in conjunction with a vertex death mechanism. This approach foregoes the expansion of inactive tree vertices to prevent the algorithm from getting stuck in concave areas. When combined with the environment-aware capability, the algorithm deftly navigates complex environments such as mazes, narrow passages, and concave regions.
(3) We also suggest a bidirectional contraction optimization strategy. Once a feasible path is identified, its points are contracted in both directions, yielding a more streamlined and efficient path.

The rest of this paper is structured as follows: Section 2 outlines the mathematical definition of the planning problem along with a brief introduction to the core principles of RRT, RRV and Fast-RRT. Section 3 offers an in-depth description of our proposed CERRT algorithm framework. Section 4 presents simulation experiments that compare the new algorithm against RRT and RRV. Finally, Section 5 concludes the paper.

2. Background

In this section, we first introduce the mathematical definition of the path planning problem and then briefly describe the RRT and RRV algorithms.

2.1. Problem Definition

Let X be the configuration space, X_{obs} be the obstacle region, and $X_{free} = X/X_{obs}$ be the feasible region. (X, X_{start}, X_{goal}) defines a path planning problem, where $x_{start} \in X_{free}$ is the initial state and $X_{goal} \subset X_{free}$ is the goal area. Let a continuous function σ:[0, n] $\rightarrow X$ of bounded variation be a path, where n is the path point number. If $\forall\tau \in [0, n]$, $\sigma(\tau) \in X_{free}$, then σ is a feasible path, defined as σ_{free}.

Definition 1. *Feasible Path Solution.*

For the (X, X_{start}, X_{goal}) problem, if $\exists\sigma \in \sigma_{free}$, where $\sigma(0) = X_{start}$ and $\sigma(n) \in X_{goal}$, then the path is called a feasible path solution σ^*; otherwise, report a path planning failure.

Definition 2. *Approximate Optimal Path Solution.*

For the (X, X_{start}, X_{goal}) problem, if $\exists\sigma^*$ satisfies $C(\sigma^*) \leq \min\left\{C(\sigma) : \sigma \in \sigma_{free}\right\} * 1.05$, then output path σ^* is the approximate optimal path solution; otherwise, report a failure.

2.2. RRT

RRT explores the configuration space by maintaining a tree T. The algorithm sets the root node of the tree as x_{start} and performs an iterative expansion. In each iteration, the sampler randomly selects a sample x_{rand} from the configuration space, finds the vertex $x_{nearest}$ in T closest to x_{rand}, and extends a step size $d_{stepsize}$ from $x_{nearest}$ toward x_{rand} to obtain the node x_{new} for expansion. If the local path from $x_{nearest}$ to x_{new} is collision-free, then x_{new} is added to the tree. The algorithm terminates either when a feasible path is obtained or when the maximum number of iterations 'N' is exceeded.

However, the randomness of the sampler often results in a low sampling probability in narrow passages, leading to fewer sampling points in such areas. Consequently, it becomes challenging for the expansion tree to detect these narrow passages. This limitation hampers the effectiveness of the RRT algorithm in complex environments.

2.3. RRV

RRV is an algorithm developed to overcome the narrow passage problem found in the RRT algorithm. It uses principal component analysis to identify the local environment type, as it can effectively map high-dimensional data to a lower-dimensional space while maximizing information retention, which facilitates the extraction of data features. Specifically, this process involves mapping sampled points within obstacles into a single feature vector that maximally retains information from the obstacle points. Subsequently, the x_{rand} point is projected onto the feature vector passing through $x_{nearest}$. The resulting projected point, $x_{projected}$, serves as a novel direction for the expansion of RRV. This strategic extension enables the random tree to circumvent obstacles, similar to the growth of a vine, along the obstacle boundaries.

As illustrated in Figure 1, the RRV algorithm generates local random sampling points (shown in red and green) and performs principal component analysis on the red obstacle points. The confidence ellipse is then used to determine the type of environment. If the ellipse does not contain a green point, it is classified as a convex obstacle environment (Figure 1a). If it contains a green point but not $x_{nearest}$, it is identified as a passage entrance environment (Figure 1b). If it contains both a green point and $x_{nearest}$, it is a passage interior environment (Figure 1c).

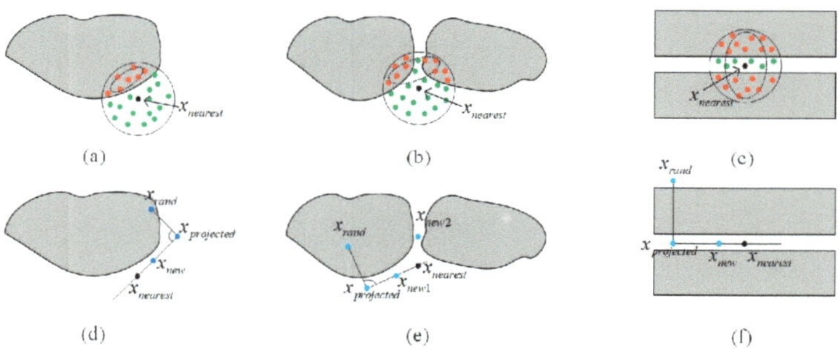

Figure 1. Environment judgment and expansion in RRV. (**a**,**d**) represent convex obstacles, (**b**,**e**) represent the entrance of the passage, and (**c**,**f**) represent the interior of the passage.

Then, the x_{rand} point is projected onto a principal component analysis feature vector passing through $x_{nearest}$ to obtain $x_{projected}$, and the tree is expanded toward this point to avoid growing toward obstacles. If the environment is identified as a convex obstacle or a passage interior, the tree is expanded along the obstacle (as shown in Figure 1d,f). If it is identified as a narrow passage entrance, as shown in Figure 1e, the tree is further expanded along the obstacle to x_{new1} and toward the interior of the passage to x_{new2}. This enables RRV to discover narrow passages more effectively than the classic RRT algorithm, and once a narrow passage is discovered, the expansion tree can grow quickly.

However, when applied to environments without narrow passages, the performance of RRV falls short compared to the original RRT algorithm. This indicates a high degree of environmental dependence in its performance.

3. CERRT

The CERRT algorithm seeks to correct the insensitivity of traditional RRT algorithms to the environment. It utilizes a novel node expansion strategy to improve expansion efficiency and incorporates new environmental awareness capabilities to address narrow passage problems. Moreover, it introduces a path optimization strategy to enhance the quality of the paths generated, making it a more efficient solution overall.

3.1. Algorithm Framework

The CERRT algorithm is an optimization of the RRT algorithm. It uses an array V to store expandable points in tree T and constantly deletes non-extensible points, known as dead nodes.

The CERRT algorithm initially stores the starting point x_{start} point as the root node in the tree and pre-allocates a corresponding set of expandable points, $x_{start}.CAND$ for it (Lines 1–3 in Algorithm 1). The sampler is adjusted to amplify the tree's growth orientation by systematically sampling points within the goal region with a certain probability. (Line 5 in Algorithm 1).

Following sampling, the algorithm selects the nearest point $x_{nearest}$ to the sampling point from the array V and queries the closest vertex to the sampling point x_{rand} from the expansion point set $x_{nearest}.CAND$ of $x_{nearest}$ to obtain the new node x_{new}. Once found, the point is removed from $x_{nearest}.CAND$. If $x_{nearest}.CAND$ is empty after removal, and the vertex $x_{nearest}$ is deemed dead and removed from the vertex array V (Lines 6–11 in Algorithm 1).

Should a collision happen during expansion, the algorithm enters the environment perception phase to determine the type of environment where $x_{nearest}$ is located. If $x_{nearest}$ is near a channel, a new x_{new} point is calculated for expansion; otherwise, resampling is performed (Lines 13–14 in Algorithm 1).

If no collision occurs during the expansion, x_{new} is added to tree T, and a corresponding set of candidate nodes for x_{new} is pre-allocated. If the set of candidate nodes overlaps with any existing nodes in the tree, no allocation is performed. If the set of candidate nodes for x_{new} is empty after allocation, it is not added to array V (Lines 17–21 in Algorithm 1).

Finally, after obtaining a feasible path, the algorithm optimizes it to reduce the cost $C(\sigma)$ (Line 24 in Algorithm 1). A more detailed description of the algorithm process is given in the subsequent section.

Algorithm 1 CERRT(x_{start}, x_{goal}, d_{step}, d_{gap}, N, Map)

1: $T = \{x_{start}\}$;
2: $V = T$;
3: $x_{start}.CAND$ = initTree(x_{start});
4: **for** $i = 1$ **to** N **do**
5: x_{rand} = $GetSample(i)$;
6: ($x_{nearest}$, $near_{idx}$) = GetNearest(V, x_{rand});
7: (x_{new}, new_{idx}) = min(distance($x_{nearest}.CAND(:)$, x_{rand}));
8: Delete($x_{nearest}.CAND(new_{idx})$);
9: **if** Empty($x_{nearest}.CAND$) **then**
10: Delete($V(near_{idx})$);
11: **end if**
12: **if** Overlap(x_{new}, T) **then** continue; **end if**
13: **if** Collision(x_{new}, $x_{nearest}$, Map) **then**
14: (x_{new}, Fig) = Aware($x_{nearest}.parent$, $x_{nearest}$, Map, d_{step}, d_{gap});
15: **if** ~Fig **then** continue; **end if**
16: **end if**
17: $x_{new}.CAND$ = SetCAND($x_{nearest}$, x_{new}, d_{step}, T, V);
18: **if** NoEmpty($x_{new}.CAND$)
19: $V = V \cup \{x_{new}\}$;
20: **end if**
21: $T = T \cup \{x_{new}\}$;
22: **if** $x_{new} \in X_{goal}$ **then** break; **end if**
23: **end for**
24: σ = GetPath(σ);
25: **return** σ = OptPath(σ);

3.2. Vertex Expansion Method

The traditional rapidly-exploring random tree (RRT) algorithm randomly samples the configuration space to guide the tree's expansion and exploration. When the RRT algorithm expands a new vertex, it calculates the distance 'd' between the point and the goal. If 'd' is less than a predetermined threshold 'r', it means the algorithm has found the goal point and the search ceases. If not, the search continues.

In this regard, each time a new node is expanded, the RRT algorithm explores the region with a radius of 'r' centered on that node. The unexplored regions of tree nodes are referred to as the 'unknown regions', while the explored areas are defined as 'exploration regions'. With each expansion, new exploration regions are created within the unknown regions.

The RRT algorithm achieves the exploration of the entire space by continuously sampling and expanding. However, this process brings about a scenario where some exploration regions are revisited, some even more than twice. The exploration efficiency is thus measured by the area of novel exploration regions explored by newly expanded vertices and the frequency of re-exploration of already known regions. Efficiency is high when the area of unexplored regions explored is large, but it is low if the area of already known regions explored is large or if they are explored multiple times.

Figure 2 illustrates the issue of redundant exploration in known regions. In this figure, the blue region represents the explored area, while the overlapped area denotes the repeated examination. The green zone denotes the exploration area for new vertices, and new vertices conduct repeated exploration on region 'T' enclosed by the red circle more than three times.

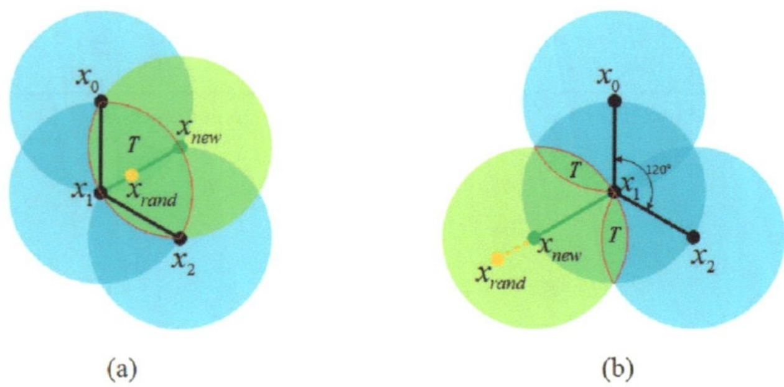

Figure 2. Issue of redundant exploration in known regions. (**a**) Redundant exploration using the traditional extension strategy. (**b**) Improved extension strategy to reduce redundancy.

Figure 2a elucidates the traditional RRT expansion process. This diagram illustrates that vertices x_0, x_1, and x_2 have already examined the 'T-region' twice. Following this, vertex x_1 extends toward the x_{rand} point and produces a fresh vertex x_{new}. Regrettably, this new vertex instigates a third redundant exploration of the 'T-region'. Multiple explorations of the same area are pointless.

In response to the problem mentioned earlier, our research suggests that setting the angle at 120 degrees between each vertex and its connected points can drastically decrease unnecessary exploration in space, as shown in Figure 2b. When a new point, x_{new}, extends from vertex x_1, the angles between the three edges x_1-x_0, x_1-x_2 and x_1-x_{new} are all 120 degrees. This decreases the 'T-region' to a mere 30% of what it is in Figure 2a. The exploration process also avoids repeating exploration areas over three times, which greatly increases efficiency.

To achieve this, we propose pre-allocated expansion points and a vertex dead strategy to ensure that the angle between each vertex's edges is 120 degrees. Algorithm 2 presents the detailed process of allocating candidate extension points. Initially, the angle ang_0 of the edge x_{new}-$x_{nearest}$ is computed within the Cartesian coordinate system. Following this, candidate extension points are created with a 120-degree bias angle (Lines 1–4 in Algorithm 2). If the candidate extension point coincides with a tree node, the allocation fails (Lines 6–10 in Algorithm 2).

Figure 3 depicts the improved vertex expansion strategy. In this figure, the black, red, green, and yellow dots, respectively, represent standard tree vertices, $x_{nearest}$, x_{new}, and x_{rand} sample points. Dashed circles are potential expansion points for each vertex, while red-crossed points signify discarded vertices removed from array V.

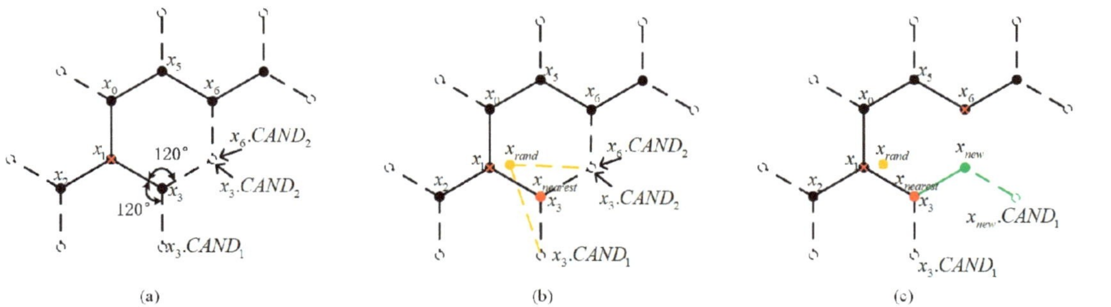

Figure 3. Improved vertex expansion process. (**a**) Before expansion. (**b**) Selection of $x_{nearest}$ after sampling. (**c**) After expansion.

Upon expanding a new vertex x_{new}, $x_{new}.CAND_j$ (where j is the candidate point index j = 0, 1, 2) is pre-assigned to maintain each vertex's edge angle at 120 degrees. Additionally, if a vertex's candidate point set is vacant, it is classified as a discarded vertex and removed from array V. Figure 3a shows that every vertex logs its corresponding candidate expansion points $x_i.CAND_j$ (where i is the vertex index in the growing tree T, i = 0, 1, ..., n). If x_i's candidate point set is void, it is deemed a discarded vertex and excluded from array V.

The expansion process of the CERRT is conveyed in Figure 3b. After sampling x_{rand} randomly, the closest vertex $x_{nearest}$ is selected from array V as the starting point. Since x_1 is a discarded vertex and has been eliminated from array V, x_3 is chosen as the nearest vertex. Subsequently, vertex x_{new} is selected from the candidate expansion points of x_3 based on its proximity to x_{rand}. After calculation, $x_3.CAND_2$ is chosen as x_{new} for expansion and $x_3.CAND_2$ is removed from the candidate set of x_3. As $x_6.CAND_2$ coincides with $x_3.CAND_2$, it is also removed from the candidate set of x_6. Since the candidate set of vertex x_6 is now empty, x_6 becomes a dead vertex and is deleted from array V.

If edge x_{new}-$x_{nearest}$ collides with obstacles, the algorithm advances to the environment awareness stage. Otherwise, x_{new} is added as a new vertex to tree T, and a set of candidate expansion points is pre-allocated for x_{new}. It is worth noting that the candidate points must not overlap with any existing points in tree T. This process is illustrated in Figure 3c.

Algorithm 2 SetCAND($x_{nearest}$, x_{new}, d_{step}, T, V)

1: ang_0 = GetCartesianAngle($x_{nearest}$, x_{new});
2: $ang_1 = ang_0 + 120$; $ang_2 = ang_0 + 240$;
3: $new_1 = x_{new} + d_{step}$*[sin($ang_1$), cos($ang_1$)];
4: $new_2 = x_{new} + d_{step}$*[sin($ang_2$), cos($ang_2$)];
5: $CAND$ = [];
6: **if** NoOverlap(new_1, T) **then**
7:     $CAND(end + 1) = new_1$;
8: **else if** NoOverlap(new_2, T) **then**
9:     $CAND(end + 1) = new_2$;
10: **end if**
11: **return** $CAND$;

3.3. Environmental Awareness

Algorithms based on Sampling often encounter difficulties when navigating narrow passages due to their lack of environmental sensitivity. However, whether a passage is deemed narrow is contingent on the extension step length. If the step length is much smaller than the narrow passage, it can be considered a spacious road. Conversely, if the step length is too small, the search accuracy will be too high, resulting in lower algorithm efficiency. To combat this, an environment perception strategy is proposed to enable the random tree to quickly identify and pass through narrow passages without decreasing the step length. When a collision occurs with an obstacle during the tree expansion process, expansion is halted and enters the environment perception stage. Algorithm 3 provides a comprehensive outline of this phase.

To begin the environmental perception process, local spatial information is collected around the $x_{nearest}$ point through local sampling. Local sampling uniformly samples n points around the vertex to be expanded using the expansion step length as the radius and stores them in a point set S. Here, n = 16 is used as an example. Set S is then separated into two subsets, S_{obs} and S_{free}, based on whether the position of S lies within the obstacle area. The boundary points between S_{free} and S_{obs} are selected and stored in S_{bdry}, where $S_{bdry} \subseteq S_{free}$ (Lines 2–3 in Algorithm 3). Figure 4 illustrates the schematic diagram of local sampling, where the red point represents the expanding vertex where a collision occurred, the blue points represent S_{obs} sampling points, the green points represent boundary points S_{bdry}, and the yellow and green points represent S_{free} points. In Figure 4, the obstacle is recognized as a wall obstacle.

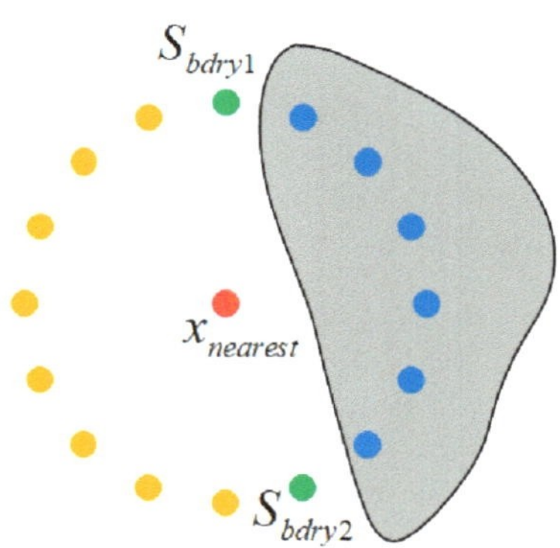

Figure 4. Local sampling process and wall obstacle types.

For the local sampling to accurately identify narrow passages, it is crucial to further clarify the quantity of local sampling points n and the extension step size d_{step}. Given the width d_{gap} of the minimum feasible driving passage and the extension step size d_{step}, to ensure that the local sampling can sample the interior of the passage, the spacing between adjacent sampling points should not exceed d_{gap}. The number of sampling points n must align with Equation (1).

$$n \geq \frac{2\pi}{\arccos\left(1 - \frac{d_{gap}^2}{2d_{step}^2}\right)} \tag{1}$$

After local sampling, environments are categorized based on the numerical relation between the point sets S_{free} and S_{bdry}. The environment is split into two primary classifications: wall obstacles and passage environments. If S_{bdry} has only two vertices and S_{free} has more than two vertices, the environment is classified as a wall obstacle, as shown in Figure 4. In this case, the algorithm exits the environment perception and performs resampling (Lines 4–6 in Algorithm 3). Otherwise, the environment is considered a narrow passage environment, which can be further classified into entrance, interior, exit, and multi-branching regions, as shown in Figure 5. In this figure, the red point represents the $x_{nearest}$ point, the yellow points represent S_{free} sample points, the green points represent the boundary points S_{bdry}, and the blue points represent the S_{obs} sample points. The green sector area R contains a continuous set of S_{free} points.

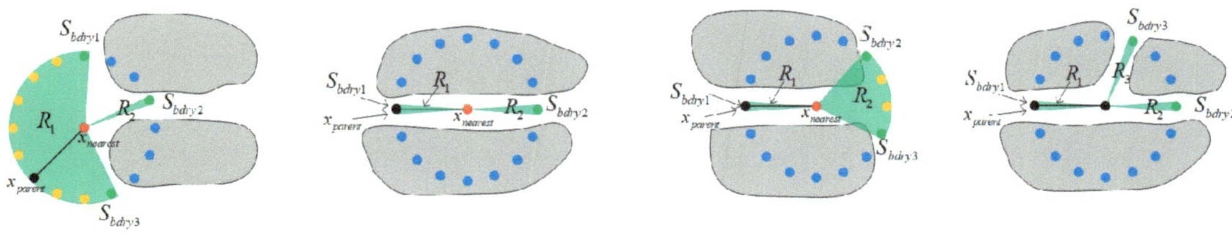

(a) (b) (c) (d)

Figure 5. Narrow passage perception. (**a**) Entrance of the passage; (**b**) Interior of the passage; (**c**) Exit of the passage; (**d**) Multiple branching passages.

To traverse the narrow passage, multiple sector regions are divided using $x_{nearest}$ as the fulcrum and the expansion step as the radius. Each sector region contains only a continuous set of S_{free} points. The resulting n sector regions are denoted as R_i (i = 1, 2, . . ., n). The sector area that does not contain the x_{parent} point is selected as the area for expansion. An S_{free} point that will not cause a collision is then selected from it as the x_{new} point for expansion (Lines 8–13 in Algorithm 3).

For example, Figure 5a illustrates the entrance of a narrow passage. After the area is divided into two sector regions (R_1, R_2), S_{bdry2} is selected from R_2 as the x_{new} point for expansion since x_{parent} is positioned in R_1. The same expansion approach is applied to the narrow passage interior shown in Figure 5b and the exit in Figure 5c.

If the number of sector areas n is greater than 2, it is considered a multi-passage branching situation, as shown in Figure 5d. After division, three sector areas (R_1, R_2, R_3) are obtained. Since R_1 contains the x_{parent} point, two S_{free} points are selected from the R_2 and R_3 regions that will not cause collisions as x_{new} for expansion.

Algorithm 3 Aware(x_{parent}, $x_{nearest}$, Map, d_{step}, d_{gap})

```
1:  x_new = []; Fig = false;
2:  (S_free, S_obs) = LocalSample(x_nearest, Map, d_step, d_gap);
3:  S_bdry = GetBoundary(S_free, S_obs);
4:  if size(S_bdry) == 2 && size(S_free) > 2 then
5:      Fig = false;
6:      return (x_new, Fig);
7:  else
8:      (R, n) = DivideRegion(S_free, S_obs, S_bdry);
9:      for i = 1 to n do
10:         if x_parent ∈ R_i then R = R−R_i end if
11:     end for
12:     x_new = ChoseNew(R, S_free, n−1);
13:     Fig = true;
14: end if
15: return (x_new, Fig);
```

3.4. Path Optimization Strategy

Sampling-based algorithms for path planning can often generate redundant nodes due to their random nature, potentially lowering the quality of path planning. In this paper, a bidirectional shrinking optimization strategy (BSOS) is proposed based on the pruning optimization strategy (POS) to further enhance the quality of the path.

In Figure 6a, moving from x_0 to x_7 requires avoiding an obstacle, and the black line path x_0-x_1-x_2-x_3-x_4-x_5-x_6-x_7 in the figure is the original planned path. This path contains a large number of redundant nodes. By using the pruning strategy and based on the triangle inequality theorem, the robot can move directly from x_0 to x_2 without passing through x_1, thus identifying x_1 as a redundant node.

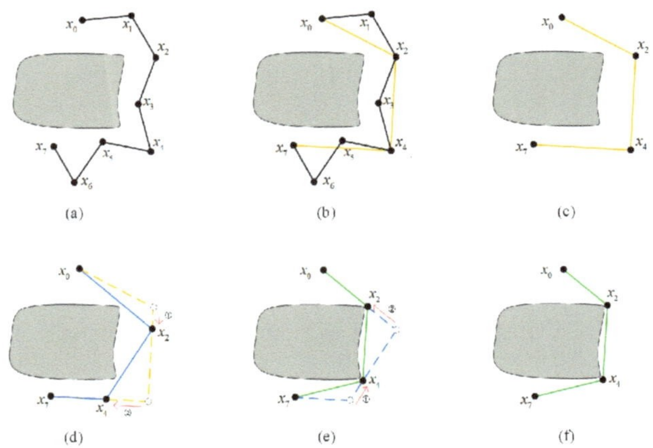

Figure 6. Path optimization strategy. (**a–c**) show pruning optimization of the original path, where the yellow path represents the pruned optimized path. (**d–f**) show bidirectional shrinkage optimization of the path, where the green path represents the final optimized path.

Figure 6b illustrates the application of the pruning operation on the entire initial path, eliminating superfluous nodes x_1, x_3, x_5, and x_6. The remaining nodes are connected to obtain the pruned and optimized yellow path x_0-x_2-x_4-x_7, as shown in Figure 6c. The yellow path obtained by POS has not only fewer path points but also a shorter path length, but it is not an approximate optimal path.

This paper proposes a bidirectional shrinking-based optimization of path points on the basis of pruning the path x_0-x_2-x_4-x_7. The endpoints x_0 and x_7 are excluded from the shrinking process, leaving the remaining points to participate in two phases of reduction.

In the first round, each path point moves toward the next point with a higher number according to the sequence of the point number. At the same time, it constantly checks whether there is any collision with the previous path point. If a collision is about to occur, the point stops moving, as shown in Figure 6d. For instance, path point x_2 moves toward x_4 until the x_0-x_2 segment is about to collide with the obstacle, and then the point stops. After that, path point x_4 moves toward x_7 until the x_2-x_4 segment is about to collide with the obstacle, and then the point stops. Once all path points have completed the shrinking movement, the blue path x_0-x_2-x_4-x_7 in Figure 6d is obtained, indicating that the first round of shrinking is completed.

In the second round of contraction, the order is reversed, with the points moving from high to low according to their vertex number. As shown in Figure 6e, path point x_4 moves toward x_2 first, and then it stops when the x_4-x_7 segment is about to collide with the obstacle. Then, path point x_2 moves toward x_0 until the x_2-x_4 segment is about to collide with the obstacle. Once all path points have completed the shrinking movement, the green path x_0-x_2-x_4-x_7 in Figure 6f is obtained, which is the final path obtained by the bidirectional search optimization strategy. The bidirectional shrinking method results in a path that is shorter and closer to the optimal configuration.

The specific process of using the bidirectional shrinking path optimization strategy to optimize the CERRT planning path is shown in Figure 7.

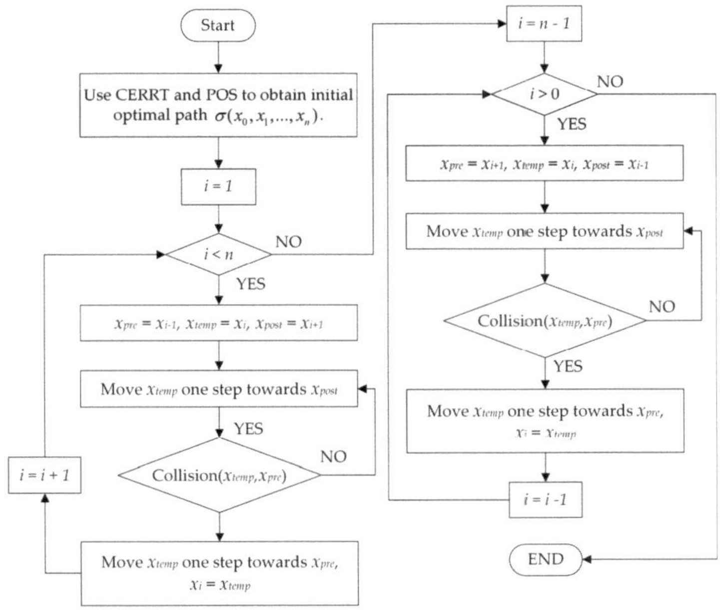

Figure 7. Bidirectional shrinking optimization strategy flowchart.

(1) Obtain the initial optimized path σ (x_0, x_1, . . ., x_n) using the CERRT and pruning optimization strategies, where x_i (i = 0, 1, . . ., n) is a path point and i is the path point number.
(2) Initialize $i = 1$.
(3) Check whether i is less than n. If it is, put x_i into the temporary variable x_{temp}, put x_{i-1} into the variable x_{pre}, and put x_{i+1} into the variable x_{post}. If not, go to step (6).
(4) Move the current variable point x_{temp} one step toward the variable point x_{post}.
(5) Check whether the path segment x_{temp}-x_{pre} collides with any obstacles. If there is a collision, move x_{temp} one step toward the opposite direction of x_{post}, update the path point x_i to x_{temp}, set $i = i + 1$, and go to step (3). If there is no collision, go to step (4).
(6) Set $i = n - 1$.
(7) Check whether i is greater than 0. If it is, put x_i into the temporary variable x_{temp}, put x_{i+1} into the variable x_{pre}, and put x_{i-1} into the variable x_{post}. If not, the optimized path is obtained, and the process ends.
(8) Move the current variable point x_{temp} one step toward the variable point x_{post}.
(9) Check whether the path segment x_{temp}-x_{pre} collides with any obstacles. If there is a collision, move x_{temp} one step toward the opposite direction of x_{post}, update the path point x_i to x_{temp}, set $i = i - 1$, and go to step (7). If there is no collision, go to step (8).

After the entire process is executed, the path points are updated by shrinking, and the obtained path $\sigma(x_0, x_1, \ldots, x_n)$ is the final optimized path. Note, that "moving one step" in the process refers to moving one pixel.

4. Simulation and Experiment

To verify the algorithm performance of CERRT, this study conducted a comparative analysis of the RRT, RRV [5], Fast-RRT [7] and CERRT algorithms in a simple environment, a maze environment, a narrow passage environment, and a bug environment. The accessible passage width was set to d_{gap} = 10 px, and the map size was 1000 px × 1000 px, as shown in Figure 8. The tree expansion step was set to d_{step} = 30 px, and the detection radius was $r = d_{step}$. The sampler probability of sampling in the target area was 0.05, and the probability of sampling in the random area was 0.95. The maximum sampling value was set to 80,000, and if the number of samples exceeded the maximum value, the path planning was considered a failure.

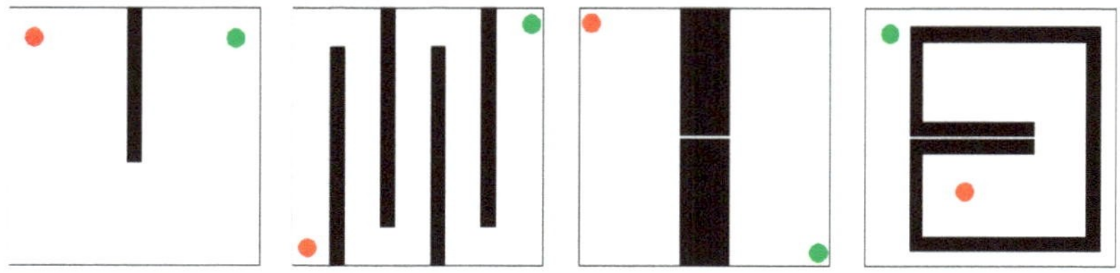

(a) (b) (c) (d)

Figure 8. Environments for the simulations. (**a**) Simple. (**b**) Maze. (**c**)Narrow. (**d**) Bug Trap.

The algorithm performance was evaluated using three criteria: execution time, number of tree vertices generated, and success rate. These evaluations were conducted by repeating each simulation 100 times. The execution time and number of vertices were only counted for successful path planning. In the path optimization experiment, this study compared the original planned path, the pruning optimized path, and the proposed bidirectional shrinking optimized path and evaluated the path quality $C(\sigma^*)$ using two indicators: path length and smoothness. All simulations were performed on a machine with an Intel (R) Core (TM) i7-12700H 2.30 GHz CPU and 16 GB of RAM. The simulation platform was MATLAB R2022a, and the function min was employed in all algorithms to find the nearest neighbor in all algorithms. The collision detection program used the linear trial method [20].

4.1. Path Planning Simulation

In the path planning simulations, we tested the performance of the RRT, RRV, Fast-RRT and CERRT algorithms in four 2D environments. The objective of using a simple environment was to assess if the new algorithm's performance was significantly impacted by additional computation. On the other hand, the maze environment, narrow environment, and Bug Trap environment were utilized to assess the algorithms' performance in complex scenarios. The red dots in the map represent the starting points, the green dots represent the target points. The entire exploration process is represented by the green lines, and the generated path is represented by the red lines. The sampling method was consistent across all experiments.

4.1.1. Simple Environment

The purpose of testing the algorithm in a simple environment was to evaluate the performance loss of the new algorithm with additional computational costs. The planning scenarios are shown in Figure 9, and the expansion shapes of the four random trees were generally similar.

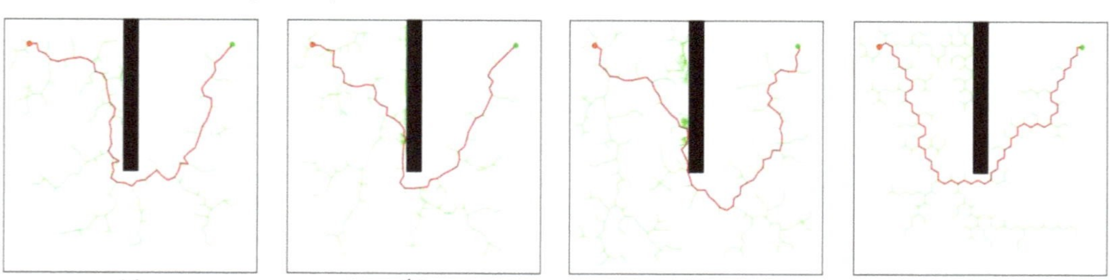

(a) (b) (c) (d)

Figure 9. Performance comparison of four algorithms in sample environment. (**a**) RRT; (**b**) RRV; (**c**) Fast-RRT; (**d**) CERRT.

Table 1 presents the performance of the algorithm in a simple environment. The success rates of the four algorithms are all 100%, and the average number of tree nodes was similar. The average running time of RRV was four times that of RRT, the average running time of Fast-RRT was 1.5 times that of RRT, and the average running time of CERRT was 1.7 times that of RRT. The time variance of CERRT was close to that of RRT and much smaller than that of RRV.

Table 1. Results for planning in a simple environment.

Algorithm	Avg. Time (ms)	Min Time (ms)	Max Time (ms)	Std	Avg. Nodes	Success Rate
RRT	9.08	5.61	14.48	2.04	263	1.00
RRV	37.04	7.33	71.00	10.90	281	1.00
Fast-RRT	13.15	6.58	32.45	3.94	355	1.00
CERRT	15.86	9.28	24.63	2.81	286	1.00

The data indicate that the performance loss of CERRT was significantly more than that of RRV, and its overall performance was almost identical to RRT, indicating that the performance loss caused by the additional computational cost of CERRT was minimal.

4.1.2. Maze Environment

The maze environment was designed to evaluate the performance of algorithms in complex environments without passages. Figure 10 illustrates the planning situations of the four algorithms, indicating that the utilization rates of tree nodes were low for RRV, RRT, and Fast-RRT, and there were numerous repeated exploration points on the left side of the map. In contrast, CERRT's tree nodes were evenly distributed, enabling the exploration of a broader area with fewer points.

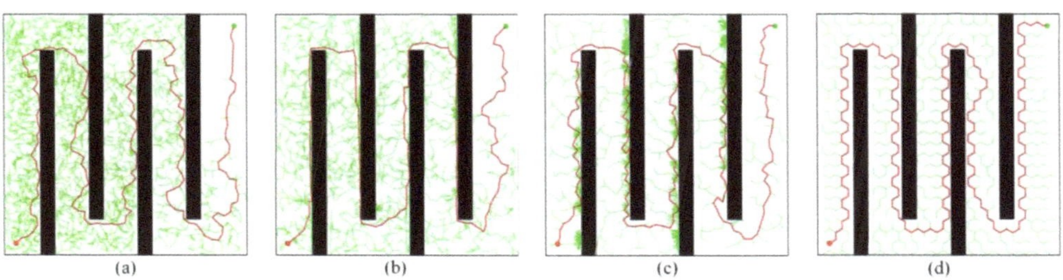

Figure 10. Performance comparison of four algorithms in maze environment. (**a**) RRT; (**b**) RRV; (**c**) Fast-RRT; (**d**) CERRT.

Table 2 presents the algorithms' performance in the maze environment. RRT's time consumption was 2.7 times that of CERRT, Fast-RRT's time consumption was 1.7 times that of CERRT, and RRV's time consumption was 15 times that of CERRT. CERRT had the shortest running time and the smallest number of tree nodes. Furthermore, CERRT's time standard deviation was significantly lower than those of RRT and RRV, further indicating that this algorithm was most stable in the Maze environment.

Table 2. Results for planning in a maze environment.

Algorithm	Avg. Time (ms)	Min Time (ms)	Max Time (ms)	Std	Avg. Nodes	Success Rate
RRT	108.29	83.66	153.99	13.97	2284	1.00
RRV	603.24	438.86	1001.21	120.67	3042	1.00
Fast-RRT	68.93	48.75	135.38	11.58	1368	1.00
CERRT	39.73	34.59	112.35	8.66	610	1.00

Based on the above, it can be concluded that the new CERRT algorithm outperformed the RRT, RRV, and Fast-RRT algorithms in various aspects in complex non-navigable environments. The main reason for this is the improved vertex expansion strategy, which effectively improved the utilization rate of vertices by pre-allocating them for expansion. This strategy reduced repeated exploration of the same region and introduced a vertex death mechanism to eliminate useless vertices, thus reducing the time cost of selecting the optimal neighboring vertex.

4.1.3. Narrow Environment

The RRT algorithm faced challenges in solving narrow passage problems, while the RRV algorithm was specifically designed to address this issue. Additionally, the Fast-RRT has also proposed solutions for narrow passages. In this study, we used a narrow environment to test the performance of the new algorithm and evaluate its environmental adaptability. Figure 11 shows the planning process of the four algorithms. It can be seen that RRT had a large number of vertices on the left side of the map, and the algorithm could not effectively detect the narrow passages on the walls. Fast-RRT expands multiple times near obstacles to find passages. RRV and CERRT were quickly able to discover and pass through the narrow passage.

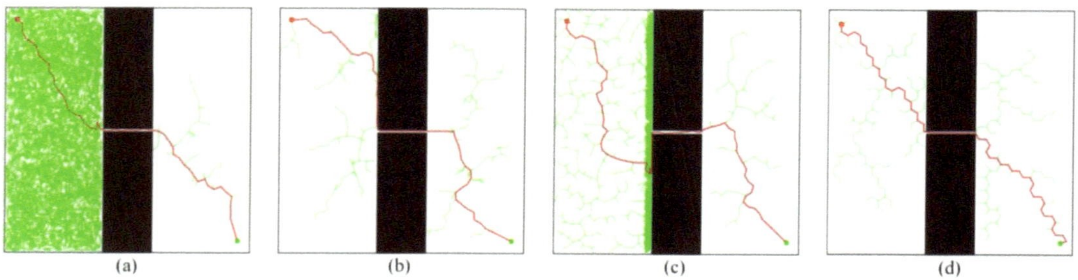

Figure 11. Performance comparison of four algorithms in narrow environment. (**a**) RRT; (**b**) RRV; (**c**) Fast-RRT; (**d**) CERRT.

Table 3 showcases the performance of the algorithms in the narrow passage environment. The planning success rate of the RRT algorithm was 0.97, while the other algorithms were both 1.00, highlighting the shortcomings of RRT in narrow environments. The average running time of RRT was 27 times that of CERRT, the average running time of RRV was four times that of CERRT, and the average running time of Fast-RRT was slightly higher than that of CERRT. The new algorithm had a shorter planning time. The tree vertex numbers of CERRT and RRV were similar, and much lower than that of the RRT and Fast-RRT algorithm, indicating that both derived algorithms can solve narrow passage problems. RRT's time standard deviation was 135 times higher than CERRT, RRV's time standard deviation was 20 times higher than CERRT, and Fast-RRT's time standard deviation was 75 times higher than CERRT, suggesting that CERRT exhibited optimal stability.

Table 3. Results for planning in a narrow environment.

Algorithm	Avg. Time (ms)	Min Time (ms)	Max Time (ms)	Std	Avg. Nodes	Success Rate
RRT	455.13	7.52	5189.01	871.53	2931	0.97
RRV	65.72	25.04	1287.83	127.12	259	1.00
Fast-RRT	58.03	4.87	3779.04	487.47	869	1.00
CERRT	16.70	11.66	74.69	6.42	255	1.00

Based on the above analysis, we can conclude that the new CERRT algorithm outperformed the RRT, RRV and Fast-RRT algorithms in narrow environments. The main reason for this is that RRV requires complex principal component analysis to determine the environment, while CERRT's environmental perception ability only requires simple sampling to determine the surrounding environment and select new expansion points without complex calculations, effectively reducing the environmental recognition cost.

4.1.4. Bug Trap Environment

The bug trap environment is created by adding concave traps to a narrow environment to assess the algorithm's ability to handle such obstacles. Figure 12 shows the planning process of the four algorithms. Notably, both the RRT and Fast-RRT algorithms performed extremely poorly in the Bug Trap environment, as evidenced by a significant accumulation of tree vertices inside the trap. The RRV algorithm can only recognize convex obstacles and mistakenly identified the concave traps as entryways, resulting in multiple attempts to expand within the trap area, which greatly reduced the algorithm's performance, with the number of tree vertices still high. The CERRT algorithm effectively utilizes each vertex to explore the space and can quickly discover and pass through the real channel.

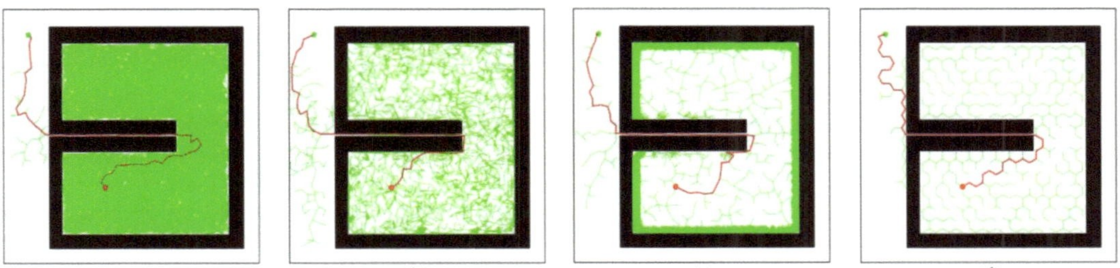

(a) (b) (c) (d)

Figure 12. Performance comparison of four algorithms in Bug Trap environment. (**a**) RRT; (**b**) RRV; (**c**) Fast-RRT; (**d**) CERRT.

Table 4 illustrates the performance of the algorithms in the Bug Trap environment. The planning success rates of RRT and Fast-RRT were only 0.90 and 0.92, respectively, while the other two algorithms achieved rates of 1.00. The average running time of RRT was 101 times that of CERRT, the average running time of Fast-RRT was 20 times that of CERRT, and the average running time of RRV was 52 times that of CERRT, with the new algorithm having the shortest planning time. The average number of tree vertices of RRT was 29 times that of CERRT, the average number of tree vertices of Fast-RRT was 21 times that of CERRT, and the average number of tree vertices of RRV was six times that of CERRT, indicating that the new algorithm had the highest vertex utilization rate. Moreover, RRT's time standard deviation was 526 times higher than CERRT, Fast-RRT's time standard deviation was 371 times higher than CERRT, and RRV's time standard deviation was 329 times higher than CERRT, underscoring the superior stability of the CERRT algorithm.

Table 4. Results for planning in a Bug Trap environment.

Algorithm	Avg. Time (ms)	Min Time (ms)	Max Time (ms)	Std	Avg. Nodes	Success Rate
RRT	2725.83	504.12	8782.20	1856.74	13902	0.90
RRV	1418.43	31.42	6234.62	1160.86	2679	1.00
Fast-RRT	1674.35	267.01	9574.50	1313.36	9798	0.92
CERRT	26.97	13.36	35.28	3.53	474	1.00

Based on the above analysis, it can be concluded that the new CERRT algorithm outperformed the RRT and RRV algorithms in the Bug Trap environment. The reason is that the vertex death mechanism can deactivate the vertices inside the traps, preventing the algorithm from becoming stuck in the concave traps. Combined with environmental awareness, the CERRT algorithm was quickly able to break through the bug trap environment.

4.2. Path Optimization Simulation

In the path optimization experiments, complex Maze and Bug Trap environments were used to test the quality of the four algorithms' paths after pruning and bidirectional shrinking optimization. To assess the stability of the optimization strategy, the experiment was repeated 100 times. The quality characteristics of the generated paths were evaluated by comparing the average length and smoothness values. The path lengths were calculated using the Euclidean distance, which refers to the straight-line distance connecting two points on a plane, and can be computed using the Pythagorean theorem, Additionally, the path smoothness values were obtained by accumulating the turning angles of each path, measured in radians.

4.2.1. Maze Environment Path Optimization

The purpose of the experiments in the Maze environment was to test the performance of the proposed optimization strategy. Figure 13 illustrates the results of pruning and bidirectional shrinking optimizations. It is evident that the original paths generated by all algorithms were convoluted and intricate. However, after the pruning optimization, the quality of the paths improved, but they were not optimal, while bidirectional shrinking optimization generated paths that were close to the optimal path. A detailed comparison of the paths is provided in Table 5. The path lengths of bidirectional shrinking optimization were the shortest, and the path smoothness values were the lowest, indicating that the generated paths were optimal. The paths generated by the four algorithms were all able to be optimized to approximate the optimal path. Hence, it can be concluded that the bidirectional shrinking path optimization strategy outperformed the pruning optimization strategy in complex environments.

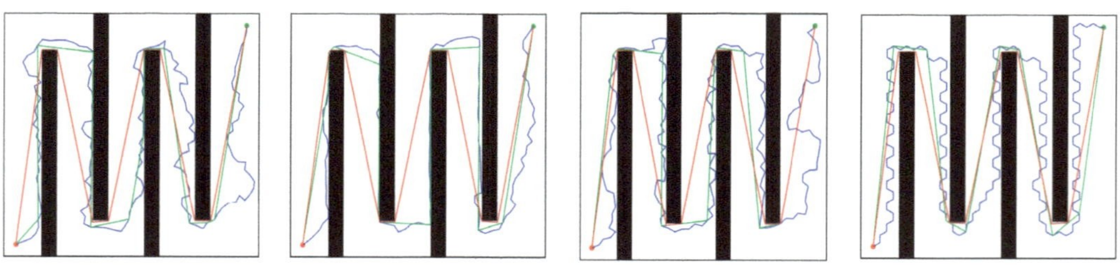

(a) RRT (b) RRV (c) Fast-RRT (d) CERRT

Figure 13. Pruning and bidirectional shrinking optimization in the Maze environment. The blue lines indicate the original paths, the green lines indicate the pruned optimized paths, and the red lines indicate the bidirectional shrinking optimized paths.

Table 5. Results for Optimization in a Maze Environment.

Algorithm	Path Cost	POS Cost	BSOS Cost	Path Smoothness	POS Smoothness	BSOS Smoothness
RRT	5346.20	4413.83	4076.84	78.24	12.20	11.16
RRV	4838.35	4392.98	4056.15	33.77	12.50	11.31
Fast-RRT	5426.26	4408.72	4062.23	65.32	11.78	11.24
CERRT	6033.47	4397.60	4071.37	226.11	11.85	11.12

4.2.2. Bug Trap Environment Path Optimization

The purpose of the experiments conducted in the Bug Trap environment was to test the performance of the proposed optimization strategies in narrow passage environments. Figure 14 displays the results of pruning and bidirectional contraction optimization. The pruned paths were not able to determine the optimal route, while the bidirectional contraction path generated the optimal paths close to the wall. A detailed comparison of the paths is provided in Table 6. The paths generated by the four algorithms all had relatively low quality, and after pruning optimization, the path smoothness values were effectively improved, but the path lengths were not optimal. After bidirectional contraction path optimization, the paths had the lowest smoothness values and the shortest lengths and they were close to the optimal path. Therefore, it is evident that the bidirectional contraction path optimization strategy still outperformed the pruning optimization strategy in narrow environments and was applicable to all initial paths generated by sampling algorithms.

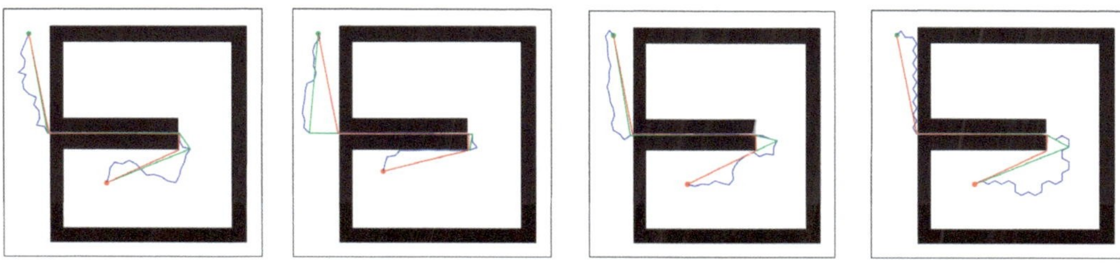

(a) RRT (b) RRV (c) Fast-RRT (d) CERRT

Figure 14. Pruning and bidirectional shrinking optimization in the Bug Trap environment. The blue lines represent the original paths, the green lines represent the pruned paths, and the red lines represent the paths optimized by bidirectional shrinking.

Table 6. Results for Optimization in the Bug Trap Environment.

Algorithm	Path Cost	POS Cost	BSOS Cost	Path Smoothness	POS Smoothness	BSOS Smoothness
RRT	1723.38	1386.87	1289.38	32.75	4.48	4.15
RRV	1618.49	1383.94	1281.70	13.58	4.78	4.34
Fast-RRT	1682.35	1375.54	1285.46	29.74	4.36	4.21
CERRT	1619.61	1323.14	1283.38	55.53	4.08	4.05

4.3. Evaluation of Algorithms in Real Environment

In order to evaluate the performance of the algorithm, we have chosen an actual map scenario, which is Tianjin Central Square with geographical coordinates of 117°04′56.88″ E, 39°05′49.19″ N, as shown in Figure 15a. The map covers an area of 175 m × 175 m and is divided into a grid of 1000 px × 1000 px, where each pixel represents an actual area of 0.03 m × 0.03 m. Figure 15b depicts the map generated based on the real scene, where black represents the obstacle areas and white represents the free space.

(a) Real map (b) Build map

Figure 15. Actual map environment.

The coordinates of the starting point are marked with a red circle at [100, 480], and the coordinates of the target point are marked with a green circle at [870, 240]. We will perform path planning tasks 100 times for a mobile robot in this scenario and evaluate the algorithm's performance by taking the average.

Algorithm Comparison

In Figure 16, the path planning results of four algorithms in a real-world environment are depicted. Notably, a curved narrow passage shortcut can be observed on the map. All four algorithms can generate feasible paths for the mobile robot, but only CERRT is able to discover and navigate through the curved passage shortcut in the actual environment. The other algorithms struggle to solve the narrow passage problem in the real environment, further confirming the practical value of the proposed algorithm. Additionally, the path quality is significantly improved after optimizing with BSOS compared to the initial paths.

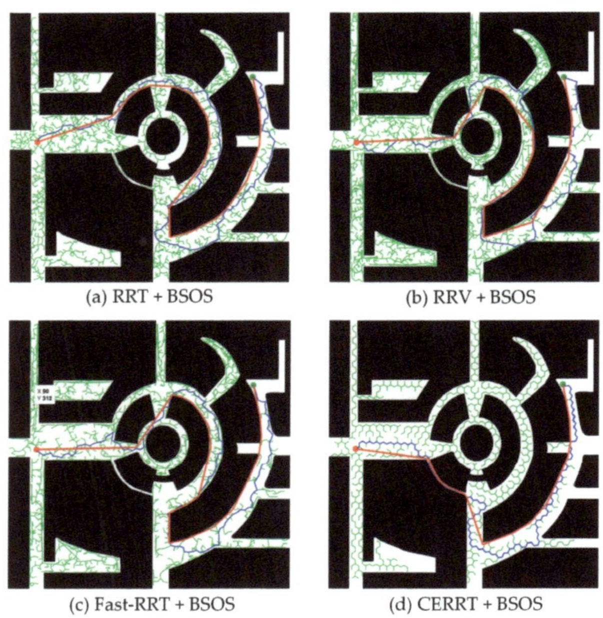

Figure 16. Performance comparison of four algorithms in real environment. The blue lines represent the original paths and the red lines represent the paths optimized by bidirectional shrinking.

Since RRT-based algorithms have probabilistic completeness, the success rate of all four algorithms in planning paths in the actual environment is 100% when the sampling limit is not restricted. Table 7 displays the performance parameters of these algorithms. The average runtime of RRT is approximately five times that of CERRT, RRV is approximately 50 times that of CERRT, and Fast-RRT is approximately seven times that of CERRT. CERRT has the shortest average runtime with a standard deviation of only 4.68, which is significantly lower than the other three algorithms, indicating excellent performance of the proposed algorithm in the actual environment.

Table 7. Results for planning in actual map.

Algorithm	Avg. Time (ms)	Std	Avg. Nodes	Path Cost	BSOS Cost
RRT	278.31	139.61	3126	2499.48	2104.93
RRV	2782.19	435.89	3149	2446.17	2036.01
Fast-RRT	365.72	138.94	2674	2545.87	2124.81
CERRT	55.23	4.68	907	1990.01	1706.98

Regarding path optimization, a horizontal comparison reveals that the path lengths after applying BSOS are all smaller than the initial paths, confirming the practicality of BSOS. A vertical comparison shows that the path length generated by CERRT is smaller than the other three algorithms. This is because CERRT is able to quickly discover and navigate through the curved passage shortcut in the actual environment, while the other algorithms struggle to pass through the curved passage efficiently.

In conclusion, the CERRT algorithm outperforms the other three algorithms in a real-world environment. The main reason is that the expansion principle of CERRT is inspired by the hexagonal honeycomb structure found in nature. Reference [21] mentions that a hexagonal honeycomb provides the least-perimeter way to enclose and separate infinitely many regions of unit area, indicating that using a hexagonal expansion strategy can greatly improve the utilization of each tree node and thus enhance the algorithm's performance.

5. Conclusions

In this paper, we propose a sampling-based path planning algorithm, CERRT, which performs better than other algorithms in complex environments and effectively solves the narrow passage problem. CERRT consists of two important parts: the first part limits the selection and expansion of tree vertices to maximize the utilization of each vertex, while the second part involves environment perception, where vertices are sampled near obstacles to ensure feasible passages are discovered. By combining the vertex selection method of the first part with the sampling strategy of the second part, the algorithm avoids redundant and useless exploration near trap-type obstacles, significantly improving planning performance. Numerical simulations comparing the proposed algorithm with others validate its effectiveness. Since sampling-based algorithms generally generate lower-quality paths, we propose the BSOS strategy to optimize the initial paths. The suitability of the algorithm has been verified through path optimization for different sampling algorithms. The optimized paths produced by the BSOS algorithm were superior to those produced by the well-known pruning optimization strategy (POS). However, for the algorithm proposed in this paper, the number of locally sampled points in three-dimensional environments may increase dramatically, limiting the environmental perception capability. Therefore, the next step is to study the performance of the algorithm in high-dimensional environments.

Author Contributions: Conceptualization, K.H. and Y.Y.; methodology, K.H. and Z.L.; software, Y.Y. and Z.L.; validation, K.H., Z.L. and Y.Y.; formal analysis, Z.L.; investigation, K.H.; resources, X.Z.; data curation, X.Z.; writing—original draft preparation, Y.Y.; writing—review and editing, K.H.; visualization, Y.L.; supervision, Y.L.; project administration, Z.L.; funding acquisition, K.H. All authors have read and agreed to the published version of the manuscript.

Funding: This work was supported in part by the Chunhui Cooperation Program of the Ministry of Education, HZKY20220590-202200265; National Natural Science Foundation of China under Grant 61902273.

Institutional Review Board Statement: Not applicable.

Informed Consent Statement: Not applicable.

Data Availability Statement: No data were used for the research described in the article.

Conflicts of Interest: The authors declare no conflict of interest.

References

1. Hichri, B.; Gallala, A.; Giovannini, F.; Kedziora, S. Mobile robots path planning and mobile multirobots control: A review. *Robotica* **2022**, *40*, 4257–4270. [CrossRef]
2. Öztürk, Ü.; Akdağ, M.; Ayabakan, T. A review of path planning algorithms in maritime autonomous surface ships: Navigation safety perspective. *Ocean. Eng.* **2022**, *251*, 111010. [CrossRef]
3. Wang, X.; Wei, J.; Zhou, X.; Xia, Z.; Gu, X. AEB-RRT*: An adaptive extension bidirectional RRT* algorithm. *Auton. Robot.* **2022**, *46*, 685–704.

4. Kuffner, J.J.; LaValle, S.M. RRT-connect: An efficient approach to single-query path planning. In *Proceedings of the Proceedings 2000 ICRA. Millennium Conference. IEEE International Conference on Robotics and Automation. Symposia Proceedings (Cat. No. 00CH37065), San Francisco, CA, USA, 24–28 April 2000*; IEEE: Piscataway, NJ, USA, 2000; pp. 995–1001.
5. Tahirovic, A.; Ferizbegovic, M. Rapidly-exploring random vines (RRV) for motion planning in configuration spaces with narrow passages. In *Proceedings of the 2018 IEEE International Conference on Robotics and Automation (ICRA), Brisbane, QLD, Australia, 21–25 May 2018*; IEEE: Piscataway, NJ, USA, 2018; pp. 7055–7062.
6. Hsu, D.; Jiang, T.; Reif, J.; Sun, Z. The bridge test for sampling narrow passages with probabilistic roadmap planners. In *Proceedings of the 2003 IEEE International Conference on Robotics and Automation (cat. no. 03CH37422), Taipei, Taiwan, 14–19 September 2003*; IEEE: Piscataway, NJ, USA, 2003; pp. 4420–4426.
7. Wu, Z.; Meng, Z.; Zhao, W.; Wu, Z. Fast-RRT: A RRT-Based Optimal Path Finding Method. *Appl. Sci.* **2021**, *11*, 11777. [CrossRef]
8. Cai, P.; Yue, X.; Zhang, H. ADD-RRV for motion planning in complex environments. *Robotica* **2022**, *40*, 136–153. [CrossRef]
9. Li, B.; Chen, B. An adaptive rapidly-exploring random tree. *IEEE/CAA J. Autom. Sin.* **2021**, *9*, 283–294. [CrossRef]
10. Chi, W.; Ding, Z.; Wang, J.; Chen, G.; Sun, L. A Generalized Voronoi Diagram-Based Efficient Heuristic Path Planning Method for RRTs in Mobile Robots. *IEEE Trans. Ind. Electron.* **2021**, *69*, 4926–4937. [CrossRef]
11. Taheri, E.; Ferdowsi, M.H.; Danesh, M. Fuzzy greedy RRT path planning algorithm in a complex configuration space. *Int. J. Control. Autom. Syst.* **2018**, *16*, 3026–3035. [CrossRef]
12. Karaman, S.; Frazzoli, E. Sampling-based algorithms for optimal motion planning. *Int. J. Robot. Res.* **2011**, *30*, 846–894. [CrossRef]
13. Islam, F.; Nasir, J.; Malik, U.; Ayaz, Y.; Hasan, O. Rrt∗-smart: Rapid convergence implementation of rrt∗ towards optimal solution. In *Proceedings of the 2012 IEEE International Conference on Mechatronics and Automation, Chengdu, China, 5–8 August 2012*; IEEE: Piscataway, NJ, USA, 2012; pp. 1651–1656.
14. Gammell, J.D.; Srinivasa, S.S.; Barfoot, T.D. Informed RRT*: Optimal sampling-based path planning focused via direct sampling of an admissible ellipsoidal heuristic. In *Proceedings of the 2014 IEEE/RSJ International Conference on Intelligent Robots and Systems, Chicago, IL, USA, 14–18 September 2014*; IEEE: Piscataway, NJ, USA, 2014; pp. 2997–3004.
15. Qureshi, A.H.; Ayaz, Y. Potential functions based sampling heuristic for optimal path planning. *Auton. Robot.* **2016**, *40*, 1079–1093. [CrossRef]
16. Jeong, I.-B.; Lee, S.-J.; Kim, J.-H. Quick-RRT*: Triangular inequality-based implementation of RRT* with improved initial solution and convergence rate. *Expert Syst. Appl.* **2019**, *123*, 82–90. [CrossRef]
17. Liao, B.; Wan, F.; Hua, Y.; Ma, R.; Zhu, S.; Qing, X. F-RRT*: An improved path planning algorithm with improved initial solution and convergence rate. *Expert Syst. Appl.* **2021**, *184*, 115457. [CrossRef]
18. Qian, K.; Liu, Y.; Tian, L.; Bao, J. Robot path planning optimization method based on heuristic multi-directional rapidly-exploring tree. *Comput. Electr. Eng.* **2020**, *85*, 106688. [CrossRef]
19. Chen, Y.; Fu, Y.; Zhang, B.; Fu, W.; Shen, C. Path planning of the fruit tree pruning manipulator based on improved RRT-Connect algorithm. *Int. J. Agric. Biol. Eng.* **2022**, *15*, 177–188. [CrossRef]
20. Hao, K.; Zhao, J.; Wang, B.; Liu, Y.; Wang, C. The Application of an Adaptive Genetic Algorithm Based on Collision Detection in Path Planning of Mobile Robots. *Comput. Intell. Neurosci.* **2021**, *2021*, 5536574. [CrossRef]
21. Morgan, F. The hexagonal honeycomb conjecture. *Trans. Am. Math. Soc.* **1999**, *351*, 1753–1763. [CrossRef]

applied sciences

MDPI

Article

Research on Multi-Sensor Data Fusion Positioning Method of Unmanned Ships Based on Threshold- and Hierarchical-Capacity Particle Filter

Yi Shen [1,2], Zeyu Zhao [1,*], Mingxin Yuan [1] and Sun Wang [1]

1 School of Mechanical Engineering, Jiangsu University of Science and Technology, Zhenjiang 212100, China; shenyi76@just.edu.cn (Y.S.); mxyuan78@just.edu.cn (M.Y.); 13115235855@163.com (S.W.)

2 Department of Intelligent Equipment Research, Zhangjiagang JUST Industrial Technology Research Institute, Zhangjiagang 215600, China

* Correspondence: 18952399326@163.com

Abstract: To improve the positioning accuracy of unmanned ships, a multi-sensor system including ZigBee, a Global Positioning System (GPS), and BeiDou Navigation Satellite System (BDS) is constructed, and an adaptive multi-sensor data fusion positioning method based on the threshold and hierarchical capacity particle filter (TCPF) is designed. First, the ZigBee-GPS/BDS multi-sensor measurement data is preprocessed to achieve a consistent space–time reference and transformed into the same coordinate system by projection. Then, the fault data is weighted and corrected through the consistency inspection of ZigBee-GPS/BDS multi-sensor positioning data, and the corresponding confidence factor is given according to the confidence distance of the positioning data; furthermore, the confidence factor is associated with stratified sampling. After that, the multi-sensor positioning data is filtered and denoised using a basic particle filter. Finally, a TCPF data fusion algorithm is designed, and the navigation positioning data of the unmanned ship is fused and filtered to obtain its positioning information. Numerical tests show that compared with other filtering algorithms, the mean square root error and standard deviation of the proposed TCPF algorithm decrease by an average of 25.0% and 28.0%, respectively, which verifies its high filtering accuracy and its advantages in suppressing particle degradation and avoiding sample scarcity. The experimental tests show that compared with other fusion algorithms, the proposed TCPF algorithm can not only realize the precise positioning during unmanned ship navigation, but also in the positioning and fault tolerance test, the average positioning error, root-mean-square error, and standard deviation of the former decrease by 36.0%, 38.0%, and 37.0%, respectively, and the corresponding performance indicators of the latter decrease by an average of 20.0%, 19.5%, and 17.5%, which verifies that it has the advantages of high data reliability and good filtering fault tolerance, and helps to improve the positioning accuracy of unmanned ships.

Keywords: unmanned ship; multi-sensor data; data fusion; positioning system; particle filter

Citation: Shen, Y.; Zhao, Z.; Yuan, M.; Wang, S. Research on Multi-Sensor Data Fusion Positioning Method of Unmanned Ships Based on Threshold- and Hierarchical-Capacity Particle Filter. *Appl. Sci.* **2023**, *13*, 10390. https://doi.org/10.3390/app131810390

Academic Editor: Jonghoek Kim

Received: 20 August 2023
Revised: 10 September 2023
Accepted: 14 September 2023
Published: 17 September 2023

1. Introduction

The water quality of rivers directly affects people's lives. In the early days, traditional water quality testing mainly relied on manual testing and reporting methods. Currently, water quality testing still relies primarily on manual testing, but data is uploaded through the Internet. Both traditional and current water quality testing face the following problems: (1) high labor input and cost; (2) high manual collection intensity and low efficiency; (3) data easily influenced by subjective factors; and (4) untimely information transmission. In recent years, unmanned ships have been widely used in the field of water quality monitoring due to their strong maneuverability and good controllability [1]. Based on this background, this paper conducts research on modern water quality testing based on unmanned ships. Unmanned ships carry various water quality sensors such as pH value, conductivity, turbidity, dissolved oxygen, etc., and can autonomously conduct regular water quality tests throughout the river area and transmit real-time data to a monitoring center via the Internet. This method has the following advantages: it (1) reduces labor costs; (2) improves detection efficiency; (3) unaffected by subjective factors, improves accuracy; and (4) ensures timely data availability. From the above analysis, it can be seen that unmanned ships play a key role in modern water quality monitoring, with their autonomous navigation positioning accuracy directly affecting safety and efficiency.

How to achieve accurate positioning of unmanned ships based on multiple combined sensors has been the focus of research [2]. Shen et al. [3] used GNSS/GPS combination sensors to obtain coordinates of unmanned ship positions while reducing noise interference caused by long-distance nodes, but they suffer from signal loss due to external environmental interference leading to reduced positioning accuracy. Wu et al. [4] realized the positioning of the unmanned ship by solving its position information collected by the GPS/INS combined sensors, but the INS has the problem of mechanical fault jumping. As time goes by, the GPS/INS combined sensor is likely to cause a large positioning deviation of the unmanned ship. Deng et al. [5] obtained the position information of the unmanned ship by fusing the GPS/IMU combined sensor data, but in the fusion process, if the sensor data is lost at a certain moment, it will cause a large fusion deviation and affect the positioning accuracy. Data fusion based on two kinds of sensors is the current mainstream technology for unmanned ship positioning, but there is a deficiency in that the positioning accuracy and reliability are affected due to the weakening or loss of a certain sensor signal. In recent years, how to improve positioning accuracy by further fusing data from three or more types of sensors has become a research hotspot. Wang et al. [6] achieved the reliable positioning of trains by using the combined positioning method of PPP-GPS/IMU. However, the built multi-sensor positioning system is prone to environmental interference, which reduces the reliability of data samples. Tang et al. [7] realized the elimination of abnormal signals by setting outlier identification in the process of IMU/ODM/UWB multi-sensor data fusion, which improved the positioning accuracy of multi-sensors, but reduced the fault-tolerant performance of the algorithm because the fault data was not weighted. Sofia et al. [8] divided the observation information of different sensors into blocks according to the size of the scale to obtain a multi-scale system model, which effectively solved the problem of measurement delay, but the efficiency and accuracy of algorithm fusion were reduced because no data confidence assignment was performed during data processing. As can be seen from the above, adding sensors helps to improve the reliability and accuracy of positioning. However, there is still a lack of research on how to further enhance the accuracy and efficiency of algorithm fusion through data confidence interval testing, as well as how to achieve effective positioning under low- or weak-signal conditions through data correction. Therefore, a ZigBee-GPS/BDS multi-sensor positioning system is built in this paper, which realizes the accurate and reliable positioning of unmanned ships by complementing and weighted-correcting signals from three types of sensors, as well as assigning confidence values. In multi-sensor data fusion, filtering algorithms are crucial, and particle filtering has been widely used due to its ability to handle various probability distribution models, but it has also become the focus of research due to its heavy dependence on initial state estimation and particle degradation. Ha et al. [9] expanded the search space of particle states by designing crossover operators and mutation operators in the particle filter calculation module, which improved the filtering accuracy of multi-sensor data but also reduced the computational efficiency. Alam et al. [10] processed the prefetched value in the weight storage of the particle-filter algorithm in parallel with the value in the random function generator, which reduced the time required for resampling and improved the fusion efficiency of sensor data, but it did not suppress the degradation of particle samples very well. Wu et al. [11] used the unscented particle-filter algorithm with constrained residuals to fuse the two sets of sensor positioning data, which effectively overcomes external environmental disturbances. However, the filtering accuracy is prone to divergence after multiple iterations. A large number of studies have shown that although the existing multi-sensor data fusion algorithms have improved the positioning accuracy of moving targets to a certain extent, their fault-tolerant performance is not good. In addition, they did not carry out pre-judgment processing on the collected data, and the problem of low filtering accuracy has not been solved when the data collected by the multi-sensor combined positioning system is fused and filtered. Therefore, a multi-sensor data fusion positioning method based on adaptive threshold and hierarchical capacity particle filter is designed in this paper. Firstly, the basic particle filter is introduced to denoise multi-sensor data. Then, the latest multi-sensor positioning data is incorporated into the proposal distribution using unscented transformation. After that, the Gaussian mixture model is constructed and the adaptive threshold is set, and the data confidence factor is associated with hierarchical sampling to further improve the fusion filtering accuracy of the algorithm. Finally, the validity of the method is verified by the numerical test of the model and the experimental test of unmanned ship navigation.

2. Unmanned Ship Positioning System and Its Multi-Sensor Data Fusion Framework

2.1. Construction of Unmanned Ship Positioning System

In order to realize the information complementarity among the multi-sensor measurement data, reduce the interference of the surrounding environment on the working state of the positioning system, improve the reliability of the multi-sensor data samples, and then improve the positioning accuracy of an unmanned ship for water quality detection, a ZigBee-GPS/BDS multi-sensor combined positioning system, as shown in Figure 1, was constructed. The entire positioning system includes a ZigBee multi-node positioning system, GPS positioning system, BDS positioning system, PC communication system, etc.

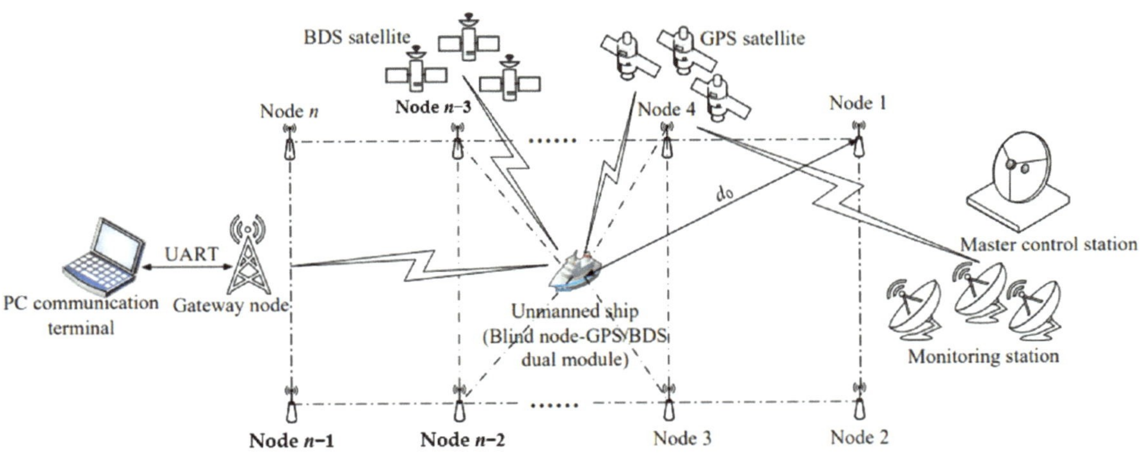

Figure 1. ZigBee-GPS/BDS multi-sensor combined positioning system for an unmanned ship.

A GPS/BDS positioning system is mainly composed of space, ground monitoring, and user receiver. First, the ground monitoring part monitors and controls the operation of each satellite through the main control station and monitoring station, traverses the navigation message, and maintains the system time. Then, the GPS and BD satellites in space continuously send their own ephemeris and time information to the GPS/BDS dual module installed in the unmanned ship. Finally, the GPS/BDS dual module calculates the latitude and longitude coordinates of the unmanned ship in real time by analyzing the signal messages sent by the respective satellites [12].

A ZigBee multi-node positioning system is mainly composed of blind nodes, reference nodes, and wireless gateway nodes. First, the communication environment of the full grid network is constructed according to the reference nodes with known position coordinates. Then, a blind node installed on the unmanned ship is placed in a communication network composed of reference nodes [13]. After that, the wireless gateway node is connected to the PC through the serial port. On the one hand, it receives the configuration data of each reference node and mobile node from the monitoring software and forwards them to the corresponding nodes. On the other hand, it receives the valid data fed back by each node and transmits them to the monitoring software. Finally, the RSSI technology is used to determine the distance or direction from the unknown node to the beacon node, and the maximum likelihood estimation method is used to calculate the position of the blind node on the unmanned ship during navigation in the river.

Figure 2 shows the self-developed unmanned ship for water quality testing. The unmanned ship is equipped with a detector with multiple water quality detection sensors connected to the launch and recovery device, as well as a positioning system including Zigbee, GPS, and BDS modules. The controller of the unmanned ship is an industrial computer with a Linux system, which is responsible for receiving and processing the data of each sensor, issuing control instructions, and transmitting signals with the PC.

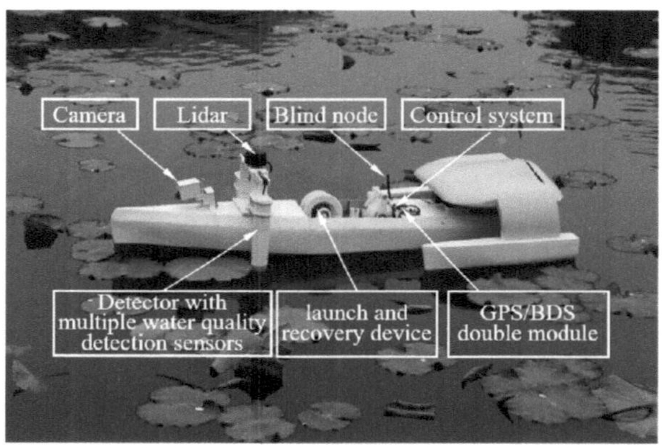

Figure 2. Test platform for unmanned-ship positioning system.

The lidar of Silan A3 is responsible for collecting the environmental information around the unmanned ship. The ATK1218-GPS-BDS dual-positioning module is selected in the GPS/BDS positioning system to obtain the latitude and longitude information of the unmanned ship. The CC2530 wireless module is selected in the ZigBee positioning system to obtain the plane coordinate information of the unmanned ship. The Hikvision DS-2 camera is selected in the vision system to be responsible for the collection of environmental information of the unmanned-ship channel.

2.2. Framework of Multi-Sensor Data Fusion

In order to improve the reliability, fault tolerance, and accuracy of multi-sensor data fusion for the unmanned ship for water quality detection, a new multi-sensor-data-fusion positioning framework based on particle filtering as, shown in Figure 3, was designed. From the figure, it can be seen that the longitude and latitude information of the unmanned ship was first unified in time and space, and projected into the local coordinate system constructed by the ZigBee multi-node module through coordinate transformation, so as to realize mutual complementarity among sensor data. Then, through the consistency inspection of the multi-sensor data, the fault data are weighted and corrected to improve the fault-tolerant performance of the data fusion algorithm. At the same time, the corresponding confidence factor was given according to the confidence distance of the positioning data, and the confidence factor was associated with hierarchical sampling to improve the fusion efficiency and accuracy of the data fusion algorithm. After that, the basic particle filter was used to de-noise the multi-sensor positioning data to improve the reliability of the data samples. Finally, an improved particle-filter algorithm was designed to fuse and filter the multi-sensor positioning data to achieve accurate positioning of the unmanned ship.

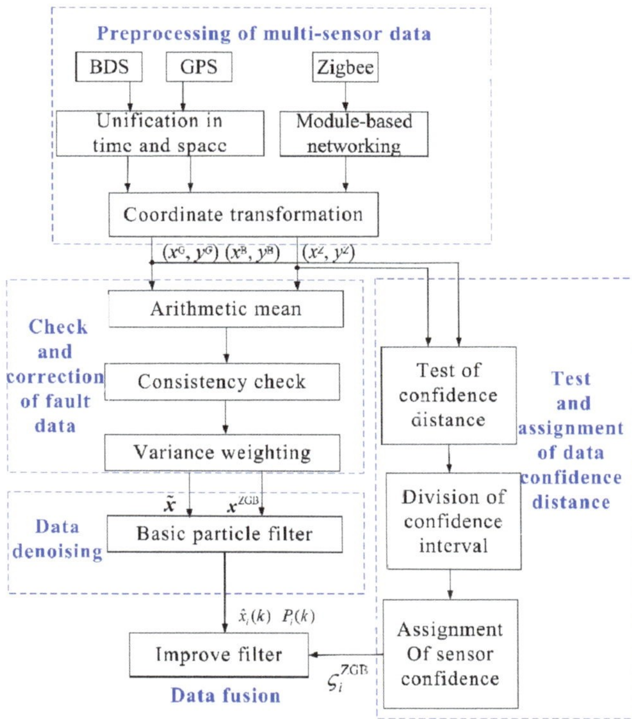

Figure 3. Particle filter-based multi-sensor data fusion positioning framework for unmanned ships.

3. Pre-Processing of Multi-Sensor Data from the Unmanned Ship

3.1. System Modeling

3.1.1. Measurement Equation of GPS/BDS

The measurement equation of GPS/BDS consists of nonlinear pseudorange measurement equations [14]. The measurement equations of the BDS pseudorange p^{B} and the GPS pseudorange p^{G} are expressed as follows:

$$\begin{aligned} p_i^{\mathrm{B}}(k) &= \left|r(k) - r_i^{\mathrm{B}}(k)\right| + c\delta t_{\mathrm{B}} + \varepsilon_i(k) + v_i(k) \\ &= \sqrt{(X^j(k) - X_i^{\mathrm{B}}(k))^2 + (Y^j(k) - Y_i^{\mathrm{B}}(k))^2 + (Z^j(k) - Z_i^{\mathrm{B}}(k))^2} \\ &+\delta t_{\mathrm{B}} + \varepsilon_i(k) + v_i(k) \end{aligned} \tag{1}$$

$$\begin{aligned} p_j^{\mathrm{G}}(k) &= \left|r(k) - r_j^{\mathrm{G}}(k)\right| + c\delta t_{\mathrm{G}} + \varepsilon_j(k) + v_j(k) \\ &= \sqrt{(X^j(k) - X_j^{\mathrm{G}}(k))^2 + (Y^j(k) - Y_j^{\mathrm{G}}(k))^2 + (Z^j(k) - Z_j^{\mathrm{G}}(k))^2} \\ &+\delta t_{\mathrm{G}} + \varepsilon_i(k) + v_i(k) \end{aligned} \tag{2}$$

where $X^j(k)$, $Y^j(k)$ and $Z^j(k)$ are the position coordinates of the receiver at time k; $X_i^{\mathrm{B}}(k)$, $Y_i^{\mathrm{B}}(k)$ and $Z_i^{\mathrm{B}}(k)$ are the position coordinates of the ith BDS satellite at time k; $X_j^{\mathrm{G}}(k)$, $Y_j^{\mathrm{G}}(k)$ and $Z_j^{\mathrm{G}}(k)$ are the position coordinates of the jth GPS satellite at time k; $r(k)$ is the position vector of the receiver; $r_i^{\mathrm{B}}(k)$ and $r_j^{\mathrm{G}}(k)$ are the position vectors of the ith BDS and the jth GPS satellite at time k, respectively; $p_i^{\mathrm{B}}(k)$ and $p_j^{\mathrm{G}}(k)$ are the pseudoranges of the ith BDS and the jth GPS satellite at time k, respectively; δt_{B}, δt_{G} is the error of the BDS- and GPS-receiving clock, respectively; $\varepsilon_i(k)$ is the non-white noise error of channel i at time k; $v_i(k)$ is the measurement noise of channel i at time k, $i = 1, 2, \ldots, n_1$, $j = 1, 2, \ldots, n_2$, $k = 1, 2, \ldots, T$. n_1 and n_2 are the number of BDS and GPS satellites received, respectively. T is the maximum duration.

When the signals of n satellites ($n = n_1 + n_2 \geq 5$) are received, the observation variables of the system are

$$z^{B,G}(k) = \left[p_1^B(k), p_2^B(k), \cdots, p_{n_1}^B(k),\ p_1^G(k), p_2^G(k), \cdots, p_{n_2}^G(k)\right]^T \tag{3}$$

Assuming that the number of GPS and BDS satellites observed at time k are equal (namely $n_1 = n_2$), the observation equation of the GPS/BDS satellite positioning system is established as

$$z^{B,G}(k) = \begin{bmatrix} \boldsymbol{p}^B(k) \\ \boldsymbol{p}^G(k) \end{bmatrix} + R(k) \tag{4}$$

where $\boldsymbol{p}^B = \{p_1^B, p_2^B, \cdots, p_{n1}^B\}$ is the BDS pseudorange, $\boldsymbol{p}^G = \{p_1^G, p_2^G, \cdots, p_{n2}^G\}$ is the GPS pseudorange, and R is the covariance matrix of observation noise.

3.1.2. Measurement Equation of ZigBee

According to the principle of ZigBee positioning networking in this paper, and taking the sampling time T_k of basic particle filter (BPF) as the time reference, the measurement equation of $RSSI$ signal collected by the ZigBee positioning system was defined [15]. The specific steps were as follows:

Step 1 Establish the signal strength power model $\widetilde{P}_i^R$ of the ith reference node according to the path loss principle.

$$\widetilde{P}_i^R = P_0 - 10\vartheta \lg(\sqrt{(x^z - x_i^R)^2 + (y^z - y_i^R)^2}/d_0) \tag{5}$$

where $\forall\ i \in [1, n]$, (x_i^R, y_i^R) is the coordinate of the ith reference node; P_0 is the received power of the reference node at d_0 from the unmanned ship; ϑ is the path loss coefficient, and (x^z, y^z) is the blind node coordinate.

Step 2 Establish the covariance matrix $\boldsymbol{R}_{RSSI}$ of the observation equation.

$$\boldsymbol{R}_{RSSI} = \mathrm{diag}(\sigma^2_{dB_1^R}, \sigma^2_{dB_2^R}, \cdots, \sigma^2_{dB_n^R}) \tag{6}$$

where $\forall\ i \in [1, n]$, $\sigma^2_{dB_i^R}$ is the initial variance of the signal strength of the ith ZigBee reference node.

Step 3 Define the observation equation for $RSSI$ measurements collected by ZigBee.

$$z_{RSSI}(k) = [\widetilde{\boldsymbol{P}}^R(k),\ \boldsymbol{R}_{RSSI}(k)]^T + v(k) \tag{7}$$

where $\widetilde{\boldsymbol{P}}^R = \left\{\widetilde{P}_1^R, \widetilde{P}_2^R, \cdots, \widetilde{P}_n^R\right\}$ is the signal strength power model; v is the covariance matrix of the observed noise.

3.1.3. Multi-Sensor Data Definition for the Unmanned Ship

In order to facilitate the description of the subsequent algorithm, the following data definition was carried out in this paper: The data collected by the sensor or the known data was defined as the measurement data, which involves the reference node coordinates (x^R, y^R) in the ZigBee system and the geodetic coordinates (B, L) collected by the GPS/BD system. The data obtained by calculation and conversion was defined as positioning data, which involves the coordinates of blind nodes (x^z, y^z) in ZigBee system, and the plane positioning coordinates (x^G, y^G) and (x^B, y^B) in GPS/BDS system. The GPS positioning data set was defined as $\boldsymbol{H}^g = \{h_1^g, h_2^g, \ldots, h_n^g\}$, the BDS positioning data set as $\boldsymbol{H}^b = \{h_1^b, h_2^b, \ldots, h_n^b\}$, and the ZigBee blind node positioning data set as $\boldsymbol{H}^z = \{h_1^z, h_2^z, \ldots, h_n^z\}$, where $h_i^z = (x_i^z, y_i^z)$, $h_i^g = (x_i^G, y_i^G)$, $h_i^b = (x_i^B, y_i^B)$ are the ith positioning data in the ZigBee, GPS, and BDS data sets, respectively, $i = 1, 2, \ldots, n$.

3.2. Unification of GPS/BDS Space-Time Reference

In order to obtain the trajectory information of the unmanned ship collected by the GPS/BDS dual-module sensor, it is necessary to unify the time of the GPS and BDS satellite systems. Due to the jump error between GPS time (GPST) and UTC time (UTCT) [16], as time goes by, the deviation between the two gradually increases, and at present the deviation has reached 16 s, that is

$$\text{GPST} \approx \text{UTCT} + 16 \tag{8}$$

BDS time (BDST) takes the national standard unit system s as the basic unit for time accumulation. Because there is no jump second in BDST, it is continuous time. The start time of BDST is converted into the number of weeks of GPST and the count of seconds in a week is 1356 weeks—14.000 s. Therefore, the following relationship exists between BDST and GPST

$$\text{BDST}_{\text{week number}} = \text{GPST}_{\text{week number}} + 1356 \tag{9}$$

$$\text{BDST}_{\text{Seconds of the week count}} = \text{GPST}_{\text{Seconds of the week count}} + 14 \tag{10}$$

Each country has established different spatial benchmarks according to its own satellite navigation system, so it is necessary to unify the spatial benchmarks when performing combined positioning of different satellite systems. WGS-84 was used for GPS, and the 2000 National Geodetic Coordinate System (NGCS2000) was used for the BDS. The origin, scale, orientation, and the scale evolved from the orientation defined by the two coordinate systems were the same, and both are closely related to the Earth Reference System (ERS). After multiple optimizations, the difference between the two is only at the centimeter level [17]. For most non-precision positioning applications such as navigation, the WGS-84 and NGCS2000 coordinate systems do not need to be converted, so they are considered as unified coordinate systems in this paper.

3.3. Network Positioning of Zigbee System

In order to meet the multi-node effective communication of the unmanned-ship Zigbee positioning system, n reference nodes were set in the positioning network coordinate system in this paper, so that the blind node on the unmanned ship is in a network composed of n reference nodes with a known signal strength and position coordinates [18]. The specific steps are as follows:

Step 1 Define the *RSSI* theoretical value of the signal strength indication of the received power, and calculate the distance between the blind node and the reference node.

$$RSSI = -(10\vartheta \cdot \lg d + A) \tag{11}$$

From Equation (11), we can obtain

$$d = 10^{-\frac{A+RSSI}{10\vartheta}} \tag{12}$$

where ϑ is the path loss coefficient, d is the distance between the blind node and the reference node, *RSSI* is the signal strength value, and A is the strength of the initial signal at the transmitter.

Step 2 Establish the following distance equations among the blind node and the reference nodes in the plane:

$$\begin{bmatrix} (x_1^R - x^z)^2 + (y_1^R - y^z)^2 \\ (x_2^R - x^z)^2 + (y_2^R - y^z)^2 \\ \vdots \\ (x_n^R - x^z)^2 + (y_n^R - y^z)^2 \end{bmatrix} = \begin{bmatrix} d_1^2 \\ d_2^2 \\ \vdots \\ d_n^2 \end{bmatrix} \tag{13}$$

where $\forall i \in [1, n]$, d_i is the distance between the blind node and the ith reference node.

Step 3 Based on the reference node's distance equation set established by Equation (13), the linear transformation equation $2A\begin{bmatrix} x^z \\ y^z \end{bmatrix} = B$ of the blind node coordinates of the unmanned ship was constructed, where A and B are expressed as follows:

$$A = \begin{bmatrix} x_1^R - x_n^R & y_1^R - y_n^R \\ x_2^R - x_n^R & y_2^R - y_n^R \\ \vdots & \\ x_{n-1}^R - x_n^R & y_{n-1}^R - y_n^R \end{bmatrix} B = \begin{bmatrix} d_n^2 - d_1^2 - (x_n^R)^2 + (x_1^R)^2 - (y_n^R)^2 + (y_1^R)^2 \\ d_n^2 - d_2^2 - (x_n^R)^2 + (x_2^R)^2 - (y_n^R)^2 + (y_2^R)^2 \\ \vdots \\ d_n^2 - d_{n-1}^2 - (x_n^R)^2 + (x_{n-1}^R)^2 - (y_n^R)^2 + (y_{n-1}^R)^2 \end{bmatrix} \tag{14}$$

Step 4 The coordinates (x^z, y^z) of the blind node in the ZigBee system network are obtained by the least-squares method.

$$\begin{bmatrix} x^z \\ y^z \end{bmatrix} = \frac{1}{2}(A^T A)^{-1} A^T B \tag{15}$$

3.4. Coordinate Transformation

In order to perform data fusion filtering on GPS/BDS positioning data and ZigBee blind node location information, it is necessary to transform the WGS-84/NGCS2000 coordinate system [19] and unify it to the local plane coordinates of the unmanned-ship navigation test with the ZigBee multi-node networking coordinate system. For this reason, the existing geodetic coordinates (B, L) were used in this paper to transform them into plane coordinates through Gauss–Kruger projection, and translate and rotate them to make them unified with the local plane coordinates of the unmanned ship navigation. The specific steps are as follows:

Step 1 convert WGS-84/NGCS2000 geodetic coordinates (B, L) to WGS-84/NGCS2000 Gaussian plane coordinates (x^s, y^s):

$$\begin{aligned} x^s = X_0^B + \tfrac{1}{2}N_p \cdot t \cdot m_0^2 + \tfrac{1}{24}(5 - t^2 + 9\eta^2 + 4\eta^4)N_p \cdot t \cdot m_0^4 \\ + \tfrac{1}{720}(61 - 58t^2 + t^4)N_p \cdot t \cdot m_0^6 \end{aligned} \tag{16}$$

$$\begin{aligned} y^s = N_p \cdot m_0 + \tfrac{1}{6}(1 - t^2 + \eta^2)N_p \cdot m_0^3 \\ + \tfrac{1}{120}(5 - 18t^2 + t^4 + 14\eta^2 - 58\eta^2 \cdot t^2)N_p \cdot m_0^5 \end{aligned} \tag{17}$$

where $m_0 = \frac{\Delta L}{\gamma}\cos(B)$, $t = \tan(B)$, $\eta^2 = (e')^2\cos^2 B$ and ΔL are the difference between the longitude (L) of the desired point and the longitude (L_0) of the central meridian, namely $\Delta L = L - L_0$. γ is the positioning coefficient error, B is the latitude of the desired point, and its unit is radians, and e' is the second eccentricity of the ellipsoid.

N_p is the curvature radius of the primary vertical circle passing through this point, which can be described as

$$N_p = \frac{a}{\sqrt{1 - e^2 \sin 2\phi}} \tag{18}$$

where a is the long radius of the ellipsoid, e is the first eccentricity of the ellipsoid, and φ is the geodetic dimension.

X_0^B is the arc length from the equator to the central meridian cut by a parallel circle passing through this point and can be described as

$$X_0^B = C_0 - \cos B(C_1 \sin B + C_2 \sin^3 B + C_3 \sin^5 B + C_4 \sin^7 B) \tag{19}$$

where C_0, C_1, C_2, C_3 and C_4 are ellipsoid parameters.

Step 2 transform the Gaussian coordinates of the unmanned ship navigation collected by GPS/BDS into the local coordinate system of the unmanned ship test through translation and rotation.

$$\begin{bmatrix} x^{G,B} \\ y^{G,B} \end{bmatrix} = \begin{bmatrix} \Delta x \\ \Delta y \end{bmatrix} + (1+m)\begin{bmatrix} \cos\theta' & \sin\theta' \\ -\sin\theta' & \cos\theta' \end{bmatrix}\begin{bmatrix} x^s \\ y^s \end{bmatrix} \tag{20}$$

where $(x^{G,B}, y^{G,B})$ are the GPS/BDS plane positioning coordinates of the unmanned ship, Δx and Δy are the translation amount of the coordinate system, m is the scale factor, and θ' is the rotation angle of the two coordinate systems.

4. Confidence Determination of Multi-Sensor Data

In order to improve the fusion efficiency and filtering accuracy of multi-sensor data of the unmanned ship, the confidence distance judgment and credibility assignment of ZigBee-GPS/BDS multi-sensor data was carried out in this paper. Firstly, the credibility of the positioning data was determined by performing a confidence interval test on the sensor positioning data. Then, the credibility was divided into the corresponding confidence intervals according to its size, and the corresponding confidence factors were assigned. Finally, the confidence factor was associated with the TCPF algorithm to improve the filtering accuracy of the algorithm.

4.1. Judgement of Confidence Distance

When the location information of the unmanned ship is positioned based on the ZigBee-GPS/BDS multi-sensor positioning system, its sensor characteristic function can be described by the Gaussian distribution curve $p(x_i)$ [20], and the positioning data x_i follows the normal distribution $N(\mu, \sigma)$, namely

$$p(x_i) = \frac{1}{\sqrt{2\pi}\sigma} e^{\frac{(x_i-\mu)^2}{2\theta^2}} \tag{21}$$

where x_i is the ith positioning data of the sensor, μ is the true value of the positioning data, θ is the measurement accuracy of the sensor information, and σ is the measurement error of the sensor information, $i = 1, 2, \ldots, n$.

Let $\boldsymbol{X} = \{\boldsymbol{H}^z, \boldsymbol{H}^g, \boldsymbol{H}^b\}$, $x_i^{ZGB} = \{h_i^z, h_i^g, h_i^b\}$, $\forall\, i \in [1, n]$, $x_i^{ZGB} \in \boldsymbol{X}$: the confidence distance of the sensor positioning data is determined using the defined $p(x_i)$. The specific steps are as follows:

Step 1 calculate the confidence distance among the positioning data of the ZigBee-GPS/BDS unmanned-ship positioning system at different times.

$$d_i(k) = p_r\left\{\frac{\left|x_i^{ZGB}(k) - x_i^{ZGB}(k+1)\right|}{\sqrt{\widehat{\tau}}\min\left(\tau_i^{ZGB}(k), \tau_i^{ZGB}(k+1)\right)}\right\} \tag{22}$$

where $p_r\{\cdot\}$ is a probability function, $x_i^{ZGB}(k)$ and $x_i^{ZGB}(k+1)$ are the observation values of the ith positioning data at time k and $k+1$, respectively, $\tau_i^{ZGB}(k)$ and $\tau_i^{ZGB}(k+1)$ are the measurement variances of the ith positioning data at time k and $k+1$, and $\widehat{\tau}$ is the mean of the measurement variance, $k = 1, 2, \ldots, T$.

Step 2 determine the credibility of the sensor information at time k according to the confidence level ε of the preset sensor support.

$$\begin{cases} \text{High data credibility}, & if\ d_i(k) \le 1-\varepsilon \\ \text{Low data credibility}, & if\ d_i(k) > 1-\varepsilon \end{cases} \tag{23}$$

Step 3 Divide the positioning data into corresponding confidence intervals according to the confidence distance of the positioning data [21].

$$\begin{cases} d_i(k) > 1-\varepsilon,\ x_i^{\mathrm{ZGB}}(k) \in [\frac{\varepsilon}{\widehat{\tau}} - \iota, \frac{\varepsilon}{\widehat{\tau}}] \\ d_i(k) \leq 1-\varepsilon,\ x_i^{\mathrm{ZGB}}(k) \in [\frac{\varepsilon}{\widehat{\tau}}, \frac{\varepsilon}{\widehat{\tau}} + \iota] \end{cases} \quad (24)$$

where ι is the variable coefficient of the sampling sample in the probability interval.

In this paper, the credibility between sensor information at different times was determined by calculating the ZigBee-GPS/BDS confidence distance, and the determined data was divided into corresponding confidence intervals according to the probabilistic measurement value between the positioning data, which caused the credible data to be close to the region with higher confidence. Giving priority to the confidence factors of the sensor positioning data is helpful in improving the efficiency of data fusion. The set confidence level ε corresponded to the mean value $\widehat{\tau}$ of the sensor measurement variance, which was used to represent the change in the confidence degree of the multi-sensor positioning data at different times. The classification of the corresponding confidence interval also changed when the mean value $\widehat{\tau}$ of the measurement variance was different, which improved the assignment accuracy of the data fusion algorithm to the sensor positioning data.

4.2. Credibility Assignment

After the confidence distance was determined for the ZigBee-GPS/BDS positioning data, the multi-sensor positioning data was divided into corresponding confidence intervals. In order to improve the filtering accuracy of the data fusion algorithm, the confidence factor was given according to the confidence distance of the positioning data, and the confidence factor was associated with the hierarchical sampling of the proposed TCPF algorithm. The higher the confidence of the positioning data, the higher the corresponding fusion bias, which helps to improve the filtering accuracy of the TCPF algorithm for the location information of the unmanned ship. The specific steps are as follows:

Step 1 define β_i as the comprehensive support degree of the ith positioning data, which is composed of several confidence weight coefficients [22].

$$\beta_i = y_1\omega_1' + y_2\omega_2' + \ldots + y_n\omega_n' \quad (25)$$

$$\omega'_i = \frac{1}{n} - \frac{1}{2a} + \frac{1}{na}\sum_{i=1}^{n} d_i \quad (26)$$

$$\sum_{i=1}^{n} \beta_i = 1 \quad (27)$$

where ω'_i is the confidence weight coefficient of the ith positioning data, n is the numericized amount of information collected by multiple sensors, and a is the adjustment parameter and can be expressed as $a = (n-1)/2$. $y_1, y_2, \ldots, y_n$ are a set of non-negative numbers.

Step 2 according to the comprehensive support degree of the defined positioning data, the norm equation of the dynamic support factor $\beta_i^o(k)$ for multi-sensor information and the probability model $p_r(x_i^{\mathrm{ZGB}}(k))$ for the positioning data is constructed.

$$\beta_i^o(k) = \frac{\left\| p_r(x_i^{\mathrm{ZGB}}(k)) \right\|_F}{\sum_{i=1}^{n} \left\| p_r(x_i^{\mathrm{ZGB}}(k)) \right\|_F} \quad (28)$$

where $||.||_F$ is the Frobenius norm, $k = 1, 2, \ldots, T$, $i = 1, 2, \ldots, n$.

Step 3 Calculate the multi-sensor measurement error w_i^{ZGB} according to the dynamic support factor $\beta_i^o(k)$:

$$w_i^{ZGB}(k) = x_i^{ZGB}(k) - A\beta_i^o(k) \quad (29)$$

where A is the state transition matrix.

Step 4 The credibility of the positioning data of the unmanned ship positioning system is assigned using w_i^{ZGB}:

$$\varsigma_i^{ZGB}(k) = \frac{(w_i^{ZGB}(k))^{-1}\omega'_i}{\sum\limits_{i=1}^{n} w_i^{ZGB}(k)} \quad (30)$$

where $\varsigma_i^{ZGB}(k)$ is the confidence factor of the ZigBee-GPS/BDS multi-sensor positioning data.

5. Inspection and Weighted Compensation of Multi-Sensor Data

In the data set collected by the ZigBee-GPS/BDS multi-sensor system of the unmanned ship, some data were valid, but some data may have caused measurement deviation due to environmental or noise interference. In order to improve the fault-tolerant performance of the data fusion algorithm, the consistency inspection of the positioning data set collected by the multi-sensor positioning system was carried out in this paper, and the inconsistency fault data was corrected by weighting.

5.1. Consistency Inspection of Sensor Data

For the multi-sensor positioning system composed of ZigBee-GPS/BDS, the ith positioning data x_i^{ZGB} can be expressed as

$$x_i^{ZGB} = x_i^t + \xi_i \quad (31)$$

where x_i^t is the true value, and ξ_i is the measurement noise, $i = 1, 2, \ldots, n$.

The obtained multi-sensor positioning data were arithmetically averaged and inspected for consistency [23]. The specific steps were as follows:

Step 1 the arithmetic mean of the positioning data, $\overline{x}$, $k = 1, 2, \ldots, T$, was calculated:

$$\overline{x}_i = (\sum_{k=1}^{T} x_i^{ZGB}(k))/T \quad (32)$$

Step 2 according to the arithmetic mean value of the positioning data, the consistency inspection of the multi-sensor positioning data was carried out.

$$\left|\boldsymbol{x}^{ZGB}(k) - \overline{x}_i\right| \leq \tau \quad (33)$$

where $\boldsymbol{x}^{ZGB}(k) = \{x_1^{ZGB}(k), x_2^{ZGB}(k), \ldots, x_n^{ZGB}(k)\}$ is the multi-sensor positioning data, and τ is the system requirement error.

Through the set system error requirements, the consistency inspection of the multi-sensor positioning data was carried out. If the difference between the sensor positioning data and the arithmetic mean value was less than the error required by the system, the positioning data was determined to be consistent, that is, credible data, and could be denoised using basic particle filtering. On the contrary, the variance of the sampled data needed weighted correction to meet the sampling requirements of the basic particle filter for data samples.

5.2. Weighted Correction for Variance

When the ZigBee-GPS/BDS multi-sensor positioning system was used to locate the navigation path of the unmanned ship at different positions in the same space, the measurement noise of each sensor followed Gaussian distribution [24]. The variance of the ith positioning data can be expressed as

$$\widehat{\sigma}_i^2 = E(x_i^{ZGB} - x_i^t)^2 \quad (34)$$

In the real navigation situation of the multi-sensor positioning system of the unmanned ship, since the true positioning value x_i^t in Equation (34) is unknown, the arithmetic mean value $\overline{x}_i$ of the positioning data in Equation (32) was taken as the unbiased estimate of the true value x_i^T, namely

$$\sigma_i'^2 = D(x_i^{ZGB} - \overline{x}_i) \quad (35)$$

In order to improve the credibility of data samples, the multi-sensor positioning system in this paper measured the position of the unmanned ship m times, and integrated the m times of positioning data into a data set. The jth positioning value of the ith sensor information was x_{ij}, and the sensor positioning data x_i^{ZGB} was replaced in Equation (35) with x_{ij} to obtain the information variance of the data set, which can be expressed as

$$\overline{\sigma}_i^2 = \frac{1}{m}\sum_{j=1}^{m}(x_{ij} - \overline{x}_i)^2 \quad (36)$$

where $j = 1, 2, \ldots, m$.

According to the variance of the data collected by the sensor, the fusion weight κ_i of the fault data was defined as follows:

$$\kappa_i = \frac{(\overline{\sigma}_i^2)^{-1}}{\sum\limits_{i=1}^{n}\frac{1}{\overline{\sigma}_i^2}} \quad (37)$$

$$\sum_{i=1}^{n}\kappa_i = 1 \quad (38)$$

According to the obtained fusion weight κ_i, the multi-sensor fault data was weighted and fused.

$$\widetilde{x}(k) = \hat{x}^o(k)\kappa \quad (39)$$

where $\widetilde{x}(k) = \{\widetilde{x}_1(k), \widetilde{x}_2(k), \ldots, \widetilde{x}_n(k)\}$ is the multi-sensor positioning data after variance weighting, $\widehat{x}^o(k) = \{\widehat{x}_1{}^o(k), \widehat{x}_2{}^o(k), \ldots, \widehat{x}_n{}^o(k)\}$ is the multi-sensor fault data, and $\kappa = \{\kappa_1, \kappa_2, \ldots, \kappa_n\}$ is the fusion weight, $i = 1, 2, \ldots, n$.

By weighting the variance of the fault data, a certain amount of information correction was realized, and the fault-tolerant performance of the data fusion algorithm was improved. Even if the ZigBee signal had measurement deviation due to environmental interference, or the GPS/BDS signal was weakened or lost due to environmental occlusion, data samples with high credibility based on the proposed method could still be obtained.

6. Denoising Processing of Positioning Data Based on Particle Filter

In order to improve the reliability of the information collected by the ZigBee-GPS/BDS multi-sensor positioning system and reduce external noise interference, the multi-sensor data after the consistency inspection and variance weighting were filtered and denoised using the basic particle-filter algorithm [25].

6.1. Principle of Particle-Filter Algorithm

The design idea of the standard particle-filter algorithm is to approximate the probability density function of the system with some discrete random sampling points and replace the integration operation with the sample mean to obtain the minimum variance estimate of the state. First, a set of random particle samples was generated according to the prior condition of the system state vector, and the weight of each sampled particle was calculated. Then, according to the system observation information, the particle weights and positions were continuously corrected, and the corrected particles were used to approximate the posterior probability density function of the target state, so that the approximate posterior probability density function can be used to estimate the target state. Finally, the particle filter state was output, namely

$$\hat{x}_i(k) = \sum_{i=1}^{n} \omega_i(k) x_i(k) \tag{40}$$

$$P_i(k) = \sum \omega_i(k)[x_i(k) - \hat{x}_i(k)][x_i(k) - \hat{x}_i(k)]^T \tag{41}$$

where $x_i(k)$ is the state of ith particle iterated at time k, $\hat{x}_i(k)$ is the weighted estimation state of the particle, $P_i(k)$ is the estimated variance, and $\omega_i(k)$ is the particle weight, $k = 1, 2, \ldots, T$, $i = 1, 2, \ldots, n$.

6.2. Denoising Processing of Sampling Data Based on Particle Filter

Particle propagation was achieved by sampling the ZigBee-GPS/BDS multi-sensor state-transition model $q(x(k)\ x(k-1),\ z(k))$ and generating new particle states $x_i(k)$. In order to reduce the interference of environmental noise on the positioning data, BPF was used to filter and denoise the multi-sensor positioning data [26], and the latest data collected by sensors was substituted into the observation equation to establish a new particle filter observation equation, which improves the reliability of multi-sensor data samples. The specific steps were as follows:

Step 1 The filtering model of the multi-sensor of the unmanned ship was established, and the state and measurement model of the multi-sensor system was defined as follows:

$$\begin{cases} x(k) = f(x(k-1)) + \lambda(k-1) \\ z(k) = h(x^{ZGB}(k), \tilde{x}(k)) + \nu(k) \end{cases} \tag{42}$$

where $x(k)$ is the system state, $z(k)$ is the sensor recursive positioning data, $f(\cdot)$ is the state transition function, $h(\cdot)$ is the measurement function, $x(k-1)$ is the system state at the last moment, $x^{ZGB}(k)$ is the sensor positioning data at time k, $\tilde{x}(k)$ is the sensor positioning data weighted by the variance at time k, $\lambda(k-1)$ is the estimation noise, and $\nu(k)$ is the measurement noise.

Step 2 initialization: the prior density $p(x_0)$ was randomly sampled and the initialization particle set $x(0)$ was generated.

Step 3 importance sampling.

(a) Randomly select n particle samples that satisfy the following distribution from th importance density function:

$$x_i(k) \sim q(x(k)|x(k-1),\ z(k)) = p(x(k)|x(k-1)) \quad (4$$

(b) Calculate the weights of the sampled particles and update them as follows:

$$\omega_i(k) = \omega_i(k-1)(p(z(k)|x_i(k))p(x_i(k)|x_i(k-1))/q(x_i(k)|x_i(k-1), z(k)) \quad ($$

where $\omega_i(k)$ is the ith particle weight at time k.

(c) Normalized the importance weights.

$$\widetilde{\omega}_i(k) = \omega_i(k)[\sum_{i=1}^{n} \omega_i(k)]^{-1} \tag{45}$$

Step 4 Resampling.

Calculate the effective particle number $N_{eff} = (\sum_{i=1}^{n} (\omega_i(k))^2)^{-1}$. If N_{eff} is less than the threshold value N_{th}, the particle set $\{x_i(k), \widetilde{\omega}_i(k)\}$ is resampled; otherwise, no resampling is required.

Step 5 Output the local estimation and covariance matrix of the filtered multi-sensor positioning data:

$$\hat{x}_i(k) = \sum_{i=1}^{n} \widetilde{\omega}_i(k) x_i(k) \; P_i(k) = \sum \widetilde{\omega}_i(k)[x_i(k) - \hat{x}_i(k)][x_i(k) - \hat{x}_i(k)]^T \tag{46}$$

where $\hat{x}_i(k)$ is the local estimation of particle information, and $P_i(k)$ is the covariance matrix, $k = 1, 2, \ldots, T$, $i = 1, 2, \ldots, n$.

7. Sensor Data Fusion Based on TCPF

When the basic particle-filter algorithm is used to filter the ZigBee-GPS/BDS sensor state-transition function $q(x(k)\; x(k-1),\; z(k))$, BPF can obtain a posterior probability that is close to the real state estimation because it is based on the Bayesian recursion of the sensor state prediction information according to the sampling idea. However, when filtering and fusing three or more input data, BPF will suffer from a low operating efficiency and lack of particle samples in the resampling process [25]. In order to meet the requirements of fusion filtering processing of the multi-input data model based on ZigBee-GPS/BDS, a particle-filter algorithm based on adaptive threshold and hierarchical capacity was proposed to perform fusion filtering processing of the multi-input data, so as to realize the precise positioning of the unmanned ship.

7.1. Principle of TCPF Algorithm

The proposed TCPF algorithm first used unscented transformation [27] to integrate the latest observation information into the proposal distribution, so that the proposal distribution was close to the real distribution of the probability density function: $\breve{x}_i(k) \leftarrow T_O(x_{i(a)}(k-1))$. Then, a Gaussian mixture model was constructed and an adaptive threshold was set to reduce the operation steps of clustering similar components in the Gaussian mixture and improve the clustering efficiency. Finally, the stratified sampling proportion capacity was set, the continuous probability density function was layered, and the combination of particle weights from the inferior layer was optimized to improve particle diversity. The specific steps are as follows:

Step 1 integrate the latest observation information into the proposal distribution.

(a) Extract the particle state and covariance matrix processed by the basic particle-filter algorithm, and calculate the sigma point set:

$$x_{i(a)}(k-1) = \left[\overline{x}_{i(a)}(k-1), \overline{x}_{i(a)}(k-1) \pm \sqrt{(n_a + \lambda)P_{i(a)}(k-1)}\right] \tag{47}$$

where $x_{i(a)}(k-1)$ is the sigma point set, $\overline{x}_{i(a)}(k-1)$ is the sigma point set after the unscented transformation, $P_{i(a)}(k-1)$ is the covariance matrix of the sigma point set, n_a is the dimension of the sigma point set, and λ is the scale parameter.

(b) Integrate ZigBee-GPS/BDS multi-sensor positioning data into the obtained Sigma sampling point set, and update the system status and covariance.

$$\breve{x}_i(k) = x_{i(a)}(k-1) + k_k z(k-1) \tag{48}$$

$$\breve{p}_i(k) = p_{i(a)}(k-1) - k_k p_{i(z(k-1))}(k-1)k_k^T \tag{49}$$

where $\breve{x}_i(k)$ is the update state of the system, $\breve{p}_i(k)$ is the update covariance of the system, k_k is the Kalman gain, and $p_{i(z(k-1))}(k-1)$ is the covariance matrix of the positioning data.

(c) Construct a proposal distribution that is closer to the target probability function using the system state and covariance, and sample from it.

$$\widehat{x}_i(k) \sim q(\boldsymbol{x}(k)\ \boldsymbol{x}(k-1),\ \boldsymbol{z}(k)) = N(\breve{x}_i(k), \breve{p}_i(k)) \tag{50}$$

where $N(\cdot)$ is the Gaussian function.

Step 2 construct a Gaussian mixture model.

(a) Generate a posterior probability density function $p_h(\boldsymbol{x}(k)|\boldsymbol{z}(k))$ with time step k according to the Gaussian mixture components.

$$p_h(\boldsymbol{x}(k)|\boldsymbol{z}(k)) = \sum_{i=1}^{C(k)} \xi_i N(\boldsymbol{x}(k)|m_i, v_i) \tag{51}$$

where $N(\boldsymbol{x}(k) \mid m_i, v_i)$ is the ith component in the mixed Gaussian model, $C(k)$ is the number of component units of discrete samples, and ξ is the number of component units for discrete samples, $k = 1, 2, \ldots, T$, $i = 1, 2, \ldots, n$.

(b) Integrate the discrete sampling points sampled by Equation (50) and their corresponding weight $\{\widehat{x}_i(k), \widehat{\omega}_i(k)\}$ into the Gaussian mixture component unit of Equation (51), and use the reconstructed continuous posterior probability density function $p_g(\boldsymbol{x}(k)|\boldsymbol{z}(k))$ to resample the discrete particles:

$$p_g(\boldsymbol{x}(k)|\boldsymbol{z}(k)) = \sum_{i=1}^{n} \widehat{\omega}_i(k) N(\widehat{x}_i(k)|\widehat{x}_i(k-1), h \cdot p(k)) \tag{52}$$

where

$$h = 0.5N^{-2/n_x} \tag{53}$$

$$\ddot{p}(k) = \sum_{i=1}^{n} \widehat{\omega}_i(k)(\widehat{x}_i(k) - \ddot{x}_i(k))(\widehat{x}_i(k) - \ddot{x}_i(k))^T \tag{54}$$

$$\ddot{x}_i(k) = (\sum_{i=1}^{n} \widehat{\omega}_i(k)\widehat{x}_i(k))/n \tag{55}$$

where $\ddot{p}(k)$ is the covariance of the discrete particle filter distribution, $\ddot{x}_i(k)$ is the mean value of the discrete particle filter distribution, h is the normalized constant, and n_x is the particle distribution dimension.

(c) Merge similar units of the Gaussian mixture in $p_g(\boldsymbol{x}(k)|\boldsymbol{z}(k))$ using cluster analysis.

Step 3 set the adaptive threshold T_c for merging similar units in cluster analysis.

(a) Take the particle x_m with the largest weight in the discrete particle sample set as the cluster center, and calculate the Mahalanobis distance D_i between the other particles i and x_m after selecting the importance sampling process.

$$D_i = \sqrt{(x_m - \beta_c^i)^T S^{-1}(x_m - \beta_c^i)} \tag{56}$$

where β_c^i is the probability density of particle i, and S is the covariance matrix.

(b) Calculate the number of effective particle samples N_e in the cluster unit.

$$N_e = \frac{n}{1 + \sigma_{\beta c}^2} \tag{57}$$

where n is the number of particle samples and $\sigma^2_{\beta c}$ is the covariance of particle probability density.

(c) Construct threshold T:

$$T = T_0 + \frac{k_e}{N_e} \cdot R \tag{58}$$

where T_0 is the initial value of the threshold, k_e is the proportion coefficient, and R is the classification times.

(d) Substitute Equation (57) into Equation (58) to obtain the adaptive threshold T_c.

$$T_c = T_0 + \frac{k_e(1 + \sigma^2_{\beta c})}{n} \cdot R \tag{59}$$

(e) Compare D_i with the adaptive threshold T_c. If D_i is less than or equal to T_c, the particle is classified into the component unit related to its probabilistic mass; on the contrary, skip the particle and cluster other particles.

(f) Select the particle with the largest weight from the remaining particle samples as the cluster center, and repeat step (e) until the clustering ends.

(g) Substitute the clustered component units into the continuous probability density function $\hat{p}(\boldsymbol{x}(k)|\boldsymbol{z}(k))$ of the constructed particle set. The $\hat{p}(\boldsymbol{x}(k)|\boldsymbol{z}(k))$ is expressed as follows:

$$\hat{p}(\boldsymbol{x}(k)|\boldsymbol{z}(k)) = \sum_{i=1}^{C(k)} \beta_i N(\widehat{x}_i(k)|\gamma_i, p_i), \ \sum_{i=1}^{C(k)} \beta_i = 1 \tag{60}$$

where β_i is the probability mass of the similar component i, γ_i is the mean of component i, and p_i is the covariance of component i, i = 1, 2, . . ., n.

Step 4 set the hierarchical sampling proportional capacity.

(a) According to the layering theory [28], the continuous probability density function $\hat{p}(\boldsymbol{x}(k)|\boldsymbol{z}(k))$ is divided into l layers, and the probability density function of each layer is defined as $p_i(x)$. According to its probability quality, the component layers are divided into a group of weight advantage layers and two groups of disadvantage layers, and are defined as l_a, l_b and l_c, respectively.

(b) Set the proportional capacity of the particle number in layers l_a, l_b and l_c as $n/4$, $n/3$ and $n/3$, respectively.

(c) The particles whose weights are less than the average value $\varpi(k)$ in l_b and l_c layers are optimized and combined to obtain the optimized particle weights $\psi'_i(k)$, and the sample data is sampled hierarchically. The $\varpi(k)$ and $\psi'_i(k)$ are calculated as follows:

$$\varpi(k) = \frac{1}{n}\sum_{i=1}^{n} \widehat{\omega}_i(k) \tag{61}$$

$$\begin{cases} \psi'_i(k) = \frac{\eta-1}{\eta}\widehat{\omega}_i(k) + \frac{1}{\eta}\varpi(k), & \begin{matrix} if\ \widehat{\omega}_i(k) < \varpi(k), \\ \text{Divide particles into } l_b \text{ layer} \end{matrix} \\ \psi'_i(k) = \widehat{\omega}_i(k), & \begin{matrix} if\ \widehat{\omega}_i(k) \geq \varpi(k), \\ \text{Divide particles into } l_a \text{ layer} \end{matrix} \end{cases} \tag{62}$$

where η is the proportional coefficient.

(d) Output the sampling results of multi-sensor data fusion.

In the process of importance sampling, the current positioning data was integrated into the design of the proposal distribution of particle sets, which made the proposal distribution closer to the real posterior probability density and improved the estimation performance of the algorithm. By constructing Gaussian mixture probability density function instead of resampling, and constructing adaptive threshold in the cluster analysis of Gaussian mixture, the discrete particle samples were merged into similar component units, which reduces the complexity of clustering operation, and improves the real-time performance and operational efficiency of system signal processing. In addition, the continuous probability density function was stratified and the proportional capacity was set to ensure that there are enough particles in the inferior layer for weight optimization combination, which improved the diversity of particles, effectively suppressed the lack of particle samples, and improved the fusion processing accuracy of the TCPF algorithm for multi-input data.

7.2. Association of Stratified Sampling with Sensor Confidence

In order to further improve the fusion filtering accuracy of the algorithm, a new design was made for the hierarchical sampling step of the TCPF algorithm in this paper. The confidence factor $\varsigma_i^{ZGB}(k)$ of Equation (30) was associated with the hierarchical sampling of the TCPF algorithm, and a new weight optimization equation was obtained. The specific steps are as follows:

Step 1 substitute the confidence factor into the calculation of the weight optimization.

$$\psi'_i(k) = \frac{\varsigma_i^{ZGB}(k) - 1}{\varsigma_i^{ZGB}(k)}\widehat{\omega}_i(k) + \varsigma_i^{ZGB}(k) \cdot \varpi(k), \widehat{\omega}_i(k) < \varpi(k) \tag{63}$$

Step 2 the data samples are stratified according to the weights of multi-sensor sampling particles after optimized combination.

$$\begin{cases} \psi'_i(k) = \frac{\varsigma_i^{ZGB}(k)-1}{\varsigma_i^{ZGB}(k)}\widehat{\omega}_i(k) + \varsigma_i^{ZGB}(k) \cdot \varpi(k), & \begin{array}{c} if\ \widehat{\omega}_i(k) < \varpi(k), \\ \text{Divide particles into } l_b \text{ layer} \end{array} \\ \psi'_i(k) = \widehat{\omega}_i(k), & \begin{array}{c} if\ \widehat{\omega}_i(k) \geq \varpi(k), \\ \text{Divide particles into } l_a \text{ layer} \end{array} \end{cases} \tag{64}$$

The confidence factor of the positioning data was associated with the stratified sampling in the TCPF algorithm, so that the sensor positioning data with a large confidence factor would be sampled first when the TCPF algorithm was used to sample and fuse the positioning data, which would improve the sampling efficiency of the TCPF algorithm. The larger the confidence factor data, the higher the credibility, which improved the reference value of samples from sensor information fusion, further improved the fusion filtering accuracy of the TCPF algorithm, and finally realized the precise positioning of the unmanned ship.

7.3. Steps of Sensor Data Fusion Based on TCPF

Step 1 generate an initial particle sample set $\{xi(0), i = 1, 2, \ldots, n\}$ by sampling from the state transition function $q(x(k)|\, x(k-1),\ z(k))$ of the ZigBee-GPS/BDS multi-sensor.

Step 2 perform basic UPF operations: $\breve{x}_i(k) \leftarrow T_O(x_{i(a)}(k-1)), k \in [1,T], i \in [1,n]$.

Step 3 sample particles from the proposal distribution $N(\breve{x}_i(k), \breve{p}_i(k))$:

$$\widehat{x}_i(k) \sim q(x(k)|\, x(k-1),\ z(k)) = N(\breve{x}_i(k), \breve{p}_i(k))$$

Step 4 construct an adaptive threshold T_c, and a Gaussian mixture continuous probability density function $\hat{p}(x(k)|z(k))$ using cluster analysis.

Step 5 the continuous probability density function is sampled hierarchically, and the sampling space is divided into l_a, l_b, and l_c layers.

Step 6 set the proportional capacity, and divide the particles into l_a, l_b, and l_c layer groups according to the weight of the confidence factor ς_i^{ZGB} of the sampled particles. At the same time, the weights of particles in the inferior layer l_b and l_c layer groups are combined and optimized, and they are added to the l_a layer to participate in sampling after the optimized particle weight $\psi_i'(k)$ is obtained.

Step 7 output the fusion sampling results.

Step 8 $k = k + 1$, and return to step 2.

8. Numerical Simulation and Experimental Testing

8.1. Simulation of Data Fusion Algorithm Based on TCPF

In order to verify the performance of the proposed fusion positioning algorithm, 30 independent tests were carried out on the simulation model shown in Equations (63)–(65), and the test results were compared with the results of an Extended Kalman Filter (EKF) [29], Unscented Kalman Filter (UKF) [30], Unscented Particle Filter (UPF) [31] and Basic Particle Filter (BPF) [32] in terms of root-mean-square error (RMSE) and standard deviation (Std).

$$f_1 = \begin{cases} x(k) = \sin(0.6\pi(k-1)) + 2x(k-1) + 6 + \sqrt{Q} \times randn() \\ z(k) = 0.5x^2(k) + \sqrt{R} \times randn() \end{cases} \tag{65}$$

$$f_2 = \begin{cases} x(k) = 2.2x(k-1) + \frac{3x(k-1)}{2+x^2(k-1)} + \cos(5(k-1)) + \sqrt{Q} \times rand() \\ z(k) = 0.35x^2(k) - 8 + \sqrt{R} \times rand() \end{cases} \tag{66}$$

$$f_3 = \begin{cases} x(k) = x(k-1) + \frac{6x(k-1)}{1+x^2(k-1)} + 4\cos(5(k-1)) + \sqrt{Q} \times randn() \\ z(k) = 0.25x^2(k) - 3 + \sqrt{R} \times randn() \end{cases} \tag{67}$$

The remaining parameters of TCPF were set as follows: N was 100, σ was 0.75, τ was 2; the process noise variance Q and measurement noise variance R were separately set according to different models, that is, the Q of the f_1, f_2 and f_3 test models were Q_1~N (0,0.5), Q_2~N (0,1) and Q_3~N(0,5), respectively; and the R of the f_1, f_2 and f_3 test models were R_1~N(0,1), R_2~N(0,5) and R_3~N(0,10), respectively. The parameters of the algorithm to be compared were taken from the corresponding references.

Table 1 shows the model test results of five algorithms. From the table, it can be seen that the mean and maximum values of RSME and Std from the proposed TCPF are smaller than the corresponding performances from the other four algorithms, which indicates that the filter fusion accuracy of TCPF was the highest. The superior performances were mainly due to the fact that in the process of importance sampling in this paper, the latest positioning data were integrated into the importance function through unscented transformation, which improved the credibility of particle samples. In addition, the algorithm in this paper clustered discrete particles by constructing an adaptive threshold in the Gaussian mixture and optimized the weights of particles in the inferior layer, which improved the diversity of data samples. The performance comparison shows that compared with the other four algorithms, the mean values of RMSE and Std of data fusion based on the proposed TCPF decreased by 25.0% and 28.0%, respectively, which fully shows that the TCPF data fusion algorithm can effectively suppress the particle shortage problem and improve the filtering accuracy.

Table 1. Model test results of five algorithms.

Test Model	Mean Value of RMSE					Mean Value of Std				
	EKF	UKF	BPF	UPF	TCPF	EKF	UKF	BPF	UPF	TCPF
f_1	0.493	0.421	0.646	0.390	0.345	0.473	0.393	0.622	0.381	0.235
f_2	4.433	2.554	3.972	3.383	2.485	4.210	2.574	2.792	2.893	2.367
f_3	5.456	3.598	4.812	4.563	3.487	5.498	3.702	3.710	4.236	3.396
Test Model	**Maximum Value of RMSE**					**Maximum Value of Std**				
	EKF	UKF	BPF	UPF	TCPF	EKF	UKF	BPF	UPF	TCPF
f_1	0.910	0.568	0.940	0.613	0.409	0.877	0.566	0.891	0.599	0.412
f_2	6.681	3.459	6.142	4.624	3.317	6.391	3.431	3.452	4.283	3.283
f_3	7.856	4.192	6.121	6.142	3.998	7.731	4.214	4.172	5.931	3.969

Figure 4 shows the results of 30 RSME and Std tests of five algorithms against three types of models (f_1, f_2, f_3). From the figure, it can be seen that when the noise variance was small, the RMSE and Std values of the TCPF data fusion algorithm were smaller than the values of the other four algorithms in each independent simulation experiment. When the noise variance increased, the RMSE and Std values of the five algorithms increased, which indicates that the filtering performance of the algorithms decreased, but the estimation accuracy of TCPF was still significantly higher than those of the other four algorithms. The test results in Figure 4 also verify that the filtering optimization performance of the TCPF data fusion algorithm was the best, and the filtering accuracy was the highest.

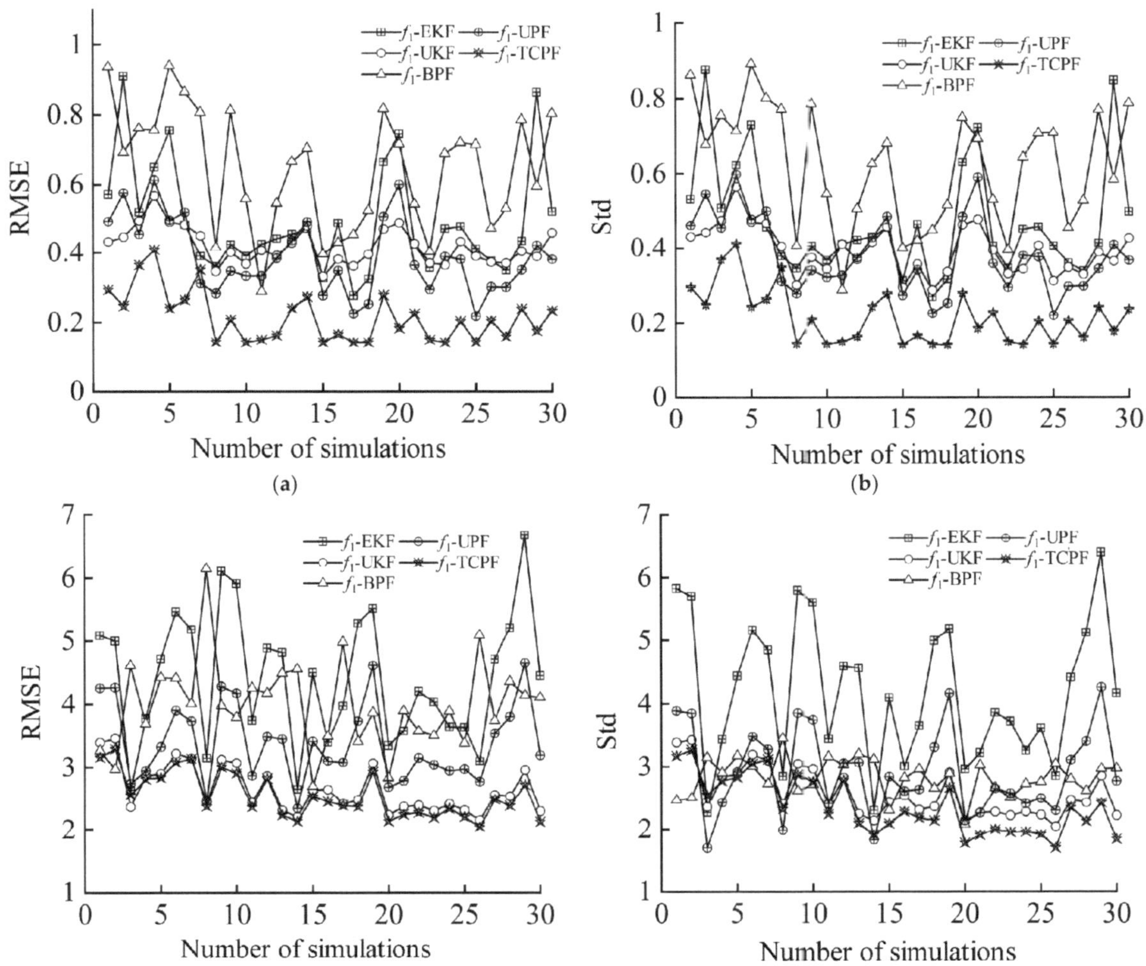

Figure 4. *Cont.*

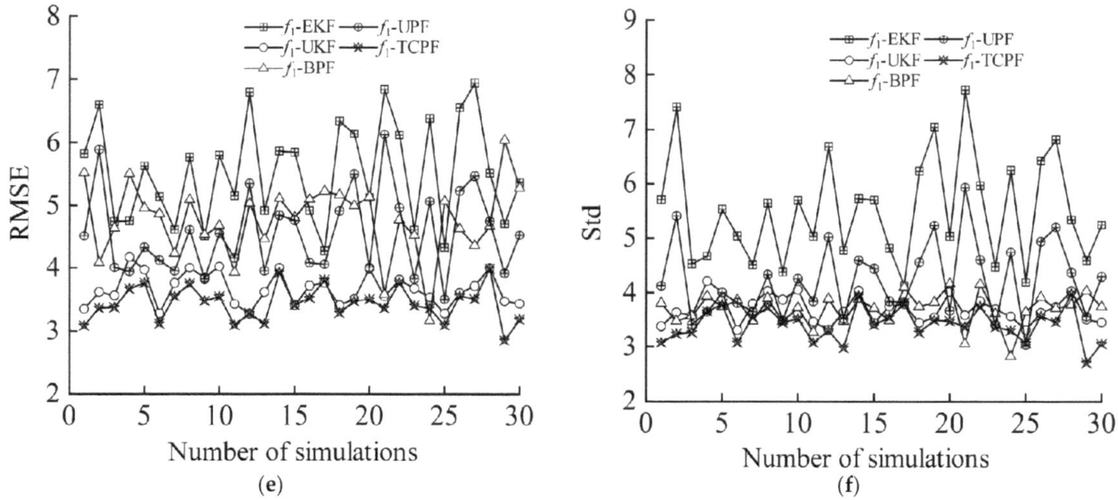

Figure 4. RMSE and Std tests of five algorithms against different models: (**a**) f_1 model, Q_1~N(0,0.5), R_1~N(0,1); (**b**) f_1 model, Q_1~N(0,0.5), R_1~N(0,1); (**c**) f_2 model, Q_2~N(0,1), R_3~N(0,5); (**d**) f_2 model, Q_2~N(0,1), R_3~N(0,5); (**e**) f_3 model, Q_3~N(0,5), R_5~N(0,10); and (**f**) f_3 model, Q_3~N(0,5), R_5~N(0,10).

8.2. Experimental Testing and Analysis

8.2.1. Establishment of Unmanned-Ship Experimental Environment

In order to further verify the effectiveness of the multi-sensor data fusion positioning method of the unmanned ship for water quality testing based on the proposed TCPF, the ZigBee-GPS/BDS multi-sensor combined positioning system test platform as shown in Figures 1 and 2 was established with the three-body unmanned ship as the carrier, and the unmanned ship navigation positioning experiment was carried out in the river channel of Zhangjiagang Campus of Jiangsu University of Science and Technology. In order to facilitate the experimental analysis and calculation, the starting point coordinate was subtracted from the preprocessed unmanned ship coordinate data, and the relative coordinate was used for testing. After calculation, the starting point coordinate of the unmanned ship was set to (3.8941, 0).

In order to evaluate the filtering accuracy of the unmanned-ship position information based on the proposed TCPF data fusion algorithm, first, the industrial computer equipped with the Linux system collected the position information of the unmanned ship using multiple sensors and outputted it from the serial port to the PC in the form of a log file through the wireless communication module. Then, the PC integrated the received position information of the unmanned ship into a data set [33]. After that, the designed TCPF data fusion algorithm was used to fuse and filter the position information data set of the unmanned ship, and finally realize the precise positioning of the unmanned ship. Figure 5 shows the experimental environment for the unmanned-ship positioning.

In order to verify the validity and superiority of the ZigBee-GPS/BDS multi-sensor positioning system for unmanned-ship navigation in different environments, the test river was divided into two sections, A and B, and two groups of ZigBee reference nodes were set along the A/B river according to the ZigBee networking principle. Among them, section A was located on the west side of the school library, which has an open water surface and was easy for the unmanned ship to navigate. However, the B section of the river was located on the west side of the school gymnasium, with dense vegetation and narrow channels, which had a certain environmental disturbance effect on the signal transmission of the ZigBee-GPS/BDS sensors. Figure 6 shows the schematic diagram of the node layout and navigation trajectory on the A/B river.

(a) (b)

Figure 5. Experimental environment for unmanned ship positioning: (**a**) Communication node; (**b**) PC terminal.

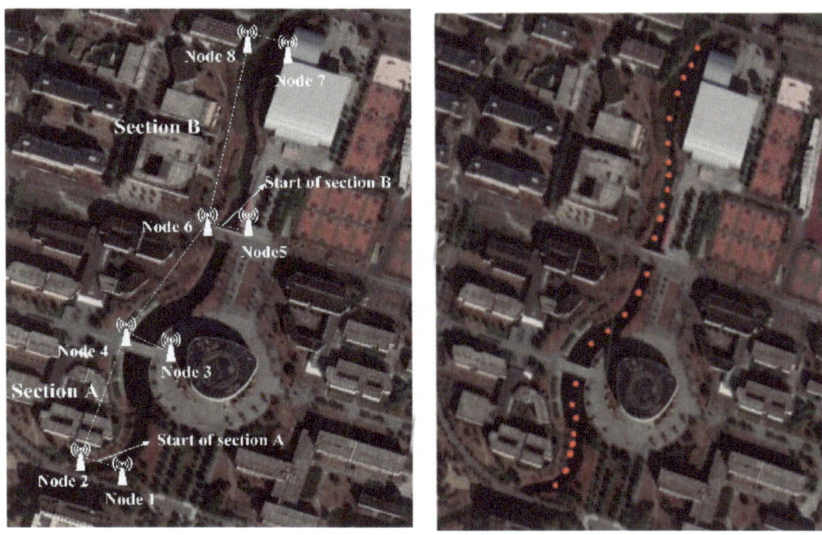

(a) (b)

Figure 6. Schematic diagram of node layout and navigation trajectory on A/B river: (**a**) ZigBee node layout; (**b**) navigation trajectory of unmanned ship.

8.2.2. Unmanned-Ship Positioning Test Based on Multi-Sensor Data Fusion

In order to verify the superior performance of the proposed TCPF algorithm, some unmanned-ship navigation and positioning experiments based on the multi-sensor positioning system were carried out. According to the unmanned-ship positioning data collected by the sensors in the A/B experimental river, the test results of the six algorithms, namely TCPF, EKF, UKF, UPF, KF, and BPF, were compared against the three performance indicators of average positioning error, RMSE and Std. The filtering data of the proposed TCPF algorithm was ZigBee-GPS/BDS positioning data, and the sensor positioning data of the other five algorithms were all ZigBee-GPS. Figure 7 shows the filtering results of the unmanned-ship navigation trajectory based on six filtering algorithms.

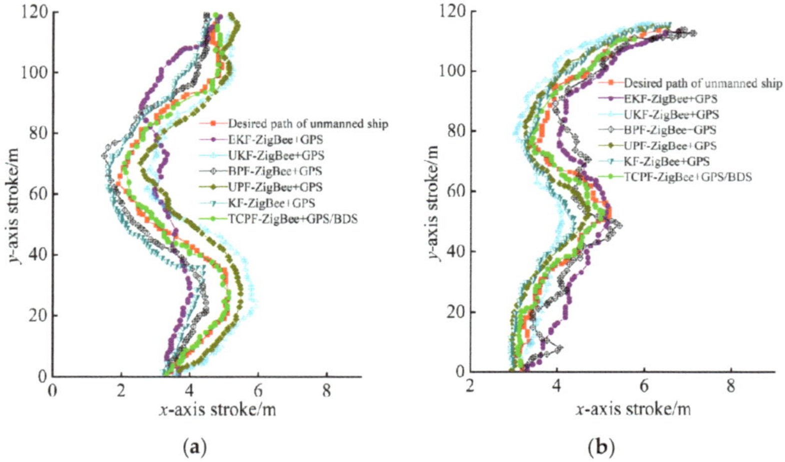

Figure 7. Filtering results of the unmanned-ship navigation trajectory based on five filtering algorithms: (**a**) positioning trajectories on section A; (**b**) positioning trajectories on section B.

From Figure 7b, it can be seen that when the unmanned ship sailed on the experimental river in section B, the sensor signal transmission was affected to a certain extent by environmental disturbances, such as dense trees and a narrow river. Except for the TCPF algorithm, the filtering and positioning results of other algorithms seriously deviated from the expected trajectory of the unmanned ship in some places; in particular, at the turning of the river, the positioning results were prone to large jumps, which caused the unmanned ship to gradually deviate from the expected trajectory and reduced the positioning accuracy and reliability. From Figure 7a, it can be seen that because the water surface of the experimental river in section A was more open than that in section B, and there were fewer shelters along the river, the positioning results of the other five algorithms showed that the unmanned ship could basically navigate according to the expected trajectory, but there was still a certain jump at the turning.

Combining Figure 7a,b, it can be seen that compared with the other five algorithms, the positioning accuracy of the unmanned ship based on the proposed TCPF algorithm was significantly higher. No matter whether at the turning of the river or in a straight line, the unmanned ship could basically navigate along the expected trajectory. This was mainly because the TCPF data fusion algorithm achieves the weighted correction of multi-sensor fault data through consistency inspection, which improves the fault-tolerant performance of the algorithm, and makes the TCPF algorithm have a high-reliability data sample for filtering. If the data are not inspected and corrected, when a type of sensor is subjected to a large environmental disturbance, the positioning data will be deviated, which will reduce the credibility of the data sample, and then reduce the positioning accuracy of the unmanned ship. Thus, this shows that the proposed TCPF algorithm of ZigBee-GPS/BDS not only ensures the accuracy of fault data inspection but also improves the fault-tolerant performance of the algorithm.

Tables 2 and 3 show the comparison results of trajectory positioning for six algorithms. From the tables, it can be seen that the test results of the proposed TCPF algorithm were smaller than those of the other five algorithms, regardless of the average positioning error or the maximum positioning error of the A/B section, which indicates that the proposed method has the highest filtering and fusion accuracy. This is mainly because the posterior probability density function representing the state was approximated as a Gaussian mixture distribution in this paper, and particle samples were extracted on it instead of resampling, which ensured the diversity of ZigBee-GPS/BDS multi-sensor sampling particles, and avoided a lack of samples.

Table 2. Trajectory positioning test results in section A.

Performance Index \ Algorithm	EKF	UKF	BPF	UPF	KF	TCPF
Average positioning error	3.265	4.110	3.538	2.821	3.856	2.352
Maximum positioning error	5.221	6.010	5.319	4.114	5.589	2.941
RMSE	3.057	3.956	3.432	2.719	3.854	2.351
Std	0.435	0.589	0.542	0.395	0.498	0.345

Table 3. Trajectory positioning test results in section B.

Performance Index \ Algorithm	EKF	UKF	BPF	UPF	KF	TCPF
Average positioning error	3.786	4.539	4.464	3.658	4.365	2.875
Maximum positioning error	6.519	8.018	7.699	6.393	7.856	4.463
RMSE	3.658	4.389	4.358	3.572	4.215	2.723
Std	0.952	1.853	1.258	1.152	1.562	0.876

From the performances of Std and RMSE in the A/B section, compared with the other five algorithms, the value of the TCPF algorithm was also the smallest, which shows that the estimation performance of the TCPF algorithm is also the best. This is mainly due to the fact that the proposed TCPF algorithm integrated the latest multi-sensor positioning data collected by the unmanned-ship positioning system into the proposal distribution of the particle set in the importance sampling process, which made the proposal distribution closer to the real posterior probability density and improved the estimation performance of the algorithm. In addition, the integration operation in Bayesian estimation was solved according to the Monte Carlo method adopted by the basic particle-filter algorithm, which reduced the interference of environmental noise on the information collected by multi-sensors, and improved the credibility of multi-sensor positioning data samples.

The calculation shows that, compared with the other five algorithms, when the unmanned ship navigated based on TCPF in section A, the average positioning error, RMSE and, Std performances of its trajectory decreased by 35.0%, 36.0%, and 35.0%, respectively. In addition, the average positioning error, RMSE and, Std performance of its trajectory in section B decreased by 37.0%, 40.0%, and 39.0%, respectively. In the experiment involving the whole river, the three performances of TCPF decreased by an average of 36.0%, 38.0%, and 37.0%, respectively, which fully shows that the proposed TCPF algorithm had a high estimation performance and fault-tolerant performance, and the positioning accuracy of the unmanned ship's navigation trajectory was improved.

8.2.3. Fault-Tolerance Test of the Unmanned-Ship Positioning Algorithm

In order to further verify the fault tolerance and filtering performance of the TCPF data fusion algorithm, two sets of fault-tolerance tests were carried out. One set of test data was from ZigBee-GPS, and the other set of selected test data was from ZigBee-BDS. In addition, the BPF algorithm was selected as the comparison algorithm for the performance test, and the positioning error, RMSE, and Std were used as performance indicators. The test data set of BPF was from ZigBee-GPS/BDS. Figure 8 shows the trajectory fusion results of the unmanned ship based on different algorithms in the A/B river.

From Figure 8, it can be seen that although the positioning results of the unmanned ship, based on the TCPF algorithm for different data sets, basically fitted the expected trajectory, the positioning accuracy of the TCPF algorithm for the ZigBee-GPS/BDS data set was significantly higher, and the positioning jump error at the river turn was also smaller, because the TCPF algorithm realized the information complementarity between the GPS and BDS data set. When a signal in the positioning system was lost, another sensor could supplement the collected position information of the unmanned ship to the sampling data set; thus, the sensor signal-loss problem caused by the large deviation in the unmanned ship position has been effectively overcome, which verifies that the constructed ZigBee-GPS/BDS multi-sensor positioning system performed the function of information complementarity between the data.

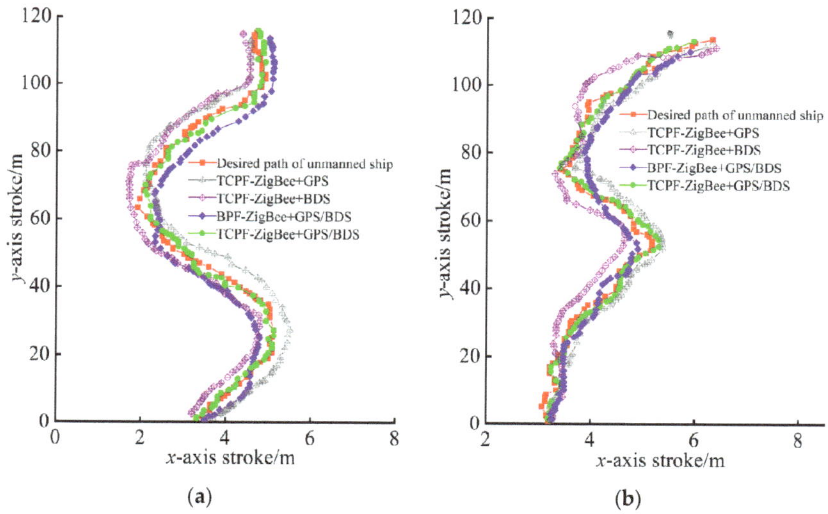

Figure 8. Trajectory fusion results of the unmanned ship based on different algorithms in the A/B river: (**a**) section A positioning trajectory; (**b**) section B positioning trajectory.

Combining the performance indicators of the average positioning error in Figure 8 and Tables 4 and 5, it can be seen that the BPF algorithm using the ZigBee-GPS/BDS data set improved the positioning accuracy of the unmanned ship to a certain extent, but there was still a certain gap compared with the positioning accuracy of the TCPF algorithm, which was mainly due to the optimization performance of the TCPF algorithm itself. Firstly, the adaptive threshold was constructed in the cluster analysis of Gaussian mixture units, and the discrete particle samples were merged into similar component units, which reduces the complexity of clustering operations and improves the real-time performance of data fusion algorithms for signal processing. Secondly, the samples of the continuous probability density function of the constructed weighted point set were stratified, and the proportional capacity of each sampling layer was set, which ensured the reasonable distribution of the number of sampling particles in the stratification. Next, the layer group l_a was sampled, and the particle weights in the layer group l_b and l_c were optimally combined, which increased the probability quality and prevented the lack of samples.

Table 4. Data positioning test results in section A.

Performance Index \ Algorithm	TCPF-Z + G	TCPF-Z + B	BPF-Z + G/B	TCPF-Z + G/B
Average positioning error	2.558	2.606	2.532	1.828
Maximum positioning error	3.438	4.013	3.663	2.676
RMSE	2.541	2.648	2.588	1.860
Std	0.368	0.605	0.381	0.332

From Tables 4 and 5, it can be seen that the RMSE and Std performance indicators of the TCPF algorithm were better than those of the BPF algorithm. This was mainly because the confidence factor of multi-sensor positioning data is associated with hierarchical sampling in the TCPF algorithm, so when using the TCPF algorithm to filter and fuse ZigBee-GPS/BDS multi-sensor positioning data, the sensor positioning data with a large confidence factor will be fused and sampled first, which improves the sampling efficiency of the TCPF algorithm. Because the confidence factor reflects the credibility of the data, the larger the confidence factor was, the higher the reliability of the data sample was, which further improved the reference value of the sensor-information fusion sample, and further improved the fusion filtering accuracy of the TCPF algorithm for multi-input data, and finally achieved the accurate positioning of the unmanned ship.

Table 5. Data positioning test results in section B.

Performance Index \ Algorithm	TCPF-Z + G	TCPF-Z + B	BPF-Z + G/B	TCPF-Z + G/B
Average positioning error	2.754	2.864	2.727	2.092
Maximum positioning error	4.561	4.875	4.945	3.631
Positioning RMSE	2.834	2.973	2.847	2.168
Positioning Std	0.572	0.717	0.751	0.506

The calculations show that, compared with the other three algorithms, the average positioning error, RMSE and Std performances of the unmanned ship, based on the TCPF algorithm during navigation in river A, decreased by 22.0%, 20.0%, and 17.0%, respectively, and the average positioning error, RMSE, and Std in river B decreased by 18.0%, 19.0%, and 18.0%, respectively. The three performances of the unmanned ship in the whole experimental river decreased by an average of 20.0%, 19.5%, and 17.5%, respectively, which fully demonstrates that the TCPF algorithm has a better optimization performance and fault-tolerant performance compared with the basic particle-filter algorithm, and improves the ship's positioning accuracy under the interference of the unmanned-ship environment.

9. Discussion

Accurate positioning is the key to achieving the high-precision autonomous navigation of unmanned ships, and the positioning of unmanned ships based on multiple combined sensors has always been the focus of research by scholars. Data fusion based on two types of sensors is the current mainstream technology for unmanned-ship positioning, but there are shortcomings in affecting positioning accuracy and reliability due to the weakening or loss of certain sensor signals. In recent years, how to improve positioning accuracy by further fusing the data of three or more types of sensors has become a research hotspot. Adding sensors can help improve the reliability and accuracy of unmanned-ship positioning, but there is still a lack of research on how to further improve the accuracy and efficiency of data fusion through testing the confidence distance of data, as well as how to achieve effective localization under low or weak signals through data correction. In view of this, for the positioning of unmanned ships, in this paper, a multi-sensor positioning system incorporating ZigBee, GPS, and BDS was first built. Then, the particle-filter algorithm was introduced for denoising multi-sensor data. Next, the latest positioning data from multiple sensors were integrated into the proposed distribution using unscented transform to cause the proposed distribution to be close to the real distribution of the probability density function. Finally, the Gaussian mixture model was constructed, the adaptive threshold was set, and the data-confidence factor was associated with the hierarchical sampling to further improve the fusion filtering accuracy of the positioning data. The numerical test results of the three groups of models and the experimental test results of the navigation path all show that the proposed method not only realizes the mutual complementation, weighted correction, and confidence assignment of multi-sensor signals, but also significantly reduces their root-mean-square error and standard deviation compared with other filtering algorithms, which verifies that the proposed method has the advantages of high data reliability and a good filtering-fault-tolerance performance, and achieves the accuracy and reliable positioning of unmanned ships.

10. Conclusions and Future Work

In this paper, a multi-sensor data fusion positioning method for unmanned ships based on a threshold- and hierarchical-capacity particle filter was proposed to address the issues of significant deviation in data fusion accuracy caused by signal weakening and loss in conventional integrated positioning systems in unmanned ships. Through algorithm simulation and experimental testing, the following conclusions could be obtained:

(1) The positioning data collected by the ZigBee-GPS/BDS multi-sensor is used to complement the information, which effectively overcomes the problem of sensor signal weakening or loss caused by environmental masking, and improves the accuracy and effectiveness of multi-sensor data fusion.

(2) By conducting consistency checks on the multi-sensor positioning data and weighted correction of the faulty data, not only does it enhance the fault-tolerance performance of the data fusion algorithm but also strengthens the credibility of the data set samples.

(3) The latest positioning data of multiple sensors are integrated into the proposal distribution of the particle set, which causes the suggested distribution to be closer to the true posterior probability density and improves the estimation performance of the algorithm. At the same time, the adaptive threshold is constructed in the cluster analysis of Gaussian mixing units, and the discrete particle samples are merged into similar component units, which improves the real-time performance and computing efficiency of the system's signal processing.

(4) Through stratified sampling and the setting of proportional capacity, a sufficient number of particles in the disadvantage layer are ensured for weight optimization and combination, and the diversity of particles is improved. Simultaneously, associating confidence factors with the TCPF algorithm's layered sampling prioritizes the selection of positioning data with larger confidence factors for sample fusion during the fusion filtering process of the unmanned ships position information to enhance the efficiency and accuracy of the data fusion algorithm.

The multi-sensor data fusion method in this paper was mainly oriented to unmanned ships that conduct water quality detection in inland rivers. In recent years, unmanned ships have also been applied in various fields such as maritime rescue, maritime monitoring, and maritime cargo transportation. The marine environment is much more complex than inland rivers. Whether the multi-sensor fusion algorithm proposed in this paper is suitable for the marine environment will be the next research topic.

Author Contributions: Methodology, Y.S. and Z.Z.; validation, Z.Z. and M.Y.; original draft preparation, Y.S. and Z.Z.; writing, Z.Z., M.Y. and S.W. All authors have read and agreed to the published version of the manuscript.

Funding: This research was funded by the 2022 Research Plan for Innovative Products in the Industrial Chain of Zhangjiagang City (ZKC2206) and the 2022 Industry-University Research Pre-research Fund Project of Zhangjiagang City (ZKYY2253).

Institutional Review Board Statement: Not applicable.

Informed Consent Statement: Not applicable.

Data Availability Statement: Not applicable.

Acknowledgments: The authors would like to thank Wei Liu of the T-SEA Marine Technology Co., Ltd. for his help in the sampling and laboratory analysis.

Conflicts of Interest: The authors declare no conflict of interest.

References

1. Nie, L.; Zhang, L.L. Key technologies and future development trends of unmanned ships. *Mar. Equip. Mat. Mar.* **2022**, *30*, 12–14. [CrossRef]
2. Liu, S.D.; Liang, T.R.; Wang, Y.; Zhang, Y. Underwater moving target localization based on multi-modal information fusion. *J. Nav. Pos.* **2022**, *10*, 14–24. [CrossRef]

3. Sheng, J.L. Application research of GNSS combined with unmanned ship technology in water system comprehensive management survey. *Sci. Tech. Inf.* **2021**, *19*, 52–54. [CrossRef]
4. Wu, S.Y.; Zhu, H.S.; Fei, X.X.; Ma, Z.J.; Tang, M. Design and implement of unmanned boat system for water quality monitoring. *Comput. Era* **2022**, *10*, 93–97. [CrossRef]
5. Deng, Z.L.; Wang, Z.G.; Sheng, W.; Liu, Y.P. Formation control of unmanned surface vehicles with position estimation algorithm. *Mech. Sci. Technol. Aero. Eng.* **2022**, *41*, 626–633. [CrossRef]
6. Wang, X.Y.; Yang, C.; Liang, S.; Wang, A. Application of PPP and DGPS in IMU/GPS aided 1:10000 aerial survey project. *Beijing Surv. Mapp.* **2016**, *6*, 45–48. [CrossRef]
7. Tang, C.; He, C.Y.; Dou, L.H. An IMU/ODM/UWB joint localization system based on modified cubature kalman filtering. *Sensors* **2021**, *21*, 4823. [CrossRef]
8. Yousuf, S.; Kadri, M.B. Information fusion of GPS, INS and odometer sensors for improving localization accuracy of mobile robots in indoor and outdoor applications. *Robotica* **2020**, *39*, 48–52. [CrossRef]
9. Hachemi, L.; Guiatni, M.; Nemra, A. Fault diagnosis and reconfiguration for mobile robot localization based on multi-sensors data fusion. *Unmanned Syst.* **2021**, *41*, 413–417. [CrossRef]
10. Alam, S.A.; Gustafsson, O. Improved particle filter resampling architectures. *J. Signal Process. Syst.* **2020**, *92*, 555–568. [CrossRef]
11. Wu, B.; Tian, Q. Improved indoor moving target localization algorithm based on UPF. *Transducer Microsy. Technol.* **2021**, *40*, 153–156+160. [CrossRef]
12. Sun, H.R.; Ge, J.L.; Niu, J.D.; Wu, W.P. Positioning precision analysis of BDS-3, GPS and combined system. *Pt. Eng. Technol.* **2022**, *59*, 117–120. [CrossRef]
13. Xu, H.Q.; Huang, K.; Yang, Y.; Zhu, S.N. Research and design of downhole positioning system based on ZigBee. *Comput. Knowl. Technol.* **2022**, *18*, 123–124+126. [CrossRef]
14. Guo, N.; Han, H.Z.; Zhang, J. Research on single frequency PPP positioning performance of BDS-3/BDS-2/GPS with high sampling rate. *Beijing Surv. Mapp.* **2022**, *36*, 935–939. [CrossRef]
15. Li, N.; Yu, M. Method of GPS coordinate transformation to Beijing 54 coordinate. *Northen Commun.* **2012**, *8*, 36–37. [CrossRef]
16. Li, H.Y. Design of general ZigBee end device based on JN5169. *Electron. Des. Eng.* **2022**, *30*, 155–158. [CrossRef]
17. Yakovlev, V.V.; Korzhenevskaya, I.V. The accounting possibilities for the earth rotation speed in satellite coordinate definitions. *Mat. Sci. Eng.* **2019**, *698*, 044007. [CrossRef]
18. Chen, H.; Xu, C.H.; Gao, J.X.; Song, X.F.; Yuan, L.S. Precision analysis of pseudorange single point positioning by BDS, GPS and combined BDS/GPS. *J. Shandong Univ. Sci. Technol.* **2015**, *34*, 72–78. [CrossRef]
19. Kimoto, R.; Ishida, S.; Yamamoto, T.; Tagashira, S.; Fukuda, A. MuCHLoc: Indoor ZigBee localization system utilizing inter-channel characteristics. *Sensors* **2019**, *19*, 1645. [CrossRef]
20. Stig, E.; Marie, L. The impact of redundancy on reliability in machinery systems on unmanned ships. *WMU J. Marit. Aff.* **2022**, *14*, 312–317. [CrossRef]
21. Zhang, M.; Lv, S.L. A method for determining confidence interval in train positioning. *Railw. Signal Commun.* **2018**, *54*, 1–4. [CrossRef]
22. Peng, X.S.; Huang, X.H.; Qiu, W.N.; Wei, K.; Liu, S.W. A weighted fusion WiFi positioning algorithm based on distance rendezvous. *Eng. Surv. Mapp.* **2017**, *26*, 72–75. [CrossRef]
23. Xu, D.; Guan, C.; Zhou, C.; Li, C. Research on consistency test of equipment maintainability pre-test data. *Aerosp. Control.* **2020**, *38*, 61–66. [CrossRef]
24. Xu, L.; Chen, X.; Ding, Z.X.; Hu, J.; Song, T.C. Outdoor positioning method combining RSSI and TOA based on lora technology. *Meas. Control Eng.* **2020**, *39*, 41–46. [CrossRef]
25. Wu, T.Z.; Zhao, T.; Xu, S.Y. Prediction of remaining useful life of the lithium-ion battery based on improved particle filtering. *Front. Energy Res.* **2022**, *10*, 95–101. [CrossRef]
26. Zeng, R.; Tian, J.; Jiang, H.; Wang, Z.N. UAV track prediction method based on improved particle filtering. *Transducer Microsyst. Technol.* **2022**, *41*, 148–151. [CrossRef]
27. Li, X.J.; Li, J.F.; Zhu, M.; Peng, N.L.; Zuo, S. Research on positioning fusion and verification algorithm based on UKF. *Automot. Eng.* **2021**, *43*, 825–832. [CrossRef]
28. Wei, J.X.; Cheng, L.; Han, P.; Zhu, Y.X.; Huang, W.D. Decision tree-based data stratification method for the minimization of the masking effect in adverse drug reaction signal detection. *Appl. Sci.* **2021**, *11*, 122–128. [CrossRef]
29. Ding, N.; Prasad, K.; Lie, T.T. State of charge estimation of a composite lithium-based Battery model based on an improved extended kalman filter algorithm. *Inventions* **2019**, *4*, 66. [CrossRef]
30. Gong, Y.J.; Xie, L.Y.; Xue, S.T.; Tang, H.S. Adaptive unscented Kalman filter for nonlinear structural identification. *J. Bldg. Str.* **2023**, *44*, 182–191+224. [CrossRef]
31. Yu, Z.; Liu, Y.; Wei, F.; Liu, L.J. Prediction of aeroengine exhaust gas temperature based on unscented particle filter algorithm. *Aeroengine* **2021**, *47*, 1–6. [CrossRef]

32. Gao, Y.; Mao, Y.H.; Yang, Y. Research on artificial fish school algorithm-particle filter. *Mod. Electron. Tech.* **2021**, *44*, 170–174. [CrossRef]
33. Rim, C.M.; Sin, Y.C.; Paek, K.H. A mobile robot localization method based on polar scan matching and adaptive niching chaos optimization algorithm. *J. Intell. Robot. Syst.* **2022**, *106*, 19. [CrossRef]

applied sciences

MDPI

Article

Application and Research on Improved Adaptive Monte Carlo Localization Algorithm for Automatic Guided Vehicle Fusion with QR Code Navigation

Bowen Zhang [1], Shiyun Li [1,*], Junting Qiu [2], Gang You [1] and Lishuang Qu [3]

[1] Faculty of Mechanical and Electrical Engineering, Kunming University of Science & Technology, Kunming 650500, China; 20212203075@stu.kust.edu.cn (B.Z.); 20212203169@stu.kust.edu.cn (G.Y.)
[2] School of Mechanical Engineering, Zhejiang University of Technology, Hangzhou 310014, China; qjt@zjut.edu.cn
[3] School of Information Engineering and Automation, Kunming University of Science & Technology, Kuming 650500, China; 20232204022@stu.kust.edu.cn
* Correspondence: 11303173@kust.edu.cn

Abstract: SLAM (simultaneous localization and mapping) technology incorporating QR code navigation has been widely used in the mobile robotics industry. However, the particle kidnapping problem, positioning accuracy, and navigation time are still urgent issues to be solved. In this paper, a SLAM fused QR code navigation method is proposed and an improved adaptive Monte Carlo positioning algorithm is used to fuse the QR code information. Firstly, the generation and resampling methods of initialized particle swarms are improved to improve the robustness and weights of the swarms and to avoid the kidnapping problem. Secondly, the Gmapping scan data and the data generated by the improved AMCL algorithm are fused using the extended Kalman filter to improve the accuracy and stability of the state estimation. Finally, in terms of the positioning system, Gmapping is used to obtain QR code data as marker positions on static maps, and the improved adaptive Monte Carlo localization particle positioning algorithm is matched with a library of QR code templates, which corrects for offset distances and achieves precise point-to-point positioning under grey-valued raster maps. The experimental results show that the particles encountered with kidnapping can be quickly adjusted in position, with a 68.73% improvement in adjustment time, 64.27% improvement in navigation and positioning accuracy, and 42.81% reduction in positioning time.

Keywords: SLAM; adaptive Monte Carlo localization; kidnapping; Gmapping; QR code template library; extended Kalman filter

Citation: Zhang, B.; Li, S.; Qiu, J.; You, G.; Qu, L. Application and Research on Improved Adaptive Monte Carlo Localization Algorithm for Automatic Guided Vehicle Fusion with QR Code Navigation. *Appl. Sci.* **2023**, *13*, 11913. https://doi.org/10.3390/app132111913

Academic Editor: Jonghoek Kim

Received: 30 August 2023
Revised: 25 October 2023
Accepted: 27 October 2023
Published: 31 October 2023

1. Introduction

An automated Guided Vehicle is a robot used for equipment handling and automatic assembly [1]. With the development of industry and the increase in the cost of human resources, material handling in industrial production has been gradually replaced by intelligent AGVs [2], which are equipped with various kinds of photoelectric sensors to network and interconnect, and issue scheduling instructions to realize the intelligent material handling system. According to the different navigation methods, AGVs can be divided into QR code, magnetic, inertial, laser, etc. [3]. When AGVs convey material along a planned route, their surroundings and ground cleanliness partially impact most navigation methods [4]. As we all know, QR code navigation requires a neat and clean floor, and the QR code must be protected to a large extent [5]. On the contrary, SLAM navigation does not require ground road conditions to build a map but has a strong need for the surrounding environment [6]. For the above situation, the navigation method is established as SLAM fused QR code navigation, and the improved AMCL positioning algorithm is used to match the QR code information to improve the positioning accuracy.

AGV navigation has been a fundamental problem for logistics robots. The key to this is the combination of algorithms with various navigation and optimization methods. Bach S H et al. [7] corrected the cumulative error in attitude estimation by combining the internal coded sensor data with the remaining external QR code data. Kulaç N. et al. [8] used RGB cameras and QR codes for mapping and localization, eliminating permanent localization errors, but the light source affects the actual localization. P. Kumar [9] used EKF for the data tracking of QR codes and the experiments proved that the positioning accuracy is improved but the quality of the QR code is required and regular maintenance is needed. C. Zhou's [10] use of QR codes for AGV navigation has been shown through experiments to reduce costs and improve positioning accuracy. Still, the system uses machine vision to recognize QR codes and therefore has an impact on the scanning time as well as precision. Adaptive Monte Carlo localization is a very effective solution for localizing the robot in a given environment [11]. Zhang, X. et al. [12] proposed an improved EKF intelligent algorithm to reduce and eliminate the deviation in the fusion of various sensor data. When in a highly similar working environment, the positioning accuracy, stability, and data processing will be significantly affected [13]. The parameter changes in the AMCL positioning algorithm also impact the algorithm response and actual positioning [14,15].

SLAM [16] is the abbreviation for "simultaneous localization and mapping," mainly used to solve the problem of localization and map construction when a robot moves in an unknown environment. It consists of four main parts: front-end scanning and matching, back-end optimization, closed-loop detection, and map construction [17]. Gmapping is a SLAM algorithm based on laser sensors, which has high map building and positioning accuracy, and is able to build accurate maps in real time and achieve more accurate positioning [18]. AMCL localization algorithms are widely used because they are highly flexible, adaptive and can handle noise and uncertainty [19]. However, in localization, AMCL is highly dependent on the accuracy of the sensors, which will affect the accuracy of the localization if the data fails. In this paper, an enhanced AMCL based on QR code information is proposed. The improved AMCL algorithm is able to automatically adjust the weights and distributions of the particles according to the estimation error through an adaptive sampling strategy, which improves the accuracy and robustness of localization. Accurate localization is still achieved in the presence of dynamic environment and sensor errors. To meet the requirements of the real-time and positioning accuracy of visual navigation, this paper uses the QR code as a landmark. A QR code is a two-dimensional matrix composed of QR code symbols and has the advantages of a fast reading speed, large amount of information, low cost, and high reliability [20]. In the process of the AGV moving, the QR code camera at the bottom of the vehicle scans the QR code. It determines the position and attitude information of the AGV by identifying the current QR code. In this process, the angular error and positional error in the motion process are analyzed and calculated. By using various algorithms for correction and compensation, the positioning accuracy of the robot is improved to a certain extent. Based on the above problems, an improved AMCL particle localization algorithm is introduced into the SLAM fusion QR code, connecting the world coordinates of the QR code to enhance the number of particles in the AMCL algorithm, optimizing the particle sampling strategy, and improving the motion model to introduce a non-linear model to improve the localization accuracy.

The AGV navigation and positioning system researched in this paper is characterized by a high accuracy, low latency, reduced maintenance cost and improved work efficiency. It can solve the navigation and positioning problems encountered by AGVs in practical work. According to the requirements of the actual production environment, this paper describes the improved methods and measures, and compares them with the unimproved methods and measures. Figure 1 shows the system framework diagram of AGV.

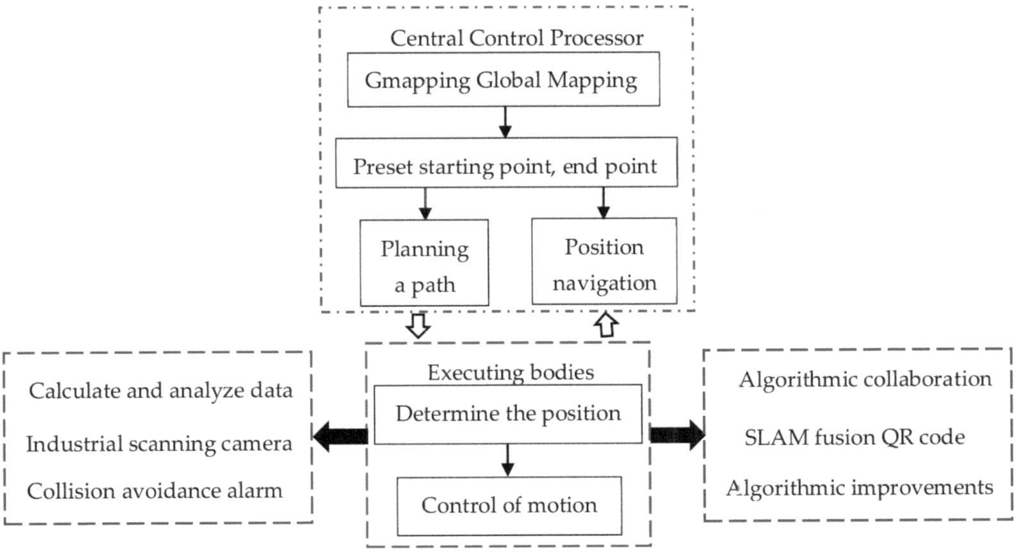

Figure 1. System Framework.

The following are the main contributions made by this paper:

(1) Improve the way of generating and resampling the initial particle swarm of the AMCL algorithm, so that the particle performance is more stable, and can effectively match the set template library to improve the positioning accuracy.
(2) In the navigation process, even if the QR code is damaged and stained, it still does not affect the navigation and positioning. Hybrid navigation can complement each other and complete the navigation task independently.
(3) The improved AMCL algorithm fuses QR code navigation. The two types of data are fused using EKF in order to improve the positioning accuracy and at the same time reduce the navigation time and improve the navigation efficiency.
(4) The improved AMCL algorithm can effectively reduce or avoid the occurrence of particle abduction events and increase the reliability of accurate AGV positioning. The superiority of the improved algorithm can be effectively proved through field experiments.

2. Improved AMCL Algorithm

The AMCL suits local and global localization problems [21,22]. The improved AMCL algorithm can effectively solve the problem of low positioning accuracy by dynamically adjusting the number of particles and incremental particle weights and improving the resampling method, which can effectively improve the accuracy and stability of positioning.

Based on the above problem, state fusion is performed by introducing an extended Kalman filter using the outputs of the two algorithms. The different data information from the improved AMCL algorithm and the Gmapping constructed graph are fused. The map position generated using Gmapping is fused with the robot position generated by the improved AMCL, matching the AGV position estimation with the map data to obtain more accurate localization results. The improved AMCL localization algorithm inputs a parameter file to subscribe to the map information scanned using Gmapping. The sensor information received by the LiDAR is input to the AMCL positioning algorithm, and the EKF enhances the positioning accuracy by fusing the output states based on the received map data information and the AMCL positioning algorithm information. Figure 2 shows the process of data fusion between the AMCL algorithm data and lidar sweep map information by the extended Kalman filter.

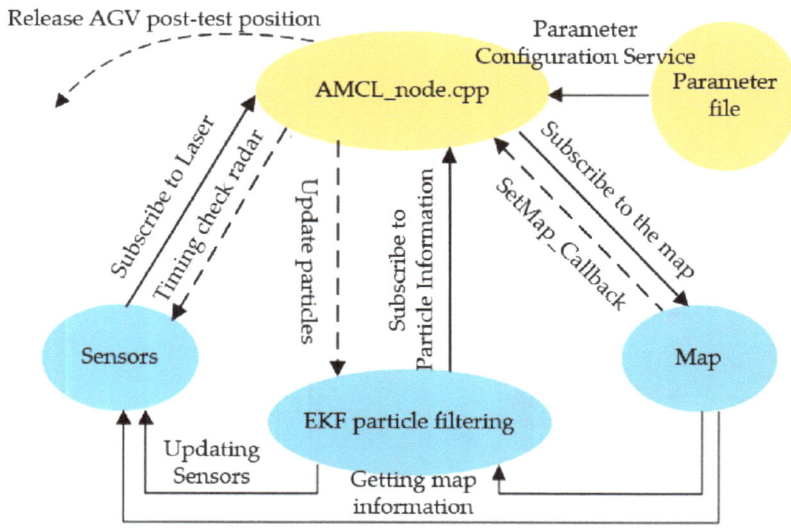

Figure 2. The overall architecture of the AMCL work.

2.1. AGV Motion Modelling and Chassis Structure Analysis

According to the drive mode, it is known that the AGV uses differential drive, so the motion model sketch of the QR code navigation AGV is shown in Figure 3.

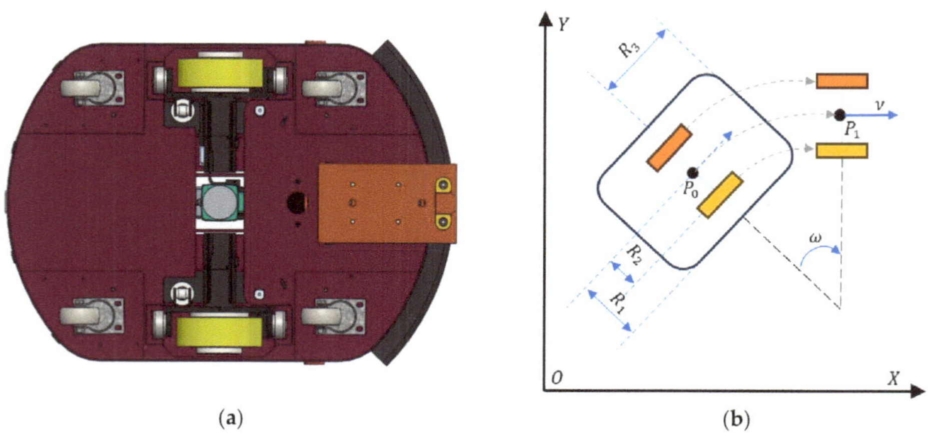

Figure 3. AGV structure and motion modelling: (**a**) chassis structure; (**b**) movement model.

As shown in the figure, the AGV dimensions R_1, R_2, and R_3 are: 38 cm, 30 cm, and 46 cm, respectively. The AGV is powered by a pair of differential wheels, with two followers at the front and rear supporting the AGV body. In Figure 3b, v is the linear velocity of the centre of the AGV, ω is the angular velocity of the AGV, and the attitude of the AGV at the moment $t + 1$ at P0. The kinematic model of the AGV is shown in Equation (1).

$$\begin{cases} x_{t+1} = x_t - \frac{v}{\omega}\sin(\theta_t) + \frac{v}{\omega}\sin(\theta_t + \omega\Delta t) \\ y_{t+1} = y_t - \frac{v}{\omega}\cos(\theta_t) - \frac{v}{\omega}\cos(\theta_t + \omega\Delta t) \\ \theta_{t+1} = \theta_t + \omega_t \end{cases} \quad (1)$$

where Δt is the sampling interval, (x_t, y_t, z_t) is the attitude of the AGV at moment t, and $(x_{t+1}, y_{t+1}, z_{t+1})$ is the attitude at moment $t+1$.

2.2. *Improving the Flow of the Algorithm*

Aiming at resampling the AMCL algorithm in the iterative computation process which will increase the computation amount, an improved AMCL algorithm combining QR code information is proposed to improve the positioning accuracy, reduce the computation amount, and shorten the navigation time. The effect of the improved algorithm is shown in Figure 4.

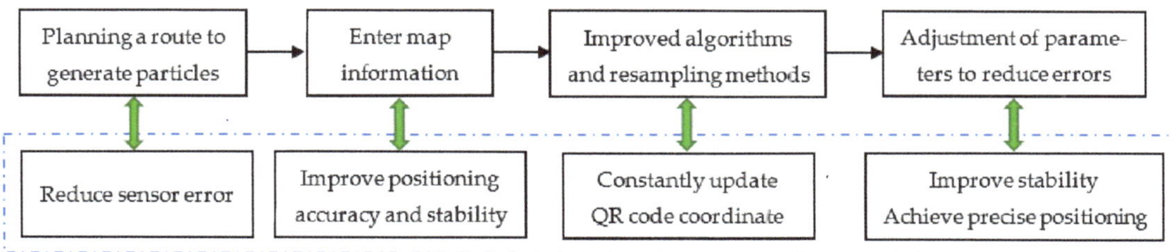

Figure 4. The effect of the improved algorithm.

2.3. *Improved Odometer Motion Model Sampling*

In the time interval $(t-1, t)$, the given motion information u_t is

$$u_t = \begin{bmatrix} \overline{x}_{t-1} \\ \overline{x}_t \end{bmatrix} \tag{2}$$

where $\overline{x}_t$ is the coordinates inside the robot; the difference between $\overline{x}_{t-1}$ and $\overline{x}_t$ enables an estimate of the difference between the two poses. The relative distances can be calculated from three steps: initial rotation δ_{rot1}, translation δ_{trans}, and secondary rotation δ_{rot2}. Calculate one translation and two rotations by the given motion information u_t.

$$\begin{cases} \delta_{trans} = \sqrt{\left(\overline{x} - \overline{x'}\right)^2 + \left(\overline{y} - \overline{y'}\right)^2} \\ \delta_{rot1} = arctan^2\left(\overline{y'} - \overline{y}, \overline{x'} - \overline{x}\right) - \overline{\theta} \quad \delta_{trans} = \sqrt{\left(\overline{x} - \overline{x'}\right)^2 + \left(\overline{y} - \overline{y'}\right)^2} \\ \delta_{rot2} = \overline{\theta'} - \overline{\theta} - \delta_{rot1} \end{cases} \tag{3}$$

Since the AGV runs for a long time, it will lead to cumulative errors. The actual values of translation and rotation can be obtained by subtracting the observed value from the interference error ε_{b^2}, the variance of which is b^2. Therefore, the motion error model can be written as:

$$\begin{cases} \hat{\delta}_{trans} = \delta_{trans} - \varepsilon_{\alpha_1 \delta^2_{trans}} + \alpha_4 \delta^2_{rot1} + \alpha_4 \delta^2_{rot2} \\ \hat{\delta}_{rot1} = \delta_{rot1} - \varepsilon_{\alpha_1 \delta^2_{rot1}} + \alpha_2 \delta^2_{trans} \\ \hat{\delta}_{rot2} = \delta_{rot2} - \varepsilon_{\alpha_1 \delta^2_{rot2}} + \alpha_2 \delta^2_{trans} \end{cases} \tag{4}$$

$\varepsilon_{\alpha_1} \sim \varepsilon_{\alpha_4}$ is the cumulative error of motion.

To obtain the actual location x_t from x_{t-1} after the initial rotation angle $\hat{\delta}_{rot1}$, follow the pan delta $\hat{\delta}_{trans}$, then add to another rotation angle $\hat{\delta}_{rot2}$:

$$\begin{pmatrix} x' \\ y' \\ \theta' \end{pmatrix} = \begin{pmatrix} x \\ y \\ \theta \end{pmatrix} + \begin{pmatrix} \hat{\delta}_{trans} \cos(\theta + \hat{\delta}_{rot1}) \\ \hat{\delta}_{trans} \sin(\theta + \hat{\delta}_{rot2}) \\ \hat{\delta}_{rot1} + \hat{\delta}_{rot2} \end{pmatrix} \tag{5}$$

Then, the final odometer error is:

$$\begin{cases} \delta_{trans} - \hat{\delta}_{trans} \\ \delta_{rot1} - \hat{\delta}_{rot1} \\ \delta_{rot2} - \hat{\delta}_{rot2} \end{cases} \tag{6}$$

2.4. Improved Resampling to Avoid Kidnapping

2.4.1. Adjustment of KLD Dynamic Resampling

The kidnapping problem refers to the fact that when the robot undergoes a drastic change in the environment or is moved to a new location, the particles in the particle filter will gather at the wrong location, resulting in inaccurate localization. In order to effectively reduce or avoid the kidnapping problem in the AMCL algorithm, this can be achieved by improving the resampling strategy of the particle filter. Traditional resampling methods, such as uniform resampling, are prone to lead to the kidnapping problem. Therefore, the kidnapping problem is avoided by using an improved KLD (Kullback–Leibler divergence resampling) sampling method to decide whether to resample or not by calculating the information entropy of the particle weights.

The Kullback–Leibler distance represents the approximation error between two probability distributions p and q, i.e.,

$$K(p,q) = \sum_{x} p(x) log \frac{p(x)}{q(x)} \tag{7}$$

The Kullback–Leibler distance is non-negative and has a value of zero if and only if the two probability distributions p and q agree.

Assuming there is a discrete distribution with k different subspaces, where the vector $X = \{X_1, X_2, X_3, \ldots, X_k\}$ represents the number of particles sampled from each subspace and the vector $P = \{P_1, P_2, P_3, \ldots, P_k\}$ represents the true probability of each subspace, the maximum likelihood estimation probability density is $\hat{P} = \frac{X}{n}$. When n satisfies a certain number, it can be ensured that the Kullback–Leibler distance $K\left(\hat{P}, P\right)$ between the true probability density and the estimated probability density is less than a threshold value ε, which ensures that the approximation error between the true probability density and the estimated probability density is minimized. At this point, according to the Wilson Ferty transformation method, the approximate calculation formula for n with the minimum approximation error can be obtained as follows:

$$n = \frac{k-1}{2\varepsilon}\{1 - \frac{2}{9(k-1)} + \sqrt{\frac{2}{9(k-1)}} z_{1-\delta}\}^3 \tag{8}$$

where $z_{1-\delta}$ is the standard normal distribution for the upper quartile $1-\delta$.

From the above equation, the number of particles required to minimize the approximation error between the estimated and true posterior distributions can be obtained. The main purpose of particle filtering is to estimate a posterior distribution, so it is only necessary to determine the number of effective subspaces k and the pre-given ε and δ to obtain the minimum number of particles required to ensure the estimation performance of particle filtering. It follows that the the improved KLD dynamically adjusted resampling process is as follows:

1. Input the collection of particles $S_{t-1} = \left\{ \left(x_{t-1}^{(i)}, w_{t-1}^{(i)} \right) \middle| i = 1, \ldots, n \right\}$ after resampling at moment $t-1$, the observation B, set ε and δ, the minimum value of the total number of particles n_{xmin}, and the number of particles n_{t-1} at moment $t-1$.
2. Set the predicted particle set at time t to $\widetilde{s}_t = \varnothing$, with a total number of particles of $n = 1500$, a number of particles on the line $n_{xmax} = 100000$, and variables k and α

both being 0. Use the dynamic adjustment method to resample particle $x_{t-1}^{(i)}$ to obtain the predicted particle $x_t^{(i)}$, and calculate the corresponding weight $w_t^{(i)}$.

3. Accumulation of weights: $\widetilde{s}_t = \widetilde{s}_t \cup \left\{x_t^{(n)}, w_t^{(n)}\right\}$, place the newly predicted particles into the set of predicted particles b. If $x_t^{(n)}$ falls into the space interval b, then $k = k + 1$, while the interval b becomes non-empty.
4. If $n \geq n_{min}$ then $n_x = \frac{k-1}{2\varepsilon}\left\{1 - \frac{2}{9(k-1)} + \sqrt{\frac{2}{9(k-1)}} z_{1-\delta}\right\}^3$, $n = n + 1$, followed by weight orthogonalization: $w_t^{(i)}$, and finally return $\widetilde{s}_t$.

2.4.2. Simulation to Verify the Analysis of the Adjusted Dynamic Results

Verify the effectiveness of KLD dynamically adjusting resampling particles to solve the kidnapping problem through the state estimation of nonlinear systems. The simulation uses a Windows 64 bit system and the simulation software is MATLAB 2021B.

Simulation research is conducted on the state estimation problem of nonlinear systems, and the nonlinear system state equation used is as follows:

$$\begin{cases} x_k = 1 + \sin(0.04\pi k) + 0.5x_{k-1} + v_{k-1} \\ y = \begin{cases} 0.2x_k^2 + n_k & k \leq 30 \\ 0.5x_k - 2 + n_k & k > 30 \end{cases} \end{cases} \tag{9}$$

where the system noise is: $v_k \sim Gamma(3,2)$, and the observation noise is: $n_k \sim N(0, 0.00001)$.

The system noise is taken as gamma noise and the observation noise is Gaussian white noise. The initial azimuthal misalignment angle is 10°; the initial horizontal alignment angle is 1°; the accelerometer constant drift is 3×10^{-5}g; the accelerometer random drift is 1×10^{-5}g; the gyroscope constant drift is 0.05°/h, and the gyroscope random drift is 0.01°/h.

From the simulation results in Figure 5a, it can be seen that the KLD and the improved KLD resampling methods conclude that only the improved KLD resampling has a more stable convergence speed and consistency, and can effectively solve the fluctuation problem of particle filtering. From Figure 5b, it can be seen that the alignment accuracy and convergence speed of the improved KLD resampling are significantly better than that of the unimproved method, and the improved simulation curve eliminates the fluctuation phenomenon present in other algorithms and has better stability.

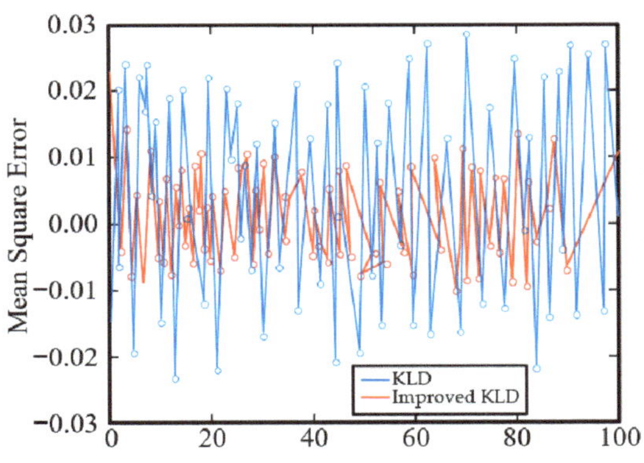

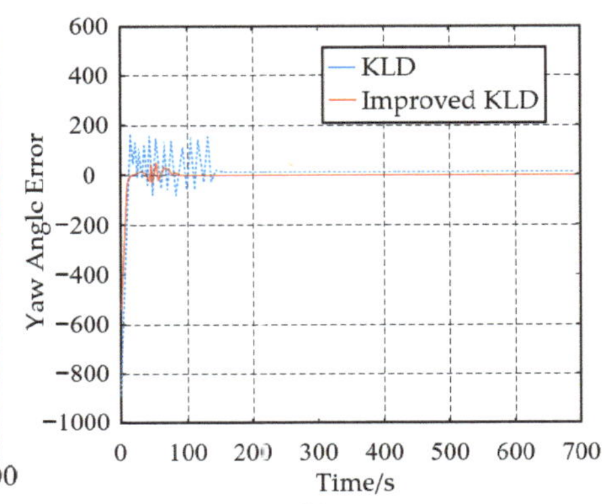

Figure 5. Simulation results: (**a**) comparison of mean square error results; (**b**) simulation results of heading angle error.

Under the same simulation conditions, the KLD and the improved KLD methods are run for 1000 steps each, and the results of calculating the mean square deviation of the state estimation of the different methods and the running time consumed by running the simulation for 1000 steps are shown in Table 1.

Table 1. Comparison of alignment accuracy and runtime.

Algorithm	Standard KLD Resampling Method	Improved KLD Resampling Method
Running Time	232	75
$\varnothing_E/(')$ Variance	1.63	0.71
$\varnothing_N/(')$ Variance	1.35	0.84
$\varnothing_U/(')$ Variance	2.59	0.93

From Table 1, it can be concluded that the estimation accuracy of the improved KLD resampling method is better than that of the KLD method, and the computation time consumed is less than 1/2 of that of the KLD method, which evidently has a better real-time processing capability and is more suitable for real-time applications.

2.5. Improvement of AMCL Algorithm Initial Particle Swarm Generation

1. Random sampling from Gaussian distribution to generate initial particles. Use the global coordinate system as the reference coordinate system, use the initial positional attitude (default 0) as the mean value of the initial particle distribution, and obtain the covariance matrix of the positional attitude from the parameter server.
2. Prediction of particle orientation. When the odometer information is received, it is sampled from the odometer model to estimate the predicted position of the particle swarm.
3. Update particle position. When receiving the measurement data, the measurement data is put under the position of each particle, to judge the possibility of the measurement data occurring, and update the weights of the particles with this possibility. Put the laser measurement data into each particle position, and then calculate the distance between the endpoint of the laser measurement and the nearest obstacle on the map, the smaller the space is, the greater the likelihood of the laser measurement data occurring, and the greater the weight of the particle. The darker the color of the particle, the greater the weight.
4. Resample. If resampling, the program will judge the variance of the weights of the particle set; the larger the variance the smaller the effective particles and the more serious the particle degradation. In this case, it is necessary to carry out the resampling. After resampling, the number of particles will remain unchanged, particles with smaller weights will be filtered out, and particles with larger weights will be copied.
5. Place the resampled particles into the histogram. The bit positions of the particles after resampling are put into the corresponding histogram. Maintain the data structure of the histogram with kd-tree, with the bit positions of the histogram as key and the weights of the particles as value. the more particles within the histogram, the darker the color of the histogram, which represents a more significant weight of that histogram.
6. Clustering statistics results. Recursively find if there are histograms containing particles at nine fixed distances around each histogram (here, the size of a histogram edge is used as the distance), and cluster them into one class if there are, and where it is clear that c1 is the highest-weighted class. Again, generate the variable particle swarm, find the particles in c1 and take the mean of their bit positions as the final result to publish.

Figure 6a,b represent the particle's position information in global coordinates and the odometer's re-prediction of the particle's position by Gaussian noise, respectively.

Figure 7a,b represent the change in the particle weights, the size of the weights based on color discrimination and the particles with larger weights are replicated, respectively. Figure 8a–c represent the process of dividing the particles based on regions, finding particles with high weights using clustering methods, and publishing particles, respectively. From the analysis of the above steps, it can be concluded that the sampling method based on the probabilistic model is used to generate the particle swarm more evenly and uniformly. According to the weights of the particles, filter out the particles with smaller weights and copy the particles with larger weights to better cover the area where the AGV is located and improve the robustness of the algorithm. For the improved initial particle swarm optimization, this can improve the initial positioning accuracy and convergence speed of the algorithm. In addition, the improved initial particle swarm generation method can also improve the accuracy of the algorithm, enabling it to effectively locate in different environments and initial conditions.

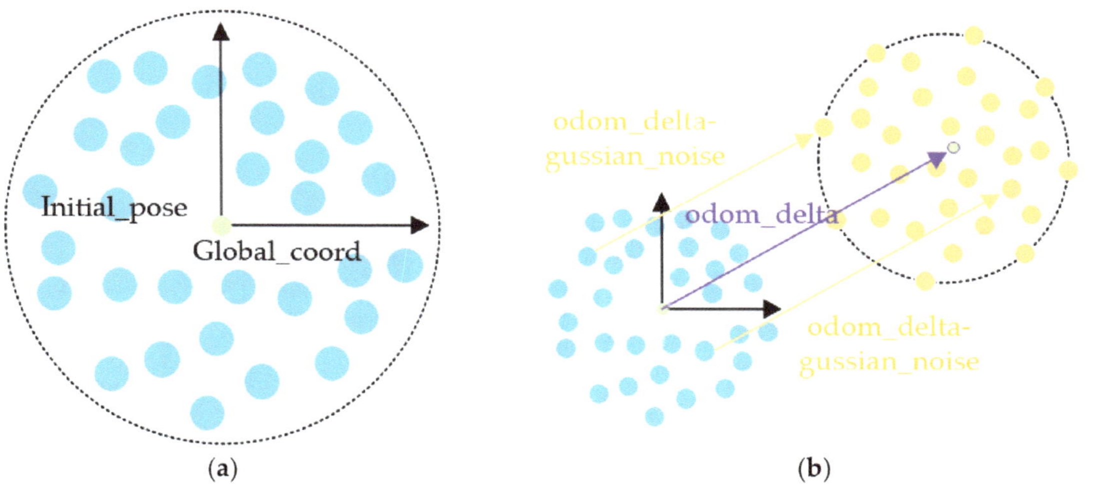

Figure 6. Initial particle generation: (**a**) obtaining the covariance matrix of the position; (**b**) updating and re-prediction of particle positions.

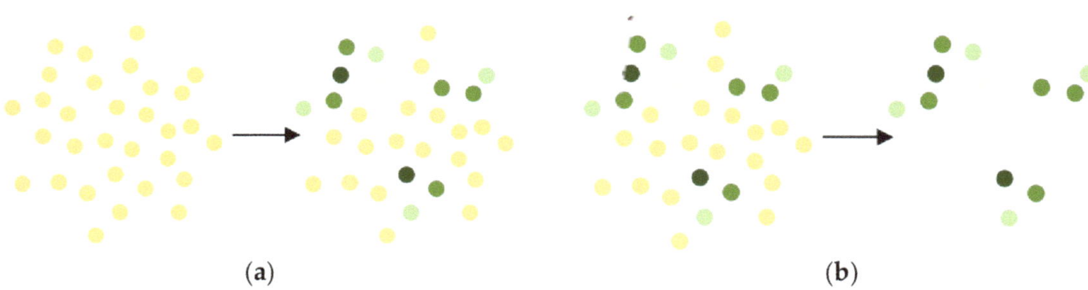

Figure 7. Particle weights and updating iterative particles, the weight of yellow particles is smaller than that of light green particles, and the weight of dark green particles is the largest: (**a**) update particle weights; (**b**) filter less weighted particles.

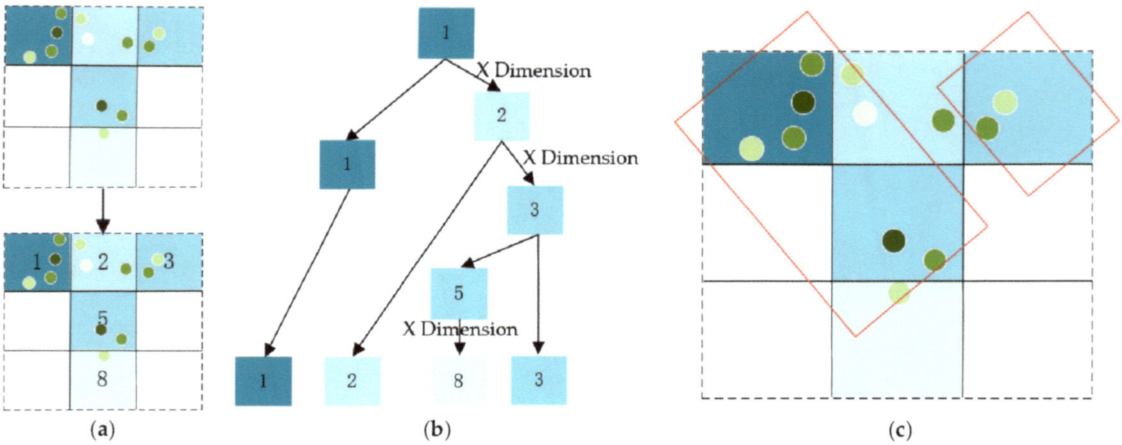

Figure 8. Analysis of clustering statistics results: (**a**) sampled particles are divided into regions based on weights; (**b**) clustering to find particles with higher weights; (**c**) Comparison of weighting information for different clustered regions.

2.6. Comparison of Performance Metrics of the Improved AMCL Algorithm

The AMCL and improved AMCL algorithms are tested in simulation experiments, respectively. The simulated AGV is started on ROS with a Gmapping raster map, and the start point and end point are set to use the two algorithms for trajectory tracking planning, respectively. The distance between the cyclic Gmapping maps is 10 m, and the walking of 10 circles is one round, ten rounds. The data of each travel is recorded using a rosbag file and then imported into Matlab for simulation analysis. As shown in Figure 9, the simulation is carried out in the ROS operating system.

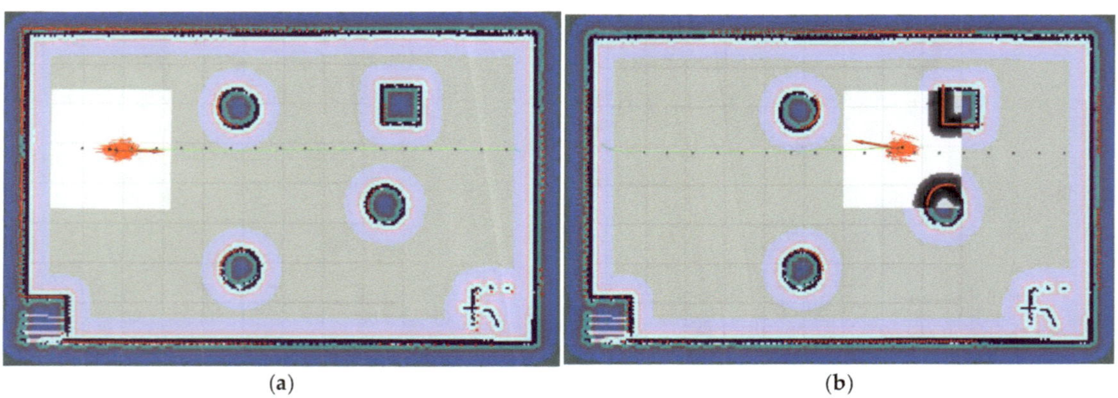

Figure 9. Simulation path planning: (**a**) from the starting point to the target point; (**b**) return from the target point to the starting point.

Where the parameters are set to $c = 0\ .357, \varepsilon = 0.2, \tau = 0.05, \delta = 0.01$. The rest of the parameters were kept unchanged.

The packet rosbag obtained on ROS was imported into Matlab 2021B, after which the data was extracted to obtain the following plot of fitted and analyzed data.

The data for each of the simulations are shown in Table 2. Table 2 represents the parameter differences between the two different algorithms in the simulation case.

Table 2. AMCL and improved AMCL fusion QR code algorithm comparison.

Parametric	Algorithm	Value
Maximum offset angle	AMCL	30°
	Improved AMCL	21°
Minimum offset angle	AMCL	24°
	Improved AMCL	10°
Average travel time for one lap	AMCL	7.3 s
	Improved AMCL	5.5 s
Maximum error from target point	AMCL	34 cm
	Improved AMCL	11 cm
Minimum error from target point	AMCL	24 cm
	Improved AMCL	5.2 cm

As can be seen from the illustration, the trace of the error is consistent with the offset error generated during AGV traveling. Still, it needs to be corrected concerning the actual speed value, a priori estimate, and a posteriori estimate. That is, the following predicted path in the process of traveling has a significant deviation from the real path, and it is easy to produce a situation in which the command route is not the same as the real route.

The improved algorithm was tested in the simulation by setting the initial yaw angle to 60°. The value change in the optimized algorithm, while ensuring that the rest of the parameters are the same, yields the silky smoothness of the real trajectory graph running over the time of (a) in Figure 10. The improved algorithm is better smoothed compared to the AMCL. (b) The grey lines in the figure serve as the measured data, with a large degree of deviation; the green lines serve as the modelled values, with a smaller degree of deviation; and the final filtered results are much closer to the true values, and the traces of error have converged to a minimum. From the figure, it can be concluded that the fluctuation is small and the performance is relatively smooth. The thin line represents the AMCL algorithm test results, and the thick line represents the improved AMCL algorithm test results. (c) and (d) in Figure 10 illustrate that the improved AMCL algorithm performs smoother for the fluctuation of the errors of the true value of the speed, the a priori estimate and the a posteriori estimate for different directions, the errors are within the permissible range, and the improvement of the algorithm meets the requirements and performs well.

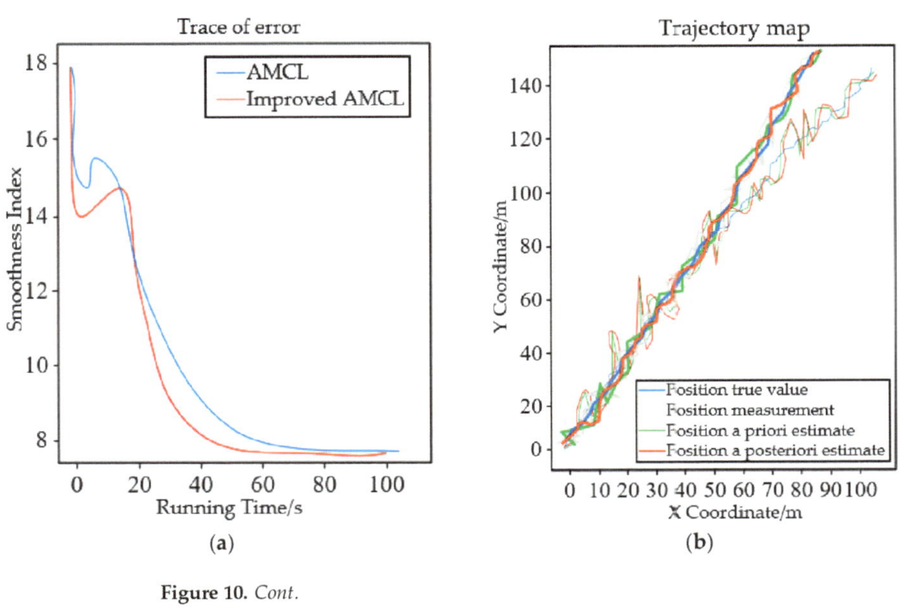

Figure 10. *Cont.*

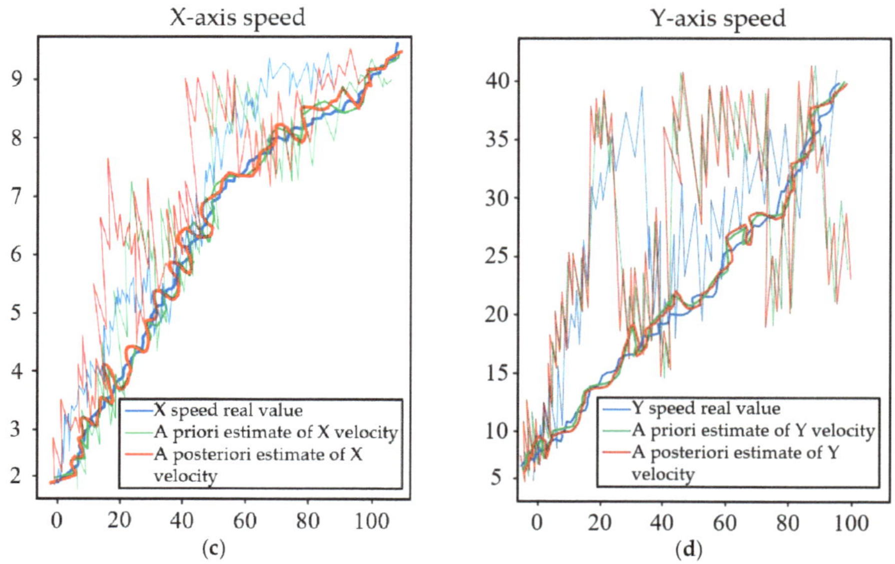

Figure 10. AMCL algorithm simulation: (**a**) trace of error; (**b**) trajectory map; (**c**) x-axis speed; (**d**) y-axis speed.

3. QR Code Navigation and Construction of a Template Matching Library

The experimental simulation verifies that the improved AMCL algorithm has better performance, yielding better data in terms of positioning accuracy, arrival time and offset angle. Therefore the improved algorithm can be verified by fusion experiments.

3.1. QR Code Navigation Process

First, the AGV uses SLAM to navigate to the area near the specified QR code. Then, the AGV opens the QR code scanning function and switches to QR code navigation after scanning the QR code setting information. Then, the AGV approaches the direction of the center of the QR code until the scanning center of the AGV chassis coincides with the center of the QR code. The vehicle-mounted camera takes the position of the QR code center as the relative position, obtains the spatial coordinates through calculation and conversion, and uses the coordinate transformation to make the AGV scan the center of the QR code and realize normal movement. Figure 11 shows the flow chart of QR code scanning and recognition.

To calculate the geometric information of objects in three-dimensional space, it is necessary to calibrate the camera. From this, the camera's internal parameter matrix A and R are obtained, and the camera's orientation relative to the world coordinate system is determined. Since the camera's focal length is not 0, and the matrices A and R are reversible, the actual coordinates of pixels in the photo in the AGV coordinate system are calculated using Formula (10).

$$\begin{cases} \begin{bmatrix} X \\ Y \\ Z \end{bmatrix} = R^{-1}.\left(A^{-1}.\begin{bmatrix} u \\ v \\ 1 \end{bmatrix}^{-1} \right) \\ (u_c, v_c) = \left(\frac{u_1+u_3}{2}, \frac{v_1+v_3}{2} \right) \end{cases} \tag{10}$$

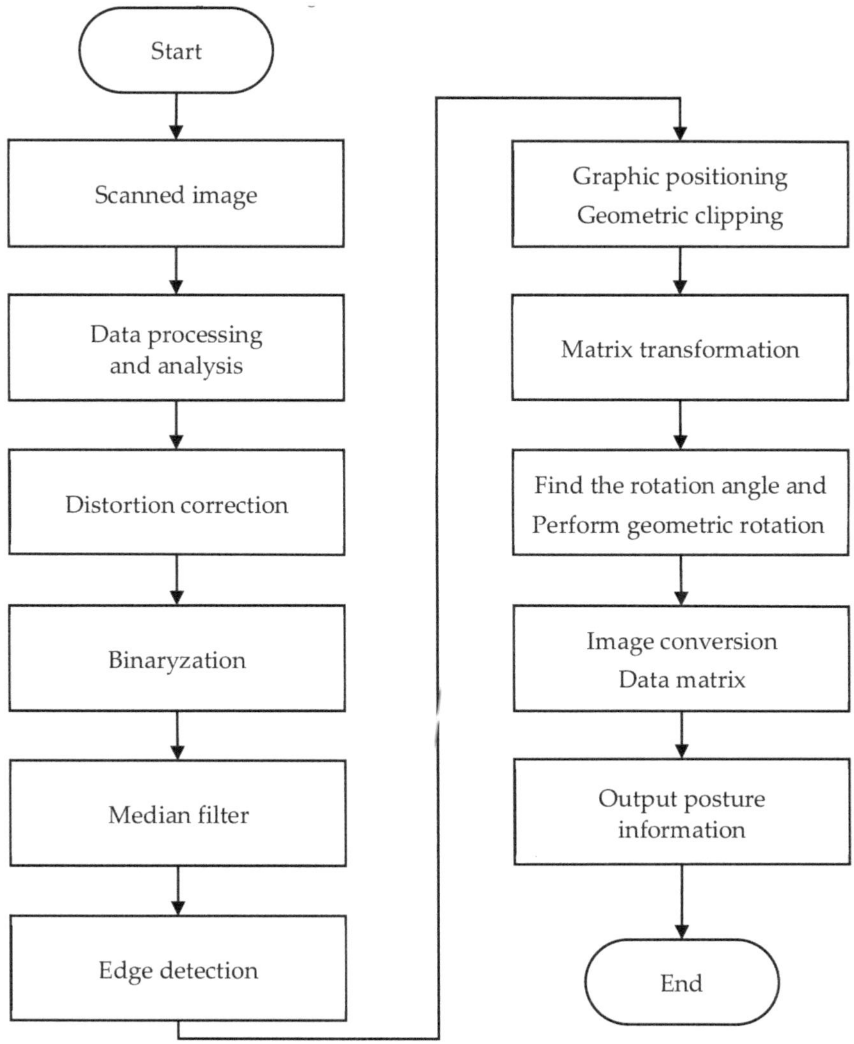

Figure 11. QR code scanning recognition process.

When the central coordinate of the QR code $p(u_c, v_c)$ is known, the coordinates $P(X_C, Y_C, Z_C)$ of the car body coordinate system can be calculated. Due to the use of QR code three-point positioning, the coordinates of the three points of the QR code in the image (u_1, v_1), (u_2, v_2), (u_3, v_3), and because the position of the required detection pattern is known, the coordinates of the center $p(u_c, v_c)$ of the QR code can be calculated, and $p(u_c, v_c)$ is substituted to obtain the coordinates $P(X_C, Y_C, Z_C)$ of the QR code in the coordinate system of the car body. Combined with the coordinate information of the current QR code in the entire map, the position of the AGV in the map can be determined. In Figure 12 (a) represents the relationship between the three coordinate systems, (b) AGV scanning for coordinate transformation.

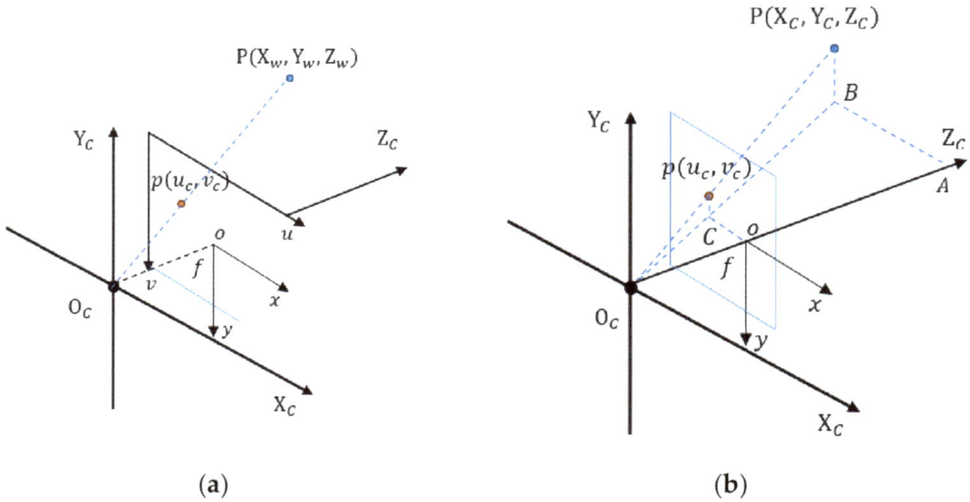

Figure 12. Coordinate transformation between QR code and AGV. $(O_W - X_W Y_W Z_W)$, represents world coordinate system describing camera position, $(O_C - X_C Y_C Z_C)$ represents camera coordinate system, $(o - xy)$ represents image coordinate system, f—the camera focal length, equal to the distance of o to the O_c. (**a**) World, camera, image physics, pixel coordinate system relationship; (**b**) the relationship between camera coordinate system and image physical coordinate system.

3.2. Construction of the Template Matching Library

The stability of QR code navigation depends on the performance of the reader (decoding speed, accuracy, and precision) and the performance of the IMU (accuracy and stability) [23,24]. The higher the decoding efficiency of the reader, the higher the AGV can travel. However, if the QR code is dirty and damaged, the navigation accuracy will significantly decrease. Therefore, we propose a library of QR code templates to improve the positioning accuracy. QR codes of the same size with different recognition areas are generated by a QR code generator, and each QR code template library contains nine independent QR codes, each of which is unique. In other words, multiple QR codes are fused into a QR code combination map, and the QR code template libraries are constructed based on the "different codes" in the QR code recognition area, and the current pose is quickly determined by triangulation matching.

To further improve the localization accuracy of QR code labels in the global graph, a particle-based localization algorithm in AMCL is used to match the QR code combination graph. The Gmapping grid map is drawn using SLAM navigation. Use the three points of the QR code identification area to form a triangle as the coordinate position of the QR code (the QR code itself also stores the position information), and save it in the map. The location of the QR code label is used as a landmark, and the AGV recognizes its ID by matching the triangles and obtains the corresponding coordinates when it detects the QR code landmark. Since each QR code tag area generates different particles, but the template library is unique, the similar triangles identified above for triangle matching can be matched with the template library. As shown in Figure 13a shows the graphic and storage information area of the QR code, Figure 13b shows the formation of QR code landmarks through multiple QR codes, and Figure 13c shows the QR code landmarks affixed to the experimental site. For example, when different numbers of particles are generated at the position of the label of the template library in the front A, then scanning the position in front can obtain a lot of similar triangles according to the different positions of the particles, at this time, the triangles generated by the particles in the fixed area are matched with the triangles of the current correct position area, which is similar but the direction is different, and thus the AGV is automatically adjusted to the angle. Therefore, the distance and offset angle between the AGV and the label can be determined. Firstly, the AGV calculates the relative position of the QR code tag in the SLAM global map based on the received QR code data, thus obtaining the approximate orientation of the detected QR code. Then, the AGV performs a self-tracking motion based on the approximate position in the map and quickly determines the part of the QR code label based on triangle matching.

Figure 13. Identification code used in navigation: (**a**) foundation template; (**b**) composite template; (**c**) paste the template in the experiment.

The matching template library formula is as follows:

$$\begin{array}{c} l_{49}, l_{48}, l_{89}, (x_4, y_4), (x_8, y_8), (x_9, y_9), 1 \\ l_{48}, l_{18}, l_{14}, (x_1, y_1), (x_4, y_4), (x_8, y_8), 2 \\ l_{33}, l_{12}, l_{26}, (x_3, y_3), (x_7, y_7), (x_5, y_5), 3 \\ \vdots \\ l_{xy}, l_{xy}, l_{xy}, (x_x, y_y), (x_x, y_y), (x_x, y_y), r \\ \vdots \\ l_{xy}, l_{xy}, l_{xy}, (x_x, y_y), (x_x, y_y), (x_x, y_y), n \end{array} \tag{11}$$

In the above equation, l_{xy} is the edge length of the triangle template library, and (x_x, y_y) is the coordinate value of each point.

In this paper, we use the estimated position of the AGV as the input to fuse the world coordinates of the QR code and increase the weight of particles in the AMCL algorithm. Many QR code landmarks were detected in the process of AGV moving, and the location of QR code landmarks was quickly determined according to the triangle matching. The relationship between the AGV and the detected QR code landmark is as follows:

$$\begin{array}{c} (x_V - x_1)^2 + (y_V - y_1)^2 = r_1^2 \\ (x_V - x_2)^2 + (y_V - y_2)^2 = r_2^2 \\ \vdots \\ (x_V - x_i)^2 + (y_V - y_i)^2 = r_i^2 \\ \vdots \\ (x_V - x_n)^2 + (y_V - y_n)^2 = r_n^2 \end{array} \tag{12}$$

where (x_V, y_V) represents the estimated position of the AGV car, which can be obtained by solving the formula. (x_i, y_i) is the ith location of the QR code detected. In addition, it can be obtained that the direction of the AGV is

$$\theta_V = \begin{cases} \frac{1}{n}\sum_{i=1}^{n}\left[\arctan\left(\frac{y_i-y_V}{x_i-x_V}\right)-\varphi_i\right] & x_i > x_V \\ \frac{1}{n}\sum_{i=1}^{n}\left[\arctan\left(\frac{y_i-y_V}{x_i-x_V}\right)-\varphi_i+\pi\right] & y_i \geq y_V,\ x_i < x_V \\ \frac{1}{n}\sum_{i=1}^{n}\left[\arctan\left(\frac{y_i-y_V}{x_i-x_V}\right)-\varphi_i-\pi\right] & y_i < y_V,\ x_i < x_V \\ \frac{1}{n}\sum_{i=1}^{n}\left[\frac{\pi}{2}-\varphi_i\right] & y_i > y_V,\ x_i = x_V \\ \frac{1}{n}\sum_{i=1}^{n}\left[-\frac{\pi}{2}-\varphi_i\right] & y_i < y_V,\ x_i = x_V \end{cases} \tag{13}$$

where θ_V represents the direction of the AGV car and φ is the observation angle of the detected QR code landmark. Therefore, the estimated pose of the AGV car can be expressed as $\mu_V = (x_V, y_V, z_V)$.

The estimated position of the AGV car can be obtained from Equations (12) and (13), and the position weight of the particle can be calculated as follows:

$$d^{[i]} = \frac{1}{(2\pi)^{\frac{3}{2}}}\mathrm{EXP}\left(-\frac{1}{2}\left(x_t^{[i]}-\mu_V\right)^{\mathrm{T}}\left(x_t^{[i]}-\mu_V\right)\right) \tag{14}$$

where the position and direction of the particle are $x_t^{[i]} = (x_i, y_i, \varphi_i)$.

The AMCL algorithm covers the whole map evenly, and then the particles converge near the QR code according to the location of the QR code landmark. As shown in Figure 14a,b. When scanning the QR code landmark at a specific location, triangular matching joins to the QR code landmark. As in Figure 14c. The coordinates of a, b, and c are (x_a, y_a), (x_b, y_b), and (x_c, y_c), respectively. The side lengths of the three measured triangles are l_{ab}, l_{ac}, and l_{bc}, respectively. First, search for similar triangles in the QR code template library. As in Figure 14d. The three sides are unequal in length but proportional, and the angles are equal. Thus, the detected Δabc matches the $\Delta 835$ in the library. As in Figure 14e,f.

Triangle detected:

$$l_{ab}, l_{ac}, l_{bc}, (x_a y_a), (x_b y_b), (x_c y_c) \tag{15}$$

Matching template library:

$$\begin{cases} l_{49}, l_{48}, l_{89}, (x_8, y_8), (x_3, y_3), (x_5, y_5), 1 \\ \vdots \\ l_{28}, l_{29}, l_{89}, (x_2, y_2), (x_8, y_8), (x_9, y_9), k \\ \vdots \\ l_{35}, l_{34}, l_{55}, (x_1, y_1), (x_3, y_3), (x_9, y_9), n \end{cases} \tag{16}$$

Triangular matching identification:

$$\begin{cases} a(x_a, y_a) = 8(x_8, y_8) \\ b(x_b, y_b) = 3(x_3, y_3) \\ c(x_c, y_c) = 5(x_5, y_5) \end{cases} \tag{17}$$

In the navigation process, the completeness and accuracy of the constructed map is one of the decisive factors in determining the accuracy of navigation and positioning [25]. Significant errors in map construction often directly lead to problems such as positioning failure and unguaranteed accuracy [26]. To improve the map-matching accuracy in navigation, the weight of the particles in the AMCL algorithm is increased in the code reading area. Then, the AGV position information is updated, and the data is estimated to be within the normal range according to the recognition of the new.

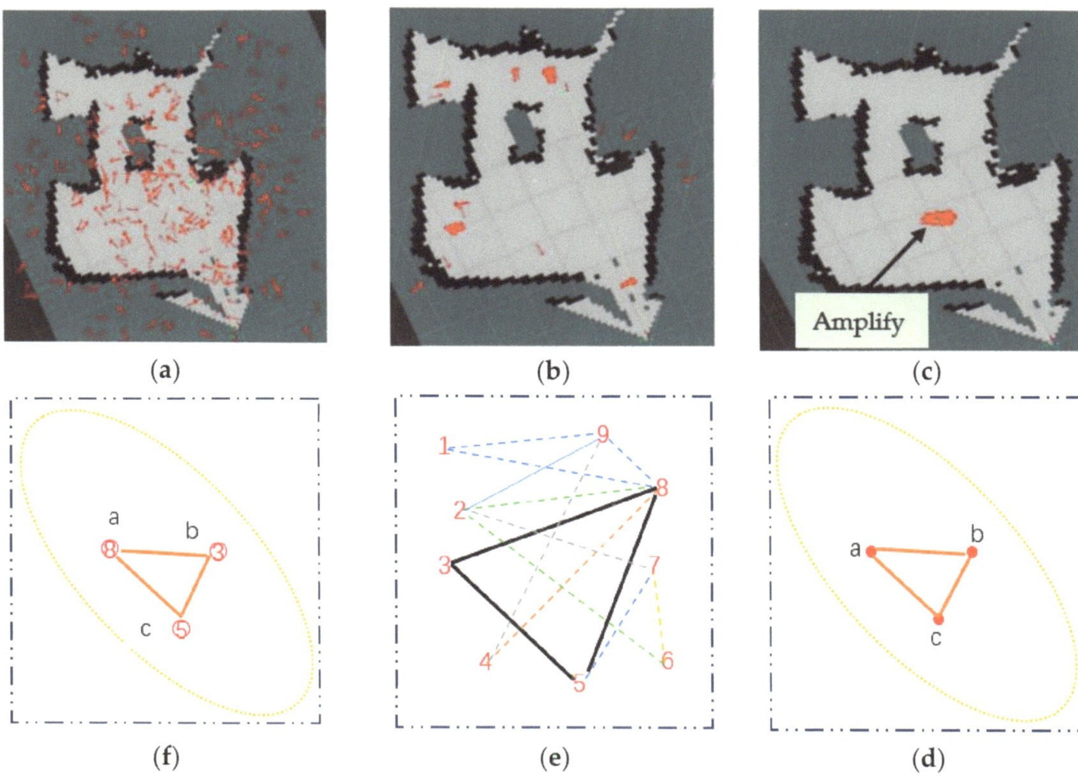

Figure 14. Process of AMCL particle distribution and triangulation matching. The number of the QR code reflection surface detected by (**a**–**c**) is the ID of the QR code landmark in the template library: (**a**) particles flatten the entire map; (**b**) the particle converges to the landmark position of the QR code; (**c**) use triangular matching positioning to increase the weight when scanning the QR code; (**d**) a, b and c are the detected QR code reflecting surfaces; (**e**) there are all kinds of triangles in the template library; (**f**) the two triangles are proportional in length and match in shape.

Establish a simulation environment and analyze and improve the various types of problems that occur in the algorithm through simulation experiments to verify the algorithm. Figures 15 and 16 represent the fusion process of QR codes in the simulated state and in the improved algorithmic positioning shown simultaneously in the RVIZ and Gazebo environments, respectively.

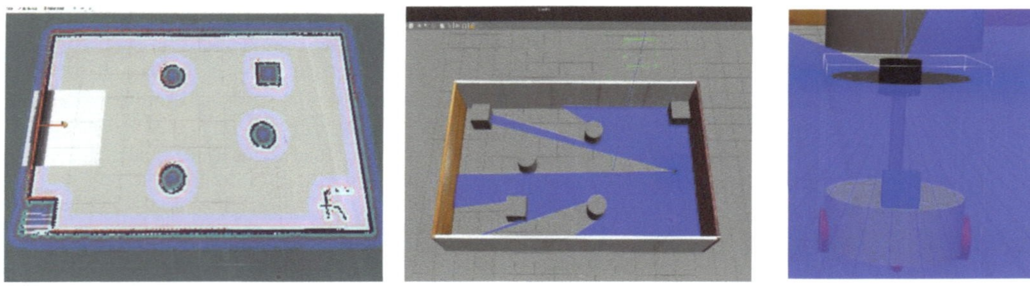

(a) (b) (c)

Figure 15. Building a simulation environment: (**a**) synchronize environment on Rviz; (**b**) synchronize environment on Gazebo; (**c**) simulation model.

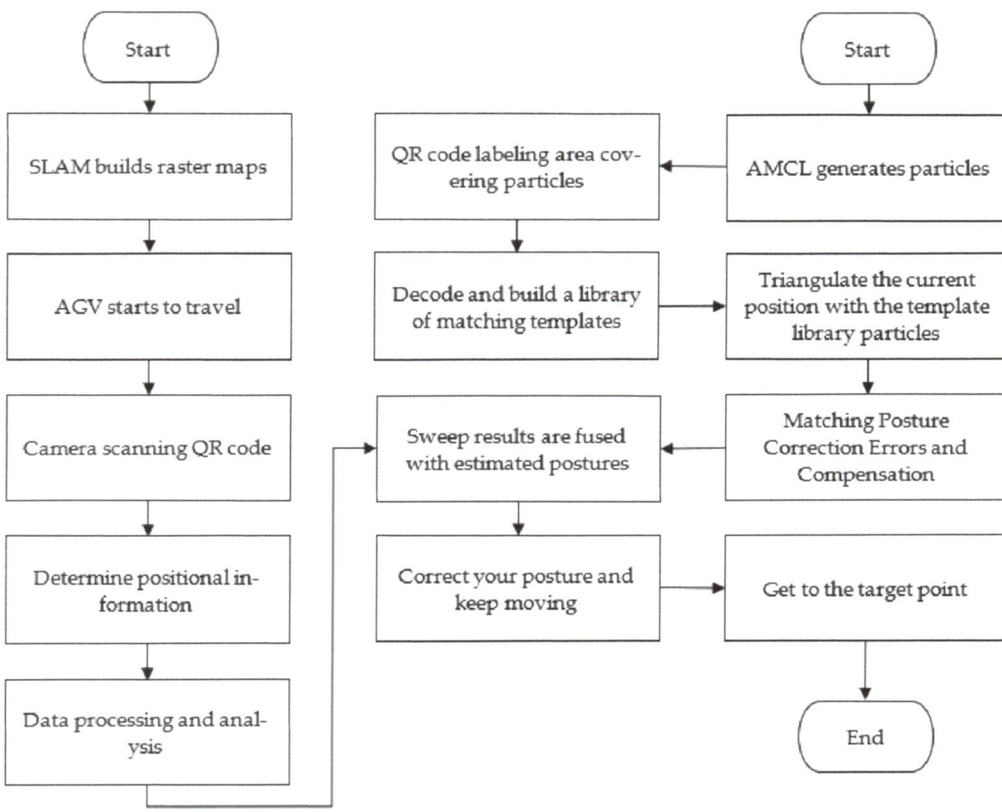

Figure 16. Improved AMCL algorithm localization process for QR code navigation fusion.

3.3. Extended Kalman Filter Fusion Data

The estimated position of the QR code navigation AGV sweep data and the AMCL algorithm can be fused by a fusion method to obtain a more accurate position estimation [27,28]. In the fusion process, it is first necessary to convert the QR code navigation AGV scanning data into positional information, which is usually obtained by decoding the QR code to obtain the position and attitude information of the QR code. Then, this position information is fused with the estimated position of the AMCL algorithm. The fusion method is as follows:

1. Prediction steps.

Establishing the equation of state:

$$X_{k+1} = \begin{bmatrix} x_k \\ y_k \\ v_k \\ \theta_k \\ \omega_k \end{bmatrix} + \begin{bmatrix} v_k \cdot \cos(\theta) \cdot \Delta t \\ v_k \cdot \sin(\theta) \cdot \Delta t \\ 0 \\ \omega \cdot \Delta t \\ 0 \end{bmatrix} + \begin{bmatrix} \frac{1}{2} \cdot \alpha_{v,k} \cos(\theta) \cdot \Delta t^2 \\ \frac{1}{2} \cdot \alpha_{v,k} \cdot \sin(\theta) \cdot \Delta t^2 \\ \alpha_{v,k} \cdot \Delta t \\ \frac{1}{2} \cdot \alpha_{\omega,k} \cdot \Delta t^2 \\ \alpha_{\omega,k} \cdot \Delta t^2 \end{bmatrix}, \omega_k \neq 0 \quad (18)$$

where x is the x-direction coordinate, y is the y-direction coordinate, v is the radial velocity, θ is the yaw angle, ω is the yaw angular velocity, $\alpha_{v,k}$ is the acceleration of the radial velocity v, and $\alpha_{\omega,k}$ is the acceleration of the yaw angular velocity ω.

Condition prediction:

$$\hat{x}_{k+1|k} = \left[f\left(\hat{x}_{k|k}\right) + F_k\left(x_k - \hat{x}_{k|k}\right) + \omega_k \right] \quad (19)$$

where $f\left(\hat{x}_{k|k}\right)$ is the state estimate, x_k is the state vector, and F_k is the higher order term derivation.

From Equations (18) and (19) the predicted covariance prediction can be derived as

$$P_{k+1|k} = [F_k\left(x_k - \hat{x}_{k|k} \cdot \cos(\theta)\Delta t\right) + \alpha_{v,k} \cdot \Delta t] \cdot [F_k\left(x_k + \hat{x}_{k|k} \cdot \sin(\theta)\Delta t\right) - \alpha_{\omega,k} \cdot \Delta t^2] \quad (20)$$

2. Update steps.

Convert the QR code navigation AGV scanning data into an observation model, i.e., the position and attitude information of the QR code is converted into a positional observation. Compare the observation model with the predicted position and attitude, and calculate the observation residuals:

$$\hat{y} = z_k - H_k \hat{x}_{k|k+1} \quad (21)$$

z_k is the state estimation transfer equation and H_k is the observation matrix.

$$\begin{cases} z_k = H_k \cdot x_k + v_k \\ x_k = F_k x_{k-1} + B_k u_k + \omega_k \end{cases} \quad (22)$$

v_k is the observation noise and B_k is the input control model acting on the controller vector u_k.

Calculate the Kalman gain:

$$K_k = \hat{P}_k \cdot H^T \cdot \left(H \cdot \hat{P}_k H^T + R\right)^{-1} \quad (23)$$

R is the measurement noise covariance matrix and H is the measurement matrix.

Based on the Kalman gain and observation residuals, the updated bit position estimate is

$$h(x') = \begin{bmatrix} \rho \\ \psi \\ \dot{\rho} \end{bmatrix} = \begin{bmatrix} \sqrt{P_x^2 + P_y^2} \\ \arctan\left(\frac{P_x}{P_y}\right) \\ \frac{P_x V_x - P_y V_y}{\sqrt{P_x^2 + P_y^2}} \end{bmatrix} \quad (24)$$

where P_x, P_y, V_x, and V_y are the positions and velocities of the fused a priori estimates.

3. Simulation result diagram.

The two were fused for simulation to test the correlation relationship.

Figure 17 shows the accuracy test of the EKF fusion data. Simulations using EKF fused data can yield close to the true values and can be used with this fusion method.

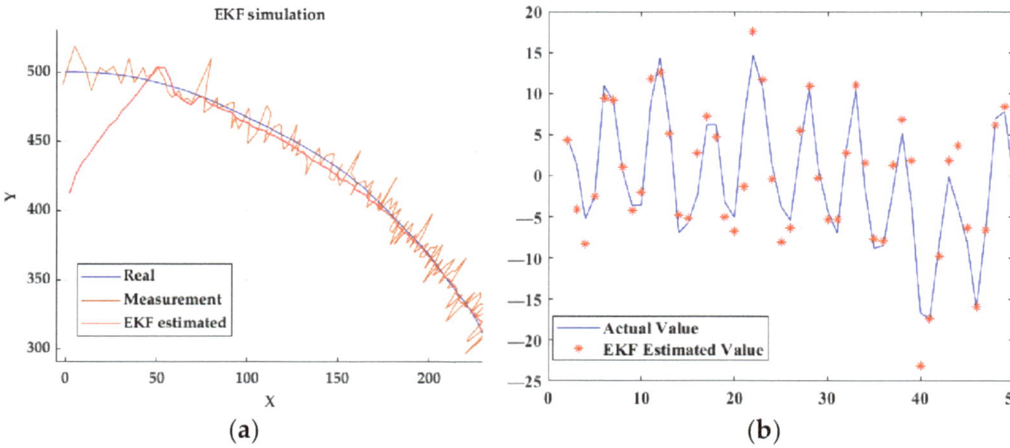

Figure 17. Simulation test: (**a**) convergence to true value; (**b**) the estimated value is near the true value.

4. Experimental Validation

In order to verify the effectiveness of the proposed combination algorithm, a simulation platform is established and real experimental scenarios are built to verify the effectiveness. The AGV used in the simulation is 0.5 m × 0.3 m × 0.2 m in size, with a mass of 10 Kg, LIDAR on top, and an inertia matrix of $[0.01,\ 0,\ 0;\ 0,\ 0.02,\ 0;\ 0,\ 0,\ 0.03]^T$. The mean value of the measurement noise is set at 0, and the covariance matrix at $[0.01,\ 0;\ 0,\ 0.01]$, which means that the variance of the measurement noise in the x- and y-directions is 0.01.

The parameters are set for simulation experiments and the rest of the data is kept constant. The data obtained in the simulation experiment platform is analyzed using MATLAB2021B. Table 3 shows the parameter settings of the AGV.

Table 3. AGV setup parameters.

Name	Number
Maximum linear speed	1 m/s
Maximum angular velocity	1 rad/s
Maximum linear acceleration	0.5 m/s^2
Maximum angular acceleration	0.5 rad/s^2
Radar scanning angle	320°
Radar scanning frequency	10 Hz
Maximum measurement distance of radar	10 m
Improved AMCL algorithm particle initial value	1000
Resampling measurement	Low variance resampling

4.1. Improved AMCL Algorithm Simulation Comparison Experiment

A raster map is generated using SLAM laser navigation, and AMCL localization particles are tiled over the raster map until the whole map is tiled. The AMCL algorithm and the improved AMCL algorithm are introduced separately, and the different situations among the two are observed and data is collected for analysis. The size of the map is 10 m × 10 m, and travelling one lap is recorded as one complete distance.

Navigation was tested using different algorithms in the same environment, and Figure 18 show the different scenarios in which the particles behaved separately.

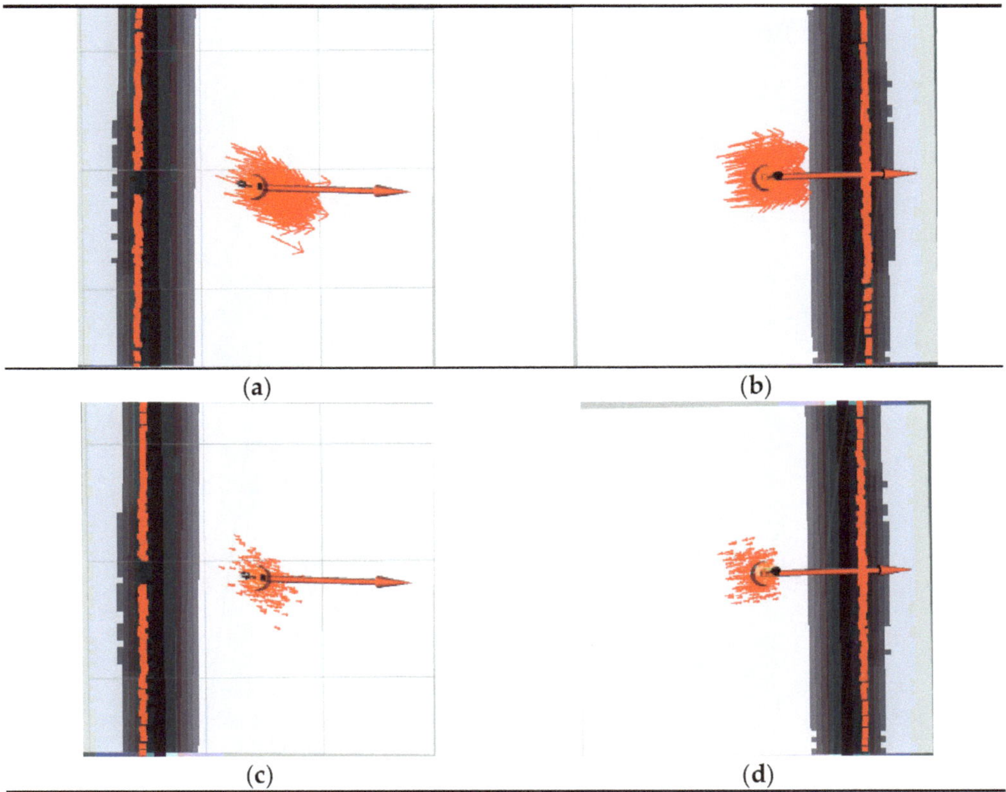

Figure 18. AMCL algorithm with improved AMCL algorithm particle changes: (**a**) AMCL algorithm for particle changes at the starting point; (**b**) AMCL algorithm changes of particles at the end point; (**c**) improved AMCL algorithm for particle changes at the starting point; (**d**) improved AMCL algorithm for particle changes at endpoints.

Photoelectric sensors were set up on both sides of the laying of the QR code and at the start-end position to detect the number of corrections and the maximum distance of error when different algorithms were used for tracking the trajectory, in order to record a week's navigation time as well as the maximum distance from the set point of the target error.

It can be learnt from the distribution of the particles that the distribution of the improved algorithm particles performs better compared to the previous algorithm, distributing on the real trajectory with less error, and the rest of the particles of the regions not involved by the AGVs disappear, only in the aggregation of the labels. From Figure 19 it can be seen that during approximately the first 30 messages released by the AMCL, the localization error is small; once 30 messages are released, the AGV is abducted to produce a large error. As the AGV moves, the positioning error gradually decreases, and after multiple messages, the error tends to 0, restoring the original positioning accuracy. Although the improved AMCL algorithm also encounters kidnapping, the adjustment time is improved by 68.73% compared to the unimproved algorithm.

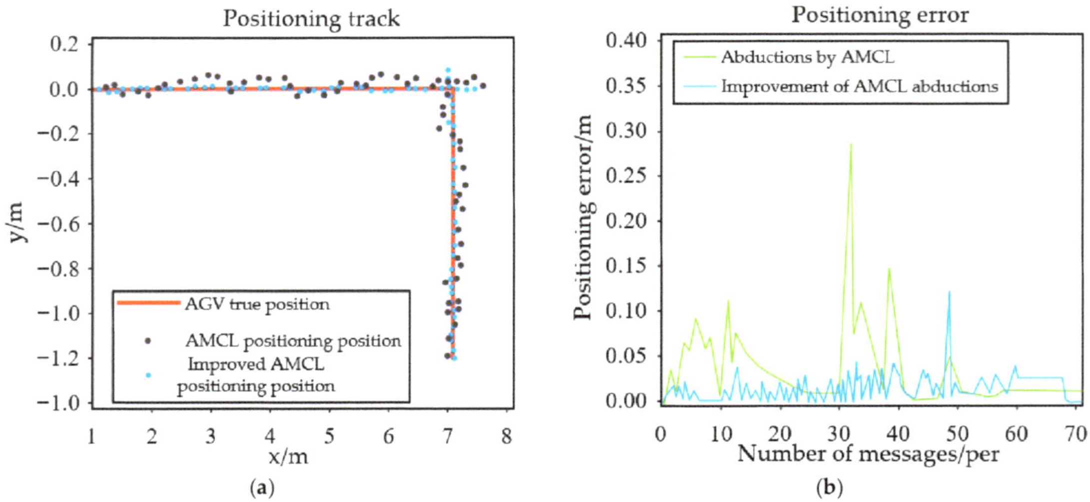

Figure 19. Particle distribution and position error: (**a**) particle distributions generated by the AMCL algorithm and improved algorithms on trajectories; (**b**) AMCL and improved AMCL algorithm for processing information in case of kidnapping.

The maximum data for the offset recorded by the photoelectric sensor in Figure 20a is (−18.79, 22.38) when using the AMCL algorithm, while the improved AMCL algorithm is (−4.79, 6.82). That is to say, adjusting the trajectory to the left exceeds the set trajectory line by a maximum of 22.83 cm and 6.82 cm, and adjusting the trajectory to the right exceeds the distance by a maximum of 18.79 cm and 4.79 cm, from which the comparison can be significantly concluded that the improved algorithm is superior. Figure 20b with the increase in the running time, the error gradually increases and reaches a peak at a certain point, and then the error gradually decreases, but the adjustment time also gradually increases. In contrast, because of the addition of the algorithm to eliminate the cumulative error link, in the improved algorithm, the error will be adjusted quickly with a slight difference between the scan data and the simulation algorithm. Because of the large deviation, the velocity variation is large, the maximum velocity cannot be reached, and the velocity variation has to be adjusted in each sweep, and the fluctuation of linear and angular velocities in Figure 20c,d are fluctuate more compared to the improved AMCL algorithm. Table 4 shows the performance of different algorithms with the same conditions.

4.2. Real Scene Experimental Test

The simulation experiments have proved that the proposed algorithm has good performance and the localization time and error distance are greatly reduced, so it can be significantly shown that the improved AMCL algorithm has better superiority. Therefore, the real effectiveness of the improved algorithm is verified, and the experiments are carried out and analyzed in real scenarios. Figure 21a,b shows the front and side view of the AGV used in the experiment. Figure 21c shows the experimental site used in the experiment, and Figure 21d shows the QR code labels pasted in the experimental site.

Real experimental scenarios are established and tested using the improved AMCL algorithm to obtain valid data for analyzing the reliability of the algorithm.

The map generated using the SLAM technique is shown in Figure 22, where the thin green line from 0 to 1 is the trajectory of the AGV. The round-trip is one lap, and the data of the AGV travelling ten laps are saved and imported into MATLAB2021B for analysis, which proves the validity of the proposed algorithm by comparing the data with the positional error of AGV arriving at the target point, the maximal corrective angle, the speed fluctuation, and line acceleration.

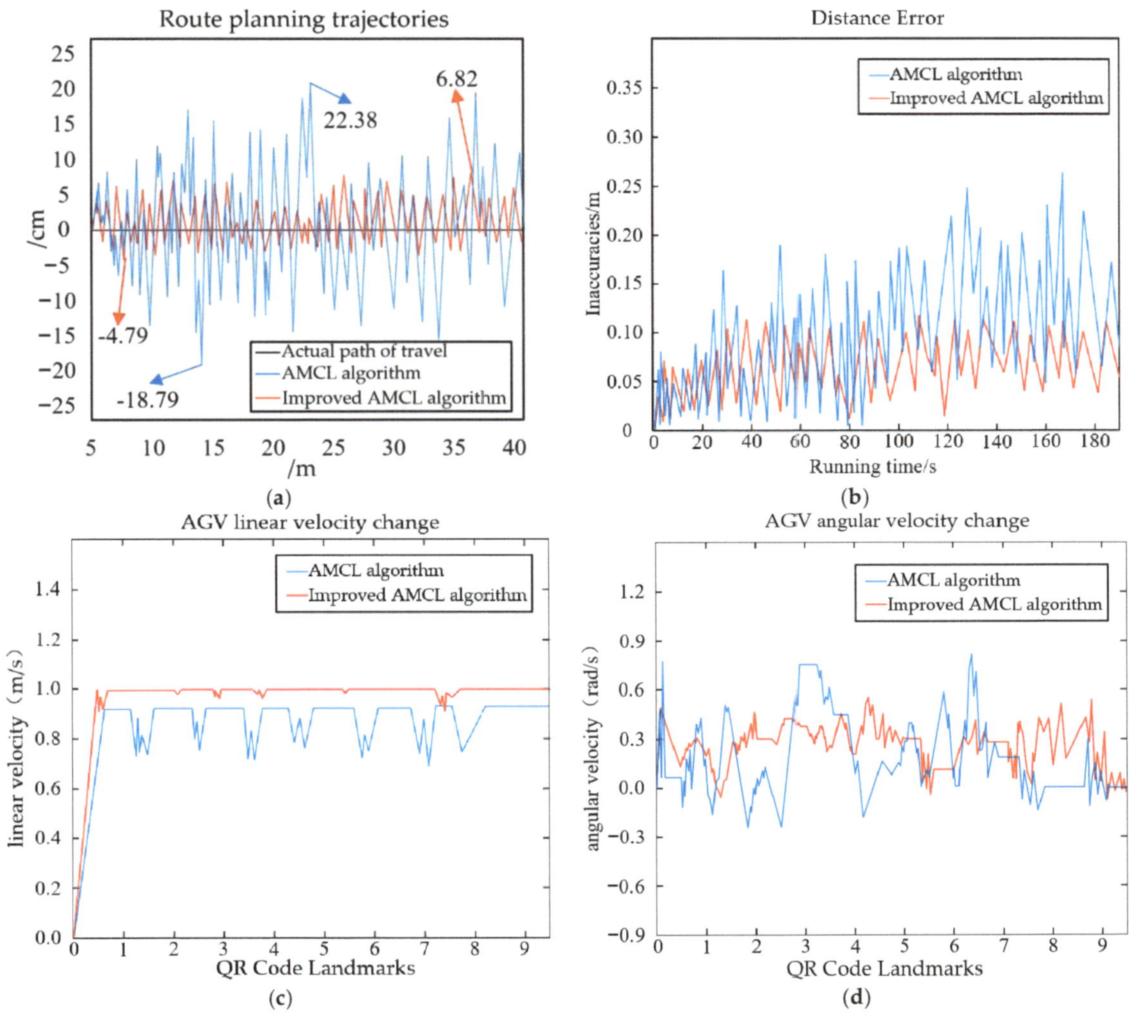

Figure 20. Comparison between the data of the simulation experiment: (**a**) distance offset on both sides when travelling; (**b**) relationship between error and time; (**c**) line speed fluctuation during scanning; (**d**) angular velocity fluctuations during sweeping.

Table 4. AMCL and improved AMCL fusion QR code algorithm comparison.

Parametric	Algorithm	Value
Starting particle	AMCL	1500
	Improved AMCL	1500
Time from start to target (10-turn average)	AMCL	12.59 s
	Improved AMCL	7.42 s
Offset maximum distance (both sides)	AMCL	(−18.79, 22.38) cm
	Improved AMCL	(−4.79, 6.82) cm
Deviation distance to target point (10-turn average)	AMCL	27.36 cm
	Improved AMCL	4.81 cm
Dissolution of kidnapping recovery time (10-turn average)	AMCL	2.2 s
	Improved AMCL	0.39 s

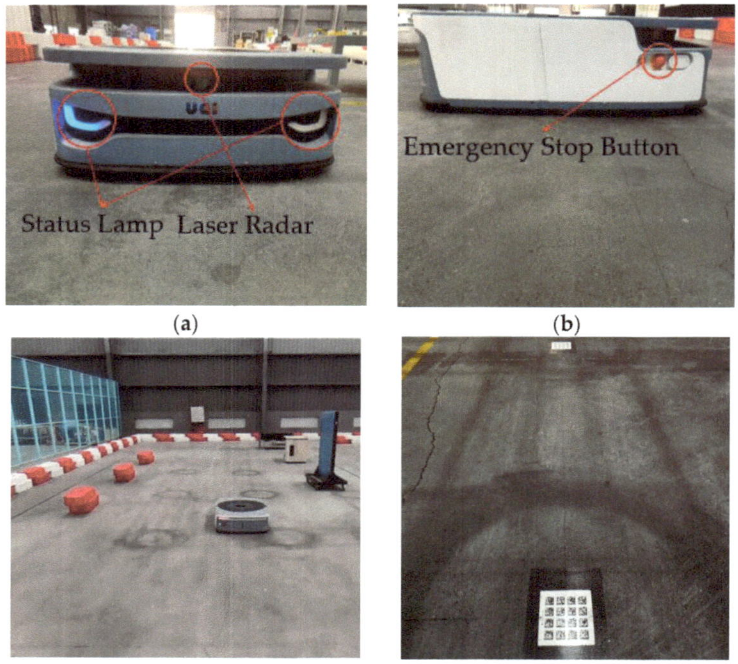

(**a**) (**b**) (**c**) (**d**)

Figure 21. Real scene testing: (**a**) AGV front; (**b**) AGV side; (**c**) experimental site and AGVs used; (**d**) paste a column of QR code landmarks.

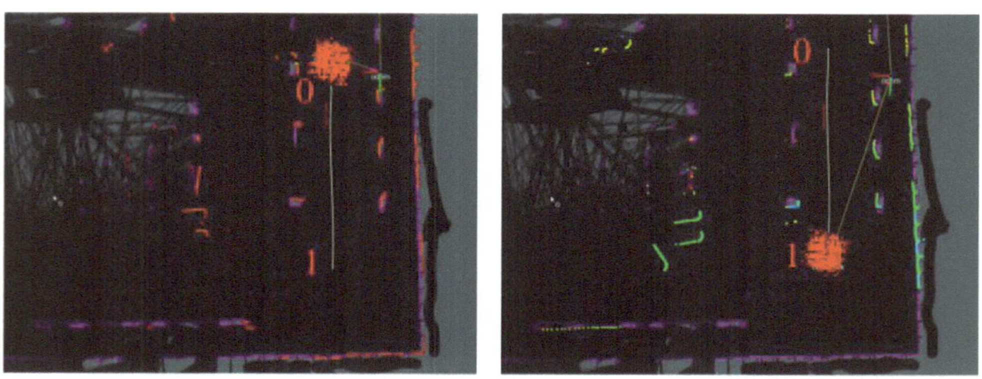

(**a**) (**b**)

Figure 22. Gmapping build synchronization action: (**a**) a half-turn from 0 to 1; (**b**) a half-turn from 1 to 0.

4.2.1. Improved AMCL Matching Template Library Accuracy Test

Use SLAM to generate particles on the QR code, mark the coordinate position of the QR code on the map, turn off the camera, use the particles to determine the coordinate position of the QR code, and the AGV travels to the particle aggregation area until it reaches the matching QR code marking point. Compare the position information with the actual coordinate position information of the QR code and analyze the error distance between the two.

Figure 23a shows that the improved AMCL particles converge on the QR1 position, at which time, Figure 23b shows that the AGV stopping position is in the area directly above the landmark. From the actual test, it is concluded that the AGV identifies the particle template on QR1 to match the exact position of the first landmark on the SLAM construction map, and then accurately travels to the first landmark area location, waiting for the next command. At this time, the industrial camera is opened to scan the code, and the real position of the AGV at this time is obtained through coordinate conversion, comparing the coordinate information and measuring the position error interval between the particle triangular matching position information using the improved AMCL algorithm and the position error interval of the QR code scanning information.

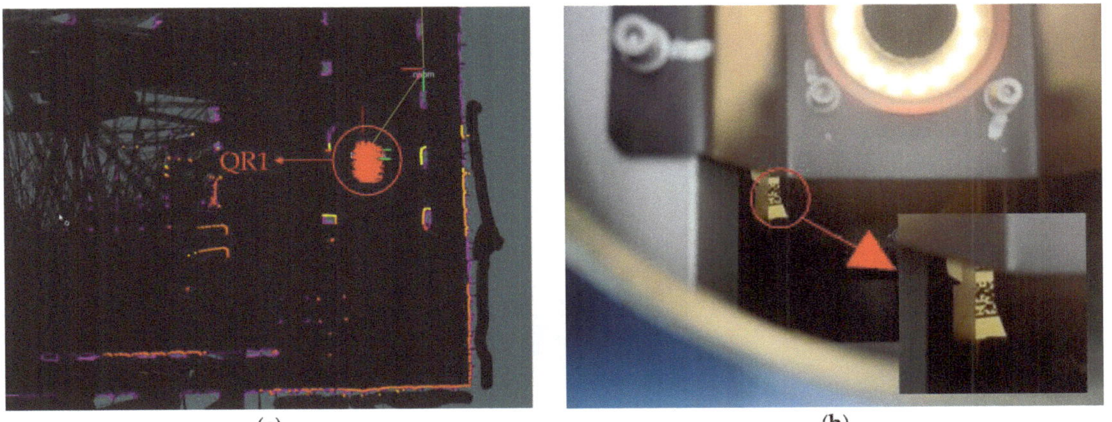

(**a**) (**b**)

Figure 23. The scanning process: (**a**) synchronised Rviz; (**b**) scanning the QR code.

Taking the forward direction of the AGV as the y-axis and the left-right offset distance as the x-axis, two photoelectric sensors are installed before and after the QR1 landmark position, and the coordinate information of each arrival position is re-recorded and weighted average. Table 5 shows that the combination of code-sweeping positioning and improved AMCL algorithm has the highest accuracy, with the maximum offset of coordinate information not exceeding 1 cm, and the reading distance before and after the combination not exceeding 0.5 cm, and the navigation and positioning accuracy has been improved by 64.27% on the original basis.

Table 5. AMCL and improved AMCL fusion QR code algorithm comparison.

Average Value	Actual Position Coordinates	Particle Triangulation Matching Position Coordinates	Industrial Camera Scanning Position Coordinates	Scanning Fusion Improved AMCL	Maximum Left Offset Maximum Right Offset
Cycle 5 times	(0, 375.00) cm	(1.96, 377.92) cm	(3.24, 373.16) cm	(0.59, 374.25) cm	(0.88, 0.53) cm
Cycle 10 times	(0, 375.00) cm	(2.88, 373.40) cm	(2.64, 377.88) cm	(0.36, 375.40) cm	(0.56, 0.62) cm
Cycle 20 times	(0, 375.00) cm	(3.72, 377.34) cm	(1.52, 377.50) cm	(0.42, 375.34) cm	(0.78, 0.62) cm
Cycle 30 times	(0, 375.00) cm	(2.48, 373.68) cm	(2.44, 373.32) cm	(0.44, 375.32) cm	(0.84, 0.66) cm
Cycle 50 times	(0, 375.00) cm	(1.54, 377.12) cm	(3.72, 377.48) cm	(0.72, 374.48) cm	(0.76, 0.64) cm

4.2.2. AGV Particle Kidnapping Experiment

In previous simulation experiments, it was found that the particles generated by the AMCL algorithm were prone to kidnapping events during the positioning process. The improvement of the algorithm as well as the improvement of resampling can effectively avoid such situations, so it is verified in real experiments that the algorithm can effectively improve the robustness and reliability of AGV positioning.

The experiments show that the particles in the AMCL algorithm will not disperse automatically after AGV is kidnapped, so it is impossible to relocate. However, the improved AMCL algorithm will scatter some particles at the kidnapped position, which can be relocated. Figure 24 shows the process of the kidnapping and location recovery of the AGV when the improved AMCL algorithm is started. Figure 24a shows the initial state of the AGV. At this time, the AGV is ready to accept the command to move forward, and the particles are in a dispersed state. Figure 24b shows the state that the AGV moves to QR2 point, at which time the particles are in a convergent state. Figure 24c shows that the AGV is kidnapped, and the particles gather in this area and do not diverge with the movement of AGV. As shown in Figure 24d, to detect the kidnapping event, the improved algorithm is adjusted and the positioning state is restored. To calculate the kidnapping time, calculate the duration of the kidnapping time based on the start time and end time of the kidnapping event. To monitor the positioning error, sensors or other positioning systems are used to monitor the difference between the actual position of the AGV and the position estimated by the AMCL algorithm. When the positioning error exceeds a certain threshold, it can be determined that an kidnapping event has occurred. Record the start time of the abduction event. When the positioning error exceeds the threshold, record the current time as the start time of the kidnapping event. To monitor the positioning recovery, continue to monitor the positioning error, and when the positioning error recovers to within an acceptable range, it can be judged that the kidnapping event is over.

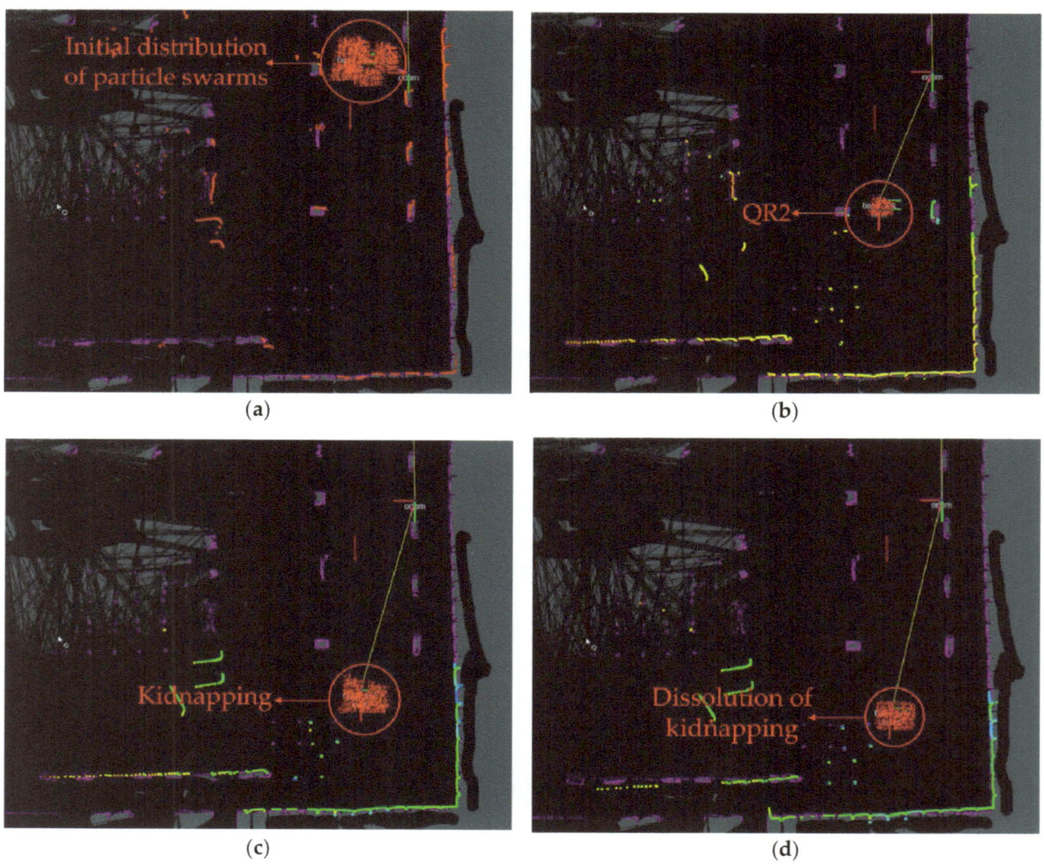

Figure 24. Kidnapping and location recovery: (**a**) initial stage; (**b**) travelling to QR2; (**c**) kidnapping situations; (**d**) dissolution of kidnapping.

The AGV was driven for 10, 20, and up to 50 laps, respectively, for the experimental testing, the time used when the kidnapping event occurred and the time after adjustment were recorded, and the data were analysed and counted in Table 6. From Table 6, it can be clearly concluded that the use of the improved AMCL algorithm can effectively avoid the occurrence of kidnapping events, and the solution of the kidnapping problem can save 42.81% of the AGV navigation time, and once such an event occurs, it can be quickly adjusted to avoid the problem of inaccurate positioning in the process of moving forward.

Table 6. Measurements of the occurrence of kidnappings.

Number of Cycles	Algorithm	Number of Kidnapping Incidents	Cumulative Time /Average Time (s)	Maximum/Minimum Recovery Time (s)	Maximum/Minimum Positioning Error (cm)
10	AMCL	17	51.64/3.04	2.59/1.75	22.81/15.69
	Improved AMCL	2	1.18/0.59	0.12/0.06	3.88/1.72
20	AMCL	31	79.04/2.54	2.78/1.16	19.26/10.71
	Improved AMCL	3	1.56/0.52	0.21/0.02	4.33/2.28
30	AMCL	44	119.68/2.72	2.13/1.54	21.16/9.31
	Improved AMCL	5	2.26/0.45	0.51/0.03	3.29/1.61
40	AMCL	60	146.28/2.44	2.88/1.02	18.74/11.26
	Improved AMCL	8	3.28/0.40	0.96/0.02	5.10/2.13
50	AMCL	71	162.64/2.29	1.35/0.97	16.89/8.92
	Improved AMCL	10	4.18/0.41	0.29/0.05	3.77/1.02

4.3. Discussion

QR code-navigated AGVs are widely used in agriculture, industrial automation, healthcare, logistics, and other fields. There are a large number of researchers contributing to QR code navigation, localization, path planning, etc. in the related literature. With the development of industrial automation, visual navigation is widely adopted because of its high accuracy and low latency. However, single navigation still faces many problems, such as a long processing time and high sensor requirements. Ref. [7] proposed to use EKF combined with an internal encoder and external QR code to correct the cumulative error generated by attitude estimation. This method can effectively solve the error problem and improve the positioning accuracy, providing a new reference for navigation methods. However, the error increases as the distance of the QR code paste increases.

In this article, EKF can only calculate the a posteriori estimate of the state through the current measurement value and the a priori estimate, and cannot directly consider the influence of historical data, so it is susceptible to the cumulative error of the sensor. EKF assumes that the noises in the system model and the measurement model are linear Gaussian distributed, but there may be nonlinear noises and non-Gaussian distributed noises in the actual application, which may lead to the filtering results of accuracy degradation. Our work meets the positioning requirements of high-precision navigation systems in terms of adaptive tuning, positioning accuracy, and navigation time.

When in a suitable environment, the navigation and positioning scheme proposed by some researchers can achieve 100% positioning and permanently eliminate accumulated errors. The method proposed by [8] can achieve 100% accuracy and complete given commands under suitable circumstances. It has achieved a qualitative leap in the field of navigation and positioning. However, due to its use of RGB sweep detection, when encountering bright light environments, or when obstacles are similar in color to the transported object, it can adversely affect the results to detect the completion of instructions.

Zhang, H., and Dong, S. et al. [29,30] proposed different fusion methods to improve the positioning accuracy and reduce the time required for navigation. However, when AGVs are in an environment similar to a long corridor, they are prone to map distortion, incorrect judgement of current position, continuing along the wrong route and collision problems.

Through the simulation experiments and scenario tests in this section, a large number of experimental comparisons can be made to conclude the superiority of the proposed method in this paper, which is better than similar methods in terms of the adaptive adjustment time, positioning accuracy, and navigation time.

5. Conclusions

This paper introduces an AGV navigation and positioning system that integrates navigation and precise positioning. The AMCL particle positioning algorithm is used to perform triangular matching, establish a library of QR code templates, and quickly identify the current position of the QR code for precise positioning. The accuracy of the proposed algorithm is verified through a large number of comparative experiments and simulation data, and the positioning and navigation tasks can be completed in a real experimental environment. It solves the problems of inaccurate positioning and low precision, which are common in the market at present, and provides reference for the accurate positioning of AGV. The focus of this research is to develop hybrid navigation AGVs based on the market scarcity to serve the public and shift from industry to service. The main contributions of this paper are as follows:

1 This article provides an in-depth analysis and research on the development status of AGV's critical domestic and international technologies and researches robot navigation, positioning, and path planning technology. The advantages and disadvantages of various methods are compared, and the overall navigation scheme and system navigation method are designed in detail. Finally, the feasibility and benefits of this choice are verified through experiments.

2 An ROS operating system was used to build the simulation environment of the AGV, and the real positioning system platform was established to prepare for the research of the AGV positioning system. The localization system proposed in this paper uses SLAM global mapping to obtain the absolute coordinates of ground punctuation. It uses the improved AMCL algorithm to combine QR codes, which improves the positioning time and accuracy in the navigation process.

3 Improving the generation of the initial particle swarm can improve the convergence speed and accuracy of the algorithm, and improve the resampling method to effectively reduce or avoid the kidnapping problem. By building the simulation model and testing the simulation using MATLAB software, the algorithm can be made to converge faster and more accurately to the AGV position, as well as improving the real-time and responsiveness of the system, and greatly reducing the time required for navigation.

4 When the particles generated by the AMCL algorithm encounter the kidnapping situation, i.e., the AGV generates too much offset and the particle state does not change with the movement of the AGV. Through comparison, it can be learnt that the improved AMCL algorithm can quickly adjust the attitude and correct the offset distance, so that it can quickly return to the original travelling route, and the resumption of the adjustment time has been improved by 68.73% compared with the unimproved algorithm.

5 During AGV navigation, the time required for navigation was reduced by 42.81% compared to the unimproved algorithm. The navigation time is greatly reduced, which speeds up the time to process the goods and improves the turnaround speed and capacity of the goods. The positioning accuracy is an important criterion to measure the accuracy of the proposed algorithm. In this paper, by comparing and contrasting, it is concluded that the positioning accuracy is improved by 64.27% compared to the previous algorithm.

In the following research, we will pay more attention to the accuracy and practicality of AGVs in positioning and navigation. For example, we will study the practical application of visual navigation in AGVs and further combine visual information with or instead of QR code data to improve the ability of AGVs to sense their surroundings. In addition, we aim to make a major breakthrough in visual autonomous navigation in AGVs, i.e., to achieve more accurate positioning and no interference throughout the autonomous navigation process, and to apply it to more complex operational tasks and environments.

Author Contributions: Conceptualization, B.Z. and J.Q.; methodology, B.Z., J.Q. and S.L; software, B.Z. and G.Y.; validation, B.Z., J.Q., L.Q. and G.Y.; formal analysis, S.L.; investigation, B.Z.; resources, J.Q. and L.Q.; data curation, B.Z.; writing—original draft preparation, B.Z.; writing—review and editing, S.L.; visualization, B.Z.; supervision, J.Q.; project administration, S.L.; funding acquisition, S.L. All authors have read and agreed to the published version of the manuscript.

Funding: This research received no external funding.

Institutional Review Board Statement: Not applicable.

Informed Consent Statement: Not applicable.

Data Availability Statement: Not applicable.

Acknowledgments: First of all, I would like to thank my supervisor, Li Shiyun, who has given me great help in writing this paper. I want to thank Qiu Junting, Zhejiang University of Technology. He often checked the technical proposal with me and confirmed the various problems encountered in the writing of the manuscript. Finally, I thank Zhejiang Wantu Sirui Robot Co., Ltd. for providing the experimental site and AGV.

Conflicts of Interest: The authors declare no conflict of interest.

References

1. Cheng, W.; Chen, J.; Yang, L.; Li, Z.; Zhou, Y. Optimization design and implementation of automatic guided vehicle. *J. Hunan Inst. Technol. Nat. Sci. Ed.* **2021**, *34*, 45–49. (In Chinese)
2. Bechtsis, D.; Tsolakis, N.; Vlachos, D.; Iakovou, E. Sustainable supply chain management in the digitalisation era: The impact of Automated Guided Vehicles. *J. Clean. Prod.* **2017**, *142*, 3970–3984. [CrossRef]
3. Okumuş, F.; Kocamaz, A.F. Cloud Based Indoor Navigation for ROS-enabled Automated Guided Vehicles. In Proceedings of the 2019 International Artificial Intelligence and Data Processing Symposium (IDAP), Malatya, Turkey, 21–22 September 2019; pp. 1–4.
4. Li, Y. Research on Multi-AGVs Cooperative Transportation Strategy in Warehouse Logistics Environment Based on HCA Algorithm. In Proceedings of the 2023 IEEE 2nd International Conference on Electrical Engineering, Big Data and Algorithms (EEBDA), Changchun, China, 24–26 February 2023; pp. 1753–1758. [CrossRef]
5. Li, W.; Zhang, N.; Yao, Y.; Zhan, H.; Ding, Y. Research on Vision Navigation and Positioning of AGV Terminal Based on QR Code. In Proceedings of the 2020 IEEE 9th Joint International Information Technology and Artificial Intelligence Conference (ITAIC), Chongqing, China, 11–13 December 2020; pp. 845–848. [CrossRef]
6. Zhang, B.; Zhu, M.; Lin, C.; Zhu, D. Research on AGV map building and positioning based on SLAM technology. In Proceedings of the 2022 IEEE 5th International Conference on Automation, Electronics and Electrical Engineering (AUTEEE), Shenyang, China, 18–20 November 2022; pp. 707–713. [CrossRef]
7. Bach, S.H.; Khoi, P.B.; Yi, S.Y. Application of QR Code for Localization and Navigation of Indoor Mobile Robot. *IEEE Access* **2023**, *11*, 28384–28390. [CrossRef]
8. Kulaç, N.; Engin, M. Developing a Machine Learning Algorithm for Service Robots in Industrial Applications. *Machines* **2023**, *11*, 421. [CrossRef]
9. Kumar, P.; Khaparde, A. QR Code Detector and Follower with Kalman Filter. In Proceedings of the 2022 International Interdisciplinary Humanitarian Conference for Sustainability (IIHC), Bengaluru, India, 18–19 November 2022; pp. 1423–1426. [CrossRef]
10. Zhou, C.; Liu, X. The Study of Applying the AGV Navigation System Based on Two Dimensional Bar Code. In Proceedings of the 2016 International Conference on Industrial Informatics—Computing Technology, Intelligent Technology, Industrial Information Integration (ICIICII), Wuhan, China, 3–4 December 2016; pp. 206–209. [CrossRef]
11. Talwar, D.; Jung, S. Particle Filter-based Localization of a Mobile Robot by Using a Single Lidar Sensor under SLAM in ROS Environment. In Proceedings of the 2019 19th International Conference on Control, Automation and Systems (ICCAS), Jeju, Republic of Korea, 15–18 October 2019; pp. 1112–1115. [CrossRef]
12. Zhang, X.; Mu, X.; Liu, H.; He, B.; Yan, T. Application of Modified EKF Based on Intelligent Data Fusion in AUV Navigation. In Proceedings of the 2019 IEEE Underwater Technology (UT), Kaohsiung, Taiwan, 16–19 April 2019; pp. 1–4. [CrossRef]

13. Trisnawan, I.K.N.; Jati, A.N. Monte Carlo method is a kind of calculation method, but it is very different from the general numerical calculation method, it is based on the theory of probability statistics. In Proceedings of the 2019 International Conference on Computer, Control, Informatics and its Applications (IC3INA), Tangerang, Indonesia, 23–24 October 2019; pp. 187–192.
14. dos Reis, W.P.N.; Morandin, O.; Vivaldini, K.C.T. A Quantitative Study of Tuning ROS Adaptive Monte Carlo Localization Parameters and their Effect on an AGV Localization. In Proceedings of the 2019 19th International Conference on Advanced Robotics (ICAR), Belo Horizonte, Brazil, 2–6 December 2019; pp. 302–307. [CrossRef]
15. Zeghmi, L.; Amamou, A.; Kelouwani, S.; Boisclair, J.; Agbossou, K. A Kalman-Particle Hybrid Filter For Improved Localization of AGV In Indoor Environment. In Proceedings of the 2022 2nd International Conference on Robotics, Automation and Artificial Intelligence (RAAI), Singapore, Singapore, 9–11 December 2022; pp. 141–147. [CrossRef]
16. Zou, Q. Research on Mobile Robot Navigation Method Based on Graph Optimization SLAM. Master's Thesis, Harbin Institute of Technology, Harbin, China, 2017. (In Chinese).
17. Zhang, F.; Li, S.; Yuan, S.; Sun, E.; Zhao, L. Algorithms analysis of mobile robot SLAM based on Kalman and particle filter. In Proceedings of the 2017 9th International Conference on Modelling, Identification and Control (ICMIC), Kunming, China, 10–12 July 2017; pp. 1050–1055. [CrossRef]
18. Kumar NT, S.; Gawande, M.; M, N.P.B.; Verma, H.; Rajalakshmi, P. Mobile Robot Terrain Mapping for Path Planning using Karto Slam and Gmapping Technique. In Proceedings of the 2022 IEEE Global Conference on Computing, Power and Communication Technologies (GlobConPT), New Delhi, India, 23–25 September 2022; pp. 1–4. [CrossRef]
19. Chung, M.-A.; Lin, C.-W. An Improved Localization of Mobile Robotic System Based on AMCL Algorithm. *IEEE Sens. J.* **2022**, *22*, 900–908. [CrossRef]
20. Tiwari, S. An Introduction to QR Code Technology. In Proceedings of the 2016 International Conference on Information Technology (ICIT), Bhubaneswar, India, 22–24 December 2016; pp. 39–44. [CrossRef]
21. Cui, G.; Bai, Y.; Li, S. AGV Research Based on Inertial Navigation and Vision Fusion. In Proceedings of the 2021 5th CAA International Conference on Vehicular Control and Intelligence (CVCI), Tianjin, China, 29–31 October 2021; pp. 1–6. [CrossRef]
22. Zheng, Z.; Yu, Y.; Chen, R.; Huang, H.; Zhao, H.; Lu, X. Localization Method Based on Multi-QR Codes for Mobile Robots. In Proceedings of the 2022 IEEE International Conference on Advances in Electrical Engineering and Computer Applications (AEECA), Dalian, China, 20–21 August 2022; pp. 1385–1391. [CrossRef]
23. Li, Z.; Huang, J. Study on the use of Q-R codes as landmarks for indoor positioning: Preliminary results. In Proceedings of the 2018 IEEE/ION Position, Location and Navigation Symposium (PLANS), Monterey, CA, USA, 23–26 April 2018; pp. 1270–1276. [CrossRef]
24. Xia, L.; Cui, J.; Shen, R.; Xu, X.; Gao, Y.; Li, X. A Survey of Image Semantics-based Visual Simultaneous Localization and Mapping: Application-oriented Solutions to Autonomous Navigation of Mobile Robots. *Int. J. Adv. Robot. Syst.* **2020**, *17*, 4158. [CrossRef]
25. Balasuriya, B.; Chathuranga, B.; Jayasundara, B.; Napagoda, N.; Kumarawadu, S.; Chandima, D.; Jayasekara, A. Outdoor robot navigation using Gmapping based SLAM algorithm. In Proceedings of the 2016 Moratuwa Engineering Research Conference (MERCon), Moratuwa, Sri Lanka, 5–6 April 2016; pp. 403–408. [CrossRef]
26. Su, Z.; Zhou, J.; Dai, J.; Zhu, Y. Optimization Design and Experimental Study of Gmapping Algorithm. In Proceedings of the 2020 Chinese Control And Decision Conference (CCDC), Hefei, China, 22–24 August 2020; pp. 4894–4898. [CrossRef]
27. Zhou, H.; Chou, W.; Tuo, W.; Rong, Y.; Xu, S. Mobile Manipulation Integrating Enhanced AMCL High-Precision Location and Dynamic Tracking Grasp. *Sensors* **2020**, *20*, 6697. [CrossRef] [PubMed]
28. Qi, X.; Ji, W.; Wang, Q. A Monte Carlo localization Algorithm based on Ranging. In Proceedings of the 2019 3rd International Conference on Electronic Information Technology and Computer Engineering (EITCE), Xiamen, China, 6–8 November 2019; pp. 1663–1666. [CrossRef]
29. Zhang, H.; Sun, C.-L.; Zeng, Y.-M.; Li, P.-P. A Fusion Localization Algorithm Combining MCL with EKF. In Proceedings of the 2018 17th International Symposium on Distributed Computing and Applications for Business Engineering and Science (DCABES), Wuxi, China, 19–23 October 2018; pp. 216–220. [CrossRef]
30. Dong, S.; Lin, R.; Zhao, W.-W.; Cheng, Y.-H. Robot Global Relocalization Based on Multi-sensor Data Fusion. In Proceedings of the 2022 2nd International Conference on Robotics, Automation and Artificial Intelligence (RAAI), Singapore, Singapore, 9–11 December 2022; pp. 35–42. [CrossRef]

applied sciences

MDPI

Article

Trajectory Planning of Shape-Following Laser Cleaning Robot for the Aircraft Radar Radome Coating

Zhen Zeng, Chengzhao Jiang, Shanting Ding *, Qinyang Li, Zhongsheng Zhai and Daizhe Chen

Hubei Key Laboratory of Modern Manufacturing Quantity Engineering, School of Mechanical Engineering, Hubei University of Technology, Wuhan 430068, China; zengzhen@hbut.edu.cn (Z.Z.); jcz@hbut.edu.cn (C.J.); qy.li@hbut.edu.cn (Q.L.); zs.zhai@hbut.edu.cn (Z.Z.); 102210080@hbut.edu.cn (D.C.)

* Correspondence: dingst@hbut.edu.cn

Abstract: At present, aircraft radome coating cleaning mainly relies on manual and chemical methods. In view of this situation, this study presents a trajectory planning method based on a three-dimensional (3D) surface point cloud for a laser-enabled coating cleaning robot. An automated trajectory planning scheme is proposed to utilize 3D laser scanning to acquire point cloud data and avoid the dependence on traditional teaching–playback paradigms. A principal component analysis (PCA) algorithm incorporating additional principal direction determination for point cloud alignment is introduced to facilitate subsequent point cloud segmentation. The algorithm can adjust the coordinate system and align with the desired point cloud segmentation direction efficiently and conveniently. After preprocessing and coordinate system adjustment of the point cloud, a projection-based point cloud segmentation technique is proposed, enabling the slicing division of the point cloud model and extraction of cleaning target positions from each slice. Subsequently, the normal vectors of the cleaning positions are estimated, and trajectory points are biased along these vectors to determine the end effector's orientation. Finally, B-spline curve fitting and layered smooth connection methods are employed to generate the cleaning path. Experimental results demonstrate that the proposed method offers efficient and precise trajectory planning for the aircraft radar radome coating laser cleaning and avoids the need for a prior teaching process so it could enhance the automation level in coating cleaning tasks.

Keywords: coating cleaning; point cloud alignment algorithm; B-spline curve; robot path planning

Citation: Zeng, Z.; Jiang, C.; Ding, S.; Li, Q.; Zhai, Z.; Chen, D. Trajectory Planning of Shape-Following Laser Cleaning Robot for the Aircraft Radar Radome Coating. *Appl. Sci.* **2024**, *14*, 1163. https://doi.org/10.3390/app14031163

Academic Editor: Jonghoek Kim

Received: 4 January 2024
Revised: 25 January 2024
Accepted: 26 January 2024
Published: 30 January 2024

1. Introduction

The maintenance and repair of aircraft radomes play a pivotal role in ensuring the integrity and functionality of aerospace systems. Radomes are subjected to various environmental factors during operation, including climate variations, flight friction, and airflows. These factors contribute to the wear and damage of the radome's surface coatings and bottom paint. Consequently, in the maintenance process, regular treatment of the radome surface coatings is an essential step for preserving its performance.

Conventional methods for aircraft paint removal include mechanical cleaning, solvent-based cleaning, and ultrasonic cleaning [1]. While these methods have reached a considerable level of maturity, they also come with significant drawbacks. For instance, chemical and mechanical cleaning methods are labor-intensive, prone to substrate damage, and can generate substantial waste, which leads to environmental pollution [2,3]. In comparison, laser cleaning technology is regarded as a greener and more promising alternative, particularly in the context of the global manufacturing industry's transformation. This technology operates by inducing a series of optical, thermal, and mechanical changes in a short period, effectively removing contaminants through their coupled effects [4–8]. Laser cleaning is considered the most promising green cleaning technology of the 21st century, as it enables minimal damage through precise control of laser parameters [9,10]. With the advancements in robotics and computer numerical control (CNC) technology, multi-joint robots have found widespread applications in various fields, including laser cleaning, processing, and welding, due to their high degrees of freedom, exceptional flexibility, and programmability [11–13]. Path planning for robot machining is a primary requirement for achieving automation in robot-based machining. Researchers have conducted relevant research on robot path planning according to the characteristics of different machining methods. Wang conducted path planning for robot machining using a parameter-optimized surface model [14]. Bian developed an offline programming system based on computer-aided design (CAD) models for robot polishing path planning [15]. Cai proposed a novel software package design based on CAD surface modeling, integrated into the offline programming software RobotStudio™ (Product of ABB Company, Sweden), which considered both simple coating models and torch kinematic parameters to generate robot trajectories automatically [16]. Morozov utilized reverse engineering techniques to reconstruct the CAD model of an aircraft wing mask and generated scanning trajectories for non-destructive testing with high positioning accuracy [17]. Most of the above studies are based on the CAD models of the workpieces. However, these methods are unable to accurately execute when the CAD model of the workpiece is missing or when the actual surface shape of the workpiece does not match the CAD model data. Point cloud information can clearly reveal the surface features of the inspected workpiece, leading many researchers to utilize point cloud data for robot path planning [18–21]. Jin et al. indirectly obtained the stereolithography (STL) model of the workpiece using a 3D scanning device and performed laser cleaning path planning through offline programming software, demonstrating the effectiveness of robot laser cleaning [22]. For simple parts, traditional robot path-planning methods can be employed for robot laser cleaning path planning. However, for large, curved surface workpieces, these methods fail to meet the requirements of precise laser cleaning operations and cannot achieve curved surface cleaning.

This work focuses on processing point cloud data of aircraft radomes in the absence of CAD models to generate motion trajectory planning for a coating laser cleaning robot. By successfully integrating point cloud data with robot actions, an automated robot coating laser cleaning system is achieved, significantly improving the cleaning efficiency in the absence of CAD models. The distance between the cleaning actuator and the surface is determined by estimating the normal vector of trajectory points and utilizing the focal length of the laser. This distance is maintained, allowing the cleaning actuator to move along the complex surface to be cleaned. Simultaneously, the laser beam remains perpendicular to the workpiece surface, achieving the desired cleaning posture planning. This approach ensures both curved surface cleaning and reduces energy loss of the laser, thereby enhancing the cleaning quality.

2. Materials and Methods

2.1. Measurement and Preprocessing of Point Cloud Data for Radar Dome Surfaces

This section presents the development of an automated robot coating laser cleaning system by incorporating 3D sensors, robots, laser cleaning actuators, and computers. The system aims to achieve efficient and precise coating removal through automated laser cleaning. The flowchart of the laser cleaning trajectory planning algorithm is as shown in Figure 1, written using the C++ (c plus plus) programming language. The operational schematic of the system is shown in Figure 2.

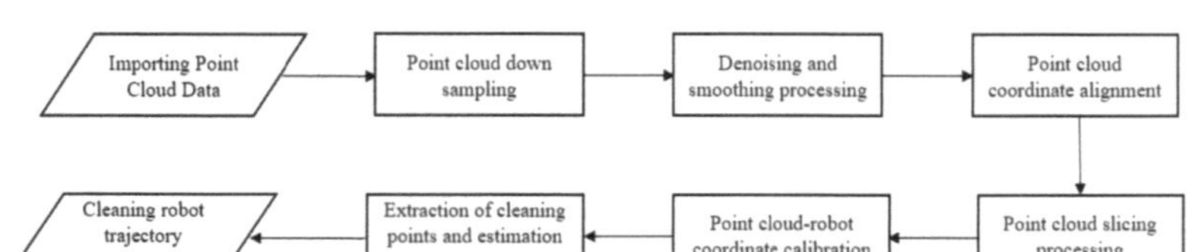

Figure 1. Flow chart of the laser cleaning trajectory planning algorithm.

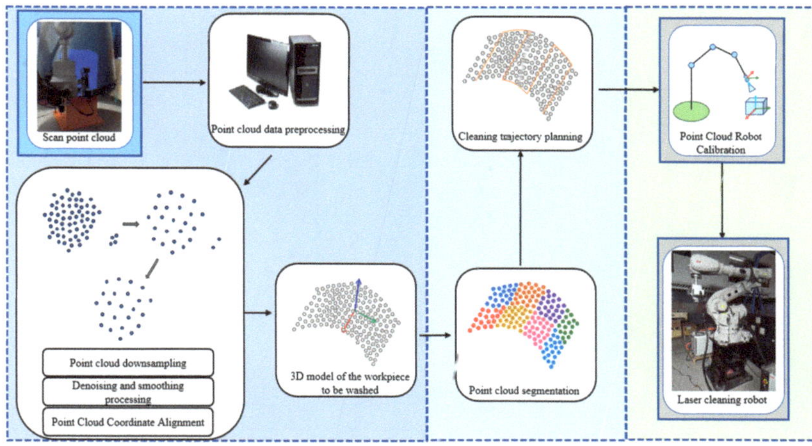

Figure 2. Development of an automated robot coating laser cleaning system.

2.1.1. Measurement of Raw Point Cloud Data

With the rapid development of 3D data acquisition technology, the utilization of 3D sensors for detecting the 3D profiles of objects has gained extensive application. In comparison to coordinate measuring machines (CMM) and articulated arm measuring machines, the advantages of 3D sensors are their non-contact measurement capability and wider detection range. In this study, measurements were performed using a 3D scanning sensor. Figure 3 presents a three-dimensional point cloud of an aircraft radome surface. The rotational symmetric radome was positioned with a base, as its vertical direction was aligned with the radome's vertical axis.

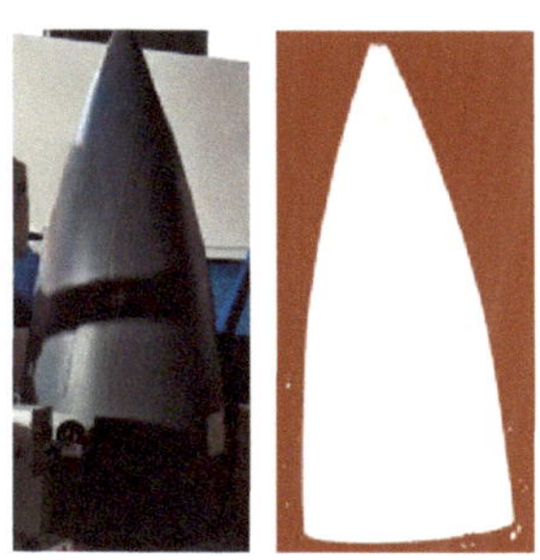

Figure 3. An aircraft radome and its 3D scanned point cloud of aircraft radome.

2.1.2. Point Cloud Data Preprocessing

During the acquisition of point cloud data on the surface of the aircraft radome, various factors may introduce disturbances. These factors include uneven lighting, vibrations, object occlusions, and aging of the scanning equipment, which could result in noise and voids in the point cloud data. Therefore, preprocessing the point cloud data becomes necessary to ensure its smoothness and uniformity. In order to handle the high-density point cloud data acquired by the 3D sensor, this study employed a voxel filtering method for down sampling [23] It is a method that effectively reduces the number of data points while preserving the details of the point cloud. Additionally, a statistical filtering technique was employed to remove outliers in the down sampled point cloud [24]. For each point Q_i in the point cloud, its 50 nearest neighboring points (N_{i1}, N_{i2}, N_{i3}, N_{i50}) are first determined.

Subsequently, the Euclidean distance d_{ij} between the point Q_i and its neighboring points was calculated using the following formula:

$$d_{ij} = \sqrt{(Q_{ix} - N_{ijx})^2 + (Q_{iy} - N_{ijy})^2 + (Q_{iz} - N_{ijz})^2} \tag{1}$$

where Q_{ix}, Q_{iy}, Q_{iz}, N_{ijx}, N_{ijy}, and N_{ijz} represent the coordinates of the points Q_i and N_{ij} along the x, y, and z axes, respectively. Next, for each point Q_i, the average distance $\overline{d}_i$ to its 50 nearest neighboring points is calculated:

$$\overline{d}_i = \frac{1}{50}\sum_{j=1}^{50} d_{ij} \tag{2}$$

Additionally, the global average of the mean distances between all points and their neighboring points was calculated, resulting in two values, denoted as μ and σ.

$$\mu = \frac{1}{N}\sum_{i}^{N} \overline{d}_i \tag{3}$$

$$\sigma = \sqrt{\frac{1}{N-1}\sum_{i=1}^{N} \left(\overline{d}_i - \mu\right)^2} \tag{4}$$

here, N represents the total number of points in the point cloud.

Lastly, a threshold value $T = \mu + k{\cdot}\sigma$ was set, where k is a constant (in this paper, k is set to 1). Any point Q_i with an average distance $\overline{d}_i$ exceeding this threshold (i.e., $\overline{d}_i > T$) was considered an outlier and was removed from the point cloud. The radome point cloud shape after preprocessing is shown in Figure 4.

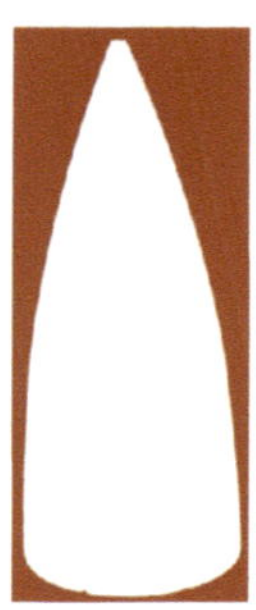

Figure 4. Radome point cloud shape after preprocessing.

2.1.3. Point Cloud Data Alignment

After preprocessing, the point cloud shape of the aligned radome became evident. However, it was necessary to align the coordinates of measuring and cleaning systems before the segmentation of the radome point cloud. To solve this issue, this study introduced the principal component analysis (PCA) algorithm for aligning the point cloud data. Specifically, the research focuses on studying the orientation determination of the principal axis vectors for point cloud alignment. A PCA-based point cloud alignment algorithm was proposed, incorporating an additional criterion for determining the main direction. The algorithm's specific steps for aligning the point cloud data are outlined as follows.

Firstly, decentralize the point cloud dataset $P = \{P_1, P_2, P_3, \cdots, P_n\}$:

$$\overline{Q}_P = \frac{\sum_{i=1}^{n} P_i}{n} \tag{5}$$

The covariance matrix of dataset P can be calculated as:

$$C_P = \frac{1}{N}\sum_{i=1}^{n} (P_i - \overline{Q}_P)(P_i - \overline{Q}_P)^T \tag{6}$$

Through eigenvalue decomposition calculation, the covariance matrix Eigenvalues and eigenvectors of C_P can be expressed as:

$$C_P = U_P D_P U_P^T \tag{7}$$

where U_P is a 3 × 3 matrix composed of the eigenvectors of the covariance matrix C_P, used to identify the main directions of the point cloud data, and D_P is a 3 × 3 diagonal matrix whose diagonal elements are the eigenvalues of the covariance matrix, indicating the variance of data in each direction.

Then a new O-XYZ right-hand coordinate system needed to be established. By taking the centroid coordinates $\overline{Q}_P$ as the origin of the new coordinate system, the first principal component of the eigenvector corresponding to the maximum eigenvalue was chosen as the *X*-axis of the new coordinate system. The second principal component of the eigenvector corresponding to the second largest eigenvalue was chosen as the *Y*-axis of the new coordinate system. Thereby, the O-XY plane was formed. By performing a cross-product operation on the *X*-axis and *Y*-axis, the new *Z*-axis coordinate was obtained and the new O-XYZ coordinate system was established.

By utilizing the newly constructed O-XYZ coordinate system, the coordinate transformation matrix (R_P, T_P) dataset could be computed to obtain the axis-aligned set P'.

$$\begin{cases} R_P = U_P^T \\ T_P = -R_P * \overline{Q}_P \end{cases} \tag{8}$$

$$P' = P * R_P + T_P \tag{9}$$

where R_P is the rotation matrix and T_P is the translation vector.

There may exist a problem with the *Z*-axis orientation, resulting in a scenario where the aligned point cloud exhibits an inverted orientation as depicted in Figure 5b. Specifically, when the *Z*-component values of the point cloud data are negative, it indicates a reversal in the *Z*-axis direction. Consequently, after the initial alignment of the point cloud coordinate system, it was necessary to perform *Z*-axis direction determination and correction.

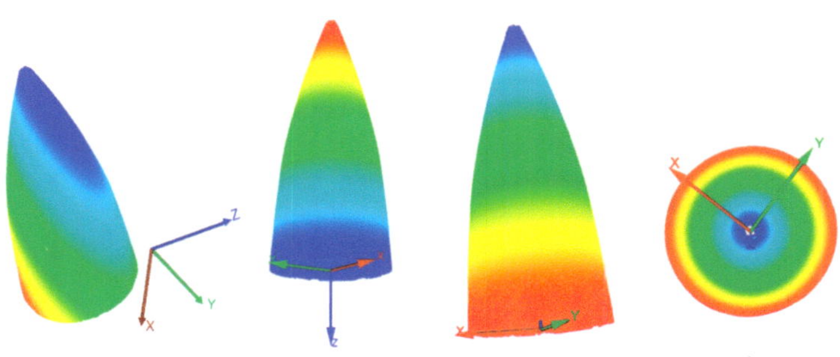

Figure 5. Alignment of radar hood point cloud coordinates (*Z*-axis rendering): (**a**) radar hood point cloud before coordinate alignment, (**b**) radar hood point cloud after initial alignment, (**c**) radar hood point cloud aligned through *Z*-axis correction, and (**d**) bottom-up view of the radar hood point cloud after alignment.

We needed to search for the maximum point P'_{k1} and the minimum point P'_{k2} in the point cloud set P'. Extract the corresponding *Z*-axis component values z_{k1} and z_{k2}, respectively. Take the absolute values of z_{k1} and z_{k2} and calculate their difference as shown in Equation (10).

From the perspective of the point cloud data, this step involves comparing the maximum and minimum points of the point cloud set to determine if the Z-axis is reversed.

$$m = |z_{k1}| - |z_{k2}| \tag{10}$$

Based on the value obtained in Equation (10), denoted as m, if m is determined to be negative, it indicates that the Z-axis is reversed. In such cases, a Z-axis correction is required, as shown in Equation (11), resulting in the aligned point cloud set P''. Figure 5b illustrates the initial alignment of the point cloud and the Z-axis direction correction, leading to the aligned point cloud depicted in Figure 5c.

$$P'' = P' * R_1 \tag{11}$$

where $R_1 = \begin{bmatrix} 1 & 0 & 0 \\ 0 & 1 & 0 \\ 0 & 0 & -1 \end{bmatrix}$.

2.2. Slicing and Visualization of Radome Point Cloud Data

2.2.1. Radome Point Cloud Data Slicing Processing

After repositioning the point cloud coordinate system at the origin, it was necessary to consider slicing the radar hood point cloud model. Due to the inherent geometric characteristics of the radar hood, vertical slicing was performed first based on its height symmetry, followed by horizontal slicing. Considering the complexity of horizontally slicing the radar hood point cloud model in three-dimensional space, a projection-based point cloud slicing technique was proposed. This technique enables rapid and accurate horizontal segmentation of the radar hood point cloud model.

Slicing in the Z-Axis Direction

The radar hood point cloud model was divided into several layers based on the z-coordinate. The division was determined by the laser cleaning range, denoted as h, and took into account an angle θ. (The choice of θ was dictated by factors such as the physical structure of the radar hood and the positioning of the cleaning equipment.) The height of each layer was calculated as $\Delta h = h * \cos\theta$. For a radar hood with a height of H, it was divided into $G = H/\Delta h$ sections. The height at which each layer's slicing line was located was denoted as $Z_k (1 \leq k \leq G+1)$. The point cloud was sorted in ascending order based on the z-coordinate, and each point was assigned to the corresponding layer based on its z-coordinate value.

Projection-Based Point Cloud Segmentation for Radar Hood Geometry

Considering the geometric characteristics of radar domes, the annular structure formed after height segmentation requires further horizontal segmentation. The projection-based point cloud segmentation technique is a method that projects the point cloud onto a 2D space for processing, aiming to reduce the complexity of horizontal slicing of the point cloud. The algorithm flow of this technique is illustrated in Figure 6.

The point cloud of the radar radome was projected onto the Z-direction and underwent normalization and binarization after height segmentation. By extracting the contour of the binarized image, the annular segments of the radar dome were obtained, as shown in Figure 7. Based on the horizontal scanning range of the laser, the segmentation dimensions were calculated and the radar dome was horizontally segmented into annular sections. The circumference of each radar dome ring was denoted as W, the horizontal cleaning range as w, the angle of the ring as α, and the number of segmentations within the ring as M. Thus, $M = W/(w * \cos\alpha)$. After partitioning, each segmentation line was assigned an angle $\alpha_g (0 \leq g \leq M)$. The regions within each ring were assigned corresponding pixel values based on the magnitude of α_g. An image with the same size as the binarized image was initialized, with all pixel values set to 0, referred to as the label image. In each segmented region of the ring, the positions of non-zero pixels within the current region were identified and the corresponding pixel values in the label image were set to the current label value. Using the label image obtained, the current original point cloud data within the segmented ring were labeled, cached, and fused, with the point cloud list being promptly updated. Upon traversing all the layers of points, the data were saved into the point cloud segmentation database, resulting in a collection of point cloud slices, denoted as $S\{s_1, s_2, s_3, \cdots, s_j\}$.

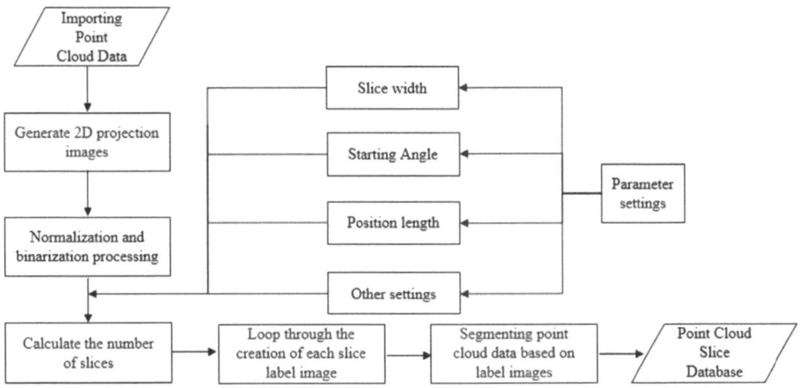

Figure 6. Projection-based point cloud segmentation process.

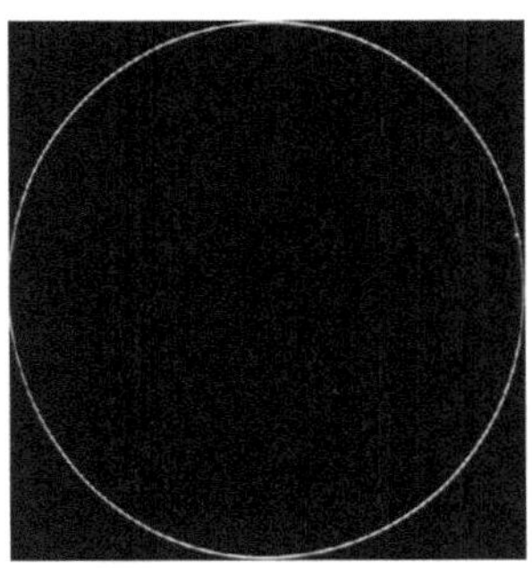

Figure 7. Projection of height-segmented image.

2.2.2. Visualization of Projected Slices

In terms of visualizing the sliced data of the radar dome point cloud, this study utilized the visualization tool based on PCL (Point Cloud Library). Prior to visualization, the radar dome point cloud model was segmented along the z-axis and the horizontal direction, with each slice sequentially stored in the collection S. Each slice was indexed based on the slicing order and assigned a unique RGB color to ensure color uniqueness and consistency. Figure 8 illustrates the unfolded view of the radar dome point cloud data slices, which facilitates a clearer observation and analysis of the sliced data.

Figure 8. Splitting expansion of radome point cloud data: (**a**) top model segmentation of radome and (**b**) expanded slice data.

2.3. Laser Cleaning Robot Trajectory Planning

The trajectory of the robot should be generated prior to the laser coating cleaning process. To achieve optimal cleaning results, two principles must be followed:

(1) The trajectory of the robot for laser coating cleaning should align with the surface height of the component to be cleaned;

(2) The central axis of the laser cleaning actuator held by the robot should always be perpendicular to the surface of the workpiece.

2.3.1. Calibration of Coordinate Relationship between Radome Measurement Point Cloud and Cleaning Robot

The aligned point cloud coordinate system differs from the robot coordinate system. In this study, a homogeneous transformation matrix was employed to perform the transformation from the point cloud coordinate system to the robot coordinate system. The transformation matrix was defined as follows:

$$\boldsymbol{M}_{4\times 4} = \begin{bmatrix} R & t \\ 0 & 1 \end{bmatrix} \tag{12}$$

where R is a 3×3 rotation matrix and t is the translation vector.

2.3.2. Calculation Model for Laser Cleaning Actuator Point Coordinates

Geometric Relationship Model between Cleaning Actuator Points and Sliced Point Cloud

The definition of the laser coating cleaning actuator point is the central point of the effective processing area of the laser spot during a single cleaning operation. In a single cleaning operation, the laser cleaning device can achieve comprehensive coverage and cleaning of the effective processing area. Therefore, all points within this area can be simplified to a single cleaning point, represented by the centroid of the point cloud slice. The calculation formula for this point is as follows:

$$u_{s_j} = \frac{1}{j}\sum\nolimits_{b=1}^{j} q_b \tag{13}$$

Among them, the set of slice points is $\{q_1, q_2, q_3, \cdots, q_j\}$. Store the cleaning point data u_{s_j} in the point set $ListU$.

The Estimation of the Normal Vector for the Cleaning Points

For a regular surface S in three-dimensional space, the normal vector u at point u_s can be approximated as the normal vector of the surface at point u_s. In the tangent plane p at u_s, vectors a and b are two non-collinear vectors, as shown in Figure 9.

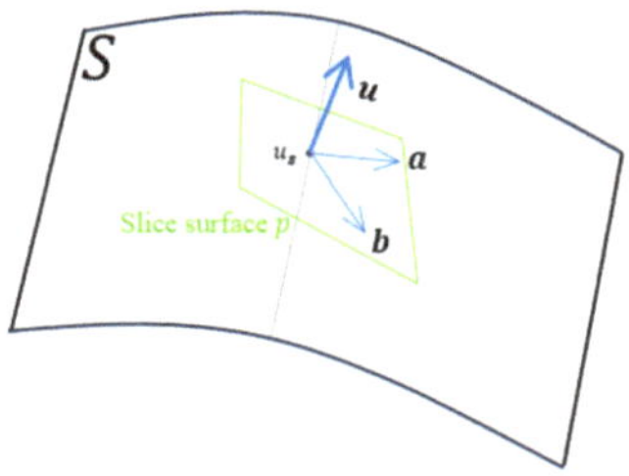

Figure 9. Illustration of point cloud normal vector estimation.

The data stored in *ListU* consist of a series of points, forming a sampling point set U. The normal vector at each data point u_{s_j} in the sampling point set represents the normal vector of the local tangent plane at that point on the surface. There are two main methods for estimating the normal vectors. One method involves the scattered data's triangulated topology, where the normal vector is obtained from the normal vectors of the relevant triangles in the point cloud data, considering the triangulated topology of the point cloud data. The other method utilizes the neighborhood information of each point and employs the least squares principle to locally fit a plane to the K neighboring points. The fitted plane can be considered as the local tangent plane at that point. The normal vector of the local tangent plane at that point is then regarded as the normal vector of the sampling point. This method provides better flexibility by adjusting the neighborhood size to balance the smoothness and details of the normal vectors. The following formula is used to calculate the plane fitting for each point's K neighborhood:

$$p_j(\boldsymbol{u}_j, d) = \underset{(u_j,d)}{\operatorname{argmin}} \sum_{l=1}^{k} (\boldsymbol{u}_j \boldsymbol{p}_l - d)^2 \tag{14}$$

where $\boldsymbol{u}_j$ represents the normal vector of the locally fitted plane p_j, while $\boldsymbol{p}_l (l = 1, 2 \cdots k)$ denotes the k neighboring points of u_{s_j} within s_j. The variable d represents the distance from the fitted plane p_j to the origin of the coordinate system.

The normal vector of plane p_j is determined through principal component analysis (PCA), by analyzing the eigenvector corresponding to the minimum eigenvalue of the covariance matrix A. This minimum eigenvector represents the normal vector of plane p_j. The covariance matrix A is defined as follows:

$$A = \sum_{l=0}^{k} (p_l - \overline{p})^T (p_j - \overline{p}) \tag{15}$$

where $\overline{p}$ denotes the centroid of the k neighborhood points around point u_{s_j}.

To fit the plane p_j, satisfying the condition of minimizing the sum of squared distances between the neighboring points and the plane, we utilized the Lagrange theorem to solve the covariance matrix A and the plane's normal vector, $\boldsymbol{u}_j$, which satisfied the following relationship:

$$A\boldsymbol{u}_j = \lambda \boldsymbol{u}_j \tag{16}$$

where λ represents the eigenvalues of matrix A. When λ reaches its minimum value, the corresponding vector $\boldsymbol{u}_j$ is the normal vector of the fitted plane p_j, which can be approximated as the normal vector of point $\boldsymbol{u}_j$, denoted as u_{s_j}.

The direction of the normal vector can be determined using the viewpoint method. Let us consider a point q within the object under inspection. If the relationship between u_{s_j} and q satisfies Equation (17), then the direction of the normal vector remains unchanged. However, if the relationship does not hold, the direction of the normal vector is reversed to ensure uniformity in the direction of the normal vectors.

$$\boldsymbol{u}_j \boldsymbol{Q} > 0 \tag{17}$$

where Q is the normal vector of point q.

2.3.3. Execute Point Fitting to Generate Cleaning Robot Trajectory

Discrete Point Fitting for Generating Cleaning Robot Trajectory

In accordance with the requirements of laser cleaning tasks performed by the cleaning robot, the vertical distance from the laser cleaning actuator to the surface of the target workpiece was denoted as L. The trajectory parameters of the laser cleaning actuator can be obtained using the following offset algorithm:

$$Q_j = u_{s_j} + L\frac{\boldsymbol{u}_j}{\|\boldsymbol{u}_j\|} \quad (18)$$

As shown in Figure 10, the point Q_j includes the position (coordinate values) and directional information (opposite to the direction of $\boldsymbol{u}_j$) of the end effector of the laser cleaning actuator.

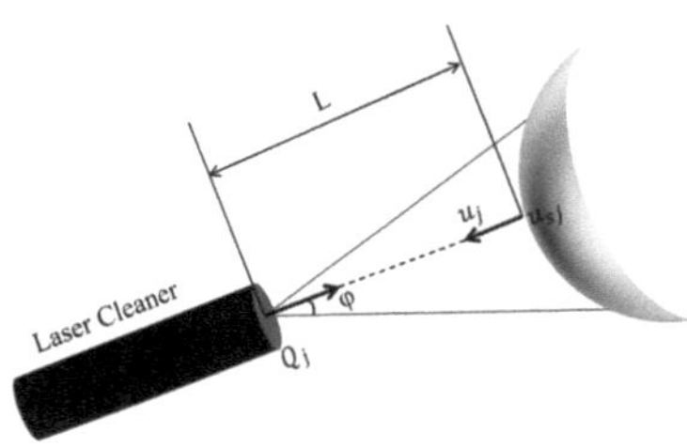

Figure 10. The position and direction of the laser cleaner at point u_{s_j}.

To obtain the offset point set Q, a similar method was applied to traverse all points in the sampling point set R. Consequently, the information contained in the entire point set Q represented the trajectory parameters (position and direction) of the laser cleaning device on the operating surface during the cleaning process.

B-spline curves are derived from the basis of Bézier curves, which were introduced by Schoenberg in the 1940s [25] and later formulated recursively by De Boor [26,27] and Cox [28]. The recursive nature of B-spline curves makes the computation straightforward and stable, leading to widespread adoption. These curves are generated by combining control points with B-spline basis functions to define the shape of the curve. In this paper, B-splines were employed for curve fitting purposes. The mathematical expression for a pth-degree B-spline curve is given by Equation (19). By connecting the respective execution points, a continuous path was formed, which can be transformed into a specific model of laser cleaning robot's motion program. This enabled automated laser cleaning of the workpiece surface for coating removal.

$$C(d) = \sum_{t=0}^{s} N_{t,p}(d)b_t \qquad l_1 \le d \le l_2 \quad (19)$$

where p represents the degree of the spline curve and $\{b_t\}$ denotes the control points of the curve.

The control polygon, formed by connecting the control points in a specific order, is referred to as the control polygon of the curve. Typically, the parameter values $l_1 = 0$ and $l_2 = 1$ are used. $N_{t,p}(d)$ represents the basis function of a pth-degree B-spline curve, as given by Equation (21), which is defined as a piecewise polynomial function determined by the knot vector $D = \{d_0, d_1, \cdots, d_e\}$. The knot vector is a non-decreasing sequence composed of all the knots. In B-spline curves, the first and last control points are typically used as the start and end points of the curve, while the intermediate control points serve as the breakpoints or knots of the curve. To ensure that the pth-degree spline curve passes through the first and last control points, the first and last knots are repeated p+1 times. In this paper, the knot vector was parameterized using the cumulative chord length method, as shown in Equation (20).

$$\begin{cases} d_0 = d_1 = d_2 = d_3 = 0 \\ d_{j+3} = d_{j+2} + \left|\Delta q_j\right|, j = 1, 2, \cdots, s-1 \\ d_{s+3} = d_{s+4} = d_{s+5} = d_{s+6} = 1 \end{cases} \tag{20}$$

where $\Delta q_j = \frac{q_j - q_{j-1}}{\sum_{j=1}^{s}\left|q_j - q_{j-1}\right|}$ represents the normalized chord length vector.

$$\begin{cases} N_{t,0}(d) = \begin{cases} 1, & d_t \leq d \leq d_{t+1} \\ 0, & others \end{cases} \\ N_{t,p}(d) = \frac{d-d_t}{d_{t+p}-d_t} N_{t,p-1}(d) + \frac{d_{t+p+1}-d}{d_{t+p+1}-d_{t+1}} N_{t+1,p-1}(d) \\ \text{set}: \frac{0}{0} \end{cases} \tag{21}$$

where the number of control points are $(s+1)$, the degree of the spline curve (p) and the number of knots $(e+1)$ satisfies $e = s + 1 + p$. Once the degree (p) and the knot vector (D) are determined, the basis functions of the B-spline curve can be computed.

Hierarchical Smooth Transition of Motion Trajectories

During laser cleaning processes, it is crucial for the robot to achieve smooth transitions between different cleaning levels. By employing B-spline curve fitting, discrete cleaning execution points have been successfully transformed into continuous path planning. However, it is necessary to ensure smooth and appropriate transitions between these paths at different levels, reducing trajectory discontinuities when switching between levels, thus ensuring stability and efficiency in the cleaning process.

Between the endpoints of each level and the starting points of the next level, the adjustment of control points in B-spline curves enables achieving consistent first-order derivatives at the connecting points of two different levels. This optimization of the connection path ensures smoothness, as shown in Equation (22). When the first-order derivatives at the connecting points of two different-level curves are equal, it indicates continuity in the tangent direction at those points. Figure 11 provides a comparative illustration, demonstrating the difference between connecting points of two different-level curves with and without the implementation of the first-order derivative consistency constraint. This approach helps avoid abrupt changes in angles at the path connections, ensuring smoother and more natural robot motion.

$$C'(d_1) = C'(d_2) \tag{22}$$

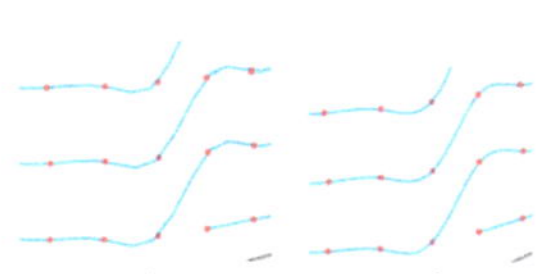

Figure 11. Comparative illustration of connecting points of two different-level curves. (**a**) Connecting points of two curves without the constraint of first-order derivative consistency. (**b**) Connecting points of two curves with the constraint of first-order derivative consistency.

In the equations, $C'(d)$ represents the first derivatives of the curve at point d. d_1 and d_2 are the parameter values at the connection points of adjacent level paths.

During the robot's transition between levels, it is essential to reduce its motion speed appropriately to avoid mechanical arm vibrations and posture deviations caused by rapid movements. Moreover, adjustments to the robot's posture are necessary to maintain an appropriate distance and angle between the laser cleaning head and the workpiece surface.

This helps decrease mechanical wear and the potential risks of malfunctions resulting from improper motion.

3. Results

The remote-control system hardware used in the experiment consists of an ABB (Asea Brown Boveri) IRB (Industrial Robot) series 6700 robot, a control cabinet, and a computer. The computer communicates with the control cabinet using the RAPID language in RobotStudio software, which controls the robot's motion. The host computer, running laser cleaning software on the Windows operating system, acts as the main control system. It establishes communication and connection with the ABB robot through the Internet options or TCP/IP protocol of the robot's PC. Additionally, it connects and communicates with the Siemens 1200 series PLC via Ethernet or other communication methods to control the robot system and laser system and monitor device status. Offline trajectory planning is performed on the 3D point cloud, which offers more freedom and accurate trajectory planning compared to online programming in RobotStudio software.

In this study, a radar dome of an aircraft was chosen as the experimental object. A 3D scanner was used to scan the radar dome and obtain point cloud data. The point cloud data were then preprocessed and aligned to generate a 3D model of the radar dome's surface. This model was imported into the offline programming software, and based on the surface curvature of the radar dome, limitations of the laser cleaning area, and focal depth requirements (in this experiment, the angle θ is 36.87 degrees), the surface of the radar dome was divided into regions, as shown in Figure 12.

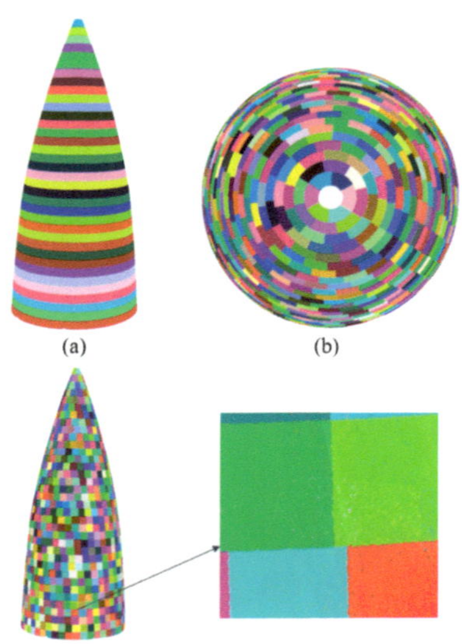

Figure 12. Radome point cloud model segmentation: (**a**) vertical segmentation effect of point cloud, (**b**) top view after point cloud segmentation, (**c**) horizontal segmentation effect of point cloud, and (**d**) local segmentation effect of point cloud segmentation.

By employing a point cloud segmentation algorithm, a series of slices were generated, as shown in Figure 12c. For each slice, a centroid calculation was performed to extract the cleansing positions of the radar radome, as shown in Figure 13a. In Figure 14a, an estimation of the normal vectors at the extracted cleaning points is presented, and numerical annotations within the figure denote the sequence of the motion trajectory during the laser cleaning process. Leveraging the B-spline curve fitting method and hierarchical stitching processing, the spatial trajectory of the laser cleaning robot was obtained, as shown in Figure 13b.

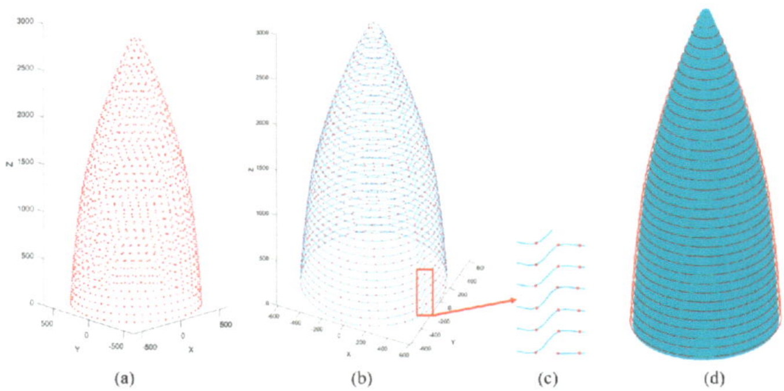

Figure 13. Execution point fitting generates cleaning robot trajectory: (**a**) extracted cleaning point positions, (**b**) trajectory planning for robot laser cleaning, (**c**) local enlarged layered connections, and (**d**) cleaning the movement trajectory of the robot.

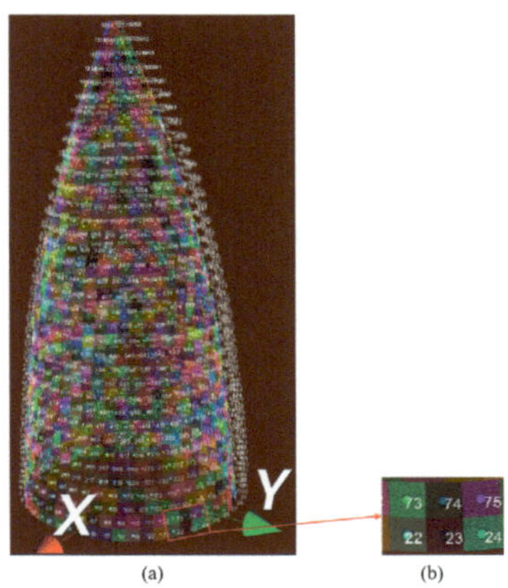

Figure 14. Estimation of normal vector for cleaning point location: (**a**) normal vector for cleaning execution point and (**b**) local enlarged display.

Before trajectory planning for laser cleaning by the robot, it was necessary to perform point cloud-to-robot calibration, which involves transforming the coordinate system of the radar radome point cloud to that of the robot. To achieve this problem, point selection calibration between the point cloud and the robot was carried out. In order to improve accuracy, a total of 10 data sets was collected and the homogeneous transformation matrix for the point cloud-to-robot calibration was computed. The computed transformation matrix is shown below.

$$M_{4\times 4} = \begin{bmatrix} -0.999999 & 0.00101104 & 0.000331104 & 2499.81 \\ -0.00101122 & -1 & 0.000034034 & -0.0950298 \\ 0.000331193 & 0.000033379 & 1 & 25.4204 \\ 0 & 0 & 0 & 1 \end{bmatrix}$$

To illustrate the practicality of the proposed method in real-world scenarios, a surface coating laser cleaning was conducted on an actual radar radome. Prior to the execution of the cleaning trials, it was imperative to fine-tune the laser cleaning parameters on a test piece composed of identical material to guarantee the attainment of thorough cleanliness in a solitary scan. The parameters for laser cleaning are delineated in Table 1. The system employs a pulsed fiber laser, characterized by a central wavelength of 1064 nm, a circular core diameter of 400 um, and a maximum output power of 1000 W. Figure 15 provides a clear demonstration of the effectiveness of laser cleaning. The efficiency of this robotic laser cleaning system was determined by calculating the ratio of cleaning duration to the area cleaned, yielding a cleaning efficiency of 2.94 m^2/h. This represents a significant improvement over the efficiency of 0.396 m^2/h reported in [29], thereby highlighting a marked enhancement in performance metrics. The trajectory planning method mentioned successfully removed the coating from the test section, meeting the experimental requirements.

Table 1. Laser cleaning parameters.

Designation	Parameters
Pulse width	100 ns
Beam diameter	340 um
Repetition rate	20–50 kHz (continuously adjustable)
Maximum single pulse energy	50 mJ

Figure 15. Laser cleaning: (**a**) the cleaning of a slice of workpieces and (**b**) the movement of the laser cleaning head between different slice layers.

4. Conclusions and Discussion

This research developed a path-planning method for the shape-following laser cleaning automated robot of aircraft radar coating. The proposed method enables accurate cleaning of coating on the radar without known surface equations or CAD models. Through the aircraft radar radome laser cleaning experiments, the effectiveness of the proposed method was validated. The final cleaning results demonstrate the wide potential and high cleaning efficiency in industrial applications.

Author Contributions: Conceptualization, Z.Z. (Zhen Zeng) and C.J.; methodology, C.J. and Z.Z. (Zhen Zeng); software, S.D., Q.L. and C.J.; validation, Z.Z. (Zhongsheng Zhai), S.D. and C.J.; formal analysis, Z.Z. (Zhen Zeng) and C.J.; investigation, Z.Z. (Zhongsheng Zhai); resources, S.D., D.C. and C.J.; data curation, Z.Z. (Zhongsheng Zhai) and S.D.; writing—original draft preparation, Z.Z. (Zhen Zeng) and C.J.; writing—review and editing, Z.Z. (Zhongsheng Zhai) and S.D.; visualization, Z.Z. (Zhen Zeng); supervision, S.D. and Q.L.; project administration, Z.Z. (Zhongsheng Zhai) and D.C.; funding acquisition, Z.Z. (Zhongsheng Zhai). All authors have read and agreed to the published version of the manuscript.

Funding: This research was funded by the Hubei Natural Science Foundation (No. 2022CFA006), and the Wuhan Key Research and Development Plan (No. 2022012202015034).

Institutional Review Board Statement: Not applicable.

Informed Consent Statement: Not applicable.

Data Availability Statement: The raw data supporting the conclusions of this article will be made available by the authors on request.

Conflicts of Interest: The authors declare no conflict of interest.

References

1. Bertasa, M.; Korenberg, C. Successes and challenges in laser cleaning metal artefacts: A review. *J. Cult. Herit.* **2022**, *53*, 100–117. [CrossRef]
2. Razab, M.K.A.A.; Jaafar, M.S.; Abdullah, N.H.; Suhaimi, F.M.; Mohamed, M.; Adam, N.; Yusuf, N.A.A.N.; Noor, A.M. A review of incorporating Nd:YAG laser cleaning principal in automotive industry. *J. Radiat. Res. Appl. Sci.* **2018**, *11*, 393–402. [CrossRef]
3. Li, X.; Huang, T.; Chong, A.W.; Zhou, R.; Choo, Y.S.; Hong, M. Laser cleaning of steel structure surface for paint removal and repaint adhesion. *Opto-Electron. Eng.* **2017**, *44*, 340–344.
4. Zhang, G.; Hua, X.; Huang, Y.; Zhang, Y.; Li, F.; Shen, C.; Cheng, J. Investigation on mechanism of oxide removal and plasma behavior during laser cleaning on aluminum alloy. *Appl. Surf. Sci.* **2020**, *506*, 144666. [CrossRef]
5. Li, Z.; Zhang, D.; Su, X.; Yang, S.; Xu, J.; Ma, R.; Shan, D.; Guo, B. Removal mechanism of surface cleaning on TA15 titanium alloy using nanosecond pulsed laser. *Opt. Laser Technol.* **2021**, *139*, 106998. [CrossRef]
6. Zhu, G.; Xu, Z.; Jin, Y.; Chen, X.; Yang, L.; Xu, J.; Shan, D.; Chen, Y.; Guo, B. Mechanism and application of laser cleaning: A review. *Opt. Lasers Eng.* **2022**, *157*, 107130. [CrossRef]
7. Ding, K.; Zhou, K.; Feng, G.; Han, J.; Xie, N.; Huang, Z.; Zhou, G. Mechanism and conditions for laser cleaning of micro and nanoparticles on the surface of transparent substrate. *Vacuum* **2022**, *200*, 110987. [CrossRef]
8. Zhao, H.; Qiao, Y.; Du, X.; Wang, S.; Zhang, Q.; Zang, Y.; Liu, X. Laser cleaning performance and mechanism in stripping of Polyacrylate resin paint. *Appl. Phys. A* **2020**, *126*, 360. [CrossRef]
9. Vorobyev, A.Y.; Guo, C. Residual thermal effects in laser ablation of metals. *J. Phys. Conf. Ser. Eighth Int. Conf. Laser Ablation* **2017**, *59*, 418–423. [CrossRef]
10. Li, X.; Zhang, Q.; Zhou, X.; Zhu, D.; Liu, Q. The influence of nanosecond laser pulse energy density for paint removal. *Optik* **2018**, *156*, 841–846. [CrossRef]
11. Bykanova, A.Y.; Kostenko, V.V.; Tolstonogov, A.Y. Development of the underwater robotics complex for laser cleaning of ships from biofouling: Experimental results. *IOP Conf. Ser. Earth Environ. Sci.* **2020**, *459*, 032061. [CrossRef]
12. Zhu, G.; Shi, S.; Fu, G.; Shi, J.; Yang, S.; Meng, W.; Jiang, F. The influence of the substrate-inclined angle on the section size of laser cladding layers based on robot with the inside-beam powder feeding. *Int. J. Adv. Manuf. Technol.* **2017**, *88*, 2163–2168. [CrossRef]
13. De Graaf, M.; Aarts, R.; Jonker, B.; Meijer, J. Real-time seam tracking for robotic laser welding using trajectory-based control. *Control Eng. Pract.* **2010**, *18*, 944–953. [CrossRef]
14. Wei, W.; Chao, Y.U.N. A path planning method for robotic belt surface grinding. *Chin. J. Aeronaut.* **2011**, *24*, 520–526.
15. Bian, Y.; Zhang, Y.; Gao, Z. A path planning method of robotic belt grinding system for grinding workpieces with complex shape surfaces. In Proceedings of the International Conference on Emerging Trends in Engineering and Technology (ICETET), Phuket, Thailand, 7–8 December 2013.
16. Cai, Z.; Liang, H.; Quan, S.; Deng, S.; Zeng, C.; Zhang, F. Computer-aided robot trajectory auto-generation strategy in thermal spraying. *J. Therm. Spray Technol.* **2015**, *24*, 1235–1245. [CrossRef]
17. Morozov, M.; Pierce, S.G.; MacLeod, C.N.; Mineo, C.; Summan, R. Off-line scan path planning for robotic NDT. *Measurement* **2018**, *122*, 284–290. [CrossRef]
18. Geng, Y.; Zhang, Y.; Tian, X.; Shi, X.; Wang, X.; Cui, Y. A novel welding path planning method based on point cloud for robotic welding of impeller blades. *Int. J. Adv. Manuf. Technol.* **2022**, *119*, 8025–8038. [CrossRef]
19. Zhang, Z.; Zhang, H.; Yu, X.; Deng, Y.; Chen, Z. Robotic trajectory planning for non-destructive testing based on surface 3D point cloud data. *J. Phys. Conf. Ser.* **2021**, *1965*, 012148. [CrossRef]
20. Hu, D.; Gan, V.J.L.; Wang, T.; Ma, L. Multi-agent robotic system (MARS) for UAV-UGV path planning and automatic sensory data collection in cluttered environments. *Build. Environ.* **2022**, *221*, 109349. [CrossRef]
21. Ye, X.; Luo, L.; Hou, L.; Duan, Y.; Wu, Y. Laser ablation manipulator coverage path planning method based on an improved ant colony algorithm. *Appl. Sci.* **2020**, *10*, 8641. [CrossRef]
22. Shuo, J.; Yuan, R.; Qingzeng, M.; Hailong, G.; Wenlong, L.; Wei, C. Off-line programming of robot on laser cleaning for large complex components. *J. Phys. Conf. Ser.* **2021**, *1748*, 022027. [CrossRef]
23. Miknis, M.; Davies, R.; Plassmann, P.; Ware, A. Efficient point cloud pre-processing using the point cloud library. *Int. J. Image Process.* **2016**, *10*, 63.
24. Gordon, W.J.; Riesenfeld, R.F. B-spline curves and surfaces. In *Computer Aided Geometric Design*; Academic Press: Cambridge, MA, USA, 1974; pp. 95–126.
25. Schoenberg, I.J. Contributions to the problem of approximation of equidistant data by analytic functions. Part B. On the problem of osculatory interpolation. A second class of analytic approximation formulae. *Q. Appl. Math.* **1946**, *4*, 112–141. [CrossRef]
26. De Boor, C. On calculating with B-splines. *J. Approx. Theory* **1972**, *6*, 50–62. [CrossRef]
27. De Boor, C. *A Practical Guide to Splines*; Springer: New York, NY, USA, 1978.
28. Cox, M.G. The numerical evaluation of B-splines. *IMA J. Appl. Math.* **1972**, *10*, 134–149. [CrossRef]

29. Pan, C.Q.; Zhu, X.W.; Yang, W.F. Design of robotic laser shape-follow cleaning control system for large freeform surface workpiece. *Appl. Laser* **2021**, *41*, 1280–1286.

applied sciences

MDPI

Article

Energy-Saving Breakthrough in the Point-to-Point Control of a Flexible Manipulator

Akira Abe

National Institute of Technology, Asahikawa College, 2-2-1-6 Shunkodai, Asahikawa 071-8142, Japan; abe@asahikawa-nct.ac.jp

Abstract: This study aims to contribute academically valuable insights into energy-efficient drives for the positioning control of flexible structures. It focuses on the point-to-point (PTP) motion control of a flexible manipulator to suppress residual vibration and reduce driving energy simultaneously. The driving energy for PTP motion is influenced by the initial deflection of the flexible manipulator. Considering this phenomenon, the study proposes a trajectory planning method for the joint angle of a flexible manipulator. In this method, the evaluation function is defined as the sum of drive torques, and its minimization through particle swarm optimization generates an optimal trajectory that minimizes drive energy and suppresses residual vibration. Numerical simulations indicate that significant energy savings can be achieved by actively deforming the manipulator. These simulation results are corroborated by experimental data, which demonstrate the practical applicability and effectiveness of the proposed method.

Keywords: trajectory planning; flexible structure; energy conservation; vibration control; flexibility utilization

Citation: Abe, A. Energy-Saving Breakthrough in the Point-to-Point Control of a Flexible Manipulator. *Appl. Sci.* **2024**, *14*, 1788. https://doi.org/10.3390/app14051788

Academic Editor: Jonghoek Kim

Received: 12 January 2024
Revised: 18 February 2024
Accepted: 20 February 2024
Published: 22 February 2024

1. Introduction

Robotic manipulators, designed for repetitive pick-and-place tasks, must be lightweight to minimize energy costs. However, their reduced weight often leads to increased vibrations due to decreased rigidity. Addressing this issue, various approaches have been developed to mitigate vibrations in flexible manipulators [1–6]. These approaches fall into two primary categories: feedback and feedforward control schemes. Feedback control is effective in managing disturbances and variable parameters in the objects being controlled. Feedforward control, which does not require sensors to measure vibrations (as demonstrated in [7,8]), is cost-effective and has received significant attention from researchers aiming to refine vibration control in flexible manipulators [9–22].

Recent studies on feedforward vibration control include the application of the finite element method to investigate vibration control in single-link flexible planar and curved manipulators, particularly focusing on triangular and trapezoidal velocity motion commands [23]. Malgaca et al. [23] reported that the residual vibrations of flexible manipulators were suppressed by selecting appropriate deceleration times in the trapezoidal velocity profile. Yang et al. [24] introduced a method for nonlinear dynamic modeling and trajectory planning in a flexure-based macro-micro manipulator, significantly reducing residual microscopic-level vibrations. Xin et al. [25] presented an optimization strategy for minimizing residual vibrations in space manipulator systems, utilizing an absolute coordinate model. Enhancements in existing input shaping techniques were reported by Kim and Croft [26], enabling the determination of multiple vibration modes in industrial robots with joint elasticity through the optimal S-curve trajectory, robust zero-vibration shaper, and dynamic zero-vibration shaper. Yoon et al. [27] examined the timing of jerks in trapezoidal motion profiles to minimize vibrations in a six-degrees-of-freedom (DOF) commercial articulated robot with a cantilevered beam. Further advancements included a trajectory planning method based on quintic polynomials for suppressing vibrations in spatial flexible manipulators with curved links [28], where a dynamic model was established as a differential algebraic equation using the absolute nodal coordinate formulation and Lagrange equation. Li et al. [29] studied vibration suppression in a planar multi-DOF serial manipulator with flexible links, proposing an online vibration-suppression path-planning method using backpropagation neural networks. A zero residual vibration position control strategy combining motion planning and optimization algorithms for two-DOF flexible systems was introduced by Meng et al. [30]. This method was applied to various systems, including a car-spring-car system, a planar two-dimensional overhead crane, and a planar single-link flexible manipulator, and its effectiveness was demonstrated through numerical simulations. Lastly, İlman et al. [31] presented an improved method for suppressing transient and residual vibrations in flexible industrial robots by adjusting acceleration/deceleration times in trapezoidal velocity profiles and incorporating a decision tree classification algorithm (C4.5) to refine input pre-shaping.

However, the aforementioned studies focused only on vibration control, and methods targeting energy conservation were limited to manipulator systems with rigid links. Soori et al. [32] conducted a comprehensive review of the literature on the optimization of energy consumption in industrial robots, in which they reviewed 136 papers published between the years 2004 and 2023. Focusing on industrial robots, autonomous vehicles, and embedded systems, Vásárhelyi et al. [33] provided a systematic review of the classification and analysis of various methodologies and solutions developed to improve the energy performance of robotic systems. Consequently, feedforward control techniques have been introduced to address both vibration control and energy conservation in flexible manipulators. Our earlier work [34,35] explored the point-to-point (PTP) motion in systems with flexible links, introducing a neural network-based trajectory planning method that simultaneously minimizes driving energy and residual vibration. Mu et al. [36] developed a trajectory planning approach for flexible servomotor systems, aiming to achieve minimal energy consumption and zero residual vibration while adhering to state constraints on velocity, acceleration, and jerk. Furthermore, the present author [37] developed an open-loop control technique to further reduce driving energy without inducing residual vibrations in the PTP motion of flexible structures, employing a combination of cycloidal and polynomial functions for optimal trajectory generation.

In the context of efforts to combat global warming, there is an increasing demand across various sectors for CO_2 emission reduction (i.e., energy conservation). This trend highlights the need for more research focusing on energy-efficient approaches in flexible manipulator systems. A common hypothesis suggests that driving a flexible manipulator along a smooth trajectory, which minimizes vibration excitation, can achieve both residual vibration suppression and energy-efficient operation. But is this assumption accurate? This study seeks to answer this critical scientific question and introduces an energy-saving feedforward control method for the PTP control problem in flexible manipulators. Simulations indicate that active deformation occurs in the manipulator during PTP motion, allowing for significant reductions in drive energy due to the interaction between the restoring force of deformation and the angular acceleration of rotation. Experimental validation further demonstrates the effectiveness and feasibility of the proposed method.

2. Single-Link Flexible Manipulator

2.1. Experimental Setup

Figure 1 shows the experimental setup of the single-link flexible manipulator used in this study. The manipulator, a brass beam, measures 550 mm in length, 50 mm in width, and 1 mm in thickness. Its one end is securely attached to a hub with a radius of 42 mm. The displacement of the manipulator was measured using a strain gauge attached at a distance of 30 mm from the clamped end. The joint angle of the flexible manipulator was measured using a serial encoder connected to an AC servomotor (SGMMJ; Yaskawa Electric Corp., Kitakyushu, Japan). For the precise tracking of the joint angle, the AC servomotor

3. Relationship between Initial Deflection and Drive Energy

Figure 4 depicts the impact of initial deflection on both the displacement (w) and the drive torque (τ) of the manipulator when it follows cycloidal motion (Equation (4)) under specific driving conditions (T_E = 0.6 s and $\theta_E = \pi/6$ rad). In the figure, the dotted, thin, and thick lines represent the results for initial deflections $w(l, 0)$ of 5, 0, and −5 cm, respectively. As depicted in Figure 4c,d, the maximum amplitude is smaller with a negative initial deflection (thick line) compared to scenarios with no initial deflection (thin line) and a positive initial deflection (dotted line). Correspondingly, the driving torque is also reduced. The driving energies, calculated using Equation (6), are listed in Table 1:

$$E = \int_0^{\theta_E} |\tau| \, d\theta. \tag{6}$$

Consistent with the observations from Figure 4, a negative initial deflection results in the lowest driving energy. Thus, it is inferred that driving energy can be minimized by inducing a substantial negative deflection in the flexible manipulator immediately after initiating the drive.

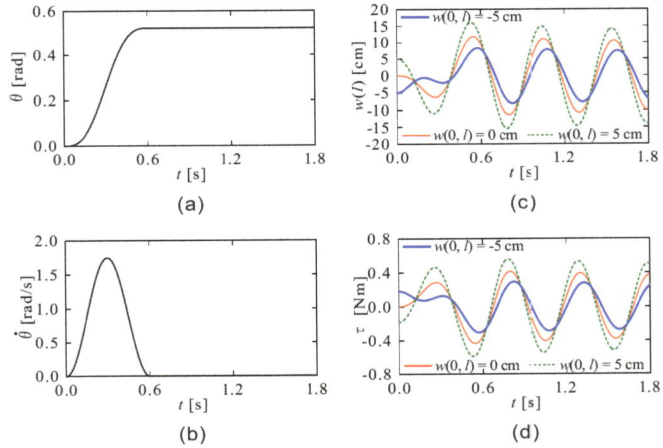

Figure 4. Effect of initial deflection on response and torque (T_E = 0.6 s and $\theta_E = \pi/6$ rad): (**a**) joint angle, (**b**) angular velocity, (**c**) tip displacement, and (**d**) motor torque.

Table 1. Effect of initial deflection on driving energy E [J].

Value of the Initial Deflection		
$w(0, l) = -5$ cm	$w(0, l) = 0$	$w(0, l) = 5$ cm
5.28×10^{-2}	1.05×10^{-1}	1.60×10^{-1}

4. Trajectory Planning Method Focused on Flexibility Characteristics

This study addresses the PTP control problem of rotating a manipulator to a target angle θ_E at time T_E. The author's previous study [37] introduced an energy-saving trajectory planning method where the joint angular trajectory of a flexible manipulator is represented by a combination of a cycloid function and a power series. Numerical simulations demonstrated that this method conserves more energy compared to neural networks [32]. However, the physical phenomena discussed in Section 3 suggest that energy savings can also be achieved by significantly deflecting the manipulator in the negative direction immediately following activation. Therefore, the following trajectory planning method is proposed.

The optimized trajectory $\theta_{opt}(t)$ of the manipulator joint angle is formulated as the sum of a cycloid curve with drive time T_0 and target angle θ_0 and a cycloid function with the input represented as a power series:

$$\theta_{opt}(t) = \theta_{cyc}(t) + \theta_f(t), \tag{7}$$

$$\theta_{cyc}(t) = \theta_0\left[\frac{t}{T_0} - \frac{1}{2\pi}\sin\left(\frac{2\pi t}{T_0}\right)\right], \ (0 \le t \le T_0), \tag{8}$$

$$\theta_f(t) = (\theta_E - \theta_0)\left\{u(t) - \frac{\sin[2\pi u(t)]}{2\pi}\right\}, \tag{9}$$

$$u(t) = \frac{t}{T_E} + (1 - T^2)\sum_{n=1}^{N} a_n T^{n-1}, \tag{10}$$

$$T = -1 + \frac{2t}{T_E}, \ (0 \le t \le T_E). \tag{11}$$

Here, the ranges of θ_0 and T_0 are defined as follows:

$$0 \le \theta_0 < \theta_E, \ \ 0 \le T_0 < T_E. \tag{12}$$

Equation (8) describes a cycloidal curve with time T_0 and angle θ_0, aiming to reduce the driving energy by significantly deflecting the flexible manipulator. Meanwhile, as indicated by Equations (9) and (10), the input $u(t)$ is given as a power series up to time T_E, generating a trajectory toward the target angle θ_E. If $\theta_0 = 0$, then $\theta_{opt} = \theta_f$, aligning with the trajectory equation from our previous study [37]. The trajectory $\theta_{opt}(t)$ generated from Equation (7) depends on T_0 and θ_0 in Equation (8) and the coefficients a_n of the power series in Equation (10). The procedure for the trajectory planning method for energy-saving and residual vibration suppression is outlined below.

Optimization parameters include T_0 and θ_0 from Equation (8) and the coefficients a_n from Equation (10), used to generate the joint angle trajectory $\theta_{opt}(t)$ from Equations (7)–(11). The numerical integration of Equation (3), based on this trajectory, reveals the dynamics of the flexible manipulator. The results from this integration are then applied to deduce the drive torque τ via inverse dynamics analysis of Equation (2). To achieve both drive energy minimization and residual vibration suppression, the evaluation function F is defined as follows:

$$F = F_1 + F_2 = \sum_{i=1}^{I}|\tau_i| + \sum_{i=I+1}^{I+J}|\tau_i| \tag{13}$$

where τ_i is the drive torque per time interval Δt = 2 ms, and $I = T_E/\Delta t$. Thus, F_1 represents the sum of the drive torque during rotation up to time T_E, with its minimization signifying drive energy reduction. F_2 represents the sum of the torque over 1 s after positioning, set to $J = 1/\Delta t$. As shown in Figure 3c,d, when residual vibration occurs after positioning, torque is required to maintain the joint angle at $\theta(t) = \theta_E$ $(t \ge T_E)$ against this vibration. F_2 is then considered to mitigate residual vibration. Hence, F is the sum of the drive torque from the start of the drive to T_E + 1 s. By considering the drive torque from the end of the drive to 1 s later, both energy savings and residual vibration suppression are achievable. We employ particle swarm optimization (PSO) [39] to optimize the search parameters for minimizing the evaluation function (Equation (13)). In our previous studies [8,38], PSO was shown to be an effective method for trajectory planning in flexible manipulators. The PSO algorithm is also described in [7]. The algorithm of the trajectory planning method was implemented in Python. The result of this optimization process generates a trajectory that conserves energy and suppresses residual vibration. By rotating the manipulator along this optimized trajectory, both drive energy and residual vibration can be reduced. Therefore, this method belongs to the category of feedforward vibration control.

tracking control and manipulator modeling in the experimental setup. The dashed lines in Figures 8c,d and 9c,d represent the experimental results for the cycloidal motion (Equation (4)). The feedforward control of the proposed method effectively mitigates residual vibration after positioning, even under varying driving conditions, confirming the method's efficacy in vibration suppression.

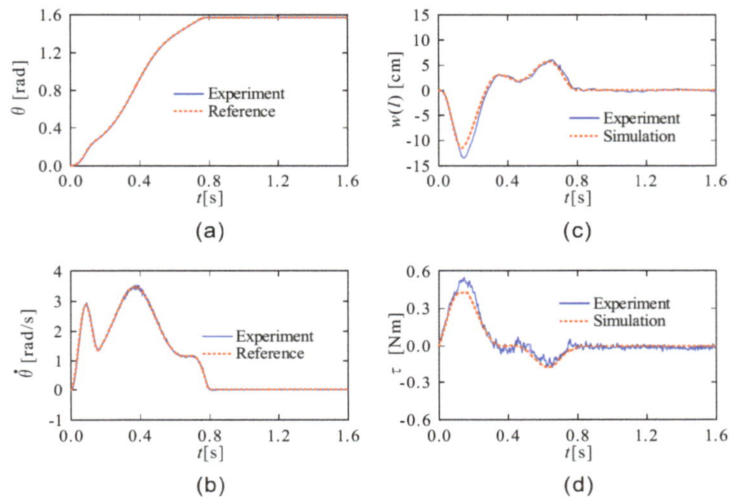

Figure 7. Comparison of simulation and experimental results obtained by the proposed method (T_E = 0.8 s and $\theta_E = \pi/2$ rad): (**a**) joint angle, (**b**) angular velocity, (**c**) tip displacement, and (**d**) motor torque.

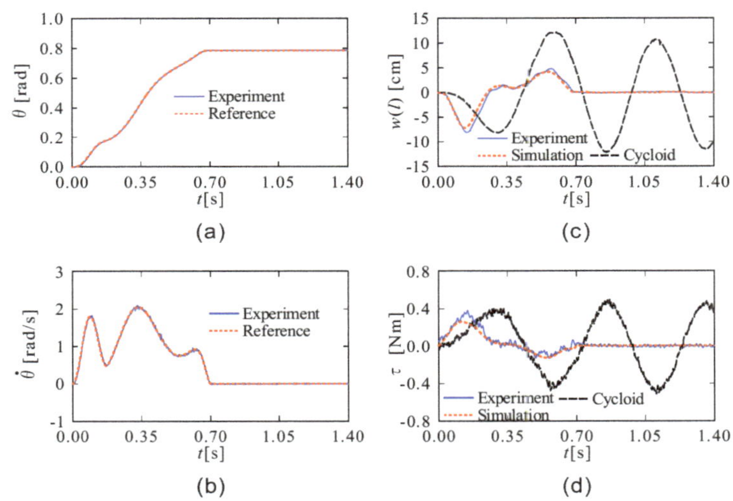

Figure 8. Comparison of the simulation and experimental results obtained by the proposed method (T_E = 0.7 s and $\theta_E = \pi/4$ rad): (**a**) joint angle, (**b**) angular velocity, (**c**) tip displacement, and (**d**) motor torque.

Next, the energy-saving effects of the proposed method are discussed. Table 2 compares the experimental drive energy values for three different driving conditions: ($\theta_E = \pi/2$ rad, T_E = 0.8 s), ($\theta_E = \pi/4$ rad, T_E = 0.7 s), ($\theta_E = \pi/6$ rad, T_E = 0.6 s). In the table, "Cyc" refers to the cycloidal motion, with simulation results presented in parentheses. The experimental results confirm that residual vibration is suppressed without inducing higher-order vibration modes under all three driving conditions in the previous method. For reference, a comparison of the simulation and experimental results from the previous method under these conditions is shown in Figures A1 and A2 in Appendix A. The experimental drive energy values were higher than the simulation values, likely due to unaccounted factors such as motor friction. As indicated in Table 2, the values for the proposed method are lower than those of the previous studies for all driving conditions, demonstrating substantial energy savings. Therefore, it can be concluded that the proposed method generates an optimal trajectory that not only suppresses residual vibration but also significantly reduces energy consumption.

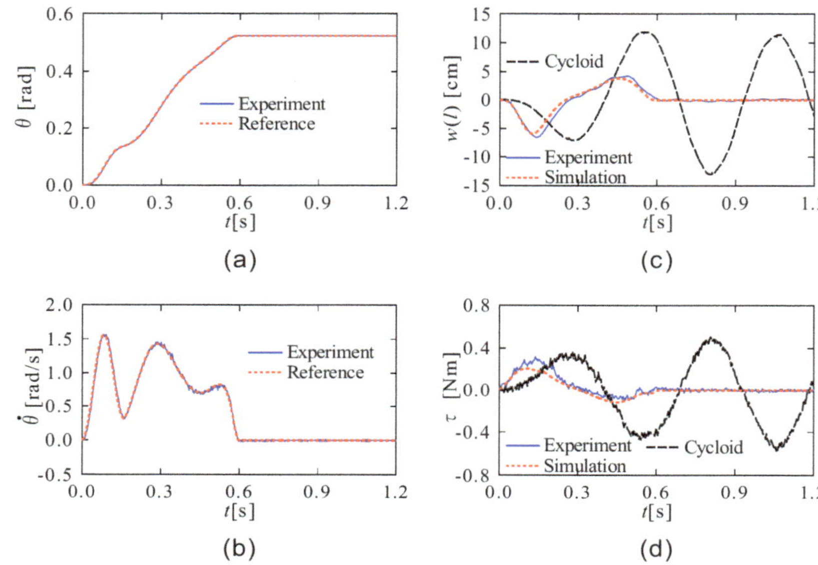

Figure 9. Comparison of the simulation and experimental results obtained by the proposed method (T_E = 0.6 s and $\theta_E = \pi/6$ rad): (**a**) joint angle, (**b**) angular velocity, (**c**) tip displacement, and (**d**) motor torque.

Table 2. Comparison of driving energy E [J].

θ_E [rad]	T_E [s]	Cyc	Previous Method	Proposed Method
$\pi/2$	0.8	5.65×10^{-1} (5.59×10^{-1})	2.98×10^{-1} (2.88×10^{-1})	2.34×10^{-1} (2.06×10^{-1})
$\pi/4$	0.7	1.98×10^{-1} (1.84×10^{-1})	9.82×10^{-2} (8.57×10^{-2})	8.06×10^{-2} (6.94×10^{-2})
$\pi/6$	0.6	1.18×10^{-1} (1.05×10^{-1})	6.58×10^{-2} (5.52×10^{-2})	5.43×10^{-2} (4.85×10^{-2})

The effectiveness of the proposed method in reducing drive energy is also explored. The drive torque, as outlined in Equatuon (2), comprises three components. To illustrate, consider the driving conditions ($\theta_E = \pi/2$ rad, T_E = 0.8 s). A comparative analysis of the time histories of these three components is shown in Figure 10. Figure 10a,b show the simulation results of the proposed method and the previous study, respectively. Figure 10a reveals that $\alpha_1\ddot{\theta}$ and $\alpha_2\ddot{W}$ in the present method have an opposite phase relationship, with $\tau = 0$ at the center. This opposite phase relationship is not present in the previous method during the time interval of approximately 0.1 to 0.7 s. Although the proposed method increases the magnitude of angular acceleration, it effectively reduces the overall drive torque due to this opposite phase relationship, thereby achieving energy savings. As illustrated in Figure 11, the proposed method can generate an optimal trajectory whereby the angular acceleration and the elastic restoring force counteract each other, enabling energy-efficient operation by utilizing the flexibility characteristics of the manipulator.

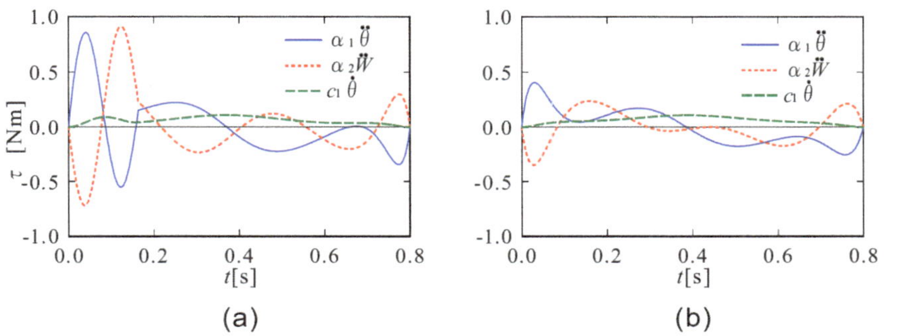

Figure 10. Time histories of torque components (T_E = 0.8 s and $\theta_E = \pi/2$ rad): (a) proposed method and (b) previous method.

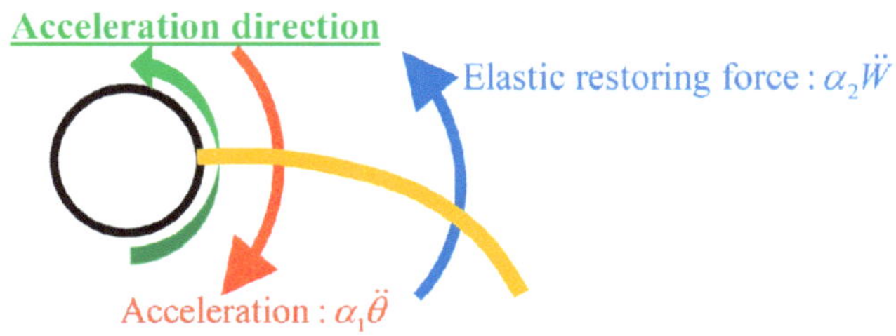

Figure 11. Schematic of the offset relationship between acceleration and elastic restoring force due to deformation.

Finally, the efficacy of the proposed trajectory planning method was validated. In a previous study [37], the amplitudes of residual vibration and driving energy were used as the two evaluated values, with trajectories for energy-saving residual vibration suppression derived from multi-objective optimization. In contrast, the present study generates trajectories through the minimization of the sum of the drive torques from the start of operation to the positioning time T_E + 1 s. In this case, driving energy was not evaluated directly. Hence, to reaffirm the validity of the proposed method's evaluation function, a comparison with simulation results from the previous study [37] was conducted. The following parameters were employed in Equations (2) and (3) [37]:

$$\left.\begin{array}{l}\alpha_1 = 2.383 \times 10^{-2}\,[\mathrm{kgm}^2],\ \alpha_2 = 9.261 \times 10^{-2}\,[\mathrm{kgm}],\ c = 3.091 \times 10^{-2}\,[\mathrm{Nms/rad}], \\ \varsigma = 9.963 \times 10^{-3}\,[-],\ \beta_1 = 2.555 \times 10^{-1}\,[\mathrm{m}],\ \beta_2 = 2.614 \times 10^{-1}\,[-],\ \omega = 10.43\,[\mathrm{rad/s}]\end{array}\right\} \quad (16)$$

Table 3 presents a comparison of driving energies under the three driving conditions. The coefficient values obtained by the proposed method and a comparative diagram of the time history data from both methods, demonstrating the effectiveness in suppressing residual vibration, are provided in Appendix B. Table 3 indicates that the proposed method achieves a significant reduction in driving energy compared to the previous study. Thus, it can be concluded that minimizing the evaluation function in Equation (13) is effective for both suppressing residual vibration and minimizing drive energy.

Table 3. Comparison of driving energy E [J] for the previous mathematical model.

T_E [s]	θ_E [rad]	Ref. [35]	Proposed Method
0.8	$\pi/6$	5.15×10^{-2}	3.80×10^{-2}
1.0	$\pi/2$	2.96×10^{-1}	1.94×10^{-1}
1.1	$\pi/2$	2.23×10^{-1}	1.60×10^{-1}

6. Conclusions

This study addressed the PTP control problem of a single-link flexible manipulator and proposed a trajectory planning method aimed at simultaneously minimizing drive energy and residual vibration. The approach was based on the physical phenomenon where drive energy decreases with initial deflection in the manipulator. The study focused on generating a trajectory that induces significant deformation immediately after activation. In the proposed method, the trajectory of the joint angle was expressed as the sum of a cycloid curve and a cycloid function, with the input represented as a power series. To achieve both drive energy minimization and residual vibration suppression, the sum of the torques was defined as the evaluation function. The parameters of the trajectory were tuned using PSO to minimize the evaluation function, and the optimal trajectory was then generated. The efficacy and feasibility of this proposed method were verified through both simulations and model experiments. The proposed method can reduce energy consumption while also suppressing residual vibrations. Generally, energy conservation in manipulator operation is often associated with the need for smooth trajectory execution to minimize vibration. However, the proposed method produced results that differed from this concept. These findings contribute new knowledge to the field of energy conservation in the context of flexible manipulators. This study, therefore, makes a significant contribution to understanding and improving energy efficiency in robotic systems. Our future challenge is to apply the utilization of flexibility for energy saving in a multi-link flexible manipulator.

Funding: This research received no external funding.

Institutional Review Board Statement: Not applicable.

Informed Consent Statement: Not applicable.

Data Availability Statement: The raw data supporting the conclusions of this article will be made available by the authors on request.

Conflicts of Interest: The author declares no conflicts of interest.

Appendix A

Table A1. Optimized parameters by the proposed method.

θ_E [rad]	T_E [s]	θ_0	T_0	a_1	a_2	a_3	a_4
$\pi/2$	0.8	1.741×10^{-1}	1.644×10^{-1}	1.672×10^{-2}	-1.401×10^{-2}	-9.818×10^{-2}	-3.069×10^{-1}
$\pi/4$	0.7	1.591×10^{-1}	1.813×10^{-1}	-3.666×10^{-2}	7.198×10^{-2}	-3.034×10^{-1}	-2.686×10^{-1}
$\pi/6$	0.6	1.309×10^{-1}	1.688×10^{-1}	-6.456×10^{-2}	6.375×10^{-2}	-3.997×10^{-1}	-1.309×10^{-1}

Table A2. Optimized parameters by the previous method [35].

θ_E [rad]	T_E [s]	a_1	a_2	a_3	a_4	a_5	a_6
$\pi/2$	0.8	1.622×10^{-2}	-6.449×10^{-2}	2.813×10^{-2}	-2.859×10^{-1}	6.359×10^{-2}	3.733×10^{-2}
$\pi/4$	0.7	1.261×10^{-2}	-1.084×10^{-1}	4.721×10^{-2}	-2.348×10^{-1}	-3.868×10^{-2}	-1.834×10^{-1}
$\pi/6$	0.6	1.197×10^{-2}	-1.666×10^{-1}	9.232×10^{-2}	-3.002×10^{-1}	-2.920×10^{-1}	6.456×10^{-2}

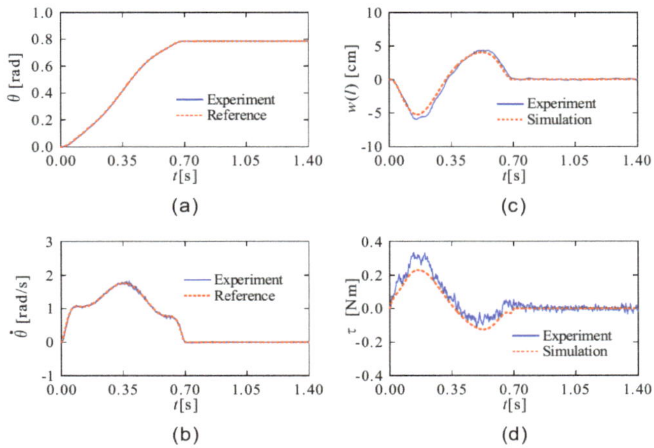

Figure A1. Comparison of the simulation and experimental results obtained by the previous method (T_E = 0.7 s and $\theta_E = \pi/4$ rad): (**a**) joint angle, (**b**) angular velocity, (**c**) tip displacement, and (**d**) motor torque.

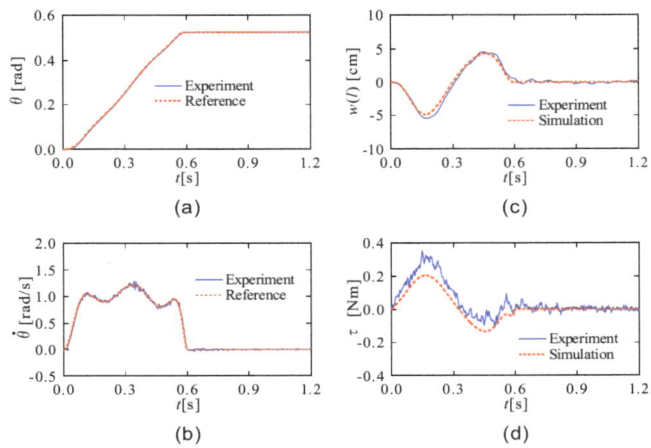

Figure A2. Comparison of the simulation and experimental results obtained by the previous method (T_E = 0.6 s and $\theta_E = \pi/6$ rad): (**a**) joint angle, (**b**) angular velocity, (**c**) tip displacement, and (**d**) motor torque.

Appendix B

Table A3. Optimized parameters by the proposed method for the previous mathematical model.

θ_E [rad]	T_E [s]	θ_0	T_0	a_1	a_2	a_3	a_4
$\pi/6$	0.8	1.150×10^{-1}	2.279×10^{-1}	-4.718×10^{-2}	-6.422×10^{-2}	-3.702×10^{-1}	-1.136×10^{-1}
$\pi/2$	1.0	1.865×10^{-1}	2.601×10^{-1}	-4.564×10^{-3}	6.426×10^{-3}	-1.281×10^{-1}	-3.045×10^{-1}
$\pi/2$	1.0	8.355×10^{-2}	2.405×10^{-1}	3.046×10^{-2}	-4.206×10^{-2}	-4.232×10^{-2}	-2.388×10^{-1}

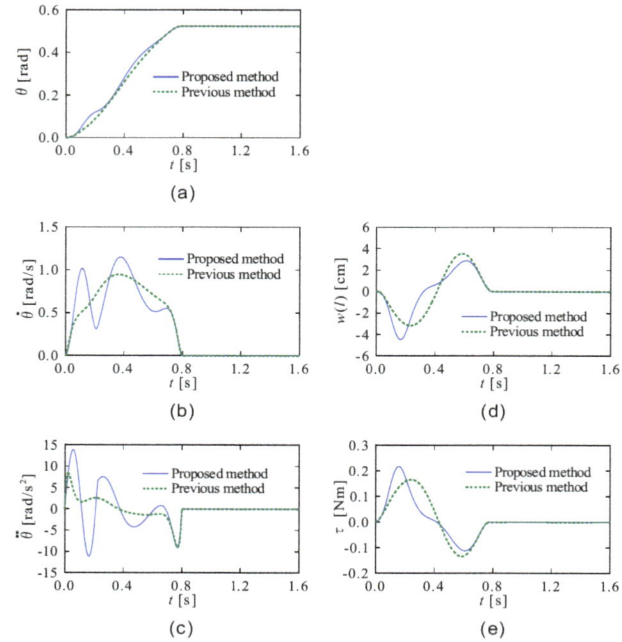

Figure A3. Comparison of the simulation results obtained by the proposed method and those obtained by the previous method (T_E = 0.8 s and $\theta_E = \pi/6$ rad): (**a**) joint angle, (**b**) angular velocity, (**c**) angular acceleration, (**d**) tip displacement, and (**e**) motor torque.

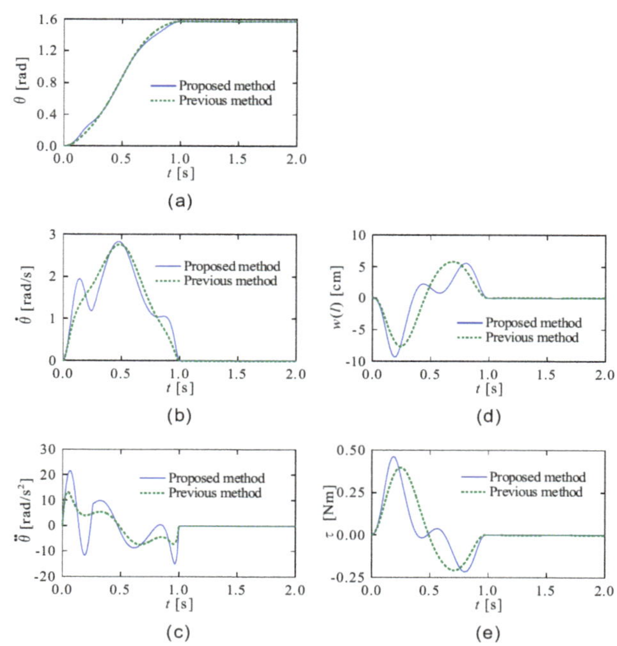

Figure A4. Comparison of the simulation results obtained by the proposed method and those obtained by the previous method (T_E = 1.0 s and $\theta_E = \pi/2$ rad): (**a**) joint angle, (**b**) angular velocity, (**c**) angular acceleration, (**d**) tip displacement, and (**e**) motor torque.

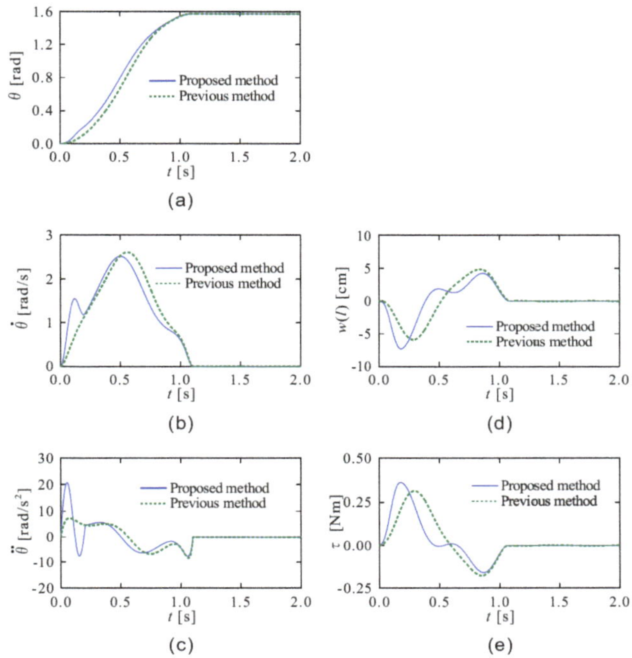

Figure A5. Comparison of the simulation results obtained by the proposed method and those obtained by the previous method (T_E = 1.1 s and $\theta_E = \pi/2$ rad): (**a**) joint angle, (**b**) angular velocity, (**c**) angular acceleration, (**d**) tip displacement, and (**e**) motor torque.

References

1. Benosman, M.; Vey, L.G. Control of flexible manipulators: A survey. *Robotica* **2004**, *22*, 535–545. [CrossRef]
2. Dwivedy, S.K.; Eberhard, P. Dynamic analysis of flexible manipulators, a literature review. *Mech. Mach. Theory* **2006**, *41*, 749–777. [CrossRef]
3. Rahimi, H.N.; Nazemizadeh, M. Dynamic analysis and intelligent control techniques for flexible manipulators: A review. *Adv. Rob.* **2014**, *28*, 63–76. [CrossRef]
4. Kiang, C.T.; Spowage, A.; Yoong, C.K. Review of control and sensor system of flexible manipulator. *J. Intell. Robot. Syst.* **2015**, *77*, 187–213. [CrossRef]
5. Lochan, K.; Roy, B.K.; Subudhi, B. A review on two-link flexible manipulators. *Annu. Rev. Control* **2016**, *42*, 346–367. [CrossRef]
6. Alandoli, E.A.; Lee, T. A critical review of control techniques for flexible and rigid link manipulators. *Robotica* **2020**, *38*, 2239–2265. [CrossRef]
7. Abe, A. Trajectory planning for residual vibration suppression of a two-link rigid-flexible manipulator considering large deformation. *Mech. Mach. Theory* **2009**, *44*, 1627–1639. [CrossRef]
8. Abe, A. Trajectory planning for flexible Cartesian robot manipulator by using artificial neural network: Numerical simulation and experimental verification. *Robotica* **2011**, *29*, 797–804. [CrossRef]
9. Park, K.J.; Park, Y.S. Fourier-based optimal design of a flexible manipulator path to reduce residual vibration of the endpoint. *Robotica* **1993**, *11*, 263–272. [CrossRef]
10. Meirovitch, L.; Chen, Y. Trajectory and control optimization for flexible space robots. *J. Guid. Contr. Dyn.* **1995**, *18*, 493–502. [CrossRef]
11. Pond, B.; Sharf, I. Experimental evaluation of flexible manipulator trajectory optimization. *J. Guid. Contr. Dyn.* **2001**, *24*, 834–843. [CrossRef]
12. Park, K.J. Path design of redundant flexible robot manipulators to reduce residual vibration in the presence of obstacles. *Robotica* **2003**, *21*, 335–340. [CrossRef]
13. Pond, B.; Vliet, J.V.; Sharf, I. Prediction tools for active damping and motion planning of flexible manipulators. *J. Guid. Contr. Dyn.* **2003**, *26*, 267–272. [CrossRef]
14. Benosman, M.; Vey, G.L.; Lanari, L.; Luca, A.D. Rest-to-rest motion for planar multi-link flexible manipulator through backward recursion. *J. Dyn. Syst. Meas. Contr.* **2004**, *126*, 115–123. [CrossRef]
15. Park, K.J. Flexible robot manipulator path design to reduce the endpoint residual vibration under torque constraints. *J. Sound Vib.* **2004**, *275*, 1051–1068. [CrossRef]

16. Kojima, H.; Hiruma, T. Evolutionary learning acquisition of optimal joint angle trajectories of flexible robot arm. *J. Robot. Mechatron.* **2006**, *18*, 103–110. [CrossRef]
17. Ramos, F.; Feliu, V.; Payo, I. Design of trajectories with physical constraints for very lightweight single link flexible arms. *J. Vib. Contr.* **2008**, *14*, 1091–1110. [CrossRef]
18. Korayem, M.H.; Nikoobin, A.; Azimirad, V. Trajectory optimization of flexible link manipulators in point-to-point motion. *Robotica* **2009**, *27*, 825–840. [CrossRef]
19. Choi, Y.; Cheong, J.; Moon, H. A trajectory planning method for output tracking of linear flexible systems using exact equilibrium manifolds. *IEEE/ASME Trans. Mechatron.* **2010**, *15*, 819–826. [CrossRef]
20. Yihuan, L.; Daokui, L.; Guojin, T. Motion planning for vibration reducing of free-floating redundant manipulators based on hybrid optimization approach. *Chin. J. Aeronaut.* **2011**, *24*, 533–540.
21. Korayem, M.H.; Rahimi, H.N.; Nikoobin, A. Mathematical modeling and trajectory planning of mobile manipulators with flexible links and joints. *Appl. Math. Model.* **2012**, *36*, 3229–3244. [CrossRef]
22. Boscariol, P.; Gasparetto, A. Model-based trajectory planning for flexible-link mechanisms with bounded jerk. *Robot. Comput. Integr. Manuf.* **2013**, *29*, 90–99. [CrossRef]
23. Malgaca, L.; Yavuz, Ş.; Akdağ, M.; Karagülle, H. Residual vibration control of a single-link flexible curved manipulator. *Simul. Model. Pract. Theory* **2016**, *67*, 155–170. [CrossRef]
24. Yang, Y.L.; Wei, Y.D.; Lou, J.Q.; Fu, L.; Zhao, X.W. Nonlinear dynamic analysis and optimal trajectory planning of a high-speed macro-micro manipulator. *J. Sound Vib.* **2017**, *405*, 112–132. [CrossRef]
25. Xin, P.; Rong, J.; Yang, Y.; Xiang, D.; Xiang, Y. Trajectory planning with residual vibration suppression for space manipulator based on particle swarm optimization algorithm. *Adv. Mech. Eng.* **2017**, *9*, 1687814017692694. [CrossRef]
26. Kim, J.; Croft, E.A. Preshaping input trajectories of industrial robots for vibration suppression. *Robot. Comput. Integr. Manuf.* **2018**, *54*, 35–44. [CrossRef]
27. Yoon, H.J.; Chung, S.Y.; Kang, H.S.; Hwang, M.J. Trapezoidal motion profile to suppress vibration of flexible object moved by robot. *Electronics* **2019**, *8*, 30. [CrossRef]
28. Cui, L.; Wang, H.; Chen, W. Trajectory planning of a spatial flexible manipulator for vibration suppression. *Robot. Auton. Syst.* **2020**, *123*, 103316. [CrossRef]
29. Li, Y.Y.; Ge, S.S.; Wei, Q.P.; Gan, T.; Tao, X.L. An online trajectory planning method of a flexible-link manipulator aiming at vibration suppression. *IEEE Access* **2020**, *8*, 130616–130632. [CrossRef]
30. Meng, Q.X.; Lai, X.Z.; Yan, Z.; Wang, Y.W.; Wu, M. Position control with zero residual vibration for two degrees-of-freedom flexible systems based on motion trajectory optimization. *Inf. Sci.* **2021**, *575*, 698–713. [CrossRef]
31. İlman, M.M.; Yavuz, Ş.; Taser, P.Y. Generalized input preshaping vibration control approach for multi-link flexible manipulators using machine intelligence. *Mechatronics* **2022**, *82*, 102735. [CrossRef]
32. Soori, M.; Arezoo, B.; Dastres, R. Optimization of energy consumption in industrial robots, a review. *Cognit. Rob.* **2023**, *3*, 142–157. [CrossRef]
33. Vásárhelyi, J.; Salih, O.M.; Rostum, H.M.; Benotsname, R. An overview of energies problems in robotic systems. *Energies* **2023**, *16*, 8060. [CrossRef]
34. Abe, A.; Kimuro, K. Minimum energy trajectory planning for vibration control of a flexible manipulator using a multi-objective optimisation approach. *Int. J. Mechatron. Autom.* **2012**, *2*, 286–294. [CrossRef]
35. Abe, A. Minimum energy trajectory planning method for robot manipulator mounted on flexible base. In Proceedings of the 9th Asian Control Conference, Istanbul, Turkey, 23–26 June 2013; pp. 1–7.
36. Mu, H.; Chen, H.; Zhu, Y. Vibration-energy-optimal trajectory planning for flexible servomotor systems with state constraints. *IET Control Theory Appl.* **2019**, *13*, 59–68. [CrossRef]
37. Abe, A. An effective trajectory planning method for simultaneously suppressing residual vibration and energy consumption of flexible structures. *Case Stud. Mech. Syst. Signal Process.* **2016**, *4*, 19–27. [CrossRef]
38. Abe, A.; Hashimoto, K. A novel feedforward control technique for a flexible dual manipulator. *Rob. Comput. Integr. Manuf.* **2015**, *35*, 169–177. [CrossRef]
39. Clerc, M.; Kennedy, J. The particle swarm—Explosion, stability, and convergence in a multidimensional complex space. *IEEE Trans. Evolut. Comput.* **2002**, *6*, 58–73. [CrossRef]
40. Parsopoulos, K.E.; Tasoulis, D.E.; Vrahatis, M.N. Multiobjective optimization using parallel vector evaluated particle swarm optimization. In Proceedings of the IASTED International Conference on Artificial Intelligence and Applications, Innsbruck, Austria, 16–18 February 2004; pp. 823–828.

applied sciences

MDPI

Article

Efficient Path Planning Based on Dynamic Bridging Rapidly Exploring Random Tree

Shulei Qiu [1,*], Baoquan Li [1], Ruiyang Tong [1], Xiaojing He [1] and Chuanjing Tang [2]

[1] School of Control Science and Engineering, Tiangong University, Tianjin 300387, China; libq@tiangong.edu.cn (B.L.); tong166159@163.com (R.T.); 2130081007@tiangong.edu.cn (X.H.)
[2] School of Computer Science and Communication Engineering, Jiangsu University, Zhenjiang 212013, China; 2222008055@stmail.ujs.edu.cn
* Correspondence: qslqer@163.com; Tel.: +86-1886-201-3989

Abstract: In the domain of mobile robotic navigation, the real-time generation of low-cost, executable reference trajectories is crucial. This paper propounds an innovative path planning strategy, termed Dynamic Bridging Rapidly Exploring Random Tree (DBR-RRT), which endeavors to enable safe and expedited path navigation. Initially, a heuristic discrimination method is engaged in the path search phase, whereby the issue of sluggish search velocity is tackled by evaluating whether sampled points reside at "bridging locations" within a free space, and by assessing the spatial–geometric relationships between proximate obstacles and auxiliary points. Subsequently, by leveraging extended speed, additional sampling points are generated in the vicinity of existing points to augment the search's efficacy. Ultimately, the path is optimized and pruned by synthesizing the local curvature of the sampling points and the proximity to obstacles, assigning varied priorities to nodes, thus ensuring that the path's quality and smoothness is upheld.

Keywords: RRT; path planning; dynamic sampling; path trimming; mobile robot

Citation: Qiu, S.; Li, B.; Tong, R.; He, X.; Tang, C. Efficient Path Planning Based on Dynamic Bridging Rapidly Exploring Random Tree. *Appl. Sci.* **2024**, *14*, 2032. https://doi.org/10.3390/app14052032

Academic Editor: Jonghoek Kim

Received: 2 February 2024
Revised: 22 February 2024
Accepted: 27 February 2024
Published: 29 February 2024

1. Introduction

Mobile robots have gained widespread application in fields such as industrial production, transportation, and the service industry, becoming a focal point in current technological research [1,2]. Path planning refers to the process by which robots navigate autonomously based on environmental perception through their sensors [3,4], and its algorithm design and optimization cannot be separated from the effective deployment of sensors [5]. The autonomous navigation of mobile robots is an intelligent system encompassing technologies such as environmental perception, path planning, and navigation positioning [6,7]. Path planning is a core research issue within this system, with its goal being to find a path from a starting point to a destination while ensuring that the robot avoids collisions with obstacles in the environment. Moreover, the path should meet one or more criteria, such as shortest distance, least time, or minimum energy consumption [8], ensuring the efficiency and reliability of pathways is also critical to achieving a highly automated urban transportation network [9]. Over the past few decades, numerous path planning algorithms have been proposed and widely applied in various fields, including, but not limited to, autonomous driving, drone navigation, and industrial robots. The study of path planning problems was initially carried out based on a grid-based approach, which first divides the environmental map into a series of grid cells. For example, graph search algorithms like A* [10,11] and D* [12], while theoretically robust and offering optimal solutions at a consistent resolution, are hampered by a significant drawback due to the challenges in selecting an appropriate a priori resolution. When the resolution is set too low, the quality of the resultant paths tends to be suboptimal. Conversely, setting a high-resolution leads to an exponential increase in computational costs, especially in the context of high-dimensional space problems, necessitating prolonged processing times. Another issue is their lack of real-time efficiency, as these algorithms require re-graphing and replanning, resulting in reduced efficiency. Optimization-based methods, such as artificial potential field methods [13], are widely used due to their ease of implementation but may fall into local minima.

With the advancement of technology, computing power has been greatly improved and more complex algorithms have been developed, so geometry-based methods have begun to be studied and applied. For example, the Voronoi diagram [14]. The Voronoi diagram divides the space into a series of polygonal regions; each region contains a generating point, and the distance from any point in the region to the generating point is less than the distance to other generating points. The Voronoi diagram guides the mobile robot to avoid obstacles by generating paths within a safe distance from the edges of obstacles, but in highly complex environments, the Voronoi diagram may generate too many Voronoi diagrams, and the generated paths tend to bypass the midpoint of the obstacle, making it difficult to find an optimal solution. This is especially true in environments with many small obstacles or obstacles with complex shapes.

Subsequently, sampling-based path planning methods have received a lot of attention. Sample-based planning methods, due to their ability to provide feasible paths for robots in the short term, have garnered considerable interest. For instance, the Rapidly Exploring Random Tree (RRT) [15] method consistently samples independent states in the search space and incrementally extends towards the goal state, offering the robot a path around obstacles. Another approach, RRT-Connect, proposed by Klemm et al. [16], improves the speed of finding feasible paths by expanding two search trees simultaneously from the starting point and the endpoint. However, although RRT exhibits good computational efficiency, the path it produces is often not optimal, causing the robot to take more time to reach the target location. In [17], Kang et al. proposed an improved RRT-Connect algorithm, which enhances the efficiency of the traditional RRT algorithm by adopting a rewire-tree method based on the triangle inequality, effectively reducing the redundancy commonly seen in paths generated by the standard RRT algorithm. Similarly, in [18], Karaman et al. introduced the improved RRT* algorithm. By updating the tree through reselecting parent nodes, it can find suboptimal paths. However, this approach is characterized by high computational costs and lower efficiency.

There are also artificial intelligence-based path planning techniques such as the ant colony algorithm [19], or the use of machine learning [20] and deep learning [21] models to predict and optimize paths. The ant colony algorithm is computationally intensive and has a slow convergence rate, although it can obtain a global optimal solution. Machine learning and deep learning approaches learn strategies for path planning from large amounts of data, and while they perform well in some scenarios, as shown Wang et al. in [22], who proposed a convolutional neural network (CNN)-based path planning algorithm that trains a model of a CNN network capable of non-uniform sampling, they typically require large amounts of training data and have poorly interpreted models.

2. Related Work

To address these challenges, researchers have gradually introduced more sophisticated strategies. In [23], Kun et al. employed the strategy of the directed expansion of new nodes, making the algorithm more efficient, combined with curvature constraints for path smoothing. However, this method's drawback is the excessive path search time. Other researchers also offered their solutions. Based on the RRT method, in [24], Karaman et al. introduced a rewire process and path cost function to improve the nodes in the existing search tree and their connections. This adjustment made RRT achieve an asymptotically optimal performance, but the process of finding the optimal path can greatly increase the time required for path planning. Gammell et al. proposed the improved algorithm Informed-RRT* in [25], which utilizes an elliptical heuristic search for optimal sampling. This approach addresses the strong randomness inherent in the traditional RRT algorithm. However, it tends to encounter local extremum issues in complex scenarios. Subsequently, in [26], Gammell et al. introduced Bi-RRT*, which constrains the sampling range and employs dual trees between the start and goal points to communicate information, thereby further enhancing the overall convergence speed, but this method still suffers from the problems of a low success rate for path planning under small samples, and the planning efficiency still needs to be improved under large samples. In [27], Yin effectively reduces the number of mobility simulations required for path planning by combining the RRT algorithm with adaptive surrogate modeling. Using the surrogate model to guide the exploration of random trees under mobility reliability constraints, and then employing these exploration trees and reliability assessments to refine the surrogate model. However, a high dependence on the accuracy of the agent model can generate problems such as, for example, insufficient preparation of the initial agent model or the inability to update it efficiently during the exploration process. Islam et al. proposed RRT*-Smart in [28], which, after generating the initial path, improves RRT*'s convergence speed by removing redundant nodes and optimizing sampling. In [29], Dai proposed a novel algorithm based on bidirectional Rapidly Exploring Random Trees and direct connection. Initially, an expansion strategy based on artificial potential fields was designed in the joint space, which was then integrated with the GB-RRT algorithm. Additionally, a direct connection strategy was developed to enhance the efficiency of expansion, ensuring larger safety margins between the obstacles and the system.

In specific environments such as narrow spaces or areas with dense obstacles issues like difficulty in node expansion and generation of convoluted paths arise. Ji proposed the Ellipsoidal Rapidly exploring Random Tree (E-RRT*) in [30]. This approach replaces line segments with ellipsoids to connect adjacent nodes and introduces a slow informed guidance method to optimize the sampling process, effectively addressing path planning challenges in confined spaces. In [31], Zhang made distinct optimizations to the sampling and proposed a flow-based VF-RRT* algorithm. This method quantifies the process of feasible path countercurrent using a parameter called the up-flow coefficient, constructs a heuristic space based on the flow function, and adaptively rejects sampling nodes outside the flow value range with an adjustable probability.

Another noteworthy method is the Fast-Marching Tree (FMT*) method [32], which combines probabilistic roadmap and RRT approaches by using a set + of sample points for tree expansion. However, the FMT method has the particular issue of redundant exploration, which leads to a decline in its path-searching performance. To solve this problem, Wu Zhen et al. proposed a direction-selective heuristic function that can evaluate the cost gradient of the samples, adjust the ordering of samples, and guide the expansion of FMT. Furthermore, in [33], the Safe Tunnel-based FMT* (ST-FMT*) method generates a preprocessed initial path before method expansion and then constructs a safe tunnel for sampling, which accelerates the convergence speed of the algorithm. Among the sampling-based path planning algorithms, the RRT planning method with a better search capability is suitable for use in complex environments, and the method only requires real-time local data to dynamically construct the search tree. However, RRT node expansion slows down its search efficiency due to the large number of collision detections required. In addition, RRT-generated paths are often redundant and convoluted, making them difficult for mobile robots to execute.

To address these issues, an efficient path planning method based on Dynamic Bridging RRT (DBR-RRT) is proposed, aiming to optimize the efficiency of the RRT-Connect algorithm, and, at the same time, to reduce the speed mutation phenomenon that occurs in the rapid operation of the mobile robot, and to achieve a good balance between the speed and the smoothness of the operation. The main contributions are as follows:

1. Firstly, a heuristic discrimination method is used in path planning to solve the problem of a slow search speed due to node extension collision detection by evaluating whether the sampling points are in free space and assessing the spatial location relationship between neighboring obstacles and setup assistance points.

2. Then, for the extended search phase, the sampling points with the slowest extension speed are searched for, and the set of sampling points is generated in their vicinity to improve the search efficiency.
3. Finally, for the path optimization and pruning phases, different priorities are assigned to the nodes by comprehensively evaluating the sampling point connectivity and proximity to obstacles, which ultimately makes the generated path smoother and easier to execute.

3. Problem Formulation

Despite significant advancements in path planning algorithms, they still encounter critical challenges. A primary issue is that tree expansion near obstacles involves extensive collision detection, which significantly hampers search efficiency. Additionally, the generated paths often lack smoothness and tend to be excessively tortuous. This shortcoming contradicts the fundamental principles of vehicle kinematics. Consequently, the pressing question is how to enhance path planning algorithms to meet the stringent requirements of robot kinematics while addressing these inefficiencies and ensuring smoother trajectories. This paper will delve deeply into this issue with the aim of balancing the generation of efficient paths as well as the stability of robot operation.

The structure of this paper is as follows: the remaining part of this section reviews the existing path planning methods and their advantages and disadvantages; the fourth part details our method and its implementation process; the fifth part presents our experimental results and compares the performance of our method with other methods; finally, the sixth part summarizes the main contributions of this paper and discusses future research directions.

3.1. Traditional Famework

This section will briefly review the RRT-Connect path planning method. The RRT-Connect algorithm, a bidirectional version of the RRT algorithm, does not require pre-sampling and storage of the entire configuration space, but explores the entire state space through function iteration.

The random tree growth process of RRT-Connect is shown in Figure 1, with the specific steps as follows:

1. Initialize the starting point q_{start}, the endpoint q_{goal}, and obstacles, and initialize the starting search tree T_{start} and the ending search tree T_{goal}, as well as the step length step.
2. A randomly selected sampling point q_{rand} in space is used to guide the expansion of the random tree. This point needs to satisfy motion constraints and capture collisions with objects in the environment. Then, iterate over all the nodes on the random tree T_{start}, compute the distance between them and the sampling point q_{rand}, and filter the node $q_{nearest1}$ that is closest to that point.
3. The node $q_{nearest1}$ on the random tree T_{start} grows a fixed number of steps in the direction of the sampling point q_{rand} to obtain a new leaf node q_{new1}. If there is no collision with an obstacle during the movement in the direction from the node $q_{nearest1}$ to the new node q_{new1}, the node q_{new1} will be added to the random tree T_{start}. Otherwise, the point will be discarded, and the process will revert to step 2.
4. Then, iterate over all the nodes on the random tree T_{goal}, and calculate their distances to the new node q_{new2}. Perform the same process as in step 3 on the random tree T_{goal} to obtain the new node q_{new2} on it.
5. Using the greedy search algorithm, step 4 is repeated on the random tree T_{goal}. When an obstacle is encountered during the growth of the random tree, the growth is stopped.
6. Then, exchange the two random trees and repeat step 2–step 6. Within the specified number of searches, when q_{new1} and q_{new2} are the same, a path connecting the initial point and the goal point can be obtained, indicating successful path planning.

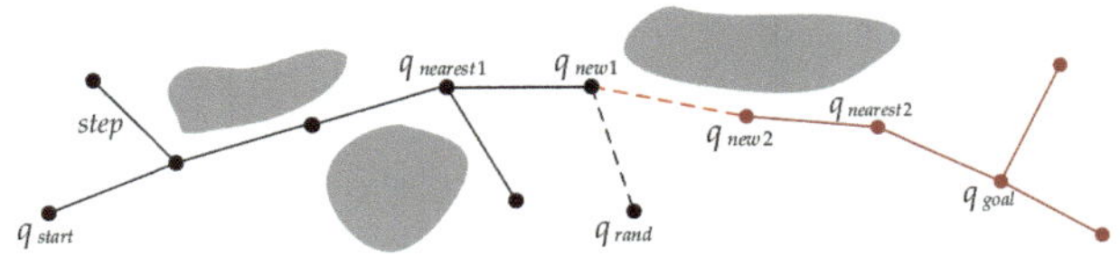

Figure 1. RRT-Connect extension steps. The nodes in T_{start} are displayed as black nodes, while the nodes in T_{goal} are displayed as red nodes.

3.2. Proposed Famework for Path Planning Algorithm

Like the RRT-Connect algorithm, the DBR-RRT algorithm uses two trees, T_{start} and T_{goal}, growing correspondingly from the starting state and target state. What differentiates our new algorithm is the use of heuristic sampling and dynamic sampling strategies.

The DBR-RRT algorithm proposed in this paper is composed of a path search (Bridge Test [34]) strategy. Firstly, DBR-RRT introduces a heuristic method to determine whether a sample point is in the free space "Bridge". This method evaluates the expansion value of the sampled point based on the relationship between the auxiliary point's location and obstacles. The tree is only expanded when the Bridge Test concept is passed, to solve the problem of slow search due to extension collision detection. Moreover, DBR-RRT introduces a dynamic sampling strategy based on expansion speed, generating sampling points set around nodes with slower expansion speed. This allows the tree to expand into obstacle-dense areas or areas with complex shapes at a faster rate, thus enhancing the search efficiency. Finally, rapid path smoothing is applied to the path generated by this technique. Through continuous iterations, the original path is optimized to produce a smoother and higher-quality path.

For initialization, two empty trees are initialized and expanded from the start node and goal node respectively. DynamicRandomNode() is used to generate an adaptive random sampling point. This function considers the current structure of the starting tree T_{start} and the goal tree T_{goal}, thereby more intelligently choosing sampling points.

The following is the Pseudocode of the DBR-RRT algorithm (Algorithm 1):

Algorithm 1. DBR-RRT

Notation: start point S, goal point G, obstacle O, vertices and edges of tree $T_{start}, T_{goal}, V_{start}, E_{start}, V_{goal}, E_{goal}$, random node X_r, nearest node X_{near}, new node X_{new}, connect node X_{con}, path P.

$(V_a, E_a) \leftarrow \{S\}, \{\}; V_b = \{G\}; E_b = \{\}$
while *LocalPlanning*() **do**
$X_r \leftarrow$ *GenerateRandomNode*()
$X_{near} \leftarrow$ *FindNearestPoint*()
$X_{new} \leftarrow$ *BridgeTest*()
if $X_{new}! = NULL$ and *IsFree*(X_{new}) **then**
$V_{start} \leftarrow V_{start} \cup \{X_{new}\}; E_{start} \leftarrow E_{start} \cup \{(X_{near}, X_{new})\}$
UpdateExpandSpeed()
$X_{con} \leftarrow$ *FindNearestPoint*()
if $X_{con}! = NULL$ **then**
$(E_{goal}, V_{goal}) \leftarrow$ *ConnectTrees*()
if *TreesConnected*() **then**
$P \leftarrow$ *ConstructPath*()
return *PathSmoothing*()
end if
end if
end if
end while
if $V_{goal}.size() < V_{start}.size()$ **then**
$Swap(V_{start}, V_{goal}); Swap(E_{start}, E_{goal})$
end if
GenerateExtraSamples()

The algorithm will attempt to identify areas in the search space that have not been fully explored (i.e., where node expansion is slow), and generate sampling points in these areas. In addition, if the starting tree and the goal tree are close to each other in space, the function may be more inclined to generate sampling points in the area between the two trees to increase the chance of tree connection.

4. Methodology

In this section, the method proposed in this paper for improving the RRT-Connect algorithm will be detailed, including the improved Bridge Test, dynamic sampling strategy, and path-smoothing techniques. Our goal is to achieve efficient, high-quality path planning.

4.1. Bridge Test

The Bridge Test is a heuristic method used to assess the likelihood of a sampling point being in the "bridge" concept of free space. The core principle of the Bridge Test is based on Voronoi diagram heuristics. Voronoi diagram is a graphical structure that partitions space into several regions, each containing a seed point, and any point within that region is closer to the seed point than to other seed points. In path planning, the Voronoi diagram can help find a path that is as far away from obstacles as possible.

The basic idea of the Bridge Test is to identify sampling points located on the edge of the free space Voronoi diagram [35], as these points are more likely to be part of an excellent path. To achieve this, a geometric analysis method is adopted, combining the spatial relationships between the sampling points, nearby obstacles, and auxiliary test points, to determine whether the sampling points align with the conditions of the Voronoi edge.

The theoretical concept of the Bridge Test is shown in Figure 2. A passage exists from the A direction to B direction. P_c represents a random sampling point on the map. P_d represents the nearest point in the known tree. P_e and P_f represent newly generated test points in two directions. Bridge Test is an algorithmic test method for quickly recognizing and passing through narrow passages (which we call "bridges") in complex environments. As shown in Figure 2, in the test we follow several steps to evaluate whether a new sampling point (P_c) can help find and enter these narrow passages. The steps in the Bridge Test are as follows:

1. Find the line between the sampled point P_c and the nearest neighbor point P_d in the search tree and calculate angle θ of this line with respect to the horizontal.
2. Depending on the angle θ, the sampling point P_c is offset by a fixed angle in each of the positive and negative directions, thus constructing two predetermined test points P_e and P_f that explore two different directions from P_c.
3. Collision detection is used to verify that the path between the test points, P_c, P_e, P_f is not blocked by obstacles.
4. If the P_c's path to one of the test points is open (i.e., unobstructed) and the path to the other test points is blocked, this indicates that the P_c is located at the entrance to a potentially narrow channel.
5. P_c is considered to have passed the Bridge Test and will be added to the current search tree.

In Figure 2b, there is an unobstructed path from P_c to P_e, which means that there are no narrow passages obstructing passage in this direction. On the contrary, the path from P_c to P_f partially enters inside the obstacle. This observation suggests that, if the path from P_c to one test point is blocked by an obstacle while the path to another test point is open, it can be determined that the point P_c passed the Bridge Test. By this means of creating handling points, the tree expansion can enter the narrow passage more quickly. This result suggests that the sampling point may be located on the Voronoi boundary of free space, and therefore has a higher expansion priority. During the search process, this priority allocation helps the algorithm to explore those possible effective paths which lead more

This method achieves effective management of the search range, allowing us to reasonably allocate and optimize unexplored areas while ensuring efficiency, avoiding invalid searches and waste of resources, and further improving the efficiency and effects of path planning. The steps for Dynamic Sampling are as follows (Algorithm 3):

Algorithm 3. Dynamic Sampling

Notation: number of extra samples N_e, the slowest expand node X_{se}, extra samples S_e, tree of samples T_s

1: Initialize $N_e \leftarrow N$
2: **if** $X_{se} \neq NULL$ **then**
3: $N_e \leftarrow StaticCast(N_e)$
4: $S_e \leftarrow GenerateExtraSamples(N_e)$
5: **for** each sample X in S_e **do**
6: **if** $\neg IsAnyObstacleInPath(X)$ **then**
7: Insert X into T_s
8: **end if**
9: **end for**
10: **else**
11: skip to the next iteration
12: **end if**

Initialize $N_e \leftarrow N$: The number of dynamic sampling points is first set to N. This is the initial setting for dynamic sampling, preparing for subsequent additional sampling.

GenerateExtraSample: The function verifies if a node with the slowest expansion speed exists (perhaps in an insufficiently explored area), if there is such a node, extra sampling points N are produced at this node. This step achieves adding more sampling points in unexplored areas, thereby improving the depth and breadth of the search.

isAnyObstacleInPath: these extra sampling points are processed individually. For each sample point, an initial check determines if obstacles exist in the path from the slowest expanding node to this sample point. If the path is clear (i.e., there are no obstacles), this sample point is added to the search tree. This step ensures that the extra sampling points generated have relevance, meaning they can be directly reached from the current node. (Algorithm 1. 5–9)

If there is no node with the slowest expansion speed, the current iteration is skipped, moving to the subsequent iteration. This means that, without a clear determination of unexplored areas, no additional sampling will be carried out, but regular search steps will continue to be performed. This step is carried out to ensure the robustness of the algorithm, which can function normally even in special circumstances. (Algorithm 1. 11–15)

While there is a need to spend time calculating the expansion speed of existing nodes in the tree to find the node with the slowest expansion speed, this also allows us to concentrate our resources on the most promising nodes. With this strategy, our tree expansion strategy can effectively avoid falling into local optimal solutions, and, to some extent, this also accelerates the search process.

In addition, this adaptive optimization method aims to dynamically adjust the sampling strategy according to the information obtained during the search process and the current environment, to improve the efficiency of the path planning algorithm in complex environments and improve the quality of the generated path. This is an optimized allocation of computational resources, aiming to maximize the possibility of global optimization, reduce search time, and provide strong support for robot navigation in complex environments. It helps to improve the efficiency of the path planning algorithm in complex environments and the quality of the generated path. This strategy can adaptively adjust the sampling distribution, making the search process more focused on difficult-to-expand areas, thereby improving the efficiency and quality of path planning given the updated node expansion speeds in each iteration, and ensuring that the search process always focuses on the most challenging area currently.

4.3. Path Smoothing

To solve the problems of redundant turns and frequent oscillations in path planning, a prominent path-smoothing strategy has been incorporated. This strategy uses an iterative optimization method to gradually optimize the original path through continuous iterations. This phase weights both local geometric contains and overarching path optimization indicators.

In the path-smoothing procedure, two non-adjacent nodes from the original path nodes are chosen. An attempt is then made to directly connect these two nodes to construct a potential new path to replace the path segment between these two nodes in the original path. This procedure must ensure that the fresh path segment does not intersect with any obstacles, preserving the path's validity. Only when the new path segment meets this criterion is it approved to replace the original path section.

Path smoothing is a key part of this research, aiming to improve the initial path generated by the search algorithm, making it more continuous, smooth, and efficient while meeting the requirements of feasibility and safety. This method helps to optimize the algorithm by directly connecting non-adjacent nodes, eliminating redundant bends in the original path, and finally generating a relatively smooth and high-quality path. The process continues until it reaches a predetermined number of iterations or the path converges. This method aims to gradually achieve global path optimization by adjusting the non-optimized parts of the path, thereby improving the navigation efficiency of robots in complex environments.

4.3.1. Curvature Calculation

It is known that the tangent vector or normal deflection angle of the curve's start and end equals the integral of the curvature length. For the polyline in Figure 4 below, the normal vector only changes at each vertex, so our goal is to find out the curvature at each vertex, but the change in the normal vector at the vertex jumps, as shown in Figure 4; the gray arrow denotes the discernible change in the vertex on the line segment, recording the turning angle of the curve.

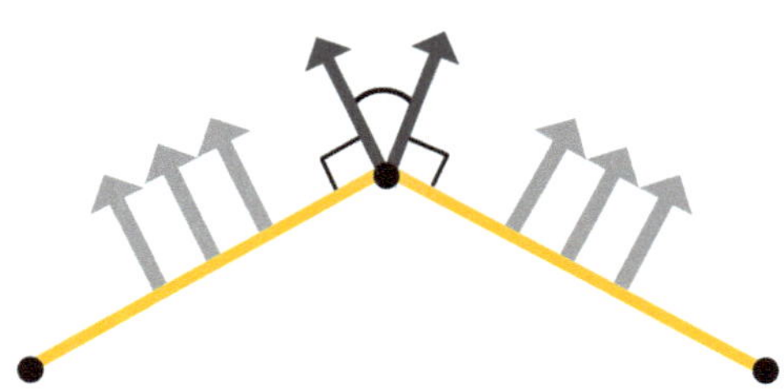

Figure 4. Normal changes at vertices record the turning angle.

The path-smoothing strategy employs a local optimization-based approach to recalibrate pivotal points in the path, minimizing its curvature. Its initial task is to scrutinize the produced preliminary path and identify possible redundant nodes and discontinuous parts. Eradicating these nodes can diminish the path's complexity while preserving essential nodes that ensure the path's feasibility.

An initial step involves assessing the local curvature between various nodes to generate our x_{new}; if the local curvature is below our preset value, then the node is deemed promising. The curvature at the vertices of a folded line is approximated by applying the Gauss–Bonnet theorem. For a simple closed curve in the plane, the total curvature and the sum of the angles at all vertices are equal to 2π. Applying this principle to the folded line model, the folded line is regarded as a polygon consisting of straight-line segments, with a corner α_i defined at each vertex, which can be found by calculating the angle between the vectors of the two neighboring sides.

Algorithm 4. PathSmoothing

Notation: node N, candidate C, candidate nodes N_c, number of candidate nodes Q_c, curcatureThreshold K.
1: **for** each N in path
2: **if** curvature of node $> K$
3: $N_c \leftarrow GenerateCandidateNodes(N, Q_c)$
4: **for** each C in N_c
5: $AssignPriority(C)$
6: $N \leftarrow HighestPriority$
7: **end for**
8: **end if**
9: **end for**
10: **return** P

5. Simulation and Experimental Results

5.1. General Framework of ROS

The system framework of ROS is divided into three main parts: the file system level, the computational graph level, and the open-source community level, with each part representing a hierarchical concept.

1. File system level: Different components in the ROS program are to be placed in different folders, and each folder also has its corresponding function. Usually we name the workspace catkin_ws, which is a folder containing function packages, compiled packages, and compiled executables. A function package is a combination of a specific file structure and a folder containing running nodes, configuration files, etc. We use the command catkin_make to compile the workspace. A function package mainly contains the files shown on the rightmost side of Figure 7, in which the src is the place where we store the source files, the build folder includes the project cache information, configurations and other intermediate files, and the devel folder is used to save the compiled program.
2. Computational graph level: The ROS creates a network to connect all of the nodes, through which any node can interact with other nodes, obtain information published by other nodes, and send its own data to the network. The basic concepts at this level include nodes, node managers, parameter servers, messages, services, topics, and message logs.
3. Open-source community level: The open-source community level of the ROS shares software and knowledge mainly through independent online communities, including ROS distributions, software repositories, ROS wiki, ROS Answer, blogs, and so on.

In Figure 8, The core of our navigation architecture is the move_base node, which fully integrates global path planning, local path planning, global cost maps, and local cost map features. These different functions are jointly implemented by subscribing to /tf,/odom, /map topics and publishing the /cmd_vel topic for motion control under specific conditions. Figure 8 shows that move_base is the core of robot navigation, providing an interface for the ROS for configuration and the operation of and interaction with navigation. The robot encounters obstacles in the environment and recalculates the path by subscribing to data from LiDAR, map information, and Monte Carlo localization to convert the path into robot velocity information and plan a new path. Figure 8 shows a diagram of the navigation function architecture which can be found on the left side of the tf information through the ROS navigation and localization module. The command amcl releases the robot pose for use with move_base; below it, odom is the robot's odometer information. The upper right corner shows the the map_server which operates through the analysis of the slam built maps released; the mainstream slam algorithms are gmapping, hector, and cartographer. The sensor in the lower right corner plays a role in local path planning. In this framework, the DWA algorithm serves as the tool for local path planning, while the DBR-RRT algorithm is utilized for global path planning.

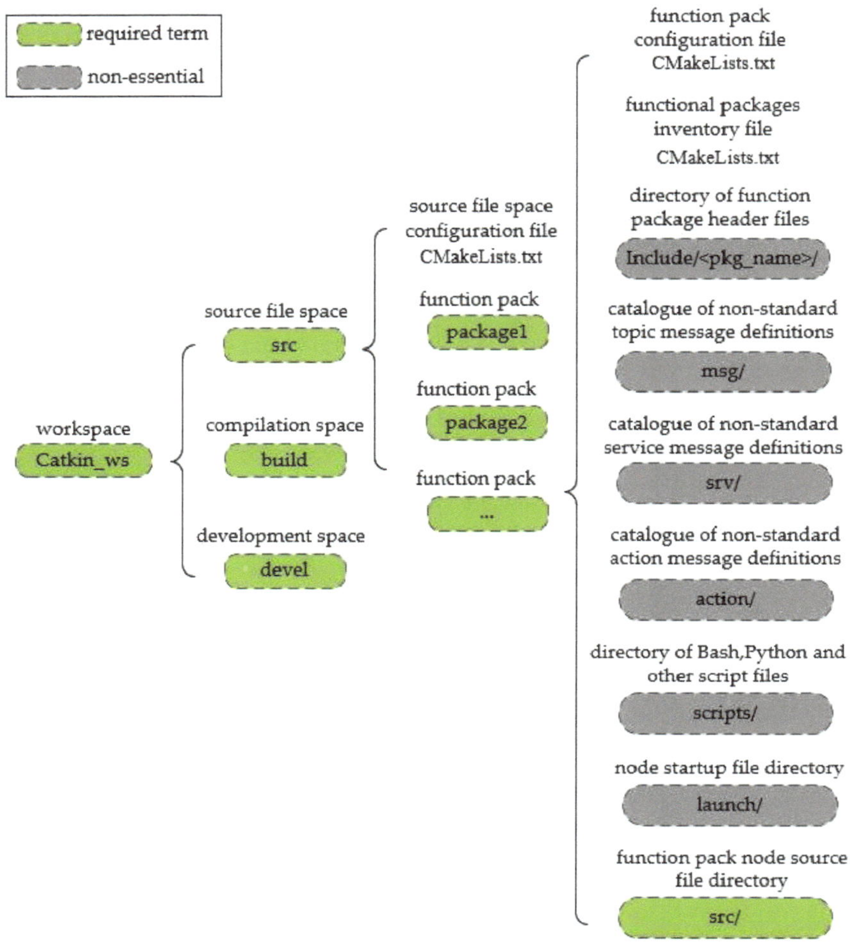

Figure 7. File system-level architecture for ROS.

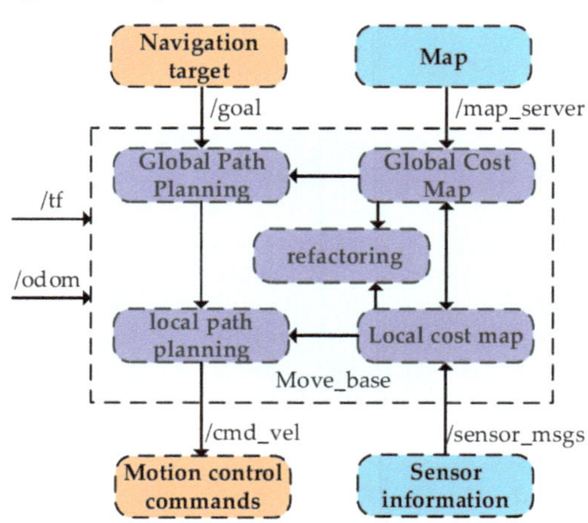

Figure 8. Navigation Framework.

5.2. Simulation I

To validate the simulation experiment, a series of simulation experiments were conducted using the above-mentioned navigation architecture on the TurtuleBot3 Waffle mobile robot platform. These experiments aim to verify whether the robot can still maintain a fast path planning speed and robustness in different obstacle environments. Simulation experiments were crafted based on the Ubuntu 20.04-ROS-Neotic system, with the establishment of a densely populated obstacle environment to simulate point-to-point navigation tasks. In simulation I, the simulation environment was 13.2 × 10.8 m in size, and two rectangular obstacles of 4.8 × 1.2 × 2 m and an L-shaped obstacle composed of two rectangular obstacles were placed, with dimensions of 8.4 × 0.2 × 2 m and 1.8 × 0.3 × 2 m, respectively. There were also a number of irregularly shaped obstacles in the environment. We used a TurtleBot3 Waffle Pi robot with dimensions of 281 × 306 × 141 mm, $v_{\max}$ = 0.26 m/s, and $w_{\max}$ = 1.82 rad/s. The model of the laser ranging sensor used is LDS-01; the scanning range of the laser radar is set to 3.0 m, the expansion radius is set to 1.0 m, and the cost scaling factor is set to 3.0. By comparing the traditional RRT-Connect algorithm with our improved algorithm, and recording the speed curve, motion trajectory, and time consumption of the mobile robot, a deeper understanding of the convergence speed and robustness of our improved algorithm emerges. The simulation experiment platform used was Gazebo, as shown in Figure 9a,b.

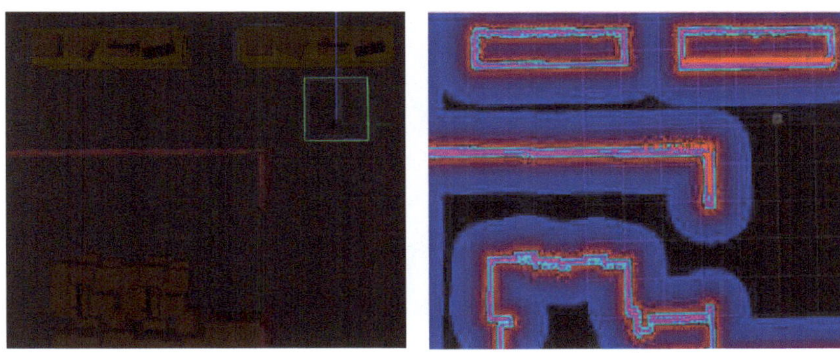

(**a**) (**b**)

Figure 9. Under the L-shaped obstacle environment. (**a**) Gazebo environment in simulation I; (**b**) RVIZ environment in simulation I.

Figure 9 shows that the physical platform is built in Gazebo; the map was constructed by the Gmapping algorithm and visualized using Rviz to simulate a real environment with different shapes of obstacles in the scene, forming a large concave obstacle. The robot had to bypass the obstacles and move to the target location in the lower left corner.

Pre-designed scripts were employed to establish and publish the target coordinates of the mobile robot, replacing the process of manually selecting the target point through the Rviz visualization tool. This ensures the consistency of the target point's position in each experiment, thereby reducing the error within the experimental results. In this experiment, our script set the target point in the lower left corner of the image. A total of 30 simulation experiments were conducted this time, and the statistical indicators used for the comparative evaluation are as follows:

Four different global path planning algorithms are thoroughly evaluated in Simulation 1: Informed-RRT*, RRT-Connect, RRT*, and the modified DBR-RRT. To ensure the fairness and accuracy of the comparison, all algorithms were run in the same test environment, and their running times and standardized speed deviation calculations are meticulously documented in Table 1, and Figure 10 shows a detailed record of each of the four speed profiles.

Table 1. Comparison between Informed-RRT*, RRT-Connect, RRT*, and proposed method in terms of time, average velocity, maximum velocity, and standard deviation of executed velocities.

Planner	Vel. (m/s)		STD Vel.	Time (s)
	Avg	Max		
Informed-RRT*	0.163	0.253	0.051	25.03
RRT-Connect	0.154	0.276	0.043	21.83
RRT*	0.157	0.256	0.039	24.16
DBR-RRT	0.189	0.301	0.037	22.53

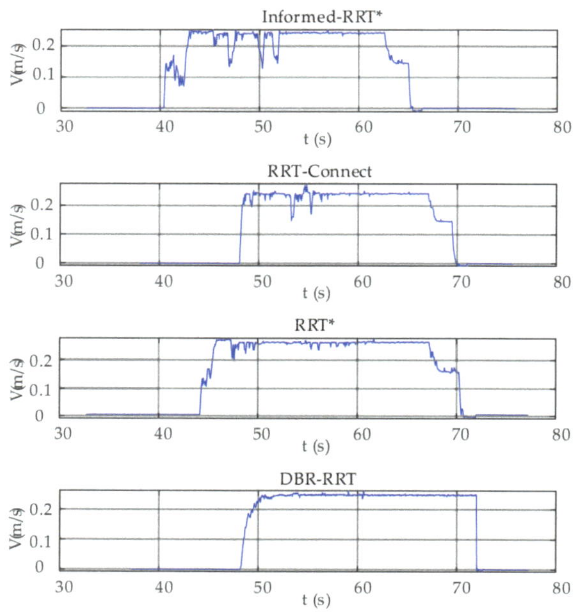

Figure 10. Speed curves generated by the Informed-RRT*, RRT-Connect, RRT*, and DBR-RRT algorithms in simulation I.

In Table 1 and Figure 10, it is shown that DBR-RRT shows an improvement of 15.95% in average speed and a 27.45% improvement in speed standard deviation compared to Informed-RRT*. Compared to RRT-Connect, DBR-RRT shows a 22.73% improvement in average speed and a 13.95% improvement in speed standard deviation. In addition, DBR-RRT results in 20.38% improvement in average speed and 5.13% improvement in standard deviation compared to RRT*. These results indicate that the proposed algorithm exhibits a high consistency and low speed fluctuations over multiple runs. In addition, DBR-RRT achieves a maximum speed of 0.301 m/s, which is slightly higher than the 0.276 m/s of RRT-Connect, indicating that the proposed algorithm maintains stability without sacrificing navigation performance.

The experimental results show that the RRT-Connect algorithm takes the least amount of time among all the reference algorithms, this is due to the bidirectional search strategy adopted by the RRT-Connect algorithm, which can quickly discover effective paths from the starting point to the end point. However, its standard deviation is large, and the speed profile shown in Figure 10 is rugged, and the excessively fast search speed sacrifices some stability. In addition, the proposed DBR-RRT algorithm inherits the bidirectional search strategy of RRT-Connect, and although the average running time is slightly longer than that of the initial RRT-Connect algorithm, the speed has a smaller standard deviation and possesses the largest maximum speed, which indicates that the speed fluctuation between each run is small, and it achieves a balance between increasing the average speed and maintaining less speed fluctuations. Although the RRT algorithm needs to consider more candidate paths in the process of finding the optimal path, and thus consumes the longest time, it is clearly unnecessary to sacrifice time to find the optimal path in a simple L-shaped environment. Compared with DBR-RRT, Informed RRT* can theoretically improve the efficiency of path planning through conducting a more informed search, but in this test environment, its efficiency does not exceed that of the other algorithms, and the large fluctuations in the velocity curves shown in Figure 10. indicate that the algorithm has a large variation in the robot's velocity when planning paths.

5.3. Simulation II

In simulation II, the simulation environment is 12 × 10.8 m in size, with three rectangular obstacles of 4.8 × 1.2 × 2 m and five circular obstacles with a diameter of 0.5 m. The robot and other environmental parameters are the same as in simulation I. Five smaller obstacles were strategically placed to surround the robot in a dense environment. The robot's task was to move to the bottom right corner of the image while avoiding these obstacles. To better simulate obstacles that may suddenly appear in the real environment, these hindrances were intentionally excluded from grid maps constructed using SLAM technology.

Figure 11 shows a path planning process using the DBR-RRT algorithm. This design makes the robot unaware of these obstacles at the beginning of the initial navigation task. However, when the robot approaches these obstacles, its internal LiDAR can effectively detect them and successfully use our algorithm to plan new paths to avoid these obstacles. Real-time robot planning trajectories were acquired and visualized by subscribing to pertinent topics, offering a clear perspective of the robot's current state. In this experiment, the target point was set in the lower right corner. In this case, the robot had to detour around five "suddenly appearing" obstacles. By incorporating such challenges, the robot was predisposed to encounter obstacles early on, positioning itself in potentially hazardous zones surrounded by impediments. This setting allows us to visually observe and compare the robustness and planning efficiency of the two algorithms.

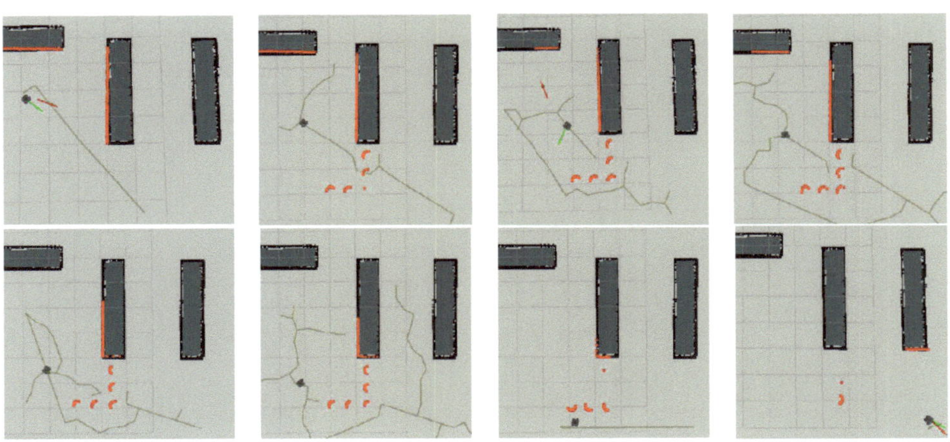

Figure 11. One experiment of DBR-RRT algorithm for path planning.

RBG color mapping was employed to vividly convey the robot's speed information on its travel path. As shown in Figure 12b, the proposed algorithm enabled the robot to have a faster average speed compared to the RT-Connect algorithm, and the speed could be smoothly transitioned during the overall path planning process. In Figure 12a, there are meanders in the robot path and there are multiple instances of sudden velocity mutations. The traditional RRT-Connect algorithm exhibits serious problems when dealing with local optimal solutions, causing the robot to fall into a state of spinning in place. In this case, the robot's speed exhibits an extremely unstable acceleration and deceleration process, even approaching zero. Therefore, the repeated rotation of the robot resulted in a significant increase in the overall path planning time.

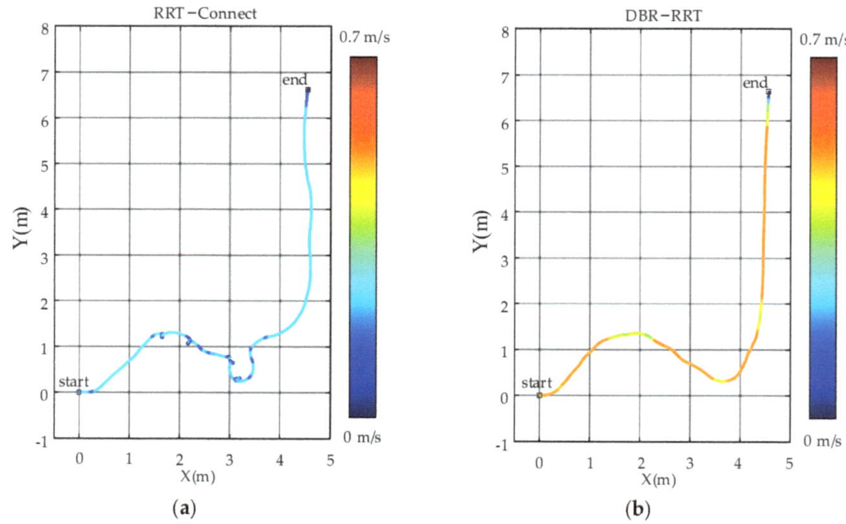

Figure 12. Trajectory velocity curves of two algorithms: (**a**) RRT−Connect and (**b**) DBR−RRT.

Figure 13 shows a comparison of the trajectories generated by the two algorithms. The robot using the RRT-Connect algorithm for path planning had three detours in the trajectory from (1.7, 1.3) to (3.5, 1.0), and there was a large slewing behavior at (3.5, 1.0), which greatly increased the path length. For unknown obstacles in a known environment, traditional algorithms perform inefficient mass collision detection and inefficiently generate paths in real time. In contrast, the algorithm proposed in this paper will first perform a Bridging Test for obstacles, select the points surrounded by obstacles, and generate a virtual node set in the iterative process to quickly guide the robot away from the dangerous area surrounded by obstacles.

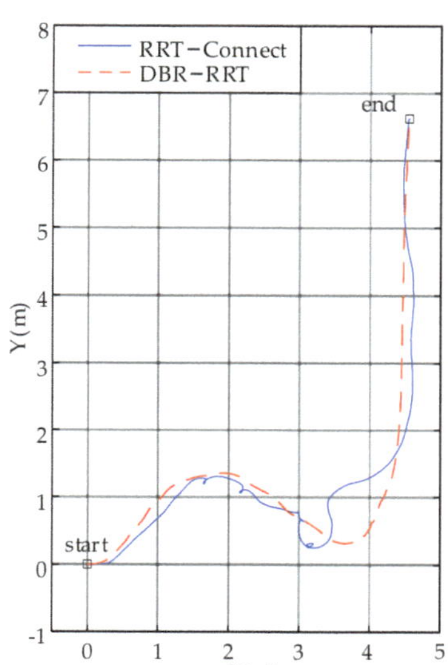

Figure 13. Comparison of trajectories of the two algorithms.

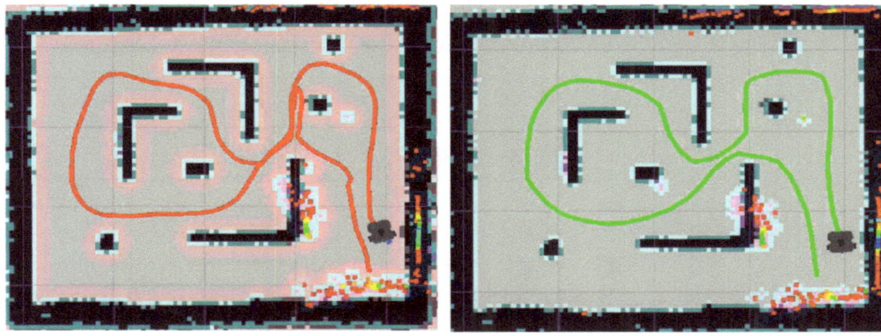

(a) (b)

Figure 18. The trajectory generated by the RRT- Connect and DBR-RRT algorithms are displayed in Rviz: (**a**) RRT-Connect. (**b**) DBR-RRT.

Upon close examination of the red trajectory, it was discerned that a judgment error occurred at the first turn. Rather than making a left at the intersection, the robot chose to proceed straight, causing an unnecessary path extension. In addition, at the point where the robot is about to enter a turn, its trajectory shows a series of obvious twists and turns, revealing that the path planning is not smooth enough. And being too close to the obstacle in the upper right corner may cause potential safety issues.

As can be seen in Figure 19, after mapping the robot's speed onto the trajectory for depth comparison, the traditional RRT-Connect algorithm exhibits significant speed fluctuations in multiple key areas. Particularly in areas requiring turns, a marked difference in robot speed was observed. Moreover, on sections devoid of significant turns, the speed presented unstable fluctuation characteristics. The speed parameters used in this experiment are as follows:

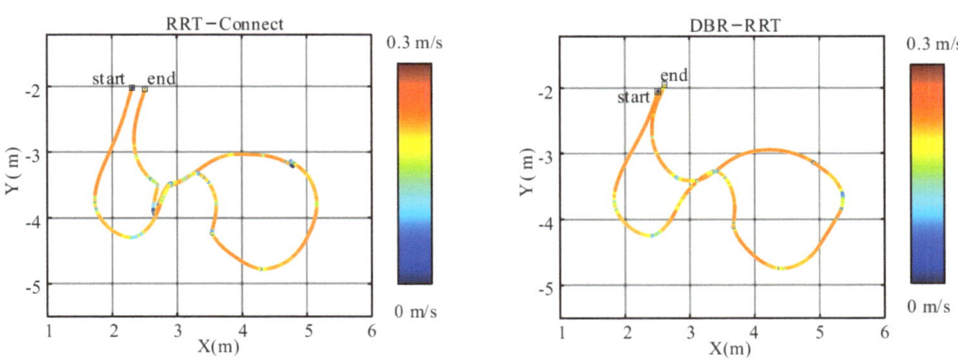

(a) (b)

Figure 19. Speed–trajectory curves of RRT−Connect and DBR−RRT. (**a**) RRT−Connect (**b**) DBR-RRT.

Following the experimental verification, Table 2 shows that the proposed DBR-RRT shows higher efficiency than the traditional RRT-Connect algorithm. Firstly, its average completion time is 9 s faster than the traditional RRT-Connect algorithm. Secondly, the average running speed of the DBR-RRT algorithm is also higher than that of the traditional algorithms, indicating that the robot can maintain higher operating efficiency throughout the entire operation process. Finally, it is worth noting that the standard deviation of the DBR-RRT algorithm is relatively low, which means that the robot's operating speed changes are more stable and less susceptible to sudden factors, thus maintaining a good driving performance in different environments and conditions.

Table 2. Comparison between RRT-Connect and proposed method in terms of time, max velocity, average velocity, and standard deviation of executed velocities.

Planner	Times (s)	Max Vel. (m/s)	Avg Vel. (m/s)	STD Vel.
RRT-Connect	106	0.230	0.100	0.103
DBR-RRT	97	0.230	0.139	0.097

6. Conclusions

In this paper, an advanced path-planning method based on Dynamic Bridging RRT was successfully developed and demonstrated, dedicated to addressing the limitations of existing strategies in complex environments and real-time requirements. The research results indicate that, based on the dynamic sampling strategy of expanding speed, key achievements include faster search speeds in regions with dense obstacles or intricate bottleneck shapes. Meanwhile, in the concluding path-generation phase, a swift path-smoothing strategy was adopted, which further optimized the path, producing a more streamlined and high-quality route. This approach makes path planning for mobile robots in complex environments more efficient and is a useful complement to existing path-planning algorithms. However, due to time and resource constraints, we were not able to independently demonstrate the potential improvement effect that the path-smoothing component may have on the RRT-Connect algorithm. This is an important dimension of comparison that will be added in future work. Future research will include applying the smoothing strategy to RRT-Connect individually under the same conditions and performing a careful comparative performance analysis to fully assess the impact of the smoothing approach. Through such work, we hoped to provide deeper insights and further validate the stability and efficiency of the methods, providing a more complete performance evaluation framework.

Author Contributions: Conceptualization, S.Q. and R.T.; methodology, S.Q.; software, S.Q. and R.T.; validation, S.Q. and B.L.; formal analysis, S.Q.; investigation, S.Q., R.T. and C.T.; resources, S.Q., B.L., and X.H.; data curation, S.Q. and R.T.; writing—original draft preparation, S.Q.; writing—review and editing, S.Q., B.L. and C.T.; visualization, S.Q. and C.T.; supervision, B.L.; project administration, S.Q. and B.L.; funding acquisition, B.L.; All authors have read and agreed to the published version of the manuscript.

Funding: This work is supported by the National Natural Science Foundation of China (grant number 61973234 and 62203326), and in part by the Tianjin Natural Science Foundation (grant number 20JCYBJC00180).

Institutional Review Board Statement: Not applicable.

Informed Consent Statement: Not applicable.

Data Availability Statement: The source code presented in this study is available on request from the corresponding author.

Acknowledgments: The authors would like to thank Tiangong University for technical support and all members of our team for their contribution to the mobile robot experiments. The authors acknowledge the anonymous reviewers for their helpful comments on the manuscript.

Conflicts of Interest: The authors declare no conflicts of interest.

References

1. Gul, F.; Mir, I.; Abualigah, L.; Sumari, P.; Forestiero, A. A Consolidated Review of Path Planning and Optimization Techniques: Technical Perspectives and Future Directions. *Electronics* **2021**, *10*, 2250. [CrossRef]
2. Zhang, H.-y.; Lin, W.-m.; Chen, A.-x. Path Planning for the Mobile Robot: A Review. *Symmetry* **2018**, *10*, 450. [CrossRef]
3. Sánchez-Ibáñez, J.R.; Pérez-del-Pulgar, C.J.; García-Cerezo, A. Path Planning for Autonomous Mobile Robots: A Review. *Sensors* **2021**, *21*, 7898. [CrossRef] [PubMed]
4. Li, X.; Tong, Y. Path Planning of a Mobile Robot Based on the Improved RRT Algorithm. *Appl. Sci.* **2024**, *14*, 25. [CrossRef]
5. Owais, M. Traffic Sensor Location Problem: Three Decades of Research. *Expert Syst. Appl.* **2022**, *208*, 118134. [CrossRef]

6. Silva Ortigoza, R.; Marcelino-Aranda, M.; Silva Ortigoza, G.; Hernandez Guzman, V.M.; Molina-Vilchis, M.A. Wheeled Mobile Robots: A Review. *IEEE Lat. Am. Trans.* **2012**, *10*, 2209–2217. [CrossRef]
7. Gao, X.; Li, J.; Fan, L.; Zhou, Q.; Yin, K.; Wang, J.; Song, C.; Huang, L.; Wang, Z. Review of Wheeled Mobile Robots' Navigation Problems and Application Prospects in Agriculture. *IEEE Access* **2018**, *6*, 49248–49268. [CrossRef]
8. Khaksar, W.; Vivekananthen, S.; Saharia, K.S.M.; Yousefi, M.; Ismail, F.B. A Review on Mobile Robots Motion Path Planning in Unknown Environments. In Proceedings of the IEEE International Symposium on Robotics and Intelligent Sensors (IRIS), Langkawi, Malaysia, 18–20 October 2015; pp. 295–300.
9. Owais, M.; Alshehri, A. Pareto Optimal Path Generation Algorithm in Stochastic Transportation Networks. *IEEE Access* **2020**, *8*, 58970–58981. [CrossRef]
10. Dong, G.; Yang, F.; Tsui, K.-L.; Zou, C. Active Balancing of Lithium-Ion Batteries Using Graph Theory and A-Star Search Algorithm. *IEEE Trans. Ind. Inform.* **2021**, *17*, 2587–2599. [CrossRef]
11. Zhang, H.; Tao, Y.; Zhu, W. Global Path Planning of Unmanned Surface Vehicle Based on Improved A-Star Algorithm. *Sensors* **2023**, *23*, 6647. [CrossRef] [PubMed]
12. Wang, S.J.; Hu, L.K.; Wang, Y.F. Path Planning of Indoor Mobile Robot Based on Improved D* Algorithm. *Comput. Eng. Des.* **2020**, *41*, 1118–1124.
13. Warren, C.W. Multiple Robot Path Coordination Using Artificial Potential Fields. In Proceedings of the IEEE International Conference on Robotics and Automation (ICRA), Cincinnati, OH, USA, 13–18 May 1990; pp. 500–505.
14. Wang, J.; Meng, M.Q.-H. Optimal Path Planning Using Generalized Voronoi Graph and Multiple Potential Functions. *IEEE Trans. Ind. Electron.* **2020**, *67*, 10621–10630. [CrossRef]
15. LaValle, S.M.; Kuffner, J.J., Jr. Randomized Kinodynamic Planning. *Int. J. Robot. Res.* **2001**, *20*, 378–400. [CrossRef]
16. Kuffner, J.J.; LaValle, S.M. RRT-Connect: An Efficient Approach to Single-Query Path Planning. In Proceedings of the 2000 ICRA. Millennium Conference. IEEE International Conference on Robotics and Automation (ICRA), San Francisco, CA, USA, 24–28 April 2000; pp. 995–1001.
17. Kang, J.G.; Lim, D.W.; Choi, Y.S. Improved RRT-Connect Algorithm Based on Triangular Inequality for Robot Path Planning. *Sensors* **2021**, *21*, 333. [CrossRef] [PubMed]
18. Karaman, S.; Walter, M.R.; Perez, A.; Frazzoli, E.; Teller, S. Anytime Motion Planning Using the RRT*. In Proceedings of the IEEE International Conference on Robotics and Automation (ICRA), Shanghai, China, 9–13 May 2011; pp. 1478–1483.
19. Qi, J.; Yang, H.; Sun, H. MOD-RRT*: A Sampling-Based Algorithm for Robot Path Planning in Dynamic Environment. *IEEE Trans. Ind. Electron.* **2021**, *68*, 7244–7251. [CrossRef]
20. Sarker, I.H. Machine Learning: Algorithms, Real-World Applications and Research Directions. *Comput. Sci.* **2021**, *2*, 160. [CrossRef]
21. Janiesch, C.; Zschech, P.; Heinrich, K. Machine Learning and Deep Learning. *Electron. Markets* **2021**, *31*, 685–695. [CrossRef]
22. Wang, J.; Chi, W.; Li, C.; Wang, C.; Meng, M.Q.H. Neural RRT*: Learning-Based Optimal Path Planning. *IEEE Trans. Autom. Sci. Eng.* **2020**, *17*, 1748–1758. [CrossRef]
23. Kun, W.; Ren, B. A Method on Dynamic Path Planning for Robotic Manipulator Autonomous Obstacle Avoidance Based on an Improved RRT Algorithm. *Sensors* **2018**, *18*, 571–586.
24. Karaman, S.; Frazzoli, E. Sampling-Based Algorithms for Optimal Motion Planning. *Int. J. Robot. Res.* **2011**, *30*, 846–894. [CrossRef]
25. Gammell, J.D.; Srinivasa, S.S.; Barfoot, T.D. Informed RRT*: Optimal Sampling-Based Path Planning Focused via Direct Sampling of an Admissible Ellipsoidal Heuristic. In Proceedings of the IEEE/RSJ International Conference on Intelligent Robots and Systems (IROS), Chicago, IL, USA, 14–18 September 2014; pp. 2997–3004.
26. Gammell, J.D.; Srinivasa, S.S.; Barfoot, T.D. Batch Informed Trees (BIT*): Sampling-Based Optimal Planning via the Heuristically Guided Search of Implicit Random Geometric Graphs. In Proceedings of the IEEE International Conference on Robotics and Automation (ICRA), Seattle, WA, USA, 26–30 May 2015; pp. 3067–3074.
27. Yin, J.; Hu, Z.; Mourelatos, Z.P.; Gorsich, D.; Singh, A.; Tau, S. Efficient Reliability-Based Path Planning of Off-Road Autonomous Ground Vehicles Through the Coupling of Surrogate Modeling and RRT*. *IEEE Trans. Intell. Transp. Syst.* **2023**, *24*, 15035–15050. [CrossRef]
28. Islam, F.; Nasir, J.; Malik, U.; Ayaz, Y.; Hasan, O. RRT*-Smart: Rapid convergence implementation of RRT* towards optimal solution. In Proceedings of the IEEE International Conference on Mechatronics and Automation (ICMA), Chengdu, China, 5–8 August 2012; pp. 1651–1656.
29. Dai, J.; Zhang, Y.; Deng, H. Novel Potential Guided Bidirectional RRT* With Direct Connection Strategy for Path Planning of Redundant Robot Manipulators in Joint Space. *IEEE Trans. Ind. Electron.* **2024**, *71*, 2737–2747. [CrossRef]
30. Ji, H.; Xie, H.; Wang, C.; Yang, H. E-RRT*: Path Planning for Hyper-Redundant Manipulators. *IEEE Robot. Autom. Lett.* **2023**, *8*, 8128–8135. [CrossRef]
31. Zhang, W.; Shan, L.; Chang, L.; Dai, Y. SVF-RRT*: A Stream-Based VF-RRT* for USVs Path Planning Considering Ocean Currents. *IEEE Robot. Autom. Lett.* **2023**, *8*, 2413–2420. [CrossRef]
32. Janson, L.; Schmerling, E.; Clark, A. Fast marching tree: A fast marching sampling-based method for optimal motion planning in many dimensions. *Int. J. Robot. Res.* **2015**, *34*, 883–921. [CrossRef] [PubMed]

33. Wu, Z.; Chen, Y.; Liang, J. ST-FMT*: A fast optimal global motion planning for mobile robot. *IEEE Trans. Ind. Electron.* **2021**, *69*, 3854–3864. [CrossRef]
34. Lee, J.; Kwon, O.; Zhang, L.; Yoon, S. SR-RRT: Selective Retraction-based RRT Planner. In Proceedings of the IEEE International Conference on Robotics and Automation (ICRA), Saint Paul, MN, USA, 14–18 May 2012; pp. 2543–2550.
35. Chi, W.; Ding, Z.; Wang, J.; Chen, G.; Sun, L. A Generalized Voronoi Diagram-Based Efficient Heuristic Path Planning Method for RRTs in Mobile Robots. *IEEE Trans. Ind. Electron.* **2022**, *69*, 4926–4937. [CrossRef]

applied sciences

MDPI

Article

Kinematic Tripod (K3P): A New Kinematic Algorithm for Gait Pattern Generation

Daniel Soto-Guerrero [1], José Gabriel Ramírez-Torres [2,*] and Eduardo Rodriguez-Tello [2,*]

1 XLIM Institute, UMR CNRS 7252, University of Limoges, 87060 Limoges, France; daniel.soto-guerrero@unilim.fr
2 Unidad Tamaulipas, Cinvestav, Km. 5.5 Carretera Victoria—Soto La Marina, Victoria 87130, Mexico
* Correspondence: grtorres@cinvestav.mx (J.G.R.-T.); ertello@cinvestav.mx (E.R.-T.); Tel.: +52-834-107-0220 (E.R.-T.)

Abstract: Insects are good examples of ground locomotion because they can adapt their gait pattern to propel them in any direction, over uneven terrain, in a stable manner. Nevertheless, replicating such locomotion skills to a legged robot is not a straightforward task. Different approaches have been proposed to synthesize the gait patterns for these robots; each approach exhibits different restrictions, advantages, and priorities. For the purpose of this document, we have classified gait pattern generators for multi-legged robots into three categories: precomputed, heuristic, and bio-inspired approaches. Precomputed approaches rely on a set of precalculated motion patterns obtained from geometric and/or kinematic models that are performed repeatedly whenever necessary and that cannot be modified on-the-fly to adapt to the terrain changes. On the other hand, heuristic and bio-inspired approaches offer on-line adaptability, but parameter-tuning and heading control can be difficult. In this document, we present the K3P algorithm, a real-time kinematic gait pattern generator conceived to command a legged robot. In contrast to other approaches, K3P enables the robot to adapt its gait to follow an arbitrary trajectory, at an arbitrary speed, over uneven terrain. No precomputed motions for the legs are required; instead, K3P modifies the motion of all mechanical joints to propel the body of the robot in the desired direction, maintaining a tripod stability at all times. In this paper, all the specific details of the aforementioned algorithm are presented, as well as different simulation results that validate its characteristics.

Keywords: hexapod robot; kinematics; gait pattern generation

Citation: Soto-Guerrero, D.; Ramírez-Torres, J.G.; Rodriguez-Tello, E. Kinematic Tripod (K3P): A New Kinematic Algorithm for Gait Pattern Generation. *Appl. Sci.* **2024**, *14*, 2564. https://doi.org/10.3390/app14062564

Academic Editor: Jonghoek Kim

Received: 20 February 2024
Revised: 12 March 2024
Accepted: 15 March 2024
Published: 19 March 2024

1. Introduction

Locomotion is the act of moving from place to place. To move forward, legged animals, as do insects, use their limbs in a gait pattern. When considering each leg individually, a cycle of the gait pattern is divided into two phases: swing and support. In the swing phase, the limb rises from the ground and moves in the desired direction of movement; the support phase begins when the limb lands and supports a fraction of the total weight of the animal. During the whole cycle, static and/or dynamic equilibrium conditions must be kept for the gait pattern to be stable.

Static stability is achieved when the projection on the ground of the robot's center of mass (CoM) falls inside the support polygon, defined as the convex hull of all feet in support phase [1,2]. Dynamic stability occurs when the zero moment point (ZMP)—the point with respect to which reaction forces at the contacts between the feet and the ground do not produce any moment in the horizontal direction—is maintained inside the support polygon throughout the gait. Gait patterns whose stability is determined by dynamic conditions allow for faster displacements of the robot because the CoM projection can be located outside of the support polygon for short periods of time [3,4]. Therefore, in order to guarantee stable locomotion, gait synthesizing algorithms must coordinate all limbs of the robot to make it move in the desired direction, while satisfying the static or dynamic equilibrium condition. In general, the stable locomotion is achieved by using precomputed gait patterns for given trajectory and terrain conditions, or by using parametric-adjusting of the gait by heuristic or bio-inspired approaches to cope with trajectory and terrain changes.

The algorithm we propose, called K3P for kinematic tripod, is a real-time kinematic gait generator capable of on-the-fly computing of the limb motions of a legged robot in order to move along an unknown arbitrary trajectory on uneven terrain, while maintaining static equilibrium, maximizing horizontal displacement of the feet contact points, and avoiding collisions between two consecutive limbs. In this paper, we describe some related approaches, analyze the performance of K3P in a virtual test scenario, and use torque estimations to measure the viability of the synthesized gait pattern.

2. Related Work

Legged robots perform different gait patterns depending on the desired horizontal speed and stability criteria [5–8]. Figure 1a shows a simplified view of the robot limb and the two phases of the gait. Complementarily, Figure 1b shows a complete gait cycle for what is known as static fast gait; as it can be observed, a minimum of three limbs (a tripod) support the robot during walking. The ratio of duration of the support phase to the total cycle duration defines the duty factor β of a gait cycle [9], Figure 1b displays gait cycle for a duty factor $\beta = 0.5$. Medium speed gaits (or ripple gait) allow for two legs on opposite sides of the robot to be in the swing phase. In slow gait (or tetrapod gait), only one limb at a time performs the swing phase while the rest support the robot; therefore, it is the most stable of the three gaits [10]. So, the problem of synthesizing a gait pattern consists of defining the best sequence of movements for all the robot limbs, where each limb features from one [11,12] up to four degrees of freedom [10,13].

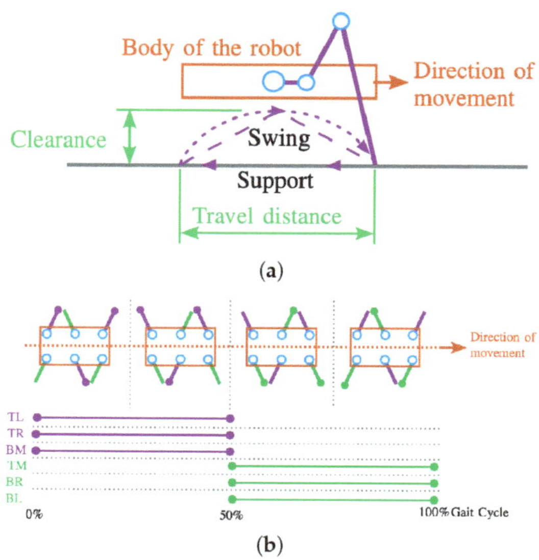

Figure 1. The gait cycle of an hexapod robot. (**a**) Side view of a leg performing a gait cycle: During the swing phase, the end side of the limb describes a curve or a triangular trajectory. Every DoF is marked with a circle; thus, the limb displayed above has 3 DoF. (**b**) Fast gait diagram: The beginning and end of the lines, marked with a circle, correspond to the landing and lifting of every limb, respectively. Every limb is marked as a combination of (T)op, (B)ottom, (M)iddle, (L)eft, and (R)ight, according to its position. When the surface is flat and every trajectory for the legs is precomputed, this gait sequence is sufficient [10].

For the purpose of this work, we have classified the related works according to the use of precomputed, heuristic, or bio-inspired approaches to generate the gait.

2.1. Precomputed Approaches

In these approaches, closed mathematical models (geometric and/or kinematic) are used to compute, in advance and for each limb, a sequential set of joint configurations that, when performed, will propel the robot body forward during the support phase, while during the swing phase, the limb tip describes a given parametric trajectory to the next support point, using most commonly sine, Bezier, and triangular trajectories [6,14–17]. During locomotion, these trajectories are repeated in a logical sequence; in some cases, the trajectory during the swing phase can be adapted to increase or decrease the horizontal travel distance and/or clearance of the gait (see Figure 1b). Furthermore, depending on the chosen mathematical model, actions such as jumping, main body orientation and clearance control may be considered [7]. Another option to a mathematical model are Probabilistic Graphical Models (PGMs) that can be trained and sampled to infer a walking gait [18].

A different paradigm consists of a footstep planning before the actual locomotion. For example, the robot ATHLETE, designed at the Jet Propulsion Laboratory (Pasadena, CA, USA), computes in advance the most useful support points across the terrain before performing any movement [19]. This planning implies that the robot must be able to accurately build a model of its surroundings, using exteroceptive sensors such as rangefinders, increasing the complexity and the total computational cost required for the robot's locomotion. In contrast, other approaches embrace the uncertainty of the terrain, not focusing on planning the position of all support points beforehand, and instead propelling low-dexterity hexapods with a fixed gait and focusing their efforts on correct state estimation under high-uncertainty circumstances [12].

These approaches allow us to obtain a precise estimation of energy consumption during the locomotion through, for example, a two-layer hierarchical cooperative control scheme [20]. A top-level controller determines the forces and torques that every limb should exert on the body of the robot, so that the robot can follow a given trajectory, while low-level controllers independently command every leg of the robot to exert such forces. Because energy consumption may vary depending on the type of surface the robot is walking on, the travel speed can be adapted by performing slow, medium, or fast gaits, depending on the energy consumed by the actuators driving every limb [15]. Another approach to hierarchical control is to use a top-level exteroceptive methodology to observe and evaluate the terrain and command a low-level routine to switch among precomputed gait patterns, with the objective of maximizing the stability of the robot when traversing uneven terrain [21].

2.2. Heuristic Methods

The major drawback of purely mathematical models is the complexity of the model itself; therefore, roboticists turned to heuristics to generate a gait pattern while still using the precomputed trajectories from an external optimization process under energetic criteria or faulty conditions.

In order to use a simpler model during the gait generation, heuristic approaches enclose a biological notion of locomotion learned from analyzing the movement of animals, expressed as simple rules for the movement of legged robots. Heuristics such as genetic algorithms (GA) can help to generate the walk pattern for a virtual legged robot [22,23], where the main criterion for fitness calculation is the stability of the robot while walking in a straight line over a flat surface, within preset borders and stability [2]. The energy efficiency, traveled distance, and deviation of trajectory from the straight line are used as feedback information for the GA [24]. The final result is a set of static gait patterns for the robot, learned without an explicit mathematical model, from which the robot can choose during locomotion.

Also, Finite State Automata (FSA) can be used to generate walking patterns [25] as well as flow charts [6], as these models can encode a sequence of movements for every limb used during locomotion maneuvers. By definition, these approaches are also static, and adaptability to faulty conditions comes at the cost of increasing the complexity of the FSA or the flow chart.

These approaches can also emulate reflexive motions using a reactive control scheme, allowing a legged robot to react to the irregularities of the terrain it is walking on. These reflexes are triggered when a limb or foot collides unexpectedly with the terrain or an obstacle. The collision detection can be performed with a set of touch sensors [14] or by measuring the electric current, forces, and torques consumed by the actuators on every joint [10].

Heuristics allow for *blind-walking*, i.e., the robot is able to walk through irregular terrain based only on proprioceptive information [26]. Neural networks can be implemented in hardware with miniaturization and energy efficiency as the main objective [27]. The Virtual Model Control [27] is one of the most relevant heuristics in order to obtain a model from experimental data, simulating the same dynamic behavior of complex mechanical systems using much simpler components such as springs, dampers, masses, etc. The resulting model is less complex but accurate enough to compute the forces acting on the robot body [28,29].

2.3. Bio-Inspired Methods

Most bio-inspired approaches for synthetic gait generation are based on central pattern generators (CPGs). CPGs are oscillators that can generate rhythmic patterns from non-rhythmic signals or no inputs at all. When applied to legged locomotion [23,30–32], the rhythmic output of CPGs corresponds to the gait pattern, in response to inputs as a gait velocity command and the proprioceptive sensor information from the limbs of the robot; this means that sensory information plays an important role in CPG-based gait generation [33]. The biggest difficulty with regard to CPGs is determining the correct range of values for the input as well as tuning the internal oscillator parameters, so that the output corresponds to the desired movement of the limbs and the transition between different gaits is smoothly performed [34]. Tuning such parameters is usually performed by trial and error, and some authors have even turned to GAs to tune CPGs [35]. Furthermore, in a decentralized scenario, where there are as much CPGs as limbs, an additional higher-level control is required [36]. In recent works, CPGs can generate a dynamic walking pattern, based on the turning radius of the desired trajectory and switch from tripod, ripple, and tetrapod gaits [37].

2.4. Main Features of K3P Algorithm

The K3P algorithm, the approach we propose for generating the walking gait for legged robots, is based on a centralized kinematic planner. This algorithm performs a steadily fast gait cycle while also being able to drive the robot at an arbitrary speed by dynamically adjusting the duty factor β. Also, K3P blurs the differentiation between slow, medium, and fast walks under static stability conditions. Basically, K3P moves the robot's limbs in tripod configurations, keeping a support tripod while swinging the second tripod to a new location, according to the actual desired speed and trajectory of the robot's center of mass.

The main differences between K3P and other approaches are as follows:

1. K3P is self-contained and makes it possible for a legged robot to walk at an arbitrary speed along an arbitrary trajectory over uneven terrain, within the physical limitations of the robot.
2. K3P does not require any precomputed limb trajectories for straight or turning maneuvers; instead, it computes the limb trajectories in real-time, to make the robot move straight ahead or in a sharp or wide curve, according to the terrain level.
3. Since K3P moves the robot's center of mass from a supporting tripod to the next one, K3P guarantees static equilibrium during the march while controlling the clearance of the robot to ground level.
4. The kinematic planner behind K3P also guarantees slip-free locomotion and collision avoidance among limbs.

Despite the mathematical complexity behind K3P, its use is fairly simple: it requires as input the physical dimensions of the robot and some PD controller gains. We must highlight that none of the limb displacements is computed beforehand, as in most of the mathematical or heuristic approaches; instead, K3P can change robot displacement at any moment. K3P takes into account the actual position and velocity of the center of mass of the robot, as well as its target trajectory, in order to compute the best position for landing the swinging legs to place the next supporting tripod. Furthermore, K3P verifies the stability of the gait by measuring how close the projection of the center of mass is to the support polygon's edges.

In comparison to precomputed approaches, K3P exhibits flexibility by allowing the dynamic adjustment of the gait pattern during execution in accordance with the specified trajectory. For example, it enables modifications to the body clearance over varied terrain. Moreover, while precomputed methods enforce minimal fixed curvature for the robot's trajectories, K3P overcomes this constraint by adapting the gait pattern to ensure that the instantaneous turning radius of the robot matches the specified trajectory. Unlike many heuristic methods that demonstrate comparable performance, K3P distinguishes itself through its ease of tuning, facilitated by the concrete nature of all its parameters.

3. Robot Description and Nomenclature

In this section, we will describe the radial hexapod robot on which the K3P algorithm was tested, as well as the nomenclature used (see Table 1). Figure 2a shows the structure of the robot, the main body is circular and the six limbs are evenly distributed along its perimeter; with the center of mass (CoM) at the origin of the B reference frame [14,19,29]. Each limb has three DoF as shown in Figure 2b. By convention, all Z axes are coaxial with the rotation's axis of each joint. With respect to B, the mounting point for the i-th limb is denoted by the M_i reference frame. The first joint provides the protraction and retraction movements, marked by the variable θ_s in the S reference frame. The L reference frame is located in the second DoF, denoted as θ_L; it provides the depression and elevation movements. The third DoF is marked as θ_k in the K reference frame, providing flexion and extension movements. The length of every link are l_c, l_f, and l_t for the coxa, femur, and tibia, respectively. The supporting point SP is at the point where the limb makes contact with the ground, supporting the body of the robot.

Table 1. Nomenclature.

Parameter	Description
	Physical parameters of a limb
$\theta_s, \theta_l, \theta_k$	The three DoF of a limb.
l_c, l_f, l_t	Length of the coxa, femur, and tibia, respectively.
	Physical parameters of the gait
l_h	Body clearance.
l_{max}	Maximum length of the gait.
l_g	Limb clearance.
	Gait state variables
η	Swing tripod is landing, moving, or taking off $\in \{-1, 0, 1\}$.
κ	Parity $\in \{-1, 1\}$.
$\mathbf{P}$	Support positions for the even and odd limbs.
ρ	Instantaneous turning radius.
$\mathbf{Z}$	Touch sensor located at SP.

Table 1. *Cont.*

Parameter	Description
	K3P input variables
$\mathbf{v}_c$	Velocity command vector.
${}^{B}\mathbf{X}_k$	The k-th current position of the legged robot.
$\mathbf{K}_p$	The proportional gains for the PD controller.
$\mathbf{K}_d$	The derivative gains for the PD controller.
	Thresholds and work ranges
l_ρ	Turning radius threshold.
φ_L	Angle threshold for two consecutive limbs.
$\boldsymbol{\theta}_l$	Range of movement for the swing joint.
$\boldsymbol{\theta}_k$	Range of movement for the knee joint.

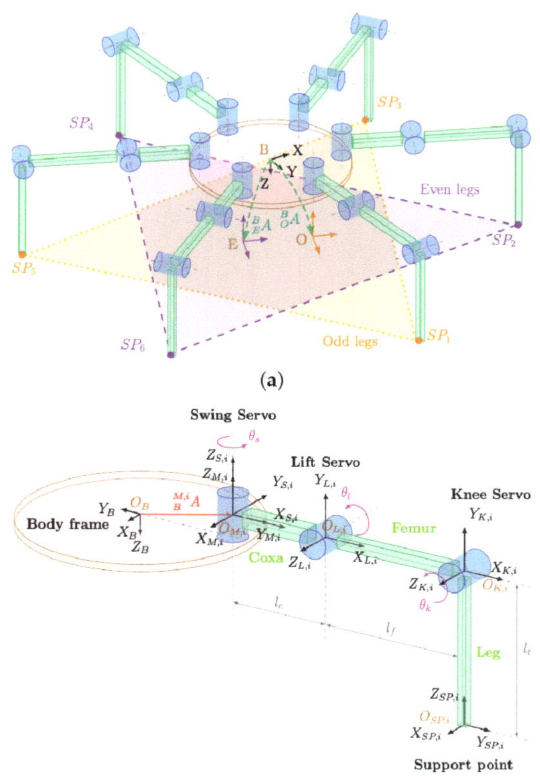

Figure 2. The mechanical description of the radial hexapod tested with K3P. (**a**) The radial hexapod as tested with the K3P algorithm. (**b**) A detailed mechanical description for one of the legs of the radial hexapod.

The six limbs are divided into two subsets, three non-contiguous limbs form the odd legs subset, while the rest are grouped in the even legs subset (see Figure 2a). Each subset defines a tripod support structure with its own reference frame, E and O, for the even and odd subsets, respectively. The subset supporting the robot defines the parity of the gait κ; if $\kappa = 1$, even limbs support the robot.

To deal with spatial relationships between two reference frames, say frame b with respect to frame a, we used rigid body transformations in homogeneous coordinates denoted as

$$ {}^{b}_{a}\mathbf{A} = \begin{bmatrix} {}^{b}_{a}\mathbf{R} & {}^{b}\mathbf{t}_{a} \\ \mathbf{0} & 1 \end{bmatrix} $$

where ${}^{b}_{a}\mathbf{R} \in SO(3)$ denotes the rotation matrix of frame b with respect to a and ${}^{b}\mathbf{t}_{a}$ is the position vector of origin of b, also with respect to a.

From the mounting point of the i-th limb, marked as M_i in Figure 2b, the links and joints of each limb form a kinematic chain. The corresponding Denavit–Hartenberg (DH) parameters [38] are summarized in Table 2. Since all limbs are equal, these parameters are valid for all limbs of the radial hexapod. These parameters allow us to compute the rigid body transformation between link n with respect to $n-1$, ${}_{n-1}^{\;\;\,n}\mathbf{A}$. Finally, the rigid body transformation from frame N with respect to the base link ($n = 0$) is given by

$$ {}^{N}_{0}\mathbf{A} = \prod_{n=1}^{N} {}_{n-1}^{\;\;\,n}\mathbf{A} $$

Therefore, with these relations, we can compute from the six sets of joint parameters $\theta_s, \theta_L, \theta_k$, the position of all six leg tips $\mathbf{SP_i}$ with respect to B. Also, the inverse kinematic model can be solved for each leg, so the joint parameters can be obtained from the position of each point $\mathbf{SP_i}$. Moreover, thanks to these physical dimensions and parameters, we can predict the maximum extension of the gait l_{max}, given a desired body clearance l_h and limb l_g clearance over the floor.

Table 2. Denavit–Hartenberg parameters for every leg.

Kinematic Chain					
n	D-H Parameter				Description
	d	θ	a	α	
0	0	0	0	0	The mounting point for the leg M_i.
1	0	θ_s	0	0	The swing DoF.
2	d_2	0	l_c	$\frac{\pi}{2}$	The length of the coxa
3	d_4	$\theta_l - \frac{\pi}{2}$	l_f	0	The lift DoF
4	d_5	$-\theta_k$	l_t	0	The knee DoF, tip of leg (O_{SP})

4. The K3P Algorithm

In this section, the K3P algorithm will be described (see Algorithm 1), beginning with the description, the objectives of the algorithm, and finally, a global overview. Subsections A to H will discuss the details of every phase of the algorithm.

The main objective of the K3P algorithm is to drive the CoM along an arbitrary trajectory at an arbitrary speed $\mathbf{v}_c$, while the two subsets of limbs perform a cyclic gait pattern, swing–support, to follow the movement of the robot's CoM and to maintain the static stability criteria. The real-time operation of the K3P algorithm is obtained by an update rate Δ_t at which the position of every limb is computed and updated. At the initial state, the two subsets of legs are landed and supporting the body of the robot. The movement begins with the odd subset starting the swing phase, while the even subset remains in the support phase of the gait cycle and propels the CoM B along the desired trajectory by updating the even tripod configuration according to the given velocity $\mathbf{v}_c$. Concurrently, K3P also drives the tripod on the swing phase forward ahead in the direction encoded by $\mathbf{v}_c$.

Algorithm 1 K3P

Require: $\dot{\mathbf{r}}_d, {}^{B}\mathbf{X}_k, l_h, l_{max}, l_g, \mathbf{K}_p, \mathbf{K}_d, l_\rho, l_\varphi, \boldsymbol{\theta}_l, \boldsymbol{\theta}_k$

if $\kappa = 1$ **then** $z_k \leftarrow \text{mean}([{}^{E}_{B}\mathbf{A}\,\mathbf{P}_e]_z)$

$\quad {}^{m}_{B}\mathbf{A}_k \leftarrow {}^{O}_{B}\mathbf{A}, \quad {}^{f}_{B}\mathbf{A}_k \leftarrow {}^{E}_{B}\mathbf{A}$

else $z_k \leftarrow \text{mean}([{}^{O}_{B}\mathbf{A}\,\mathbf{P}_o]_z)$

$\quad {}^{m}_{B}\mathbf{A}_k \leftarrow {}^{E}_{B}\mathbf{A}, \quad {}^{f}_{B}\mathbf{A}_k \leftarrow {}^{O}_{B}\mathbf{A}$

if $\eta = 0$ **then**

$$\quad {}^{D}\mathbf{X}_k \leftarrow {}^{B}\mathbf{X}_k + \left[\dot{\mathbf{r}}\Delta_t,\ l_h - z_k,\ 0,\ 0,\ \dot{\psi}\Delta_t,\ v_x,\ v_y,\ \frac{l_h - z_k}{\Delta_t},\ 0,\ 0,\ \dot{\psi}\right]^T$$

else ${}^{D}\mathbf{X}_k \leftarrow {}^{B}\mathbf{X}_k$

$\mathbf{e}_k = {}^{D}\mathbf{X}_{W,k} - {}^{B}\mathbf{X}_{W,k}$

$\mathbf{u} \leftarrow \mathbf{K}_p\mathbf{e}_k + \mathbf{K}_d\frac{\partial \mathbf{e}_k}{\partial t}, \quad {}^{B}\mathbf{X}_{k+1} = {}^{B}\mathbf{X}_k + \mathbf{u}\Delta_t$

${}^{B}_{W}\mathbf{A}_k \leftarrow \Upsilon({}^{B}\mathbf{X}_k), \quad {}^{B}_{W}\mathbf{A}_{k+1} \leftarrow \Upsilon({}^{B}\mathbf{X}_{k+1}), \quad {}^{m}\mathbf{X}_{B,k} \leftarrow \Upsilon^{-1}({}^{m}_{B}\mathbf{A}_k)$

${}^{f}_{B}\mathbf{A}_{k+1} \leftarrow {}^{B}_{W}\mathbf{A}^{-1}_{k+1}\{{}^{B}_{W}\mathbf{A}\,{}^{f}_{B}\mathbf{A}\}_k$

$\rho \leftarrow \frac{\|\dot{\mathbf{r}}\|}{[\dot{\mathbf{q}}]_\psi}, \quad \theta_r \leftarrow \frac{l_{max}}{2\rho}$

$${}^{L}\mathbf{X}_{B,k} \leftarrow \begin{cases} \left[\frac{1}{2}l_{max},\ 0,\ l_g,\ \mathbf{0}_{1\times 9}\right]^T \text{ if } |\rho| \geq l_\rho \\ \left[\frac{1}{2}l_{max}\cos\theta_r,\ \text{sign}(\theta_r)\frac{1}{2}l_{max}\sin\theta_r,\ l_g,\ \mathbf{0}_{1\times 9}\right]^T \end{cases}$$

$\mathbf{e}_{L,k} \leftarrow {}^{L}\mathbf{X}_{B,k} - {}^{m}\mathbf{X}_{B,k}$

$\mathbf{u}_L \leftarrow \mathbf{K}_p\mathbf{e}_{L,k} + \mathbf{K}_d\frac{\partial \mathbf{e}_{L,k}}{\partial t}$

${}^{m}\mathbf{X}_{B,k+1} \leftarrow {}^{m}\mathbf{X}_{B,k} + \mathbf{u}_L\Delta_t$

$\Delta_z \leftarrow [{}^{m}\mathbf{X}_{B,k+1}]_z - [{}^{m}\mathbf{X}_{B,k}]_z$

if $\eta = -1 \wedge \text{any}(\mathbf{Z}_i)$ **then** $[\mathbf{p}_i]_z \leftarrow [\mathbf{p}_i]_z + \Delta_z$

else if $\eta = -1 \wedge \text{all}(\mathbf{Z}_i)$ **then** $\kappa \leftarrow (-1)\kappa, \quad \eta \leftarrow 1$

else if $\eta = 1$ **then** $\text{highest}([\mathbf{p}_m]_z) \leftarrow \text{highest}([\mathbf{p}_m]_z) - \Delta_z$

else if $\eta = 1 \wedge \text{none}(\mathbf{Z}_i)$ **then** $\eta \leftarrow 0$

if $\kappa = 1$ **then** ${}^{O}_{B}\mathbf{A}_{k+1} \leftarrow \Upsilon({}^{m}\mathbf{X}_{B,k+1}), \quad {}^{E}_{B}\mathbf{A}_{k+1} \leftarrow {}^{f}_{B}\mathbf{A}$

else ${}^{E}_{B}\mathbf{A}_{k+1} \leftarrow \Upsilon({}^{m}\mathbf{X}_{B,k+1}), \quad {}^{O}_{B}\mathbf{A}_{k+1} \leftarrow {}^{f}_{B}\mathbf{A},$

$\Theta_s, \Theta_l, \Theta_k \leftarrow$ **Inverse kinematics** $\left({}^{O}_{B}\mathbf{A}_{k+1}, {}^{E}_{B}\mathbf{A}_{k+1}, \mathbf{P}\right)$

${}^{m}_{f}\mathbf{A} \leftarrow {}^{m}_{B}\mathbf{A}\,{}^{f}_{B}\mathbf{A}^{-1}, \quad l_{gait} \leftarrow \|{}^{m}\mathbf{r}_f\|$

$\mathbf{r}_i = {}^{O}_{B}\mathbf{A}\,\mathbf{p}_{\{1,3,5\}}, \quad \mathbf{r}_i = {}^{E}_{B}\mathbf{A}\,\mathbf{p}_{\{2,4,6\}} \quad \varphi_i = \arccos(\mathbf{r}_i \bullet \mathbf{r}_{i-1})\ \forall i \in [1,6]$

if $l_{gait} \geq l_{max} \vee \text{any}(\varphi_i \leq l_\varphi) \vee \text{any}(\Theta_L \notin \boldsymbol{\theta}_L) \vee \text{any}(\Theta_K \notin \boldsymbol{\theta}_K)$ **then**

$\quad \eta \leftarrow -1$

Execute $(\Theta_s, \Theta_l, \Theta_k)$

The act of landing the swing tripod to receive the robot's weight and to allow the other tripod to take off is called a phase shift of the gait cycle. These phase shifts must be performed in such a way that the projection of the CoM on the ground remains at the interior of the supporting polygon at all times, and the total number of phase shifts along the trajectory is kept at minimum, so the step length is maximum. To decide a phase shift of the gait cycle, the algorithm K3P makes use of three different criteria, named $\text{K3P}_1, \text{K3P}_2$, and K3P_3, in order to minimize the number of phase shifts during walking while guaranteeing a static stable gait. These criteria are described in subsection F.

The KP3 algorithm uses different frames to describe the configuration of the hexapod robot and to compute the gait: the main reference frame B attached to the robot's body and CoM, a frame E to describe the tripod formed by the even subset of limbs, and a frame O to describe the odd subset. While in the support phase, any of the two subsets defines a tripod-supporting structure, standing on the ground, so the support polygon corresponds to a triangle at ground level. All support points SP_i are described with respect to either E or O reference frames, depending on whether they belong to the even or odd subset (see Figure 2a). As will be shown later, the actual configuration of the robot can be computed from these frames at any time.

In the following subsections, we will introduce some key aspects that give shape to K3P, starting by describing how the CoM is propelled forward along the desired trajectory, given an arbitrary speed command $\mathbf{v}_c$. We will also cover how K3P seamlessly distinguishes the forward and turn maneuvers and how K3P performs the phase shift of the gait cycle using the three aforementioned criteria.

4.1. Gait Cycle

From the odd and even subsets of limbs, the supporting tripod of the gait cycle is determined by the parity κ of the gait cycle. If $\kappa = 1$, the even subset is fixed to the ground, supporting the body of the robot, while the odd subset is moving over ground level, further ahead of the CoM, in a swing motion; the opposite occurs for $\kappa \neq 1$. For each case, the rigid body transformations at time instant k of the swing ${}^{m}_{B}\mathbf{A}$ and supporting legs ${}^{f}_{B}\mathbf{A}$ can be obtained using the position of frames E and O, both with respect to B, using the inverse kinematic models of the limbs (lines 1 and 3 of Algorithm 1).

4.2. Propelling Forward the Body of the Robot

The position and orientation of the robot's body frame B, as well as their corresponding derivatives, with respect to the world reference frame W, describe the desired trajectory of the robot. This target trajectory can be expressed by the state vector:

$$^{B}\mathbf{X}_{W,k} = [\mathbf{r}, \mathbf{q}, \dot{\mathbf{r}}, \dot{\mathbf{q}}]^T$$

where $\mathbf{r} = (x, y, z)^T$ corresponds to the three-dimensional coordinates of B, while vector $\mathbf{q} = (\theta, \phi, \psi)^T$ contains the three Euler angles that define the orientation of the robot's body, both with respect to the world frame W.

K3P works at a fixed rate, so after every time step Δ_t, the desired position $^{D}\mathbf{X}_{W,k}$ for the CoM of the robot is determined by the commanded velocity vector $\dot{\mathbf{r}}_d = (\dot{\mathbf{r}}, \dot{\mathbf{q}})^T$ and the current position of the robot (see line 7 of Algorithm 1). This is only carried out if the robot is moving $\eta = 0$, otherwise $^{D}\mathbf{X}_{W,k}$ remains the same. The $[\dot{\mathbf{r}}]_z$ component is updated to manage any elevation changes of the terrain. To determine $[\dot{\mathbf{r}}]_z$ with respect to B, the difference between the average height of the leg tips SP of the supporting tripod (see lines 1 and 3 of Algorithm 1), and the commanded clearance height is divided by Δ_t.

The spatial difference between $^{B}\mathbf{X}_{W,k}$ and $^{D}\mathbf{X}_{W,k}$ defines an error metric (line 10 of Algorithm 1), which is fed to a PD controller; the result is a control command $\mathbf{u}$ to propel B in the direction encoded in $\mathbf{v}_c$ (line 11 of Algorithm 1). Diagonal matrices $\mathbf{K}_p$ and $\mathbf{K}_d$ contain the proportional and derivative gains.

4.3. Tripod in Support Phase

Given the desired position of the center of mass $_{B}\mathbf{X}_{k+1}$ (line 11 of Algorithm 1) and the current tripod supporting the robot ${}^{B}_{f}\mathbf{A}_k$ (lines 1 and 3 of Algorithm 1), K3P defines the new configuration for the support tripod (line 13 of Algorithm 1) constrained by

$$^{W}_{f}\mathbf{A}_k = {}^{W}_{f}\mathbf{A}_{k+1}. \tag{1}$$

Such a constrain implies that all limbs corresponding to the support tripod remain fixed to the ground with respect to W. In consequence, the gait generated is slip-free and none of the limbs loses contact with the ground. We can solve for ${}^{B}_{f}\mathbf{A}_{k+1}$:

$$\{{}^{W}_{B}\mathbf{A}\,{}^{B}_{f}\mathbf{A}\}_k = \{{}^{W}_{B}\mathbf{A}\,{}^{B}_{f}\mathbf{A}\}_{k+1}$$
$${}^{B}_{f}\mathbf{A}_{k+1} = {}^{W}_{B}\mathbf{A}^{-1}_{k+1}\{{}^{W}_{B}\mathbf{A}\,{}^{B}_{f}\mathbf{A}\}_k$$

where ${}^{W}_{B}\mathbf{A}_k$ and ${}^{W}_{B}\mathbf{A}_{k+1}$ are determined by the present and desired poses of B. From these matrices, the inverse kinematic model for the supporting legs can be solved.

4.4. Tripod in Swing Phase

As described earlier, K3P defines dynamically the desired position for the CoM; thus, K3P also defines dynamically the desired position for the tripod during the swing phase. The target position L for the swing tripod is defined with respect to B based on three different parameters: the maximum gait distance l_{max} (see Figure 3a); the instantaneous turning radius of the hexapod ρ; and the swing tripod clearance l_g.

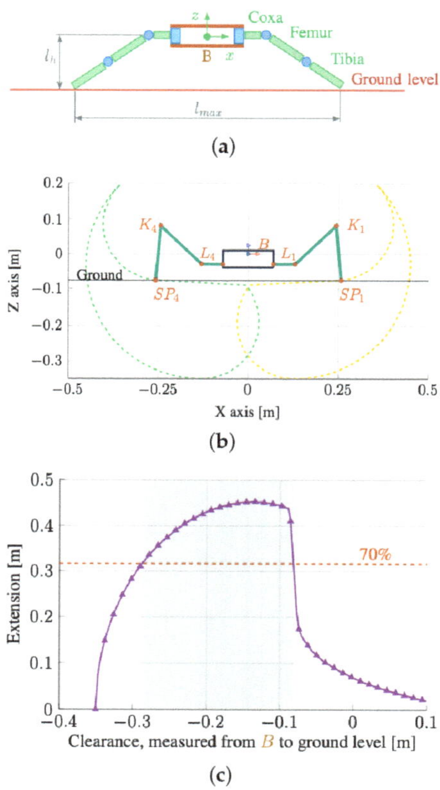

Figure 3. The position in the XY_W plane and the orientation ψ of the robot, while traversing the lemniscate on uneven terrain. (**a**) The theoretical maximum gait for the radial hexapod. (**b**) The dexterous work envelope of two opposite limbs (represented with dotted lines). (**c**) The shaded area represents the best trade-off between clearance and longest horizontal travel.

The first parameter l_{max} is determined by the dexterous work envelope of two opposite limbs. Figure 3b displays such a dexterous work envelope when $\theta_k \in [0, -3\pi/4]$ and $\theta_L \in [-\pi/4, \pi/2]$. The envelope defines the horizontal travel distance that a limb can perform. In order to keep phase shifts at minimum, a good compromise between body clearance l_h and the maximum gait distance l_{max} has to be found. In Figure 3c, we plot the maximum limb extension vs. the clearance; the shaded area represents the range of walk heights for which the K3P algorithm can guarantee a horizontal travel distance of 70% of maximum limb extension. Therefore, K3P considers l_{max} to be equal to the 70% of the theoretical maximum extension of the legged robot (see Figure 3a), with a walking height within the shaded range in Figure 3c.

The second parameter that determines the location of frame L is based on the instantaneous turning radius ρ of the CoM B. Such a turning radius is given by the ratio between the current velocity $\dot{\mathbf{r}} = \| \dot{x}, \dot{y}, \dot{z} \|$ and the angular velocity around the Z_B axis $[\dot{\mathbf{q}}]_\psi$ (line 12 of Algorithm 1). If ρ is above a given threshold l_ρ, then L is located straight ahead $\frac{1}{2}l_{max}$ meters from B in the direction of $\dot{\mathbf{r}}$ (see Figure 4a); otherwise, it is located over the instantaneous circular trajectory such that the arc length from B to L is equal to $\frac{1}{2}l_{max}$ (see Figure 4b). Determining the most suitable position for L based on the magnitude of the instantaneous turning radius ρ is how the algorithm differentiates from straight and turning maneuvers.

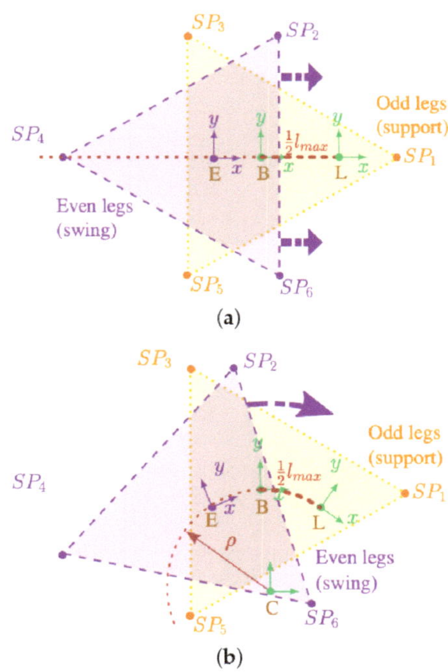

Figure 4. The two situations that may occur when driving the subset of legs during the swing phase of the gait. (**a**) Gait forward, $\rho \geq l_\rho$, reference frame E moves towards L, which is located straight ahead from B, along the x axis, d_{max} meters. (**b**) Turning gait, $\rho < l_\rho$. Reference frame L is located d_{max} meters ahead over the momentaneous trajectory around the turning point C.

The third and last parameter l_g represents how high the tripod will be raised during the swing phase. This parameter is expressed as a percentage of clearance l_h and it can be changed dynamically, depending of the roughness of the terrain.

The desired position for the tripod in the swing phase is computed in line 13 of Algorithm 1. Similar to Section 4.2, an error metric $\mathbf{e}_{L,k}$ is fed to a PD controller to generate a control command $\mathbf{u}_L$ over the subset legs in the swing phase, lines 14 and 15 of Algorithm 1, respectively. Then, the position of the tripod performing the swing phase at time instant $k+1$ can be computed (line 16 of Algorithm 1).

4.5. Joint Reconfiguration

In lines 9 and 16 of Algorithm 1, the new positions of the body body frame B and both tripods O and E are obtained, using PD controllers on their target positions.

After the corresponding location for the tripods in the support and swing phases have been updated to follow the movement of the CoM B, their corresponding state vectors ${}^{B}\mathbf{X}_{W,k+1}$, ${}^{O}\mathbf{X}_{B,k+1}$ and ${}^{E}\mathbf{X}_{B,k+1}$ define the spatial relationships between reference frames W, B, O, and E. From here, it is possible to obtain the positions of each leg tip SP_i with respect to the robot's body and, using the inverse kinematics of the 3 DoF RRR limb (line 24 of Algorithm 1), we can compute the values for all articular joints $[\theta_s, \theta_l, \theta_k]_{1,2,\ldots,6}$ [6].

At this moment, K3P tests the stability of the arrangement at time instant $k+1$ by measuring the Euclidean distance from the projection of B on the support polygon to all edges; in case the minimum distance falls below a predefined threshold l_{thd}, K3P will command the robot to come to a halt. This rarely occurs because the phase shift tests $K3P_{1,2,3}$, which will be introduced in the following section, were designed to guarantee a stable gait.

4.6. Phase Shift

While the robot is walking, a phase shift occurs when the two tripods toggle the phase of the gait they are in. The tripod in the support phase takes off to begin the swing phase and the swing tripod lands to start supporting the main body of the robot. In order to maintain the robot in static equilibrium throughout the walking cycle, the K3P algorithm determines when to shift phases based on three different criteria, named $K3P_1$, $K3P_2$, and $K3P_3$. In particular, K3P1 and K3P2 ensure a collision-free gait pattern.

The first condition $K3P_1$ ensures that every step is as long as mechanically possible for the robot, before the CoM approaches the border of the support polygon too closely, and the gait becomes unstable. For the tripod in the swing phase, $K3P_1$ measures the horizontal traveled distance from the starting point to the current position of its origin, if the traveled distance reaches the maximum travel distance l_{max} (see Figure 5a), $K3P_1$ will trigger a phase shift (line 25 of Algorithm 1).

The second condition $K3P_2$ avoids collisions between two consecutive limbs during a turning maneuver (line 26 of Algorithm 1). $K3P_2$ works by measuring the angle φ, formed by the projections on the XY_B plane of two consecutive position vectors ${}^{SP,i}\mathbf{t}_B$. If φ is smaller than a certain threshold, this criterion triggers a phase shift (see Figure 5b).

Together, the criteria $K3P_1$ and $K3P_2$ reduce the number of phase shifts when the robot is walking, allowing it to make big steps while advancing and/or turning; additionally, these criteria are enough to control the robot in open-loop blind-walking if all position controllers driving each joint are accurate enough. That said, K3P incorporates the inherent position information of all limbs as a third criterion, named $K3P_3$, to make sure they are working properly. $K3P_3$ tests the stability of the inverse kinematics solution; so all joint angles of every limb $[\theta_s, \theta_l, \theta_k]_{1,2,\ldots,6}$ remain within a certain range of operation, avoiding the proximity to the mechanical limits and singularities that otherwise could lead to an unstable walking pattern.

If any of these criteria are met, then the algorithm K3P commands a shift phase of the gait cycle (line 27 of Algorithm 1). The CoM stops its motion to wait for the subset of legs in the swing cycle to land on the ground and begin the stand phase; while the opposite occurs for the subset of legs in stand phase.

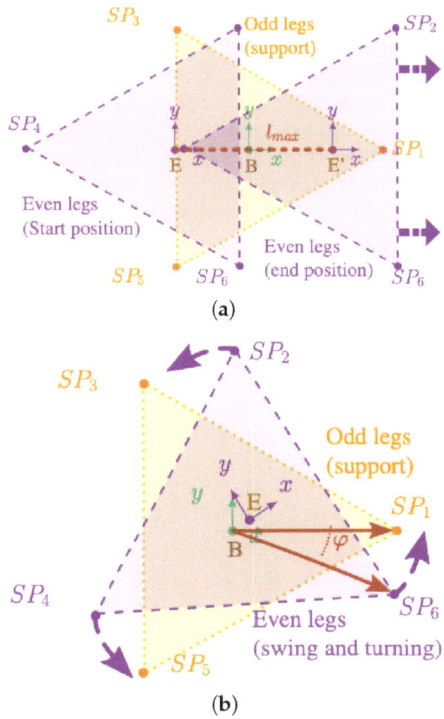

Figure 5. First two criteria that trigger a phase shift. (**a**) K3P$_1$. The linear distance between centroids at start and end positions is limited to l_{max} meters. (**b**) K3P$_2$. As the even legs turn counter-clockwise, the angle φ cannot be smaller than φ_m.

4.7. Uneven Terrain

When a phase shift has been triggered, the tripod in the swing phase has to land. If the surface is uneven, the limbs have to adapt to the elevation changes of the surface. In contrast to the tripod control strategy, where the three legs are commanded simultaneously, during landing, each limb is controlled individually and the three landing events are treated separately. By utilizing interoceptive information, the position of each leg is always known. K3P considers that the swing phase has ended only when all three limbs of the swinging tripod have touched the surface, and therefore, it can begin the support phase of the gait cycle (line 19 of Algorithm 1). To adapt to changes in elevation, K3P must receive information when every *SP* has touched the surface and updates their height with respect to its corresponding reference frame (*E* or *O*). The update process is carried out according to the displacement Δ_z performed by the tripod in swing phase while landing (lines 18 and 20 of Algorithm 1). The displacement update Δ_z for the *i*-th limb (line 17 of Algorithm 1) is expressed as the difference of the *z* components of two consecutive state vectors of the moving tripod.

The swing phase ends when all limbs have landed, and at this moment, their positions with respect to *B* have been adapted to the elevation of the terrain below the robot. In order to perceive ground contact, we consider using inexpensive ToF distance sensors (for example, VL53L0X by ST semiconductor). Eventually, when a tripod restarts its swing phase, the limbs take off, starting from the lowest limb (line 20 of Algorithm 1). The update process Δ_z is applied only to those limbs at the same elevation with respect to *B*, making the limbs separate from the ground in the opposite order on which they landed and move at unison once all the limbs of the tripod have taken off.

Using this approach, the K3P algorithm does not require prior information about the terrain texture. In the Results section, simulations obtained with predefined values of l_h and l_g are shown. However, it is possible to adapt these parameters during execution based on information obtained, for example, from visual data regarding terrain conditions, allowing a higher-level trajectory planning module to modify these parameters.

However, it is important to highlight the limitations of the algorithm: the gait pattern generator will fail when encountering obstacles of significant height, such as large debris or stairs. Additionally, the terrain must be rigid; hence, viscous terrains cannot be considered either.

4.8. Torque Estimation

To test the mechanical viability of the K3P algorithm driving the hexapod robot, we estimated the torques exerted on the knee K, swing S, and lift L joints, as they represent the electric actuators which exert a torque to drive every limb in the commanded direction. Considering that every limb consists of one or more concentrated masses m_j, the way to estimate the torque $\boldsymbol{\tau}_{S,i}$ for the any given joint, say ϑ, is

$$\boldsymbol{\tau}_{\vartheta,i} = \sum_{j=1}^{J} {}^{CoG,j}\mathbf{t}_{\vartheta} \times m_j \mathbf{g} \tag{2}$$

where $\mathbf{g}$ is the gravity vector with respect to W, m_j is the mass for the j-th link, and ${}^{CoG,j}\mathbf{t}_{\vartheta}$ is the position vector for the CoM of the j-th link with respect to reference frame ϑ. Table 3 lists the three joints of interest along the CoM coordinates for every concentrated mass m_j that exerts a torque on the j-th joint. Using Equation (2), the torques on the knee, swing, and lift joints were estimated, and the results will be shown when we describe the simulation process.

Table 3. Parameters for torque estimation with respect to listed reference frames.

Torque Variables			
J	**Reference Frame**	**Link**	**CoG**
1	K, Knee	Tibia	$[1/2l_t, 0, 0]^T$
2	L, Lift	Femur	$[1/2l_f, 0, 0]^T$
3	S, Swing	Coxa	$[1/2l_c, 0, 0]^T$

5. Test Results

In this section, we show the test results of the K3P algorithm when commanding an hexapod robot with a radial base of 0.65 m in a virtual uneven terrain. The numerical values for all physical dimension and parameters are listen in Table 4. The results shown in this section were obtained from computations on Matlab in order to simulate the kinematics of the robot.

During the simulation, the speed commands $\mathbf{v}_c$ for the robot were generated so that it describes a lemniscate trajectory. We chose the lemniscate trajectory because it defines two turns in opposite directions and two almost straight segments for the robot to travel. The parametric equations of the lemniscate $\mathbf{r}_d(s)$ is shown in Equation (3):

$$\mathbf{r}_d(s) = \begin{bmatrix} x_d(s) \\ y_d(s) \\ z_d(s) \\ \psi_d(s) \end{bmatrix} = \begin{bmatrix} a\sin(\frac{s}{\epsilon}) \\ b\sin(\frac{2s}{\epsilon}) \\ c\sin(\frac{3s}{\epsilon}) \\ \arctan 2(\dot{y}_d, \dot{x}_d) \end{bmatrix} \tag{3}$$

Table 4. Simulation parameters.

Hexapod			
θ_l	$[-2\pi/9, 2\pi/9]$	θ_k	$[-\pi/4, \pi/4]$
l_{max}	0.165 m	φ_L	15°
l_c	0.06 m	l_f	0.16 m
l_t	0.16 m	v_B	0.15 m/s
$^B\mathbf{X}_{W,0} = [0,0,0.2,0,0,0,0,0,0,0,0,0]^T$			
Controller gains			
$\mathbf{K}_p = diag(2,2,2.5,0,0,0.9,0,0,0,0,0,0)$			
$\mathbf{K}_d = diag(0.05,0.05,0.1,0,0,0.05,0,0,0,0,0,0)$			
Lemniscate parameters			
a	1.75	b	1.15
c	0.0	ϵ	30

For this numerical example, the robot is commanded to follow a lemniscate covering a rectangular region of 3.5 m long and 2.3 m wide; the values for all parameters of the lemniscate equations are listed in Table 4. For a given speed value v_B and time step Δ_t, we iteratively computed the increment for the parameter Δ_s that yields an equal incremental displacement $\Delta_{\mathbf{r}_d} = v_B \Delta_t$. This incremental displacement is then used as the input command for the K3P algorithm $\mathbf{v}_c$ (Section 4.2) as

$$\mathbf{v}_c = \begin{bmatrix} v_x \\ v_y \\ v_z \\ \dot{\psi} \end{bmatrix} = \begin{bmatrix} \dot{x}_d(s+\Delta_s) \\ \dot{y}_d(s+\Delta_s) \\ \dot{z}_d(s+\Delta_s) \\ \arctan 2(\dot{y}_d+\Delta_s, \dot{x}_d+\Delta_s) \end{bmatrix}$$

Figure 6a shows how K3P drove the hexapod around the lemniscate trajectory over uneven terrain, as well as the trail of all limbs; the initial state of the robot was

$$^B\mathbf{X}_{W,0} = [0,0,0.16,\mathbf{0}_{1\times 9}]^T$$

Figure 6b shows the trajectory described by the CoM of the robot, overlapping the desired trajectories in the XY_W plane. Figure 6c shows the transient response at the beginning of the trajectory when the hexapod aligns itself with the lemniscate trajectory from its initial state; the shaded areas represent the moments at which the even subsets of legs are at the swing phase of the gait cycle. The reader can verify that after every phase shift is triggered, the desired angle of orientation $\psi_d(t)$ equals $\psi(t)$; this causes the robot to stop spinning while the swinging tripod lands. Figure 6d displays how K3P adapts to the elevation of the terrain $z_d(t)$ as measured from right below B. K3P tries to maintain a constant walking height $z(t)$ with respect to the surface elevation ≈0.16 m (Section 4.7). Figure 7 shows several footprints of the hexapod right after a phase shift occurs, and both subsets of legs are touching the ground. The corresponding locations of $B_{1,2,\ldots,15}$ are shown to display that the gait is stable because B is within the support polygon of the vehicle. Furthermore, Figure 7 shows where the turning radius is smaller than the threshold $\rho_m = 0.8$ m used in the simulation to better determine the desired location L for the swing tripod, as discussed in Section 4.4.

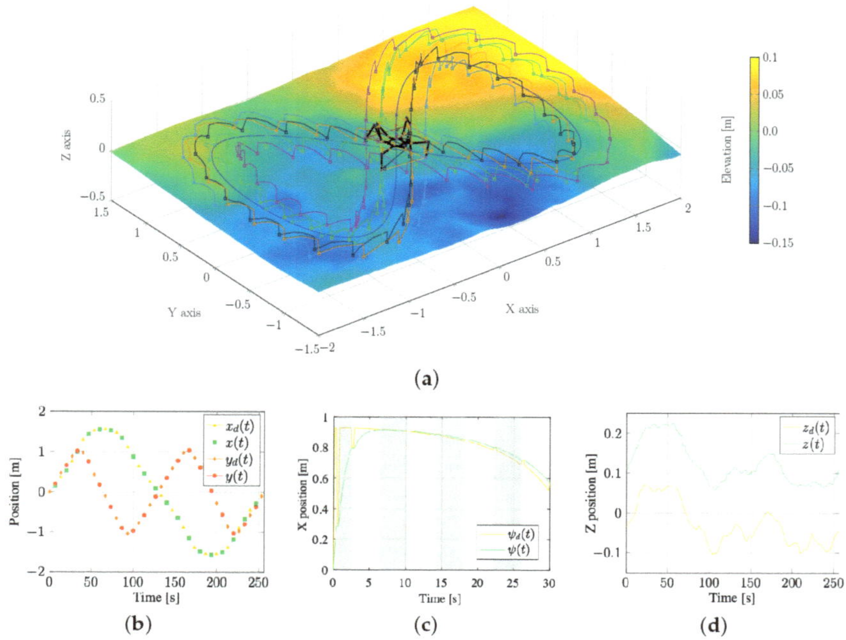

Figure 6. K3P driving the hexapod robot over uneven terrain, describing the lemniscate trajectory. (**a**) Walk gait around the lemniscate over uneven terrain (lighter colors indicate higher elevations of the terrain). (**b**) Desired ($x_d(t)$ and $y_d(t)$) and actual ($x(t)$ and $y(t)$) trajectories. (**c**) The transient ψ response. (**d**) Walking height $z(t)$ vs. terrain elevation $z_d(t)$.

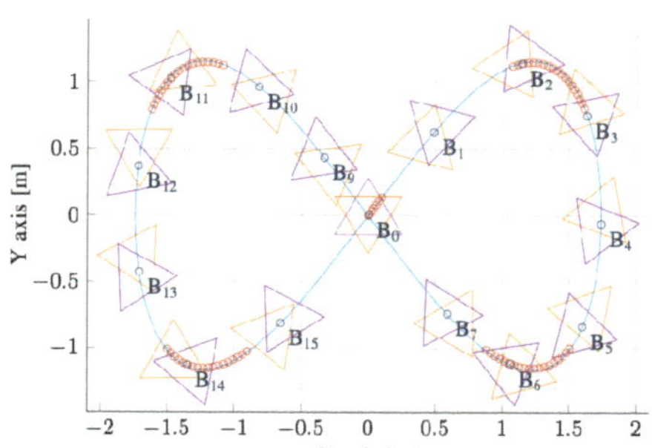

Figure 7. The stability of the walking gait. Consecutive orange dots over the trajectory represent the location of B where $\rho < 0.8$ m.

Figure 8a represents the first 40 s of simulation when the hexapod traverses the lemniscate trajectory; the shaded areas show when the even tripod is in the swing phase, while the white areas show where the odd tripod is performing the swing phase of the gait cycle. At every change in shading, the graph shows the criterion triggering the phase shift at the specific moment in time it occurred: number 1 for K3P$_1$, number 2 for K3P$_2$, and 3 for K3P$_3$. At the beginning of the simulation, when the hexapod robot aligns with the lemniscate, K3P$_2$ triggers the phase shift because the legs were getting too close to each other during the turning maneuver; this corresponds to the transient response in the ψ angle as displayed in Figure 6c. Then, K3P$_1$ triggers the phase shift because the traveled distance of the swinging tripod is longer than l_{max}. The dotted lines correspond to the thresholds $\frac{1}{2}l_{max}$ and φ_m, for the maximum gait distance and minimum angle between two consecutive legs, respectively. Moreover, to display that K3P is capable of commanding the robot to move at an arbitrary velocity through an arbitrary trajectory, the graph in Figure 8a displays a change in velocity, commanded right after the fourth phase shift at $t \approx 22$ s. The velocity is doubled from $v = 0.02$ m/s to $v = 0.04$ m/s; the change in velocity can be observed from the duration of the swing phase, where they become narrower after $t \approx 22$ s because it takes less time for the robot to cover the maximum traveled distance l_{max}. Note, however, that the change in speed can be commanded at any given time, changing immediately the duty factor β.

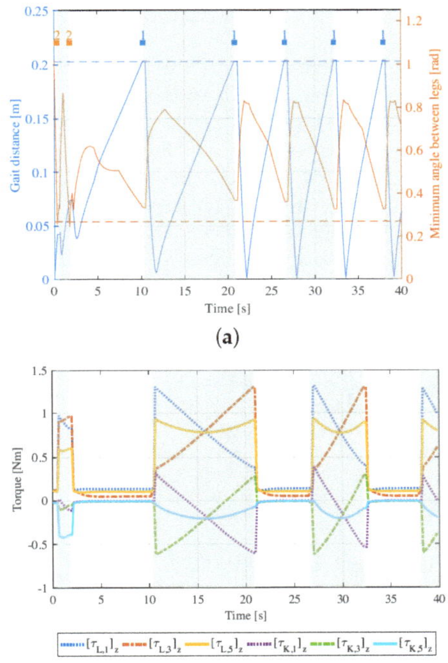

Figure 8. The phase shifts and torques generated when the hexapod robot traverses the lemniscate. Only the first 40 s of simulation are shown. (**a**) The phase shifts triggered during the first 40 s of the simulation. The commanded speed starts at $v = 0.02$ m/s; after the fourth phase shift ($t \approx 22$ s), it was changed to $v = 0.04$ m/s. (**b**) For the odd tripod, the torques exerted on the L and K joints around the z axis.

After running the simulation, we estimated the torques exerted on every joint of the robot. We modeled the robot as a set of discrete masses, listed in Table 5, whereby the overall mass of the robot is approximately 1.6 kg. Figure 8b shows the resulting torques on the knee and lift joints for the odd subset of limbs during the same period of time and phase shifts as previously discussed for Figure 8a. Torques for the even subset of limbs are in the same order of magnitude, since the robot is symmetrical. Because the axis of rotation of the knee and lift joints are coaxial with the two axes $[K]_z$ and $[L]_z$, in Figure 8b, we only show $[\boldsymbol{\tau}_{L,i}]_z$ and $[\boldsymbol{\tau}_{K,i}]_z$. We omitted the torque exerted on the swing joint, because $[\boldsymbol{\tau}_{S,i}]_z \approx 0$. As it can be observed, the maximum torques exerted on the lift and knee joints occur when the even tripod is in the support phase of the gait cycle; furthermore, the maximum absolute values were 1.36 Nm and 0.60 Nm for the $[\boldsymbol{\tau}_{L,i}]_z$ and $[\boldsymbol{\tau}_{K,i}]_z$ axes, respectively, which are manageable for commercially available servo motors.

Table 5. Discrete masses that form the robot.

Mass Distribution of the Robot			
Qty.	**Link**	**Unit Mass [gr]**	**Subtotal**
1	Main body	640	640
6	Coxa	80	480
6	Femur	53	318
6	Tibia	26	156
		Total weight	1594

As mentioned in Section 4.4, the K3P algorithm can command the swing tripod to increase or decrease the maximum clearance of the robot. Figure 9 displays two different values of clearance when the robot travels in a straight line uphill: the first (see Figure 9a) with a clearance equal to 50% of l_h and the second with a clearance of 90%. The latter causes the support point SP to travel almost as high as the CoM B. If required, the walk clearance can be updated at any moment to better adapt to the terrain's changes in elevation. Section 4.4 also shows the profile of every gait cycle that the K3P algorithm describes. Right after a phase shift is triggered, the limbs are taken to land to begin the support phase of the gait cycle.

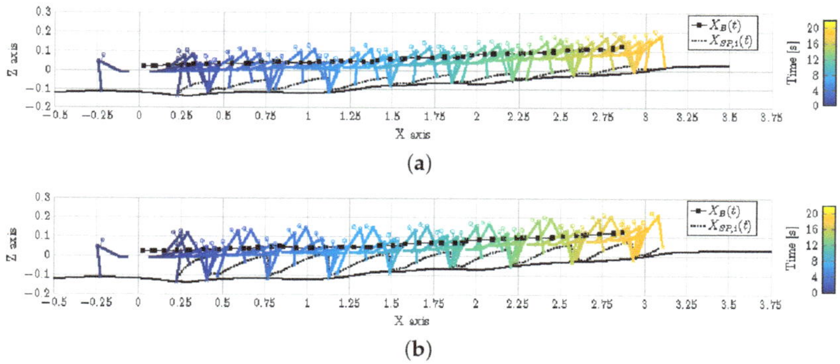

Figure 9. The trajectories of SP_1 when K3P drives the hexapod uphill (black solid line) at two different settings for the maximum height. The hexapod robot moves from left to right, the trajectory of B is also shown. For simplicity, we only show the trajectory for leg number 1. (**a**) The limbs are swinging at 50% of clearance. (**b**) The limbs are swinging at 90% of clearance.

6. Conclusions

In this article, the K3P algorithm is proposed as a novel approach for dynamic gait generation for hexapod robots. This new algorithm is based on a kinematic planner for the legs organized as tripods. The core of K3P are three shift phase conditions, $K3P_{1,2,3}$, that ensure the static slip-free stability of the robot throughout its operation, without requiring any precomputed paths or trajectories whatsoever.

The methodology and numerical results are presented for a radial hexapod traversing a lemniscate trajectory, shown as a versatile methodology when commanding an hexapod. Compared to other approaches, K3P does not require any precomputed information from the trajectory to be followed, nor the trajectory for every support point SP_i, nor precomputed gait patterns. Instead, all trajectories for every tripod were dynamically generated in real-time and made possible that B described a smooth arbitrary trajectory at an arbitrary velocity while the support points remained still over the uneven surface that the robot was walking on. Additionally, K3P is able to dynamically change the clearance of the robot, and we studied the trade-off between clearance vs. step length, given the physical dimensions of the robot.

The K3P algorithm is executable on commercially available embedded computers, using fast linear algebra computation libraries, such as Lapack. Modern CPUs support instruction sets enabling parallelization, such as Single Instruction Multiple Data (SIMD), while GPUs can further enhance the algorithm's execution speed. The possibility of implementing the algorithm on an FPGA can also be considered. Consequently, K3P algorithm finds application in real-time scenarios for robot control. However, it can also be utilized in offline contexts. For example, K3P could serve as a teaching tool for neural networks, with K3P criteria employed to reinforce the learning process of walking.

Because the K3P algorithm performs at real-time and under static stability, it can change the direction of movement of the robot at any given moment. This can be useful if the robot performs in an ever-changing environment with static and dynamic obstacles, e.g., humans or other mobile robots. Such is the case in collaborative robotics; in this emerging research field, robots perform alongside humans or other robots [25]. K3P can offer a development opportunity in collaborative robotics because it can make the robot stop or perform an immediate change in direction of movement when close to a moving obstacle or in a dangerous situation.

The viability of the algorithm is proven by estimating the torques exerted on the knee and lift joints, which are below the maximum torque of commercially available electric servo motors.

As it was shown in the previous section, the K3P algorithm can drive a hexapod robot over irregular terrain without planning in advance every step of the robot at an arbitrary speed. This key design choice for K3P has an important implication: a higher-level trajectory planner can determine the most suitable path for the robot to follow, so that all traversed portions of the terrain can support a footstep. Therefore, K3P can be described as a low-level kinematic planner for a hexapod robot operating in open loop. Its features would allow us to use it alongside different abstraction models of a hexapod to make the robot change its shape when walking in confined environments [39]. Changing the shape of the robot when walking can be useful to adapt to not only uneven terrain as shown here, but to also adapt to constrained and unstructured environments such as a tunnel.

7. Future Work

As future work, we plan to test K3P with an actual robot. The main purpose will be to integrate this algorithm as a low-level feature, allowing for higher-level algorithms to plan the desired trajectory for the robot. Additionally, we plan to integrate an Inertial Measurement Unit as a loop back sensor to work in a closed-loop scheme for the position and pose of the robot; this would make it possible for the robot to display some level of adaptation to sudden external perturbations (mud, gliding, external agents, etc.).

Author Contributions: Conceptualization, J.G.R.-T.; methodology, J.G.R.-T. and D.S.-G.; software, D.S.-G.; validation, J.G.R.-T., D.S.-G. and E.R.-T.; formal analysis, J.G.R.-T. and D.S.-G.; investigation, J.G.R.-T. and D.S.-G.; resources, J.G.R.-T. and D.S.-G.; data curation, D.S.-G.; writing—original draft preparation, D.S.-G.; writing—review and editing, J.G.R.-T., D.S.-G. and E.R.-T.; visualization, D.S.-G. and E.R.-T.; supervision, J.G.R.-T. All authors have read and agreed to the published version of the manuscript.

Funding: The research of the second and third authors was partially funded through Conahcyt-SNII grants 70840 and 44223, respectively.

Institutional Review Board Statement: Not applicable.

Informed Consent Statement: Not applicable.

Data Availability Statement: Code and tests can be found in the K3P repository, available online https://github.com/djaniel/k3p (accessed on 15 March 2024).

Conflicts of Interest: The authors declare no conflicts of interest.

References

1. Zhang, C.; Jiang, X.; Teng, M.; Teng, J. Research on gait planning and static stability of hexapod walking robot. In Proceedings of the 8th International Symposium on Computational Intelligence and Design (ISCID), Hangzhou, China, 12–13 December 2015; Volume 2, pp. 176–179. [CrossRef]
2. Manglik, A.; Gupta, K.; Bhanot, S. Adaptive gait generation for hexapod robot using genetic algorithm. In Proceedings of the IEEE 1st International Conference on Power Electronics, Intelligent Control and Energy Systems (ICPEICES), Delhi, India, 4–6 July 2016; pp. 1–6. [CrossRef]
3. Jalal, A.; Behzad, M.; Fariba, B. Modeling gait using CPG (Central Pattern Generator) and neural network. In Proceedings of the Biometric ID Management and Multimodal Communication (BioID 2009), Madrid, Spain, 16–18 September 2009; pp. 130–137. [CrossRef]
4. Smaldone, F.M.; Scianca, N.; Modugno, V.; Lanari, L.; Oriolo, G. ZMP constraint restriction for robust gait generation in humanoids. In Proceedings of the IEEE International Conference on Robotics and Automation (ICRA), Paris, France, 31 May–31 August 2020; pp. 8739–8745. [CrossRef]
5. Booysen, T.; Marais, S. The development of a remote controlled, omnidirectional six legged walker with feedback. In Proceedings of the 2013 Africon, Pointe aux Piments, Mauritius, 9–12 September 2013; pp. 1–6. [CrossRef]
6. Isvara, Y.; Rachmatullah, S.; Mutijarsa, K.; Prabakti, D.E.; Pragitatama, W. Terrain adaptation gait algorithm in a hexapod walking robot. In Proceedings of the 13th International Conference on Control Automation Robotics Vision (ICARCV), Singapore, 10–12 December 2014; pp. 1735–1739. [CrossRef]
7. Zhai, Y.; Gao, P.; Sun, Y.; Zhao, S.; Jiang, Z.; Li, B.; Hu, Y.; Zhang, J. Gait planning for a multi-motion mode wheel-legged hexapod robot. In Proceedings of the IEEE International Conference on Robotics and Biomimetics (ROBIO), Qingdao, China, 3–7 December 2016; pp. 449–454. [CrossRef]
8. Wang, B.; Zhang, K.; Yang, X.; Cui, X. The gait planning of hexapod robot based on CPG with feedback. *Int. J. Adv. Robot. Syst.* **2020**, *17*, 1729881420930503. [CrossRef]
9. Nishii, J. Legged insects select the optimal locomotor pattern based on the energetic cost. *Biol. Cybern.* **2000**, *83*, 435–442. [CrossRef] [PubMed]
10. Ji, W.S.; Cho, B.K. Development of a walking algorithm for stair formed obstacle for the hexapod walking robot LCR200. In Proceedings of the 14th International Conference on Control, Automation and Systems (ICCAS 2014), Gyeonggi-do, Republic of Korea, 22–25 October 2014; pp. 1614–1616. [CrossRef]
11. Chou, Y.C.; Yu, W.S.; Huang, K.J.; Lin, P.C. Bio-inspired step crossing algorithm for a hexapod robot. In Proceedings of the IEEE/RSJ International Conference on Intelligent Robots and Systems, San Francisco, CA, USA, 25–30 September 2011; pp. 1493–1498. [CrossRef]
12. Lin, P.; Komsuoglu, H.; Koditschek, D.E. Sensor data fusion for body state estimation in a hexapod robot with dynamical gaits. *IEEE Trans. Robot.* **2006**, *22*, 932–943. [CrossRef]
13. Kriengkomol, P.; Kamiyama, K.; Kojima, M.; Horade, M.; Mae, Y.; Arai, T. New tripod walking method for legged inspection robot. In Proceedings of the IEEE International Conference on Mechatronics and Automation, Harbin, China, 7–10 August 2016; pp. 1078–1083. [CrossRef]
14. Marais, S.T.; Nel, A.L.; Robinson, P.E. Reflex assisted walking for a hexapod robot. In Proceedings of the Pattern Recognition Association of South Africa and Robotics and Mechatronics International Conference (PRASA-RobMech), Stellenbosch, South Africa, 30 November–2 December 2016; pp. 1–6. [CrossRef]
15. Kottege, N.; Parkinson, C.; Moghadam, P.; Elfes, A.; Singh, S.P.N. Energetics-informed hexapod gait transitions across terrains. In Proceedings of the IEEE International Conference on Robotics and Automation (ICRA), Seattle, WA, USA, 26–30 May 2015; pp. 5140–5147. [CrossRef]
16. Sun, Y.; Jing, Z.; Dong, P.; Chen, W.; Huang, J. Locomotion Control for a Land-Air Hexapod Robot. In Proceedings of the 6th IEEE International Conference on Advanced Robotics and Mechatronics (ICARM), Chongqing, China, 3–5 July 2021; pp. 887–892. [CrossRef]
17. Rahme, M.; Abraham, I.; Elwin, M.L.; Murphey, T.D. Dynamics and domain randomized gait modulation with Bezier curves for sim-to-real legged locomotion. *arXiv* **2020**, arXiv:2010.12070. [CrossRef]
18. Chavali, R.A.; Kent, N.; Napoli, M.E.; Howard, T.M.; Travers, M. Inferring Distributions of Parameterized Controllers for Efficient Sampling-Based Locomotion of Underactuated Robots. In Proceedings of the American Control Conference (ACC), Philadelphia, PA, USA, 10–12 July 2019; pp. 5767–5773. [CrossRef]
19. Hauser, K.; Bretl, T.; Latombe, J.; Wilcox, B. Motion planning for a six-legged lunar robot. In *Algorithmic Foundation of Robotics VII*; Selected Contributions of the Seventh International Workshop on the Algorithmic Foundations of Robotics; Springer: Berlin/Heidelberg, Germany, 2008; Volume 47, pp. 301–316. [CrossRef]
20. Stoian, V.; Vladu, I.C. A control algorithm for hexapod mobile robot gait in fault conditions. In Proceedings of the 20th International Conference on System Theory, Control and Computing (ICSTCC), Sinaia, Romania, 13–15 October 2016; pp. 349–354. [CrossRef]
21. Chen, G.; Han, Y.; Li, Y.; Shen, J.; Tu, J.; Yu, Z.; Zhang, J.; Cheng, H.; Zhu, L.; Dong, F. Autonomous gait switching method and experiments of a hexapod walking robot for Mars environment with multiple terrains. *Intell. Serv. Robot.* **2024**, 1–21. [CrossRef]

22. Currie, J.; Beckerleg, M.; Collins, J. Software Evolution of a Hexapod Robot Walking Gait. In Proceedings of the 15th International Conference on Mechatronics and Machine Vision in Practice, Auckland, New Zealand, 2–4 December 2008; pp. 305–310. [CrossRef]
23. Wang, B.; Cui, X.; Sun, J.; Gao, Y. Parameters optimization of central pattern generators for hexapod robot based on multi-objective genetic algorithm. *Int. J. Adv. Robot. Syst.* **2021**, *18*, 17298814211044934. [CrossRef]
24. Seljanko, F. Hexapod walking robot gait generation using genetic-gravitational hybrid algorithm. In Proceedings of the 15th International Conference on Advanced Robotics (ICAR), Tallinn, Estonia, 20–23 June 2011; pp. 253–258. [CrossRef]
25. Liu, M.; Li, M.; Pang, J. Fault-tolerant gait implementation of hexapod robot based on finite state automata. In Proceedings of the 29th Chinese Control And Decision Conference (CCDC), Chongqing, China, 28–30 May 2017; pp. 6800–6805. [CrossRef]
26. Mrva, J.; Faigl, J. Tactile sensing with servo drives feedback only for blind hexapod walking robot. In Proceedings of the 10th International Workshop on Robot Motion and Control (RoMoCo), Poznan, Poland, 6–8 July 2015; pp. 240–245. [CrossRef]
27. Kurosawa, M.; Sasaki, T.; Ohara, M.; Tanaka, T.; Hayakawa, Y.; Kaneko, M.; Uchikoba, F.; Saeki, K.; Saito, K. Gait Pattern Generation of Hexapod-Type Microrobot Using Interstitial Cell Model Based Hardware Neural Networks IC. In Proceedings of the International Conference on Electronics Packaging (ICEP), Niigata, Japan, 17–20 April 2019; pp. 316–319. [CrossRef]
28. Sun, Q.; Gao, F. An online gait planner of hexapod robot to safely pass through crowded environment based on tactile sense and virtual dynamic model. In Proceedings of the 9th International Conference on Human System Interactions (HSI), Portsmouth, UK, 6–8 July 2016; pp. 176–182. [CrossRef]
29. Liu, Y.; Ding, L.; Gao, H.; Liu, G.; Deng, Z.; Yu, H. Efficient force distribution algorithm for hexapod robot walking on uneven terrain. In Proceedings of the IEEE International Conference on Robotics and Biomimetics (ROBIO), Qingdao, China, 3–7 December 2016; pp. 432–437. [CrossRef]
30. Ijspeert, A. Central pattern generators for locomotion control in animals and robots: A review. *Neural Netw.* **2008**, *21*, 642–653. [CrossRef] [PubMed]
31. Zhong, B.; Zhang, S.; Xu, M.; Zhou, Y.; Fang, T.; Li, W. On a CPG-Based Hexapod Robot: AmphiHex-II with Variable Stiffness Legs. *IEEE ASME Trans. Mechatron.* **2018**, *23*, 542–551. [CrossRef]
32. Thor, M.; Manoonpong, P. A Fast Online Frequency Adaptation Mechanism for CPG-Based Robot Motion Control. *IEEE Robot. Autom. Lett.* **2019**, *4*, 3324–3331. [CrossRef]
33. Jose Hugo Barron-Zambrano, C.T. FPGA implementation of a configurable neuromorphic CPG-based locomotion controller. *Neural Netw.* **2013**, *45*, 50–61. [CrossRef] [PubMed]
34. Yu, H.; Gao, H.; Ding, L.; Li, M.; Deng, Z.; Liu, G. Gait Generation with Smooth Transition Using CPG-Based Locomotion Control for Hexapod Walking Robot. *IEEE Trans. Ind. Electron.* **2016**, *63*, 5488–5500. [CrossRef]
35. Li, W.; Chen, W.; Wu, X.; Wang, J. Parameter tuning of CPGs for hexapod gaits based on Genetic Algorithm. In Proceedings of the IEEE 10th Conference on Industrial Electronics and Applications (ICIEA), Auckland, New Zealand, 15–17 June 2015; pp. 45–50. [CrossRef]
36. Morantes, G.; Cappelleto, J.; Fernández, G.; Clotet, R.; Torrealba, R.; Guerrero, S. Comparison of CPG topologies for bipedal gait. In Proceedings of the IEEE Ecuador Technical Chapters Meeting (ETCM), Auckland, New Zealand, 15–17 June 2015; pp. 1–6. [CrossRef]
37. Čížek, P.; Milička, P.; Faigl, J. Neural based obstacle avoidance with CPG controlled hexapod walking robot. In Proceedings of the International Joint Conference on Neural Networks (IJCNN), Anchorage, AK, USA, 14–19 May 2017; pp. 650–656. [CrossRef]
38. Corke, P.I. A Simple and Systematic Approach to Assigning Denavit-Hartenberg Parameters. *IEEE Trans. Robot.* **2007**, *23*, 590–594. [CrossRef]
39. Russell, B.; Tirthankar, B.; Marko, B.; Lorenz, W.; Marco, H.; Navinda, K. Walking Posture Adaptation for Legged Robot Navigation in Confined Spaces. *IEEE Robot. Autom. Lett.* **2019**, *4*, 2148–2155. [CrossRef]

applied sciences

MDPI

Article

Research on Path Planning for Robots with Improved A* Algorithm under Bidirectional JPS Strategy

Fujie Wang, Wei Sun, Pengfei Yan, Hongmei Wei and Huishan Lu *

School of Mechanical Engineering, North University of China, Taiyuan 030051, China;
wangfujie@nuc.edu.cn (F.W.); 18235908468@163.com (W.S.); ypf5321@163.com (P.Y.); weihmchina@163.com (H.W.)
* Correspondence: 13934597379@139.com

Abstract: Aiming to address the A* algorithm's issues of traversing a large number of nodes, long search times, and large turning angles in path planning, a strategy for multiple improvements to the A* algorithm is proposed. Firstly, the calculation of the heuristic function is refined by utilizing the Octile distance instead of traditional distance, which more accurately predicts the optimal path length. Additionally, environmental constraints are introduced to adaptively adjust the weight of the heuristic function, balancing the trade-off between search speed and path length. Secondly, the bidirectional jump point search method is integrated, allowing simultaneous path searches from both directions. This significantly reduces path search times and the number of nodes traversed. Finally, the path undergoes two rounds of smoothing using a path smoothing strategy until the final path is generated. To validate the effectiveness of the improved A* algorithm, simulations are conducted on ten types of grid maps. Results demonstrate that the improved A* algorithm markedly decreases path search times while maintaining path length, with greater speed improvements observed as the map size increases. Furthermore, the improved algorithm is applied in experiments with mobile robots, achieving significant reductions in average path search times of 79.04% and 37.41% compared to the traditional A* algorithm and the JPS algorithm, respectively. This enhancement effectively meets the requirements for rapid path planning in mobile robotics applications.

Keywords: robot; path planning; A* algorithm; heuristic function; jump point search

Citation: Wang, F.; Sun, W.; Yan, P.; Wei, H.; Lu, H. Research on Path Planning for Robots with Improved A* Algorithm under Bidirectional JPS Strategy. *Appl. Sci.* **2024**, *14*, 5622. https://doi.org/10.3390/app14135622

Academic Editor: Jonghoek Kim

Received: 18 May 2024
Revised: 20 June 2024
Accepted: 24 June 2024
Published: 27 June 2024

1. Introduction

In today's society, with the continuous development of intelligent manufacturing technology, robots have received widespread attention and have permeated various aspects of human life. Currently, research on robot autonomous navigation is primarily divided into three parts: perception, planning, and execution. Among these, path planning, referred to as "planning", serves as the bridge connecting the other two components and is also the core focus of robotics research [1,2]. The importance of path planning can be reflected in the framework diagram of navigation, as shown in Figure 1. The navigation framework is mainly composed of the following key components. The 'Move_base' node is the core of the navigation framework, being responsible for coordinating and managing other modules. It receives the target position, calls Global_Planner and Local_Planner to generate a path, and sends control commands to Base_controller. The 'Global_Planner' node calculates the optimal path from the start point to the target point based on the static map and the target position. The 'Local_Planner' node is responsible for generating local paths suitable for the current environment during the execution of the global path, performing real-time obstacle avoidance and path adjustment. The 'Base_controller' node transforms high-level path planning instructions into actual motion control signals to ensure that the robot can navigate accurately and safely to the target position. The 'Costmap' node consists of the Global_costmap node and Local_costmap node, representing the traversal costs of different locations in the environment. Global_costmap is based on a static map while Local_costmap is based on real-time sensor data. The 'Recovery_behaviors' node is used to handle faults in various modules during navigation. Recovery behaviors are activated when there are failures in global planning, local planning, or oscillations. The 'Amcl' node is responsible for robot localization based on sensor data and the map. The 'Map_server' node provides the static environment map, usually generated by SLAM. This map is used for path planning and localization. The 'Odometry source' node provides the robot's position information and motion state in the environment, estimating the robot's posture and position changes relative to its starting position. The 'Sensor sources' node collects data about the robot's surroundings using sources such as LiDAR, depth cameras, and ultrasonic sensors. These data are used to build and update the costmap. The 'Sensor transforms' node converts the data collected by the sensors into a unified coordinate system or reference frame for consistent processing and decision making. Additionally, by using communication mechanisms such as topic communication with topics like "/map", the nodes work collaboratively. The navigation framework achieves key autonomous navigation functions through the coordinated work of Global_Planner and Local_Planner. Path planning specifically means that the robot plans an optimal path from the start position to the goal position without any collision with obstacles in the space when the environment is known or unknown. Path planning according to the degree of mastery of the environment information is divided into two forms. One is global path planning with full knowledge of the environment; at present, the commonly used global path planning algorithms are the Dijkstra algorithm [3], A* algorithm [4–8], ACO algorithm [9], etc. The other is local path planning with only partial or no knowledge of the environment; local path planning algorithms mainly include the DWA algorithm [10,11], artificial potential field method [12], etc.

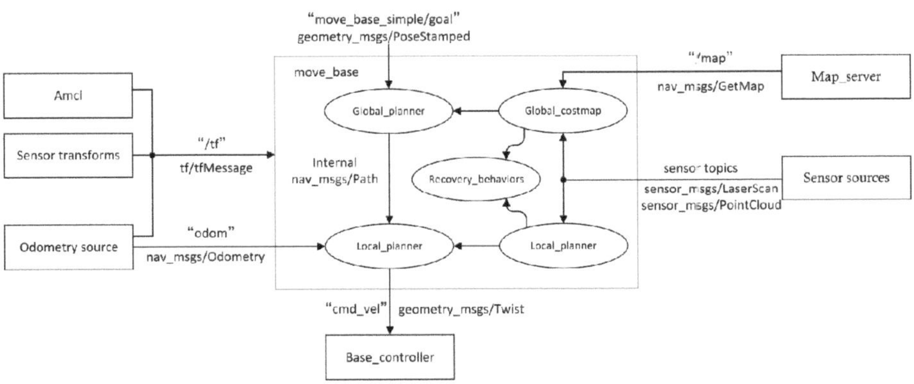

Figure 1. Framework diagram of navigation.

The A* algorithm is a path planning algorithm based on graph search that is widely used in global path planning due to its fast computation speed, high efficiency, and maturity, being the most established path planning algorithm known so far. However, the A* algorithm still has many shortcomings. For example, it constantly visits the surrounding eight neighboring nodes during pathfinding, adding them to OpenList; selects the node with the smallest cost as the current node for expansion after calculating its evaluation function value; and adds it to CloseList after expansion, resulting in a large number of redundant nodes being stored and accessed. When the map scene is large, it will lead to the wastage of both storage space as well as time due to the increase in computation, resulting in inefficient pathfinding. Therefore, many scholars have made a series of improvements to the A* algorithm through research based on these problems. Zhang Mo et al. [13] reduced the number of turns in diagonally planning paths by changing the equation for total cost, i.e., increasing the influence of the predicted cost on the total cost, and smoothed the paths by introducing the five-point cubic smoothing method, but the algorithm still suffers from the problem of having too many redundant nodes. Gong Yunxin et al. [14] used an extension to include convex corner points that satisfy the adjacency relationship as the pattern of adjacency extension for the A* algorithm. This modification reduces access to unnecessary nodes and effectively improves the algorithm's search efficiency. However, the improved A* algorithm's efficiency is still not high in scenarios with too many convex corner points, and it fails to meet real-time demands. Harabor et al. [15] proposed a jump point search algorithm that removes non-essential nodes through screening and pruning rules, enables jumping between nodes with variable step sizes, and improves the pathfinding speed. However, the path planned by this method still suffers from the problem that the turning angles are not smooth and cannot satisfy the robot's motion characteristics. In Zhao Xiao et al.'s [16] method, which combines the A* algorithm with the JPS algorithm to improve the A* algorithm, a series of representative nodes are screened for expansion by preprocessing the nodes, which not only saves memory but also enhances the algorithm's efficiency. In this paper, we analyze and study the existing improved A* algorithms and a new strategy for the improvement of the A* algorithm is proposed. It involves the following steps:

1. Improve the heuristic function of the A* algorithm: firstly, utilize the Octile distance formula instead of the commonly used distance formula to approximate the predicted cost of the heuristic function to the actual path length. Afterward, environmental constraints, i.e., the percentage of obstacles, are introduced to make the algorithm adaptive to the changes of the environmental map;
2. Incorporating a bidirectional JPS search strategy: the searched jump points are added to OpenList for access and expansion. This replaces the need to access and compute a large number of neighboring nodes in the A* algorithm, thereby reducing memory consumption and significantly accelerating the path search speed through a bidirectional search strategy;
3. Further optimize the generated paths: first reduce the number of redundant nodes on the paths, then eliminate unnecessary inflection points on the paths, and, finally, smooth the paths further using the dynamic circular cutting method to improve the turning angles of the paths.

2. Traditional A* Algorithm

2.1. Modeling of Grid Maps

Establishing a map model of the environment is the primary prerequisite for path planning, and commonly used methods for map modeling include the grid method, visual graphic method, topological map method, Voronoi diagram, and others [17]. The grid method is widely adopted for its intuitive representation, simple operation, and ease of implementation. In this paper, 2D grids of uniform size are used to abstractly represent information about the actual environment, with each grid corresponding to a unique coordinate. Each grid is classified into two states based on whether it is occupied by an obstacle [18]. One is the free state, represented by white grids in Figure 2, indicating areas where the robot can freely pass; the other is the occupied state, represented by black grids in Figure 2, indicating areas impassable for the robot.

Commonly used neighborhood expansion patterns in A* algorithms include four-neighborhood expansion and eight-neighborhood expansion [19]. In actual robot operations, the model's complexity is increased due to uncertainties in the robot's operational state, which affect its movement direction. To mitigate this complexity, this paper selects an eight-neighborhood expansion pattern when there are no boundaries or obstacles around the robot, as depicted in Figure 3.

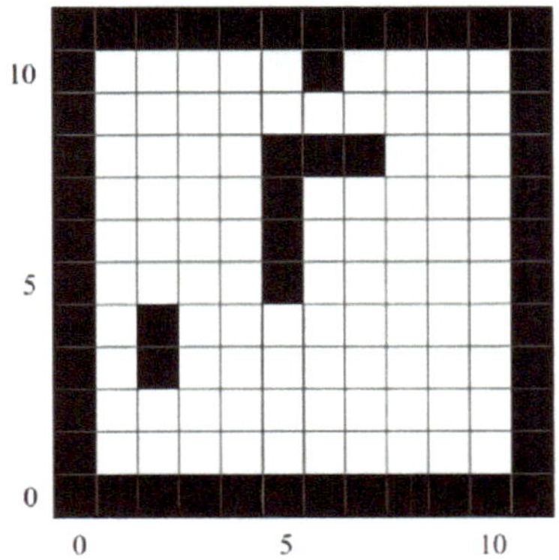

Figure 2. Grid map.

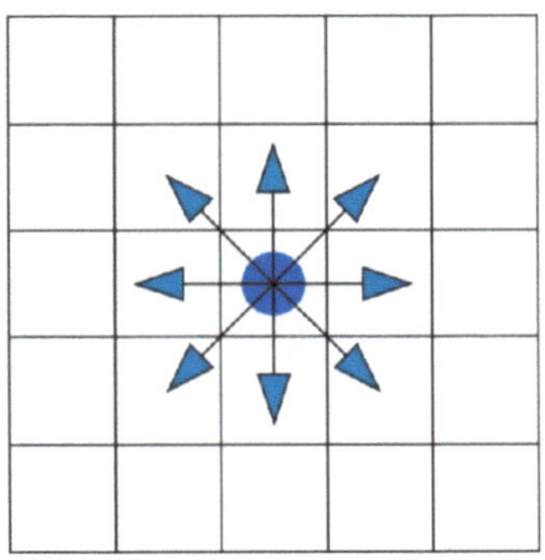

Figure 3. Eight-neighborhood expansion.

2.2. Traditional A* Algorithm

The A* algorithm is a heuristic path planning algorithm designed for static scenarios that introduces a Best First Search algorithm, i.e., heuristic function, based on the Dijkstra algorithm, which enhances the heuristic of the algorithm, makes the algorithm more purposeful in searching towards the goal point, and reduces the algorithm's traversal of the redundant nodes. The evaluation function of the A* algorithm is given in the following equation:

$$f(n) = g(n) + h(n) \tag{1}$$

In Equation (1), $g(n)$ denotes the total actual cost accumulated from the start node to the current node n; $h(n)$ is a heuristic function representing the predicted cost from the current node n to the target node; and $f(n)$ is an evaluation function representing the total predicted cost for the start node to reach the target node via node n.

The heuristic function $h(n)$ is crucial in determining the efficiency of the algorithm. When $h(n)$ is smaller than the actual cost $h^*(n)$, the algorithm can definitely plan the optimal path, but it may traverse more nodes, slowing down the search speed; when $h(n)$ is 0, $h(n)$ must be less than or equal to $h^*(n)$, i.e., when the algorithm degenerates into the Dijkstra algorithm; when $h(n)$ is greater than the actual cost $h^*(n)$, the algorithm reduces the number of traversed nodes, speeding up the search, but it will ensure that the path planning result is not the optimal solution; when $h(n)$ is much larger than $h^*(n)$, the algorithm evolves to the Best First Search algorithm; the A* algorithm needs to consider the search speed during the search process while ensuring that the optimal path is found. Therefore, for optimal efficiency, $h(n)$ should closely approximate $h^*(n)$.

Commonly used representations of heuristic functions are Euclidean distance and Manhattan distance. Let the coordinates of the current node be (x_n, y_n) and the coordinates of the target point be (x_g, y_g); then, the Euclidean distance formula is as below:

$$h(n) = \sqrt{(x_g - x_n)^2 + (y_g - y_n)^2} \tag{2}$$

The Manhattan distance formula is as follows:

$$h(n) = |x_g - x_n| + |y_g - y_n| \tag{3}$$

Due to the presence of obstacles in the actual environment and the selected expansion method in this paper being eight-neighborhood expansion, when using a Euclidean distance of $h(n)$, the distance value must be smaller than the actual cost $h^*(n)$; when using a Manhattan distance of $h(n)$, the distance value is larger than the actual cost $h^*(n)$. In this paper, we need to select an appropriate distance formula that makes $h(n)$ closer to $h^*(n)$.

The A* algorithm will add the node with the smallest value of the current evaluation function as the current node to CloseList during the path search and, at the same time, visit the neighboring nodes reachable by this node and add them to OpenList, then calculate the node with the smallest value of the evaluation function in OpenList to add it to CloseList. It will repeat this process until the target point becomes the current node, completing the pathfinding. Finally, the final path is generated by backtracking the parent–child relationships of the nodes. The pathfinding process is illustrated in Figure 4, which highlights that the A* algorithm generates a large number of redundant nodes, represented by the light blue grids in the figure. Maintaining and accessing these nodes increases the computational workload of the algorithm and wastes storage space, thereby impacting the efficiency of path planning.

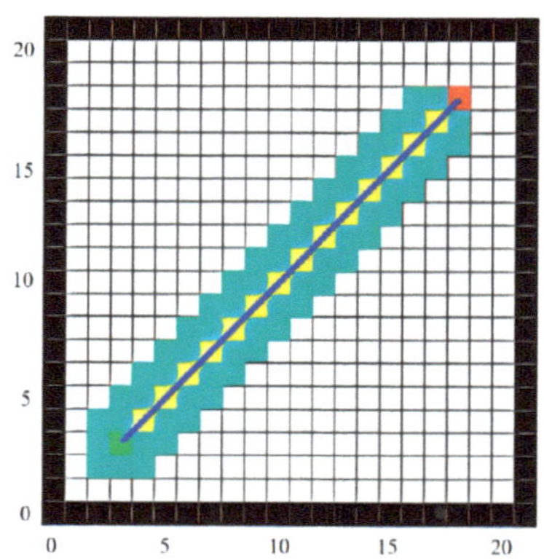

Figure 4. Pathfinding process of A* algorithm.

3. Improved A* Algorithm

3.1. Improvement of the Heuristic Function

3.1.1. Selection of Heuristic Function

The heuristic function, as the core of the A* algorithm, significantly impacts the result of the planned path. In response to the issues with the general heuristic function's expression mentioned above, this paper proposes a new distance expression for the heuristic function, namely Octile distance. Octile distance is a combination of Euclidean distance and Manhattan distance that brings the predicted distance closer to the actual distance, thus improving the efficiency of the algorithmic search. Octile distance is illustrated in Figure 5, where the yellow grid indicates the starting point, the purple grid indicates the target point, the gray grid indicates the actual path, the blue straight line indicates the Octile distance,

3.2. Combining the Bidirectional JPS Algorithm

3.2.1. Traditional JPS Algorithm

Jump Point Search (JPS) is a more efficient path planning algorithm that screens natural neighbor nodes and forced neighbor nodes through specific screening rules and further screens to obtain the jump points needed for path planning, which well solves the problem of the A* algorithm accessing more redundant nodes.

Natural neighbor nodes are nodes that extend directly along the current search direction in the hopping point search algorithm; using these nodes can efficiently skip intermediate invalid nodes, reduce the search space, and improve the speed of path planning.

Forced neighbor nodes are the nodes that must be considered in the jump point search algorithm in order to bypass obstacles. By checking these nodes, JPS can effectively avoid obstacles in path search and ensure that it remains the optimal path.

The screening rules are as follows.

(1) No obstacles around the node

As shown in Figure 8, node x represents the current node, node $P(x)$ is the parent of node x, and n is a neighboring node of x. The length of the path taken by the parent node $P(x)$ to reach node n through x, when moving horizontally, must be shorter than the lengths of other paths that do not pass through x to reach node n. Similarly, the length of the path taken by the parent node $P(x)$ to reach node n through x, when moving diagonally, must be less than or equal to the lengths of other paths that do not pass through x to reach node n.

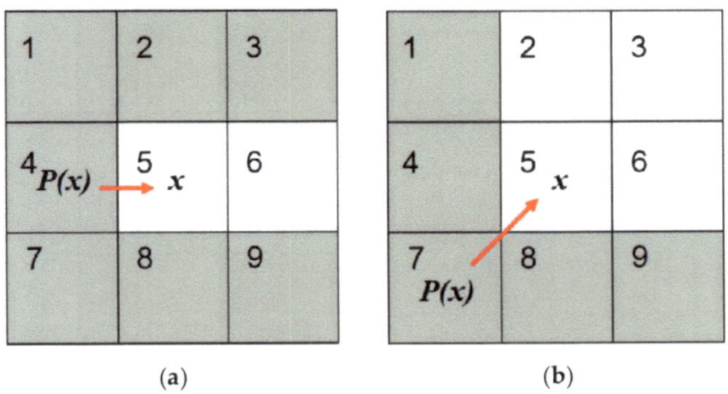

Figure 8. Screening rules without adjacent obstacles: (**a**) horizontal move; (**b**) diagonal move.

Therefore, the following screening rules for natural neighbor nodes can be derived:

- When moving horizontally, the natural neighbor node n satisfies the following equation:

$$L(< P(x), x, \ldots, n >) < L(< P(x), \ldots, n > | x) \tag{8}$$

Here, $L(<P(x), x, \ldots, n>)$ denotes the length of the path from the starting point $P(x)$ through x to reach n and $L(<P(x), \ldots, n> | x)$ denotes the length of the path from the starting point $P(x)$ to reach n without passing through x.

- When moving diagonally, the natural neighbor node n satisfies the following equation:

$$L(< P(x), x, \ldots, n >) \leq L(< P(x), \ldots, n > | x) \tag{9}$$

(2) Presence of obstacles around the node

As shown in Figure 9, the black grid indicates the obstacles. At this time, in addition to screening the natural neighbor nodes, it is also necessary to screen out the forced neighbor nodes. The forced neighbor nodes are shown in the green grid in the figure.

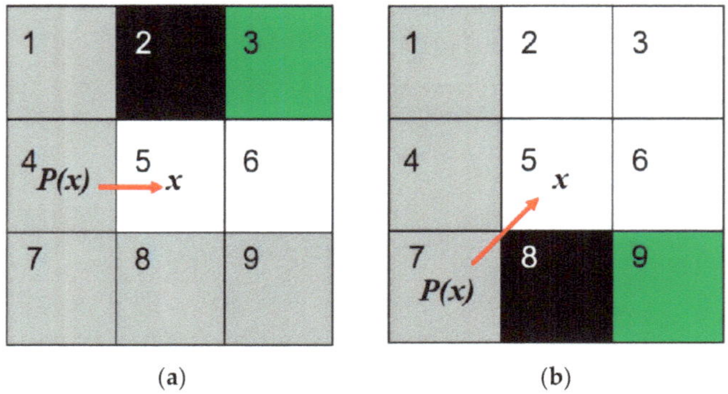

Figure 9. Screening rules with adjacent obstacles: (**a**) horizontal move; (**b**) diagonal move.

The screening rules for natural neighbor nodes are the same as the screening rules when there are no obstacles around a node. The screening rules for the forced neighbor nodes are as follows:

- n is not a natural neighbor of node x;
- n satisfies the following equation:

$$L(< P(x), x, \ldots, n >) < L(< P(x), \ldots, n > | x) \tag{10}$$

For the natural neighbor nodes and forced neighbor nodes obtained by the above screening rules, it is necessary to further screen them to obtain the jump point. The jump point screening rules are as follows:

- If node n is a starting or goal point, then n is a jump point;
- If node n has at least one forced neighbor, then n is a jump point;
- A node n is a jump point if the parent node $P(x)$ is in a diagonal direction from node n and n has nodes in the horizontal or vertical directions that satisfy conditions (1) and (2) above.

The purpose of the integration of the A* algorithm with the JPS algorithm is to first use the JPS algorithm to preprocess to obtain representative jump points, then add these jump points to OpenList and select the node with the smallest cost as the current node for expansion, then continue to screen its jump points to be added to OpenList. At the same time, the jump points after the expansion is completed are added to CloseList. This process is repeated until the current node is the target point, at which point the path planning is complete.

3.2.2. Bidirectional JPS Algorithm

The traditional JPS algorithm can better deal with the problems of the A* algorithm accessing more redundant nodes and having a high computation and memory consumption. However, in larger scenes or with more obstacles, a substantial number of jump points can still be generated, leading to slower pathfinding. Therefore, this paper proposes to improve the A* algorithm by using the bidirectional JPS algorithm on the basis of the traditional JPS algorithm.

Using the bidirectional JPS algorithm means that both the starting point and the goal point are used as the starting point and each takes the other as the goal point for path planning. Typically, the search from the starting point to the goal point is referred to as the forward search while the search from the goal point to the starting point is known as the reverse search. A successful path search is indicated when the currently expanded node of one party is the currently expanded node of the other party or exists in the CloseList of the other party.

The pathfinding process of the A* algorithm incorporating the bidirectional JPS algorithm is shown in Figure 10. Node s represents the start node, node g represents the target node, the black grids denote obstacles, and the red arrows and green arrows represent the paths for the forward and reverse searches, respectively. The flow of the algorithm is as follows:

- Create four lists, *OpenList1*, *OpenList2*, *CloseList1*, and *CloseList2*. Place node *s* into *OpenList1* and node *g* into *OpenList2*;
- Start the bidirectional search; during the forward search, place jump point *a* into *OpenList1*; during the reverse search, place jump point *b* into *OpenList2*. Place node *s* into *CloseList1* and node *g* into *CloseList2;*
- Continuing the forward search with *a* as the current node yields jump points *c* and *d*, which are added to *OpenList1*. In the reverse search, jump point *e* is found for *b* and added to *OpenList2*. Nodes *a* and *b* are placed into *CloseList1* and *CloseList2*, respectively;
- Since *d* is the node with the smallest value of *f(n)* in *OpenList1*, the forward search expands *d* to obtain jump points *e* and *f*, which are added to *OpenList1*. In the reverse search, jump point *h* is found for *e* and added to *OpenList2*. Nodes *d* and *e* are placed into *CloseList1* and *CloseList2*, respectively;
- Node *e* has the smallest value of *f(n)* in *OpenList1*, so *e* is selected as the current node for forward expansion. Since *e* exists in *CloseList2*, the path search is successful.

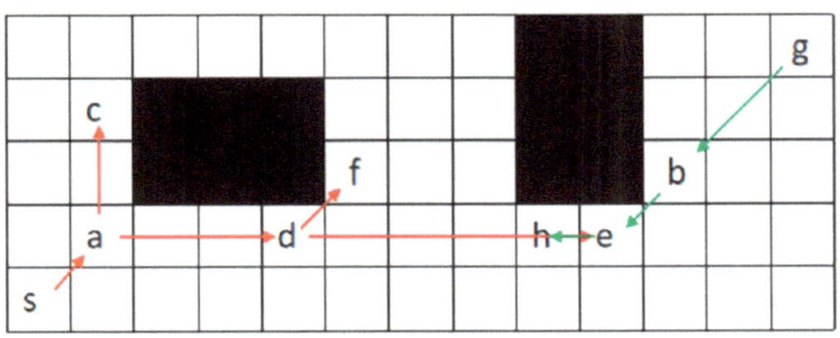

Figure 10. Schematic diagram of bidirectional JPS algorithm.

According to the process diagram of pathfinding using bidirectional JPS shown in Figure 10, theoretically, bidirectional JPS can double the efficiency compared to unidirectional JPS. However, bidirectional JPS also has drawbacks. When there is symmetry in the environment map, bidirectional JPS may plan two symmetric paths. This results in the paths searched in the forward direction not intersecting with the paths searched in the reverse direction. At this time, each search may independently plan a path, which not only fails to improve the efficiency of path planning but may even reduce it. To address this issue, this paper proposes an improved method where each search considers the other's current extended node as the target node.

3.3. Path Smoothing

3.3.1. Primary Smoothing

The paths planned by the improved algorithm still exhibit issues such as having numerous redundant nodes and excessive turning points. Therefore, this paper employs a redundant point removal strategy to enhance path smoothness by reducing both the number of redundant nodes and excessive turning angles. The steps of the redundant point removal strategy, i.e., primary smoothing, are as follows:

- Plan the optimal path using the improved algorithm and save information such as the path length and number of nodes;
- Create a collection of path nodes, $R = \{N_i\}$, $i \in [0, n]$, where N_0 is the start node, N_n is the goal node, and $N_1, \ldots, N_{n-1}$ are the intermediate nodes;

- Determine whether the vectors $\overrightarrow{N_0N_1}$ and $\overrightarrow{N_1N_2}$ formed by points N_0, N_1, and N_2 are parallel using the principle of vector parallelism. If they are parallel, the three points are collinear, and the redundant node N_1 is removed. If they are not parallel, then the three points are not collinear, and there is no redundant node. If node N_1 is removed, then continue to evaluate nodes N_2, N_3, and N_4 to determine whether they are collinear. Otherwise, evaluate nodes N_1, N_2, and N_3 for collinearity. Repeat this process until no more collinear nodes exist in the path. Create a new set P of path nodes and add the reserved nodes to set P. Set $P = \{p_j\}, j \in [0, m]$, where p_0 is the start node, p_m is the goal node, and $p_1, \ldots, p_{m-1}$ are the intermediate nodes;
- Create the set K of nodes that the path must pass through. Add p_0 to set K and consider p_0 as the current point for evaluation. First, connect nodes p_0 and p_m to determine whether the straight line segment p_0p_m passes through obstacles. If it does not, p_0p_m is considered a viable path, and one should add p_m to set K. If p_0p_m is obstructed, connect p_0 to p_{m-1} and continue the evaluation. If p_0p_{m-1} is clear, add p_{m-1} to set K. If p_0p_{m-1} is obstructed, connect p_0 to p_{m-2} and continue evaluating. Continue this process until the must-pass node p_{t1} is identified and added to set K. Then, use p_{t1} as the current point and repeat the evaluation process. Repeat these steps until p_m is added to set K;
- Connect the nodes extracted from set K sequentially; the resulting path is the path after one smoothing. The schematic diagram of the redundant node removal process is shown in Figure 11.

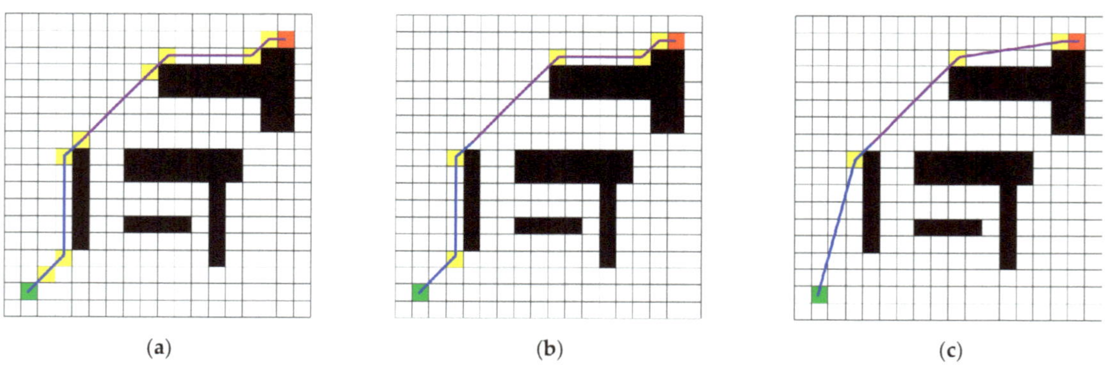

Figure 11. Schematic diagram for removing redundant nodes: (**a**) original path; (**b**) removing collinear nodes; (**c**) removing redundant turning nodes.

3.3.2. Quadratic Smoothing

Although the path after primary smoothing reduces the number of redundant nodes and inflection points, it still exhibits problems such as unsmooth turning angles, which do not satisfy the motion characteristics of the robot. Therefore, to enhance the smoothness of the path, this paper adopts the dynamic circular cutting method for secondary smoothing. The schematic diagram of the dynamic circular cut method is shown in Figure 12.

For the turning corner shown in Figure 12 ∠ABC, the steps of the dynamic circular cut method are as follows:

Compare the lengths of the sides of ∠ABC. Choose the endpoint A of the shorter side AB as the tangent point and draw a perpendicular line through A to intersect the angle bisector BO_0 of ∠ABC at point O_1. Then proceed to step 2;

Make a circle with O_1 as the center and AO_1 as the radius. Judge whether the arc AQ crosses obstacles or not. If it does, perform step 3; otherwise, use the arc AQ instead of the straight lines AB and BQ;

The tangent point A is moved upward along AB by a length λ (the value of λ is chosen according to the actual situation), denoted as A_1. Use A_1 as the tangent point to perform step 1;

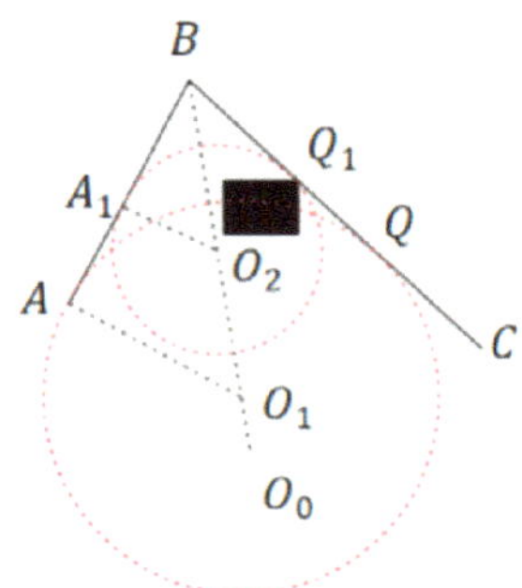

Figure 12. Schematic diagram of the dynamic circle cutting method.

- Determine whether all the turning corners have been smoothed. If yes, the secondary smoothing is completed; otherwise, continue with step 1.

4. Simulation Analysis and Experimental Validation

4.1. Simulation Analysis

In order to verify the effectiveness of the improved algorithm, this paper simulates and tests the A* algorithm, the JPS algorithm, and the improved A* algorithm on ten different grid maps. The computer we use is configured with a Windows 11 operating system, an R7-5800H processor with 3.2 GHz, and 16 GB of RAM.

The results of the three algorithms tested on a grid map of size 40 × 40 with a 30% obstacle ratio are shown in Figure 13. In the figure, the green grid indicates the start node, the red grid indicates the target node, the black grids indicate obstacles, the blue grids indicate nodes added to OpenList, and the yellow grids indicate nodes added to CloseList. The blue solid line represents the planned path, the green solid line represents the path obtained by primary smoothing, and the red solid line represents the final path obtained after secondary smoothing. As can be seen from Figure 13, the length and the number of turns of the optimal path obtained by the improved A* algorithm are better than those of the other two algorithms in the same environment. Other data obtained from the simulations are shown in Tables 1 and 2.

Table 1 compares the three algorithms in terms of the nodes traversed, path lengths, numbers of turns, and search times in grid maps of different sizes with a consistent obstacle percentage of 30%. It is evident that the number of nodes traversed by the improved algorithm is better than that of the A* algorithm and worse than that of the JPS algorithm. However, due to the nature of bidirectional search, it is guaranteed that the search time will outperform the other two algorithms. In 20 × 20 maps, the improved algorithm's search time efficiency improves by 50.29% and 26.37%, respectively, relative to the other algorithms; in 40 × 40 maps, it improves by 62.07% and 30.59%, respectively; in 60 × 60 maps, it improves by 68.66% and 37.13%, respectively; in 80 × 80 maps, it improves by 72.62% and 42.41%, respectively; and in 100 × 100 maps, it improves by 78.37% and 50.43%, respectively. From these data, it can be concluded that as the map size increases, the algorithm's improvement in search speed becomes more significant. However, in some cases, this improvement may lead to longer path lengths. Additionally, the improved algorithm's paths also exhibit significantly fewer turns compared to the other two algorithms.

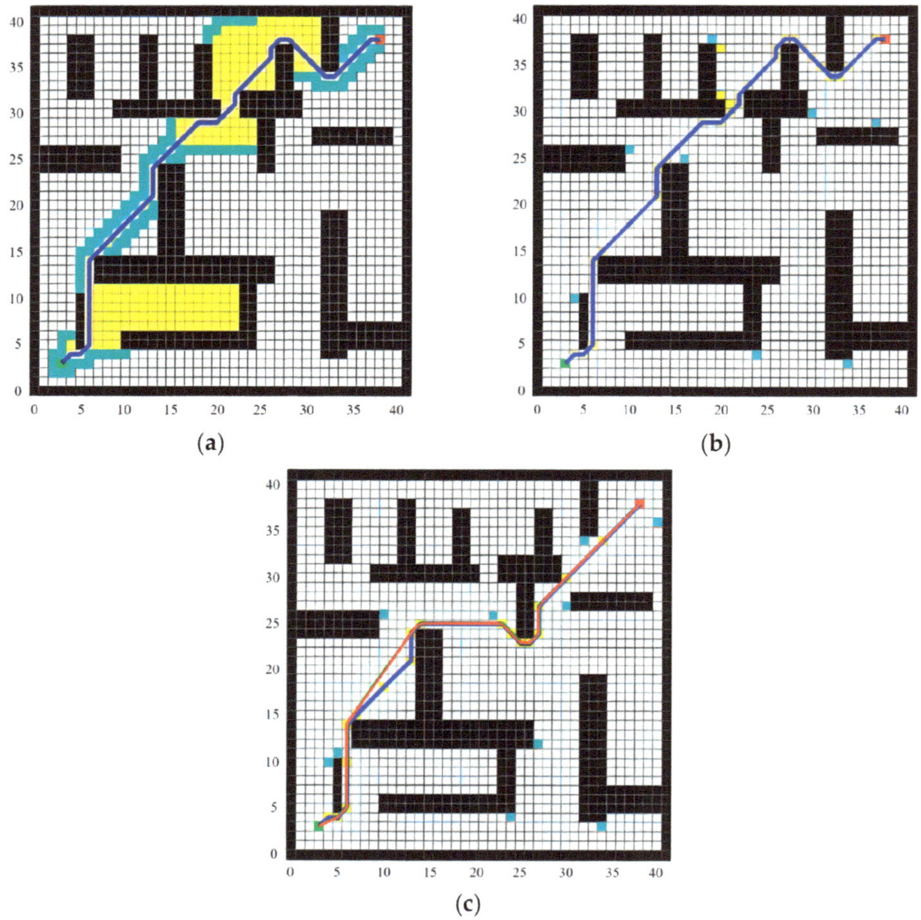

Figure 13. Comparison diagram of simulation results for three algorithms: (**a**) traditional A* algorithm; (**b**) traditional JPS algorithm; (**c**) improved algorithm.

Table 2 compares the three algorithms in terms of the nodes traversed, path lengths, numbers of turns, and search times in grid maps of varying sizes with a consistent obstacle percentage of 70%. It shows that the improved algorithm can generally plan optimal paths, as evidenced by shorter path lengths and fewer turns compared to the other two algorithms. According to Table 2, the search time efficiency of our algorithm improves by 26.14% and 12.70% relative to the other two algorithms in 20 × 20 maps, 39.25% and 20.80% in 40 × 40 maps, 52.47% and 32.10% in 60 × 60 maps, 58.21% and 39.79% in 80 × 80 maps, and 66.58% and 48.22% in 100 × 100 maps. It can be seen that the larger the size of the map is, the more effective the algorithm is in improving search speed. But in cases where the proportion of obstacles is higher, the improvement in search speed is not as significant as when the proportion of obstacles is lower. However, the optimal path is guaranteed to be obtained when the proportion of obstacles is higher.

Table 1. Comparison of three algorithms on grid maps with low proportions of obstacles.

Map Size	Experimental Algorithms	Number of Nodes Traversed (Number)	Path Length (Meters)	Number of Turns (Times)	Search Time (Seconds)	Percentage of Obstacles
20 × 20	A* algorithm	77	24.7279	7	0.233068	30%
	JPS algorithm	20	24.7279	7	0.157325	
	Improved algorithm	26	24.1381	6	0.115842	
40 × 40	A* algorithm	326	61.0122	17	0.707153	
	JPS algorithm	39	61.0122	17	0.386459	
	Improved algorithm	32	58.0070	10	0.268241	
60 × 60	A* algorithm	290	88.3259	11	1.101714	
	JPS algorithm	38	88.3259	11	0.549135	
	Improved algorithm	30	93.5291	6	0.345265	
80 × 80	A* algorithm	836	131.6396	21	3.334070	
	JPS algorithm	51	131.6396	21	1.585116	
	Improved algorithm	52	134.3342	15	0.912868	
100 × 100	A* algorithm	876	160.5097	23	5.128762	
	JPS algorithm	55	160.5097	23	2.237949	
	Improved algorithm	48	159.7931	16	1.109351	

Table 2. Comparison of three algorithms on grid maps with high proportions of obstacles.

Map Size	Experimental Algorithms	Number of Nodes Traversed (Number)	Path Length (Meters)	Number of Turns (Times)	Search Time (Seconds)	Percentage of Obstacles
20 × 20	A* algorithm	131	29.3137	12	0.448762	70%
	JPS algorithm	40	30.7279	9	0.379653	
	Improved algorithm	57	30.2587	9	0.331437	
40 × 40	A* algorithm	183	62.6274	25	1.247487	
	JPS algorithm	79	62.6274	25	0.956823	
	Improved algorithm	125	60.3749	18	0.757804	
60 × 60	A* algorithm	280	100.7696	34	2.161697	
	JPS algorithm	86	100.7696	32	1.513218	
	Improved algorithm	128	99.3118	29	1.027475	
80 × 80	A* algorithm	694	136.9117	31	4.919608	
	JPS algorithm	113	136.9117	31	3.414556	
	Improved algorithm	132	134.7762	24	2.055904	
100 × 100	A* algorithm	791	173.3970	35	7.117738	
	JPS algorithm	127	173.3970	35	4.593951	
	Improved algorithm	144	169.9316	27	2.378748	

The simulation results indicate that the efficiency of the improved A* algorithm is influenced by the size of the environment map and the ratio of obstacles. Specifically, the improved algorithm demonstrates greater efficiency in improving the search speed with larger maps. Conversely, with smaller maps, the efficiency of the improved algorithm in enhancing the search speed diminishes. When the ratio of obstacles in the map varies, due to the introduction of environmental constraints, if the percentage of obstacles is low, the improved algorithm shows less noticeable improvement in path length, and in some cases, the path length may even worsen. However, the improvement in search time is significant. Conversely, when the percentage of obstacles is higher, the improved algorithm demonstrates more pronounced improvement in path length, but the improvement in search time deteriorates compared to the scenario with a lower obstacle percentage. However, in general, compared to the other two algorithms, the improved A* algorithm has significantly enhanced both the efficiency of pathfinding and algorithm optimization.

4.2. Experimental Validation

In order to verify the feasibility of the improved A* algorithm in the actual operation of the mobile robot, the algorithm was applied to the two-wheel-drive differential robot built by our team, as shown in Figure 14a.The differential robot utilizes the RPLIDAR A1 laser rangefinder to obtain the 2D point cloud data of the external environment. It integrates odometer data for localization and constructs a 2D map using the Amcl and Gmapping function packages. Finally, it performs global path planning using the global planner in the Move Base function package [20]. The experimental scene described in this paper was a section of aisles in the laboratory, as shown in Figure 14b.

(a) (b)

Figure 14. Mobile robots and experimental environments: (**a**) differential robot; (**b**) laboratory.

The two path plannings performed by the differential robot at the experimental site are shown in Figure 15. The results of the planning have been displayed using the ROS visualization tool Rviz, where the orange square represents the model of the cart and the green solid line depicts the path planned by the improved A* algorithm.

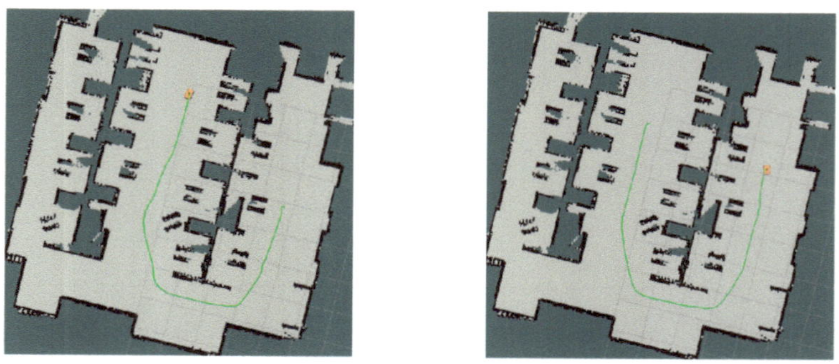

(a) (b)

Figure 15. The results of path planning using the improved algorithm: (**a**) planning for Path 1; (**b**) planning for Path 2.

The search times of the three algorithms for the two paths are shown in Tables 3 and 4, respectively. It can be concluded that the improved algorithm proposed in this paper further enhances search speed while essentially producing optimal paths. Therefore, the improved algorithm outperforms the other two algorithms significantly.

Table 3. Comparison of pathfinding times of the three algorithms for Path 1.

	A* Algorithm	JPS Algorithm	The Algorithms in This Paper
Path length/m	16.8	16.7	16.5
Search time/ms	267.89	84.43	56.25

Table 4. Comparison of pathfinding times of the three algorithms for Path 2.

	A* Algorithm	JPS Algorithm	The Algorithms in This Paper
Path length/m	14.7	14.7	15.0
Search time/ms	217.43	78.10	45.47

5. Conclusions

In this paper, for the traditional A* algorithm in path planning, there are many redundant nodes, which lead to problems such as high memory consumption and long search time, etc.; an improved A* algorithm under the combination of bidirectional JPS strategy is proposed. The effectiveness of the improved algorithm has been demonstrated by analyzing and comparing the results of the three algorithms for path planning under different grid maps. The simulation results in different grid maps demonstrate that while ensuring basic path length guarantees, the algorithm proposed in this paper not only significantly improves path search speed but also notably reduces the number of turns in a path. Especially in larger scenarios, the effect of improving the search speed is more pronounced. The improved algorithm has been applied to the robot platform for experiments, and the experimental results fully demonstrate that the improved algorithm meets the requirements for fast path planning, with noticeable optimization effects.

Author Contributions: Project administration, F.W.; writing—original draft preparation, W.S.; writing—review and editing, P.Y. and H.W.; supervision, H.L. All authors have read and agreed to the published version of the manuscript.

Funding: "This research was funded by the Fundamental Research Program of Shanxi Province, grant number 20210302123054" and "Supported by the Opening Foundation of Shanxi Provincial Key Laboratory for Advanced Manufacturing Technology (No. XJZZ202307)".

Institutional Review Board Statement: Not applicable.

Informed Consent Statement: Not applicable.

Data Availability Statement: Data are contained within the article. The dataset is available on request from the authors.

Conflicts of Interest: The authors declare no conflicts of interest.

References

1. Qu, D.K.; Du, Z.J.; Xu, D.G.; Xu, F. Research on Path Planning for a Mobile Robot. *J. Robot.* **2008**, *30*, 97–101+106.
2. Cui, W.; Zhu, F.Z. Review of Path Planning Algorithms for Robot Navigation. *J. Comput. Eng. Appl.* **2023**, *59*, 10–20.
3. Xu, Y.; Guan, G.F.; Song, Q.W.; Jiang, C.; Wang, L.H. Heuristic and random search algorithm in optimization of route planning for Robot's geomagnetic navigation. *J. Comput. Commun.* **2020**, *154*, 7–12. [CrossRef]
4. Zhi, C.B.; Zhang, A.J.; Du, X.Y.; Peng, P. Research on Global Path Planning of Mobile Robot Based on Improved A* Algorithm. *J. Comput. Simul.* **2023**, *40*, 486–491.
5. Qi, F.L.; Wang, X.Q.; Zhang, W.Y. Research on AGV Obstacle Avoidance Path Planning Based on Improved A* Algorithm. *J. Mach. Tool Hydraul.* **2023**, *51*, 34–39.
6. Chen, C. Research on Robot Shortest Path Planning with Improved A* Algorithm. *J. Comput. Digit. Eng.* **2023**, *51*, 1697–1701.
7. Zhong, X.Y.; Tian, J.; Hu, H.S.; Peng, X.F. Hybrid Path Planning Based on Safe A* Algorithm and Adaptive Window Approach for Mobile Robot in Large-Scale Dynamic Environment. *J. Intell. Robot. Syst.* **2020**, *99*, 65–77. [CrossRef]
8. Sang, T.T.; Xiao, J.C.; Xiong, J.F.; Xia, H.Y.; Wang, Z.Z. Path Planning Method of Unmanned Surface Vehicles Formation Based on Improved A* Algorithm. *J. Mar. Sci. Eng.* **2023**, *11*, 176. [CrossRef]

9. Huang, C.J. Robot Path Planning Design Based on Improved Ant Colony Algorithm. *J. Value Eng.* **2023**, *42*, 51–53.
10. Fu, Q.; Ning, Y.K.; Ji, Y.F.; Sun, X.Y. Path Planning Based on Improved RRT and DWA Fusion Algorithm. *J. Comput. Simul.* **2023**, *40*, 429–435.
11. Wang, Y.; He, T.; Silva, P. Wide-band high-accuracy ADC using segmented DAC with DWA and mismatch shaping. *J. Electron. Lett.* **2017**, *53*, 713–714. [CrossRef]
12. Gao, F.X.; Hao, W.J.; Wu, Y.; Sun, C. Research on Robot Obstacle Avoidance Path Planning Based on Improved Artificial Potential Field Method. *J. Comput. Simul.* **2023**, *40*, 431–436+442.
13. Zhang, M.; Wu, Y.Z. Optimization of transfer robot path planning based on A* algorithm. *J. Mod. Electron. Tech.* **2023**, *46*, 135–139.
14. Gong, Y.X.; Liu, G.H.; Zhang, W.K.; Yu, D.Y.; Cui, Y.X.; Shen, Z.B. Path Planning Method of Improving A* Algorithm Using Convex Corner. *J. Comput. Eng. Appl.* **2023**, *59*, 309–315.
15. Harabor, D.; Grastien, A. The JPS pathfinding system. In Proceedings of the 5th Annual Symposium on Combinatorial Search, Niagara Falls, ON, Canada, 19–21 July 2012.
16. Zhao, X.; Wang, Z.; Huang, C.K.; Zhao, Y.W. Mobile Robot Path Planning Based on an Improved A*Algorithm. *J. Robot.* **2018**, *40*, 903–910.
17. Lv, T.Z.; Zhao, C.X.; Xia, P.P. Global path planning based on simultaneous visibility graph construction and A* algorithm. *J. Nanjing Univ. Sci. Technol.* **2017**, *41*, 313–321.
18. Jeddisaravi, K.; Alitappeh, J.R.; Pimenta, A.C.L.; Guimarães, G.F. Multi-objective approach for robot motion planning in search tasks. *J. Appl. Intell.* **2016**, *45*, 305–321. [CrossRef]
19. Zhang, W.M.; Zhang, Y.; Zhang, H. Path planning of coal mine rescue robot based on improved A*algorithm. *J. Coal Geol. Explor.* **2022**, *50*, 185–193.
20. Harabor, D.; Grastien, A. Online graph pruning for pathfinding on grid maps. In Proceedings of the 25th AAAI Conference on Artificial Intelligence, Menlo Park, CA, USA, 7–11 August 2011.

applied sciences

MDPI

Article

Path Planning Based on Artificial Potential Field with an Enhanced Virtual Hill Algorithm

Hyun Jeong Lee [1], Moon-Sik Kim [2,*] and Min Cheol Lee [1,*]

1 School of Mechanical Engineering, Pusan National University, Pusan-si 46241, Republic of Korea; lhjeong98@gmail.com
2 Department of Intelligent Mobility Engineering, Kongju National University, Cheonan-si 31080, Republic of Korea
* Correspondence: mskim2@kongju.ac.kr (M.-S.K.); mclee@pusan.ac.kr (M.C.L.)

Abstract: The artificial potential field algorithm has been widely applied to mobile robots and robotic arms due to its advantage of enabling simple and efficient path planning in unknown environments. However, solving the local minimum problem is an essential task and is still being studied. Among current methods, the technique using the virtual hill concept is reliable and suitable for real-time path planning because it does not create a new local minimum and provides lower complexity. However, in the previous study, the shape of the obstacles was not considered in determining the robot's direction at the moment it is trapped in a local minimum. For this reason, longer or blocked paths are sometimes selected. In this study, we propose an enhanced virtual hill algorithm to reduce errors in selecting the driving direction and improve the efficiency of robot movement. In the local minimum area, a dead-end algorithm is also proposed that allows the robot to return without entering deeply when it encounters a dead end.

Keywords: mobile robots; artificial potential field; local minima problem; enhanced virtual hill; dead-end algorithm

Citation: Lee, H.J.; Kim, M.-S.; Lee, M.C. Path Planning Based on Artificial Potential Field with an Enhanced Virtual Hill Algorithm. *Appl. Sci.* **2024**, *14*, 8292. https://doi.org/10.3390/app14188292

Academic Editor: Christos Bouras

Received: 27 August 2024
Revised: 9 September 2024
Accepted: 12 September 2024
Published: 14 September 2024

1. Introduction

Unlike complex global path planning [1], which requires a vast amount of prior information, the local method allows for path planning, including obstacle avoidance, in an unknown environment with a small amount of computation. Local path planning using the APF (artificial potential field) technique has been widely used in real-time path planning due to its easy implementation and high computational efficiency [2–9]. APF is formed by the sum of the attractive potential field to reach the destination and the repulsive potential field to avoid obstacles [10]. Due to the potential at each point, the robot travels to the destination without colliding. In this process, when the sum of potentials approaches '0', local minima may occur, in which the robot stops operating [11,12]. In order to apply the APF technique, this problem must be solved, and various techniques have been studied so far. There are methods of modifying the potential field function or the repulsive force function [6,13–15] or adding vortex fields or local attractors [16–18]. In [17], vortex fields prevent collisions between obstacles and robots and enable smooth running when the robot moves through narrow corridors. However, it is difficult to determine when to apply vortex potential for optimal performance, and it may not be effective in complex environments. References [7,19] propose a short and optimal path planning method using bacterial foraging optimization. In [7], virtual particles are initially moved randomly, the cost is calculated at each location, the best particle is found, and the robot's driving path is created. The optimal path can be found even in complex environments. Reference [19] shows that when a robot is trapped in a local minimum, a virtual obstacle suitable for the location condition is created, and the robot is guided to move to a new path. However, both of these techniques require an iterative learning process to optimize the path and virtual obstacles. In the bacterial potential field method [20], the problems of the APF are solved by applying the bacterial evolutionary algorithm (BEA) in addition. There are also methods to plan the optimal path while solving the local minimum using simulated annealing (SA), a genetic algorithm (GA), or reinforcement learning [21–24]. Reference [24] introduces an algorithm that converges to the optimal path using reward and value functions. The optimal path can be found through automatic learning without prior information. However, since it starts randomly, it can be unstable and requires a lot of calculations to optimize the path. References [25,26] propose new algorithms to be combined with APF. In [26], the improved A* algorithm is fused with APF, and search accuracy and efficiency are improved using a heuristic method. To escape the local minimum, a virtual target point is set in [27,28], and a grid-based obstacle cell is used in [29,30]. In [30], efficient computation is possible using a discrete grid, and local minima can be avoided by creating a new repulsive potential between adjacent obstacles. However, if the resolution of the grid is low, precise path planning is difficult, and if the resolution is high, the amount of calculation increases. A perpendicular approach based on APF has been proposed as a local path plan for unknown obstacles along with a global path plan performed before the departure of the robot [31]. This method can solve the local minima problem with a relatively short calculation time, but it is difficult to apply in an unknown environment because it requires prior information about obstacles.

Recently, research on learning-based optimization techniques has been introduced to overcome the shortcomings of APF. These methods can find the optimal path for robots even in complex environments and efficiently avoid and escape local minima. However, there are issues such as the robot being unstable at the beginning of the learning process, results varying depending on the initial conditions, and high computational load. Additionally, in unknown environments, the robot's responsiveness is reduced until the learning process is completed, making immediate path planning difficult. On the other hand, the artificial potential field algorithm with a virtual hill used in this paper has the advantage of being able to avoid obstacles, escape local minima, and reach the destination without a learning process in unknown environments.

In [32], a method of escaping from the local minimum using a virtual hill is introduced. The virtual hill generates extra force instead of attractive force to repel the robot from a local minimum. The virtual hill technique does not require prior information about space and obstacles, modeling [33] and learning [23,24] processes for generating potential fields may be omitted, and a new local minimum is not created. It also has the advantage of being applicable to dynamic environments and being easy to implement. And in most cases, a robot can escape from the local minimum. This is an easy and safe way for a robot to move to a destination while avoiding obstacles in an unknown environment without being trapped in a new local minimum. The virtual hill technique is a method of moving along the outline of the obstacle that caused the local minimum. The robot determines which direction to follow among the left and right obstacles to drive as soon as it is trapped in the local minimum. Efficient and reliable path planning is possible if the robot selects the obstacle that leads to the shortest distance. In the previous study [32], the direction of movement was selected only in the relative positional relationship between the robot, the closest obstacle, and the destination without considering the shape of the path. As a result, the robot may go a long distance or choose a dead end. Therefore, there is a need to reduce errors in the selection of the direction of driving regardless of obstacle shape and to improve the efficiency of the robot's movement.

This paper proposes 'the enhanced virtual hill' technique, a method of determining the driving direction that considers the shape of surrounding obstacles. At the moment the robot is trapped in a local minimum, the distance to obstacles is measured, and it is determined which of the two directions is the more open path. The driving efficiency of the robot can be improved by blocking closed-path driving in advance. Additionally, a dead-end algorithm is proposed that allows the robot to come back without going deep when it encounters a dead end. Through simulations, it is confirmed that the robot's moving distance is reduced when the proposed methods are applied.

2. Simulation Environments

For evaluating the path planning algorithm, a mobile robot model, LiDAR sensor, visualization function, and maps provided by MATLAB 2022b's Mobile Robotics Simulation Toolbox are utilized. Figure 1a,b shows the directions of the linear velocity and angular velocity of the robot and the environment of the Mobile Robotics Simulation Toolbox, respectively. The forward/inverse geometry of the robot is provided, and the position, posture, movement path, sensor measurement distance(the dashed blue lines), etc. of the robot are shown through the visualization function, as in Figure 1b. For the simulation, the radius of the wheel of the mobile robot is set to 0.1 [m], and wheelbase, which is the distance from the left wheel to the right wheel, is set to 0.5 [m]. The local coordinate system of the robot is fixed to the center of the robot. In the simulation environment, the LiDAR sensor is configured in a radial form, with the center of the robot as the origin. According to the user settings, the robot obtains obstacle information of up to 4 m at 19 points through the LiDAR sensor. LiDAR sensors are widely used to create maps or model and track objects using numerous measurement points. However, in this paper, since large amounts of data in the form of a cloud are not required, the LiDAR sensor is simulated by a low-cost ultrasonic sensor widely used in mobile robots. In the reference paper [32], 5–7 robot skeleton points are set, and the distance from each point to the obstacle is calculated and simulated. Although this paper also uses the LiDAR sensor supported by MATLAB's Toolbox, it receives distance data by dividing 360 degrees into 19 points as if it were equipped with 19 ultrasonic sensors. Computation time can be saved by reducing the amount of processed data. The maximum measurement distance is set to 4 m. This study targets mobile robots that provide services in offices or commercial facilities. Accordingly, 4 m is considered the distance that people can feel as an open space in general indoor office environments. At too far a distance, the spacing between each sensor data increases, making it inappropriate to treat it as information about continuous obstacles. Reference [32] is referred to for the path planning and control process of the robot. All simulations are saved as figures, data files, and videos.

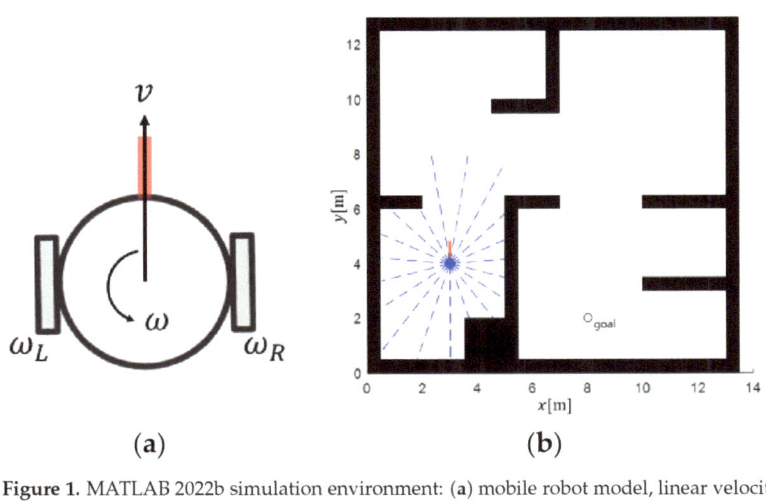

Figure 1. MATLAB 2022b simulation environment: (**a**) mobile robot model, linear velocity v, angular velocity ω; (**b**) visualizer of the Mobile Robotics Simulation Toolbox.

3. Virtual Hill Concept and Open Path Indicator, *New* e_b

The virtual hill concept allows the robot to escape from the local minimum by generating extra force instead of attractive force when in the local minimum area. As explained in

Section 3.1, the extra force follows the direction of the unit tangent vector $\mathbf{e}_t$ of γ, which represents the trajectory of the nearest obstacle while the robot is moving within the local minimum. The unit tangent vector is determined as the cross product of the unit normal vector $\mathbf{e}_n$ and the unit binormal vector $\mathbf{e}_b$. Since $\mathbf{e}_n$ is the vector between the nearest obstacle and the robot, the decisive factor that determines the robot's driving direction is $\mathbf{e}_b$. Unlike the previous $\mathbf{e}_b$ determined according to the relative positions of the robot, the closest obstacle, and the goal, the *new* $\mathbf{e}_b$ proposed in this paper operates as an open path indicator. *new* $\mathbf{e}_b$, which is determined by the shape of surrounding obstacles, makes the robot travel on a more open path, improving path efficiency.

3.1. Virtual Hill Concept

APF can be expressed as the sum of the attractive potential corresponding to the goal and the repulsive potential generated by the obstacle [10]. The artificial forces that determine the control output of the robot, that is, the linear velocity and angular velocity, are the negative gradients of the potentials [10,12]. The attractive potential U_{att}, force $\mathbf{F}_{att}$, and the repulsive potential U_{rep}, force $\mathbf{F}_{rep}$, are obtained as follows.

$$U_{att} = \begin{cases} k_a d^2, & d \le d_0 \\ k_a\left(2d_0 d - d_0^2\right), & d > d_0 \end{cases} \tag{1}$$

$$\mathbf{F}_{att} = -\nabla U_{att} = \begin{cases} -2k_a\left(\mathbf{P} - \mathbf{P}_{goal}\right), & d \le d_0 \\ -2k_a d_0 \frac{\mathbf{P}-\mathbf{P}_{goal}}{d}, & d > d_0 \end{cases} \tag{2}$$

$$U_{rep} = \begin{cases} \frac{1}{2}k_r\left(\frac{1}{\rho} - \frac{1}{\rho_0}\right)^2, & \rho \le \rho_0 \\ 0, & \rho > \rho_0 \end{cases} \tag{3}$$

$$\mathbf{F}_{rep} = -\nabla U_{rep} = \begin{cases} k_r\left(\frac{1}{\rho} - \frac{1}{\rho_0}\right)\frac{\mathbf{P}-\mathbf{P}_{co}}{\rho^3}, & \rho \le \rho_0 \\ 0, & \rho > \rho_0 \end{cases} \tag{4}$$

d_0 is the radius of the quadratic range, k_a is the proportional gain of the function, and $d = \left\|\mathbf{P} - \mathbf{P}_{goal}\right\|$. In the quadratic area, the attractive force is proportional to distance between the goal and the robot. If d_0 is large, a robot smoothly and slowly stops at the goal. $\mathbf{P}$ and $\mathbf{P}_{goal}$ are the position vectors of the robot and the goal. ρ_0 is a potential field's distance limit of repulsive potential influence, and ρ is the shortest distance between the robot and its closest obstacle, where $\rho = \|\mathbf{P} - \mathbf{P}_{co}\|$. $\mathbf{P}_{co}$ is the position vector of the closest obstacle to the robot. The entire artificial force $\mathbf{F}_{total}$ is $\mathbf{F}_{att}$ + $\mathbf{F}_{rep}$. The robot is controlled by this force, which is converted to speed and angular velocity according to the force-to-velocity conversion method [32].

However, before arriving at the goal, the robot is trapped in local minima and stops driving when the attractive force and the repulsive force are the same or very similar. In [32], the virtual hill allows the robot to escape from the LM (local minimum) by generating an extra force instead of an attractive force when in the LM area. The artificial force, $\mathbf{F}_{total}$, applied to the robot from the moment of being trapped in the LM to the moment of escape, is $\mathbf{F}_{rep} + \mathbf{F}_{ext}$. After the robot is out of the LM, $\mathbf{F}_{total}$ = $\mathbf{F}_{att}$+$\mathbf{F}_{rep}$ is applied to move to the goal. Extra force can be applied even in complex environments and does not create a new local minimum. Extra potential is applied from the moment the robot is trapped in the LM until the escape is completed and is defined as follows.

$$U_{ext} = -k_{e1}\Psi + k_{e2}\rho^2 \tag{5}$$

k_{e1} and k_{e2} are proportional constants. ρ is the distance between the robot and the closest obstacle. Ψ is the path integral, which is the line integral for γ and expressed as

$$\Psi = \int_{\gamma} d\Psi, \qquad \gamma(t) : \{\mathbf{Q}(t^*) : t_0 \leq t^* \leq t \text{ and } t_0 \leq t \leq t_k\} \tag{6}$$

$$\mathbf{Q}(t_i) = \mathbf{P}_{co}(t_i) \qquad \text{where } i = 0, 1, \cdots, K \tag{7}$$

γ is defined as the path of the closest obstacles, indicating the trajectory of the closest obstacle while the robot is moving within the LM. The concept of γ is shown in Figure 2a [34]. t_0 is the time when the robot is trapped in the LM, and t_k is the time when it escapes from the LM. $\mathbf{Q}$ is the position vector of the closest obstacle on the continuous trajectory. $\mathbf{P}_{co}$ is the position vector of the closest obstacle measured discretely by the range sensor. t_i represents a discrete sampling time of the range sensor. Assuming that $\mathbf{Q}$ is continuously interpolated and differentiable, Ψ can be written as follows.

$$\Psi(t) = \int_{\gamma(t)} d\Psi = \int_{t_0}^{t} \frac{d\Psi}{dt} dt = \int_{t_0}^{t} \|\dot{\mathbf{Q}}\| dt \qquad \text{where } \dot{\mathbf{Q}} = d\mathbf{Q}/dt \tag{8}$$

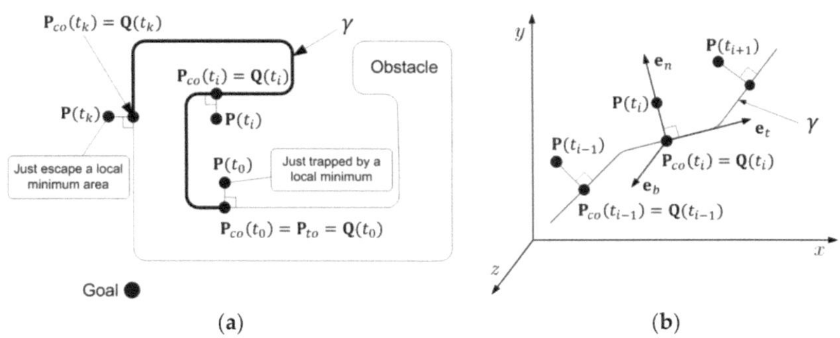

Figure 2. Within the local minimum area: (**a**) concept of γ; (**b**) concept of $\mathbf{e}_t$, $\mathbf{e}_n$, and $\mathbf{e}_b$ with respect to γ.

$\Psi(t)$ denotes a distance from $\mathbf{Q}(t_0)$ to $\mathbf{Q}(t)$ along γ. The extra force is the negative gradient of extra potential, and it is shown in Equation (9).

$$\mathbf{F}_{ext} = -\nabla U_{ext} = k_{e1}\nabla\Psi - k_{e2}\nabla\left(\rho^2\right) \tag{9}$$

This can also be expressed as Equation (10) [34].

$$\mathbf{F}_{ext} = k_{e1}\mathbf{e}_t - 2k_{e2}\rho\mathbf{e}_n \tag{10}$$

In Figure 2b, $\mathbf{e}_t$ is defined as the unit tangent vector of γ, and the derivative of $\mathbf{Q}$ has the direction of $\mathbf{e}_t$.

$$\mathbf{e}_t = \frac{\dot{\mathbf{Q}}}{\|\dot{\mathbf{Q}}\|} = \mathbf{e}_n \times \mathbf{e}_b \tag{11}$$

$\mathbf{e}_t(t_0)$ can be obtained by the cross product of the unit normal vector $\mathbf{e}_n$ and unit binormal vector $\mathbf{e}_b$.

$$\mathbf{e}_n = \frac{\mathbf{P} - \mathbf{P}_{co}}{\|\mathbf{P} - \mathbf{P}_{co}\|} = \frac{\mathbf{P} - \mathbf{P}_{co}}{\rho} \tag{12}$$

$$\mathbf{e}_b = \mathbf{e}_t \times \mathbf{e}_n = \mathbf{e}_t \times \frac{\mathbf{P} - \mathbf{P}_{co}}{\rho} \tag{13}$$

$\mathbf{P}_{co}$ is the closest point from the robot to γ, and $\overline{\mathbf{PP}_{co}}$ is always vertical to γ. The initial direction of γ is determined by $\mathbf{e}_t(t_0)$. In order for the initial direction of γ to be directed toward goal, $\mathbf{e}_b$ is defined as follows.

$$\mathbf{e}_b = \frac{\left(\mathbf{P}_{goal} - \mathbf{P}_{to}\right) \times (\mathbf{P}(t_0) - \mathbf{P}_{to})}{\left\| \left(\mathbf{P}_{goal} - \mathbf{P}_{to}\right) \times (\mathbf{P}(t_0) - \mathbf{P}_{to}) \right\|} \qquad where \quad \mathbf{P}_{to} = \mathbf{P}_{co}(t_0) \tag{14}$$

$\mathbf{P}(t_0)$ is the position vector of the robot when it is trapped in the LM. As shown in Figure 3, $\mathbf{e}_b$ is set to $\mathbf{K}$ or $-\mathbf{K}$ at the time of t_0 according to the relative positions of the robot, the closest obstacle, and the goal. At the moment the robot is trapped in the LM, $\mathbf{e}_t(t_0)$, which is the direction for the robot to move, is determined according to the sign of $\mathbf{e}_b$.

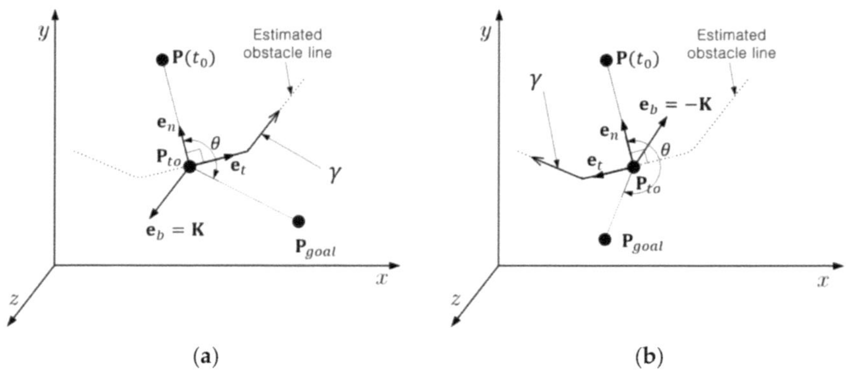

Figure 3. Direction of γ: (**a**) when $\mathbf{e}_b = \mathbf{K}$; (**b**) when $\mathbf{e}_b = -\mathbf{K}$.

3.2. Problems with Previous e_b

$$F_{total} = F_{rep} + F_{ext} = \begin{cases} k_{e1}e_t + \left(k_r\left(\frac{1}{\rho} - \frac{1}{\rho_0}\right)\frac{1}{\rho^2} - 2k_{e2}\rho\right)e_n, & \rho \le \rho_0 \\ k_{e1}e_t - 2k_{e2}\rho e_n, & \rho > \rho_0 \end{cases} \tag{15}$$

From the moment the robot is trapped in the LM to the moment of its escape, the artificial force applied to the robot is $\mathbf{F}_{rep} + \mathbf{F}_{ext}$. In Equation (10), the first term $k_{e1}\mathbf{e}_t$ of $\mathbf{F}_{ext}$ acts for a robot to move along the obstacle. The second term $-2k_{e2}\rho\mathbf{e}_n$ acts in the opposite direction to the repulsive force so that the robot does not move too far away from the obstacle. The extra force operates to move the robot along the outline of the obstacle until the robot escapes the LM without being too far away from the obstacle. In the previous virtual hill algorithm, $\mathbf{e}_b$ was determined by the location relationship, and the form of the surrounding obstacle was not considered. Therefore, depending on the form of the map, the robot sometimes set an inefficient path. Figure 4 shows the moment when the robot is trapped in the LM, and $\mathbf{P}_{to}$ is the position of the closest obstacle at the moment when it is trapped in the LM. According to Equation (14) used in the previous algorithm, $\mathbf{e}_b$ has the $-\mathbf{K}$ direction. In the LM section, since $\mathbf{e}_t$ indicating the driving direction is $\mathbf{e}_n \times \mathbf{e}_b$, the robot moves in the negative direction (clockwise). Eventually, it goes around the blocked path, as shown in Figure 5 (Point E is the moment it is determined that the robot has escaped the LM). Depending on which direction is selected at the moment when the robot is trapped in the LM, it may take quite a long path, or a dead path may be selected. Determining $\mathbf{e}_b$ by considering only the relative positions, as in Equation (14), may complicate the movement path, as in Figure 5. Therefore, a more effective path planning algorithm that considers the form of obstacles is required.

3.3. Open Path Indicator, New e_b

new e_b is proposed, in which the obstacle form is considered. As soon as the robot is trapped in the LM, it is necessary to select in which of the two directions to move. This is divided into the (+) direction and (−) direction based on $\mathbf{P}(t_0)$. For each direction, the rate of change in the obstacle contour is calculated. The direction with the larger rate of change is considered the more open path, that is, the driving direction. This direction setting helps the robot avoid a closed path and reach the destination with a shorter moving distance. More efficient path planning becomes possible when the driving direction is determined in consideration of the shape of the obstacle.

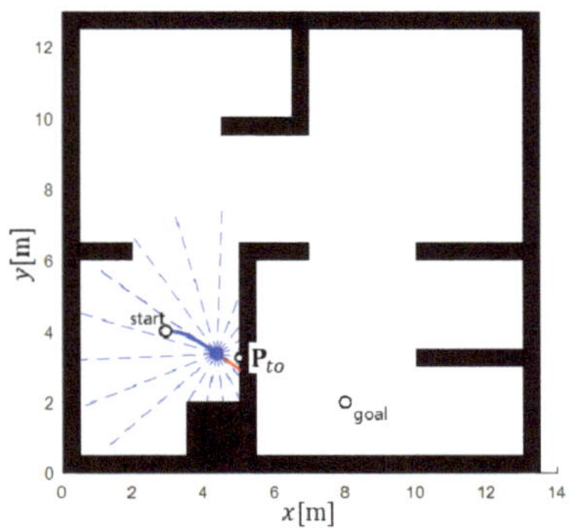

(a)

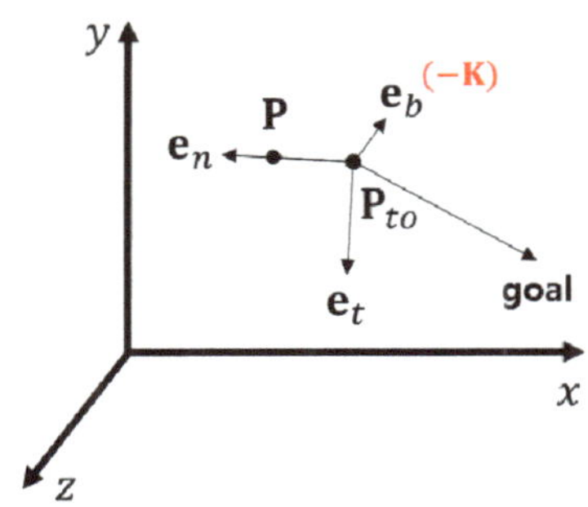

(b)

Figure 4. Direction determined by *previous* $\mathbf{e}_b$: (**a**) the moment the robot is trapped in the local minimum; (**b**) $\mathbf{e}_t$ determined by *previous* $\mathbf{e}_b$.

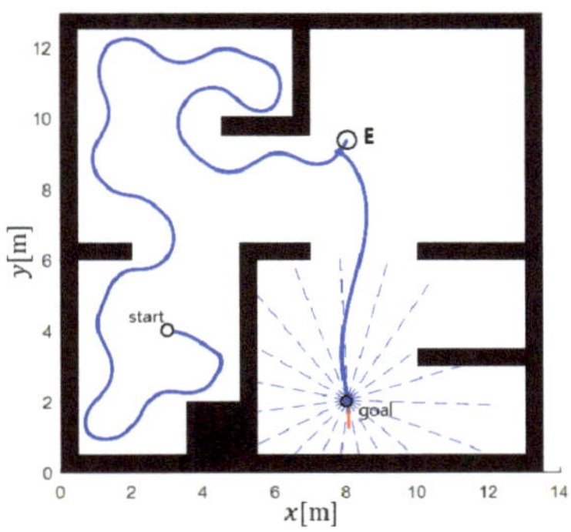

Figure 5. Driving path determined by *previous* $\mathbf{e}_b$. The path length is 44.26 [m].

3.3.1. Two-Directional Obstacle Measurement and *New* $_{e_b}$

In the simulation environment, the range sensor is configured in a radial form with the center of the robot as the origin. The degree of openness in both directions is estimated using the rate of change in the detection distance according to the measurement angle of each sensor data. In order to calculate the rate of change in the obstacle detection distance, it is expressed as a polar coordinate system fixed to the center of the robot. Figure 6 shows the polar coordinate system with the origin at the center of the robot and the angle of each sensor data set in the simulation. The angle and detection distance for each sensor data are expressed as θ_i and r_i. At the moment when the robot is trapped in the LM, the rate of change in the outline of the obstacle is calculated for both the left and right directions based on the n-th sensor data that measured $\mathbf{P}_{to}$. The one with the larger value is considered the open path and becomes the direction for the robot to move. When the contour of the two-directional obstacle is expressed as continuous functions χ_p and χ_m, the moment of the LM is shown in Figure 7. The differential lengths $d\chi_p$ and $d\chi_m$ in the polar coordinate system can be written as follows.

$$d\vec{\chi_p} = dr\ \mathbf{a}_r + r\ d\theta\ \mathbf{a}_\theta \qquad d\vec{\chi_m} = dr\ \mathbf{a}_r + r\ d\theta\ \mathbf{a}_\theta \tag{16}$$

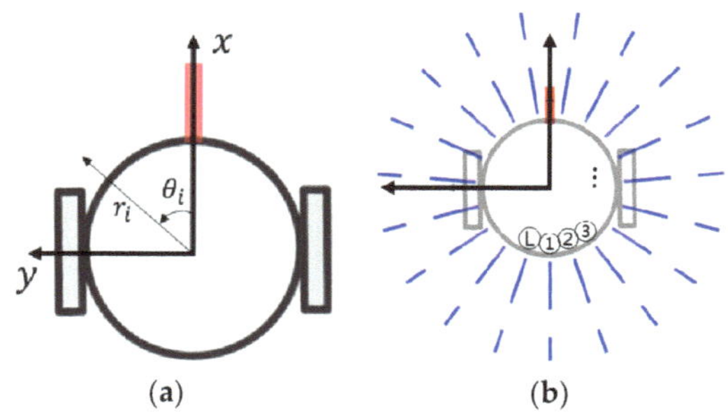

Figure 6. Local polar coordinate system with an origin at the center of the robot: (**a**) sensor measurement angle θ_i and detection distance r_i; (**b**) sensor data numbers $(1 \cdots L)$ and measurement angles.

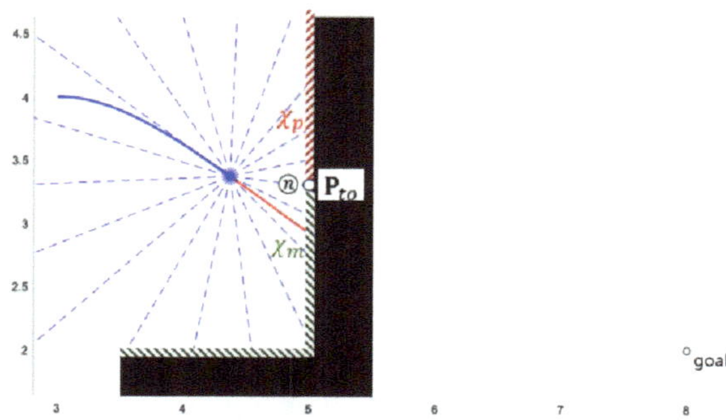

Figure 7. The $(+)$ directional obstacle χ_p and $(-)$ directional obstacle χ_m based on the n-th sensor data in which $\mathbf{P}_{to}$ is measured (enlargement of Figure 4a).

The rate of change in the obstacle contour for θ, $d\chi_p/d\theta$ and $d\chi_m/d\theta$, is as follows.

$$\left|\frac{d\chi_p}{d\theta}\right| = \sqrt{\left(\frac{dr}{d\theta}\right)^2 + r^2} \qquad \left|\frac{d\chi_m}{d\theta}\right| = \sqrt{\left(\frac{dr}{d\theta}\right)^2 + r^2} \tag{17}$$

The obstacle contour is discretized as distance values received from the L sensor data, $\{\mathbf{R}_1(r_1,\theta_1),\ \mathbf{R}_2(r_2,\theta_2),\ \ldots,\ \mathbf{R}_i(r_i,\theta_i),\ \ldots \mathbf{R}_L(r_L,\theta_L)\}$. If $dr/d\theta$ is replaced with $\Delta r/\Delta\theta$, the sum of the power of the rate of change in each of the two directions based on the n-th sensor data may be written as follows.

$$P_{sum} = \sum_{i=n}^{n+L/2} \left(\frac{r_{i+1} - r_i}{\Delta\theta}\right)^2 + r_i^2 \qquad M_{sum} = \sum_{i=n-L/2}^{n} \left(\frac{r_{i+1} - r_i}{\Delta\theta}\right)^2 + r_i^2 \tag{18}$$

The angle and obstacle measurement distance of the i-th and $(i+1)$-th sensor data are expressed in the polar coordinate system, as in Figure 8. The angle change $\Delta\theta$ with the neighboring sensor data is determined according to the sensor resolution L and is calculated as $2\pi/L$. P_{sum} is the sum of squares of the rate of change in the obstacle contour in the $(+)$ direction based on the n-th sensor data, and M_{sum} is the sum of squares of the rate of change in the $(-)$ direction. The direction with the further distance to the obstacle and the greater rate of change is regarded as an open path. That is, $\mathbf{e}_b$ is set to $+\mathbf{K}$ when $P_{sum} > M_{sum}$ and to $-\mathbf{K}$ when $P_{sum} < M_{sum}$. In the map of Figure 4, *previous* $\mathbf{e}_b$ is $-\mathbf{K}$, and the robot rotates in the direction of $-\theta$. *new* $\mathbf{e}_b$, determined by the new algorithm, is $+\mathbf{K}$, and the robot rotates in the direction of $+\theta$, as shown in Figure 9. More efficient path planning is possible. If the degree of opening in both directions is similar by satisfying the following conditions, *previous* $\mathbf{e}_b$ is maintained. *threshold* is determined experimentally.

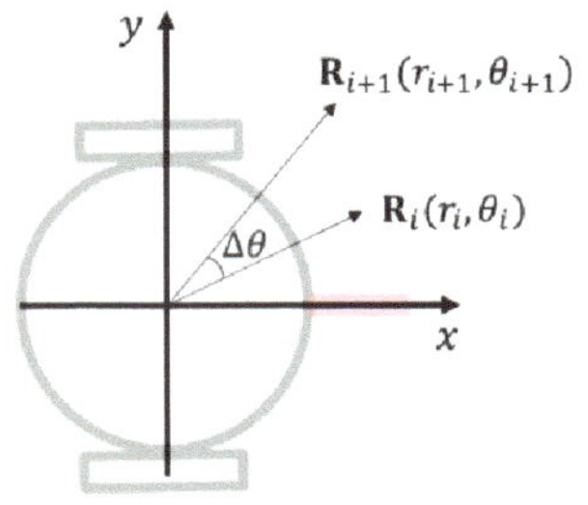

Figure 8. The i-th and the $(i+1)$-th vectors of the L sensor data.

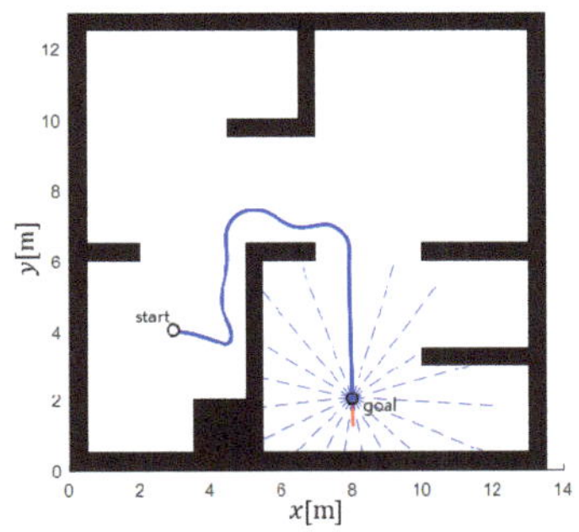

Figure 9. Driving path determined by *new* $\mathbf{e}_b$ (in comparison to Figure 5). The path length is 13.82 [m].

$$if \quad P_{sum} > threshold \ \& \ M_{sum} > threshold \quad then \quad new \ \mathbf{e}_b = previous \ \mathbf{e}_b \tag{19}$$

3.3.2. Dead-End Algorithm

When the robot encounters a dead end while traveling to the destination, it is efficient to avoid it unless the destination is inside the dead end. When a robot encounters a dead end, if it is not on the way to its destination, it can avoid the dead end using the dead-end algorithm. Figure 10a shows the path planning results when $\mathbf{e}_t$ is determined by *previous* $\mathbf{e}_b$. This is the path where the robot trapped in the LM at point $\mathbf{P}(t_0)$ moves to the destination along the obstacle ABCD section using the virtual hill technique. The ABCD section is a dead end and unnecessarily increases the robot's travel distance and time. In this case, the robot can also come back through the dead-end algorithm just as people come back after confirming that it is a dead-end path.

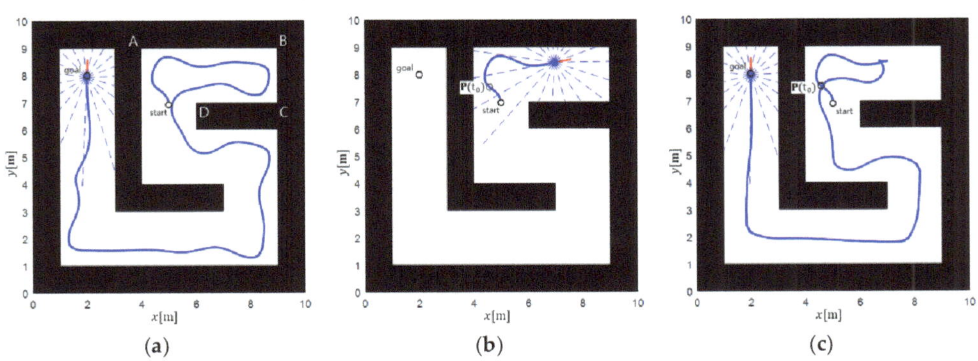

Figure 10. Dead-end algorithm: (**a**) driving path determined by a virtual hill with *previous* $\mathbf{e}_b$. The path length is 32.87 [m]. (**b**) The moment the robot notices a dead end. (**c**) Driving path determined by the dead-end algorithm. The path length is 28.99 [m].

The robot returns to the point $\mathbf{P}(t_0)$ and then moves in the opposite direction. The robot travels along obstacles closer to the destination. If the dead-end algorithm is applied when encountering a dead end, the robot can avoid the dead end without completely entering it, as shown in Figure 10c. This algorithm is applied between the beginning and the end of the LM. The flowchart of the dead-end algorithm is shown in Figure 11. The criteria for determining that the robot is at a dead end are as follows.

$$if r(\theta) < maxRange, -\frac{\pi}{2} \leq \theta \leq \frac{\pi}{2} then\ the\ path\ is\ a\ dead\ end. \tag{20}$$

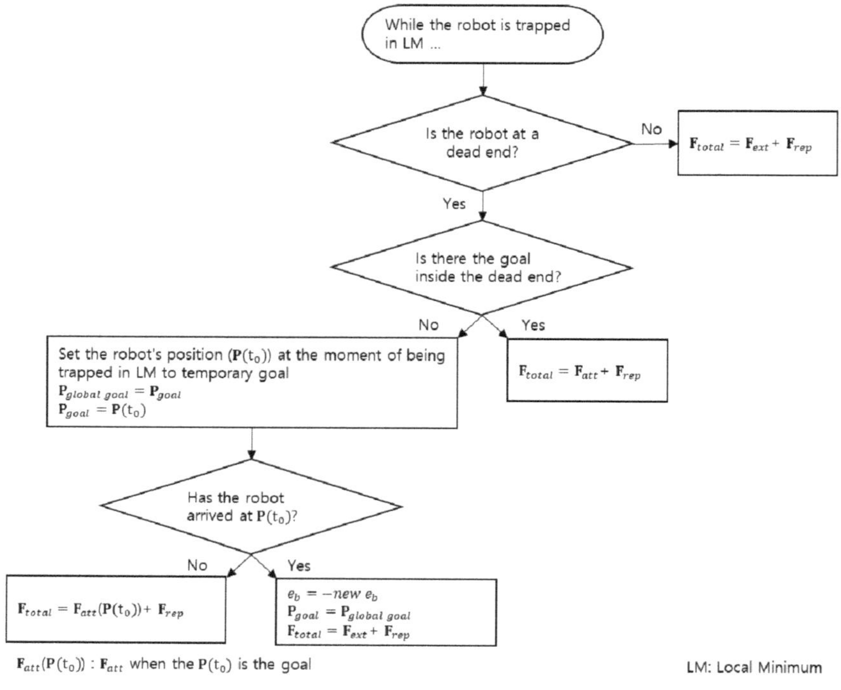

Figure 11. Flowchart of the dead-end algorithm.

$r(\theta)$ is the distance of the obstacle detected by the sensor, and $maxRange$ is the sensor's maximum detection distance. If the obstacle detection distance of all sensor data in the $[-\pi/2, \pi/2]$ section corresponding to the front portion of the robot is shorter than $maxRange$, it is determined that the robot has entered a dead end. If the robot is at a dead end, whether to enter or avoid further depends on the location of the destination. The following two conditions are examined to determine whether the destination is inside or outside the dead end.

$$\begin{cases} \left|\mathbf{P}^{\mathbf{L}}_{goal}\right| < r_k \\ -\frac{\pi}{2} \leq \alpha \leq \frac{\pi}{2}, \quad \mathbf{P}^{\mathbf{L}}_{goal} = \left|\mathbf{P}^{\mathbf{L}}_{goal}\right| \angle\alpha \end{cases} \tag{21}$$

$\mathbf{P}^{\mathbf{L}}_{goal}$ represents the location vector of the destination for the local coordinate system with an origin at the center of the robot, as shown in Figure 6a. r_k is the measured distance of the k-th sensor data, which is at the angle closest to the vector $\overrightarrow{\mathbf{P}_{goal} - \mathbf{P}}$ among all sensor data. $\left|\mathbf{P}^{\mathbf{L}}_{goal}\right| < r_k$ means that there is no obstacle between the robot and the destination. α is the angle of $\mathbf{P}^{\mathbf{L}}_{goal}$ in the local coordinate system. $-\pi/2 \leq \alpha \leq \pi/2$ means that the goal is at the front portion of the robot. If both conditions are satisfied, it is determined that the destination is inside the dead end. If the destination is inside the dead end, the robot is controlled to the destination with $\mathbf{F}_{total} = \mathbf{F}_{att} + \mathbf{F}_{rep}$. This $\mathbf{F}_{att}$ is an attractive force heading to the initially set goal (destination). If the destination is outside the dead-end road, the robot returns to $\mathbf{P}(t_0)$ and moves in the opposite direction. In order for the robot to return to $\mathbf{P}(t_0)$, $\mathbf{P}_{goal}$ is temporarily changed to $\mathbf{P}(t_0)$, and $\mathbf{F}_{total} = \mathbf{F}_{att}(\mathbf{P}(t_0)) + \mathbf{F}_{rep}$ is applied to the control. $\mathbf{F}_{att}(\mathbf{P}(t_0))$ described an attractive force generated when $\mathbf{P}(t_0)$ is set as $\mathbf{P}_{goal}$. When the robot arrives at the position $\mathbf{P}(t_0)$, $\mathbf{P}_{goal}$ is reset to the initial goal, and it moves in the opposite direction to $e_t(t_0)$. $\mathbf{F}_{total} = \mathbf{F}_{ext} + \mathbf{F}_{rep}$ is applied until the robot escapes from the LM.

4. Simulations

Some conditions for simulation are referred to in [32]. The virtual hill algorithm has several parameters for controlling the robot, such as the constant coefficients of the extra force, the application distance of the reactive force, and the like. Depending on the setting, the degree to which the robot approaches or moves away from obstacles, the straightness of the driving, and similar parameters may vary. Some maps provided by MATLAB are applied in sizes of 10 [m] × 10 [m]. The total artificial force is converted into linear velocity and angular velocity by the force-to-velocity conversion method [32]. The path length is indicated in each figure description.

Figures 12 and 13 show the results of the enhanced virtual hill algorithm. In the previous algorithm, when the robot is trapped in an LM (local minimum), the direction of progress is determined by the relative positions of $\mathbf{P}(t_0)$, $\mathbf{P}_{to}$, and the goal. According to this algorithm, as shown in Figure 12a, the robot moves along the obstacle BCDE. On the other hand, when the enhanced virtual hill algorithm including *new* $\mathbf{e}_b$ is applied, the robot moves along the obstacle BA, as shown in Figure 12b. When the robot reaches the goal by following the obstacle BA, the efficiency of the driving path may be improved. In the case of Figure 13, according to the direction determination by the previous virtual hill algorithm, that is, *previous* $\mathbf{e}_b$, the robot goes around in the (+) direction, as shown in Figure 13a. The robot moves along the obstacle ABCDEFGH until it escapes the LM. Since the path determined by *new* $\mathbf{e}_b$ goes to the goal along obstacle OA, unnecessary driving may be avoided. In Figure 13, the driving distance varies considerably depending on the direction determination at the LM.

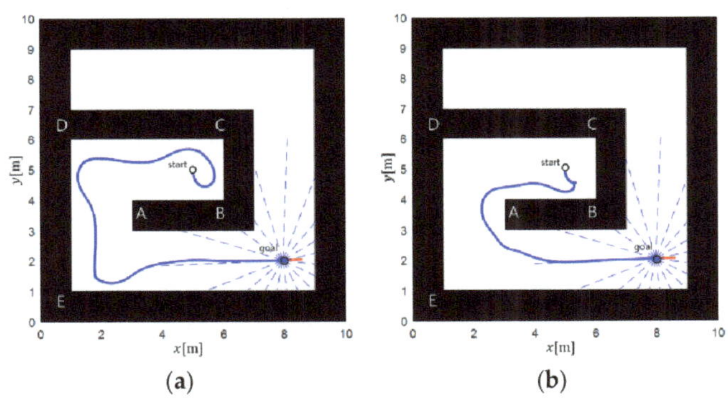

Figure 12. Driving paths: (**a**) Determined by *previous* $\mathbf{e}_b$. The path length is 16.51 [m]. (**b**) Determined by *new* $\mathbf{e}_b$. The path length is 11 [m].

Figures 14–19 show the simulations of the driving path by *new* $\mathbf{e}_b$ and the dead-end algorithm. If the robot recognizes that it is at a dead end while driving in the LM section, it returns to the robot's position $\mathbf{P}(t_0)$. At $\mathbf{P}(t_0)$, the robot moves in the opposite direction and follows a closer obstacle to its goal. In the case of Figure 14a, at the point of t_0, it travels in the (−) direction and travels along the obstacle ABCDEFGHI for a long distance. Figure 14b travels along the (+) direction according to *new* $\mathbf{e}_b$ and travels the shortest distance along the obstacle BA. Figure 14c shows the result of applying the dead-end algorithm during driving, as shown in Figure 14a. After recognizing the obstacle BCDE as a blocked road, the robot returns to $\mathbf{P}(t_0)$ and moves in the direction of obstacle BA. In Figure 17c, as the robot rotates in the $(+)$ direction, the recognition of the blocked road, arrival at $\mathbf{P}(t_0)$, and setting of the opposite direction occur almost simultaneously, drawing a path similar to that in Figure 17b without any special driving.

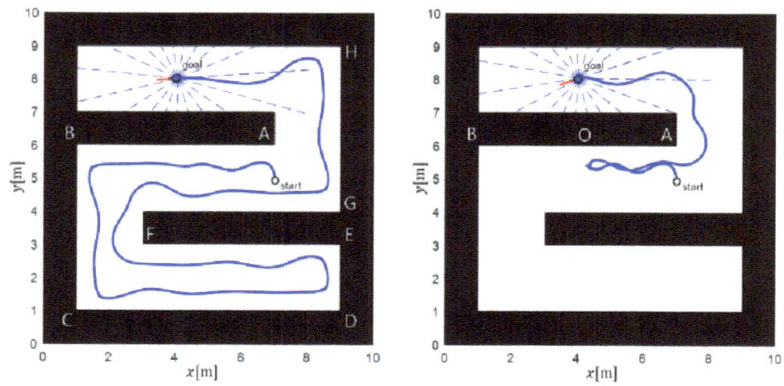

(a) (b)

Figure 13. Driving paths: (**a**) Determined by *previous* $\mathbf{e}_b$. The path length is 41.08 [m]. (**b**) Determined by *new* $\mathbf{e}_b$. The path length is 12.91 [m].

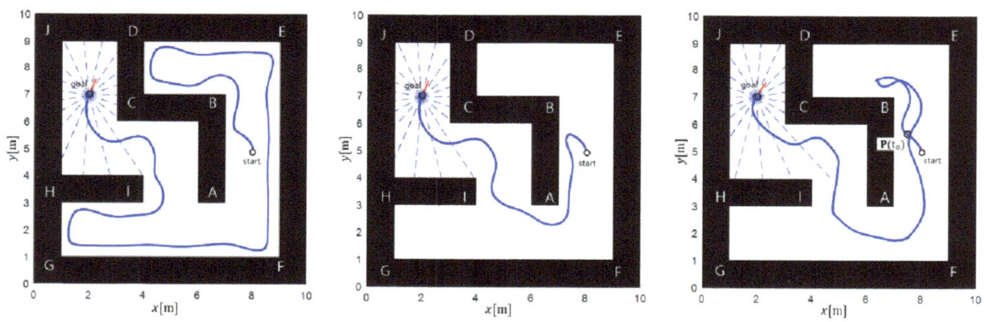

(a) (b) (c)

Figure 14. Driving paths: (**a**) Determined by *previous* $\mathbf{e}_b$. The path length is 36.66 [m]. (**b**) Determined by *new* $\mathbf{e}_b$. The path length is 12.87 [m]. (**c**) Determined by *previous* $\mathbf{e}_b$ and the dead-end algorithm. The path length is 21.7 [m].

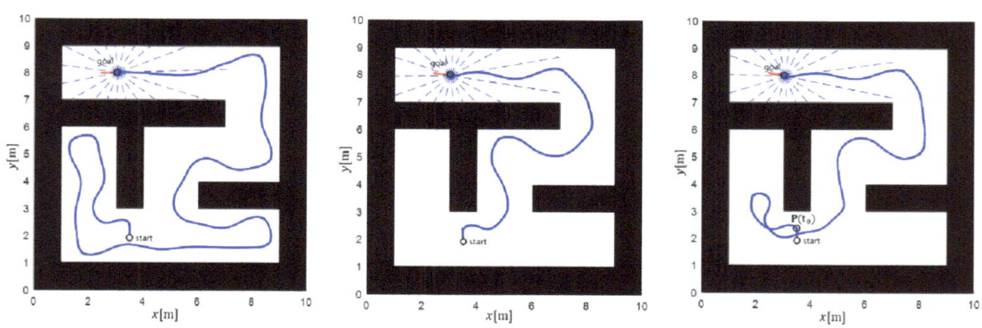

(a) (b) (c)

Figure 15. Driving paths: (**a**) Determined by *previous* $\mathbf{e}_b$. The path length is 35.95 [m]. (**b**) Determined by *new* $\mathbf{e}_b$. The path length is 15.98 [m]. (**c**) Determined by *previous* $\mathbf{e}_b$ and the dead-end algorithm. The path length is 21.72 [m].

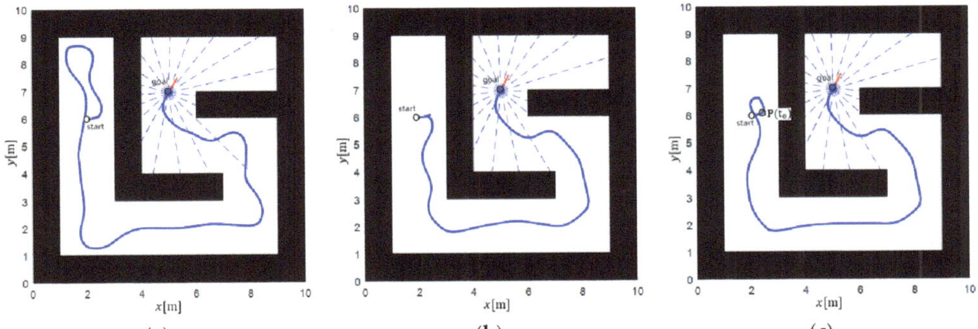

(**a**) (**b**) (**c**)

Figure 16. Driving paths: (**a**) Determined by *previous* $\mathbf{e}_b$. The path length is 25.6 [m]. (**b**) Determined by *new* $\mathbf{e}_b$. The path length is 17.33 [m]. (**c**) Determined by *previous* $\mathbf{e}_b$ and the dead-end algorithm. The path length is 19.02 [m].

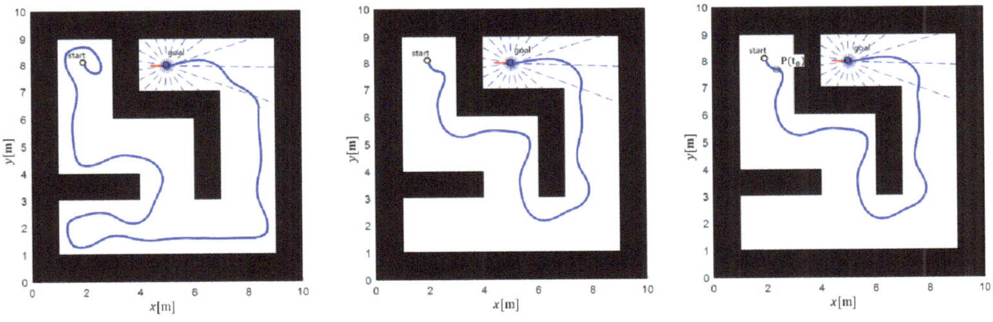

(**a**) (**b**) (**c**)

Figure 17. Driving paths: (**a**) Determined by *previous* $\mathbf{e}_b$. The path length is 32.34 [m]. (**b**) Determined by *new* $\mathbf{e}_b$. The path length is 18.96 [m]. (**c**) Determined by *previous* $\mathbf{e}_b$ and the dead-end algorithm. The path length is 19.14 [m].

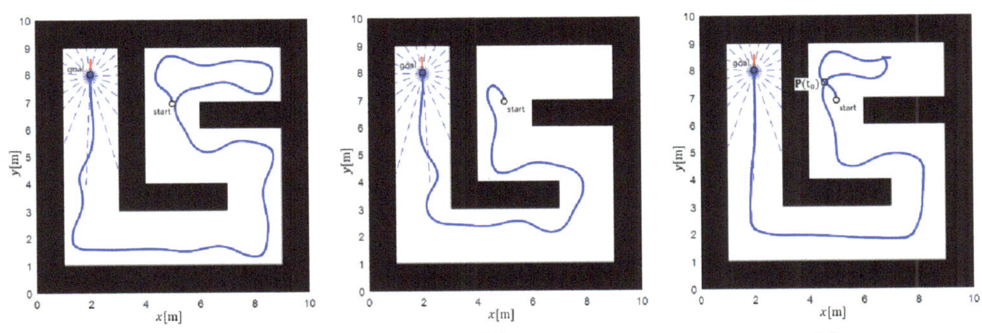

(**a**) (**b**) (**c**)

Figure 18. Driving paths: (**a**) Determined by *previous* $\mathbf{e}_b$. The path length is 32.87 [m]. (**b**) Determined by *new* $\mathbf{e}_b$. The path length is 20.17 [m]. (**c**) Determined by *previous* $\mathbf{e}_b$ and the dead-end algorithm. The path length is 28.99 [m].

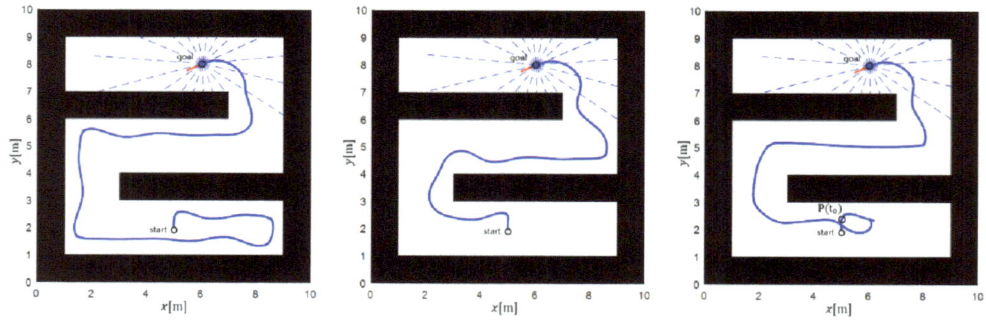

(a) (b) (c)

Figure 19. Driving paths: (**a**) Determined by *previous* $\mathbf{e}_b$. The path length is 26.12 [m]. (**b**) Determined by *new* $\mathbf{e}_b$. The path length is 16.61 [m]. (**c**) Determined by *previous* $\mathbf{e}_b$ and the dead-end algorithm. The path length is 19.48 [m].

In Figure 20, (c) shows the moment of the LM, (a) shows the driving path determined by *previous* $\mathbf{e}_b$, and (b) shows the driving path determined by the *new* $\mathbf{e}_b$. Using *new* $\mathbf{e}_b$, $\mathbf{e}_t$ is determined so that the robot may travel on a more open path. However, depending on the map, the direction that was the more open path may be blocked, as shown in Figure 20. The more open direction is the (+) direction when comparing (+) and (−) directions from the robot's position shown in Figure 20c, but the robot eventually encounters a dead end. Nevertheless, there is no significant difference in path length between Figure 20a,b. In Figure 20a, the robot avoids the dead end using *previous* $\mathbf{e}_b$ but follows obstacles far from the destination. In Figure 20b, the robot enters the dead end, but the path length is reduced by following obstacles closer to the destination due to *new* $\mathbf{e}_b$.

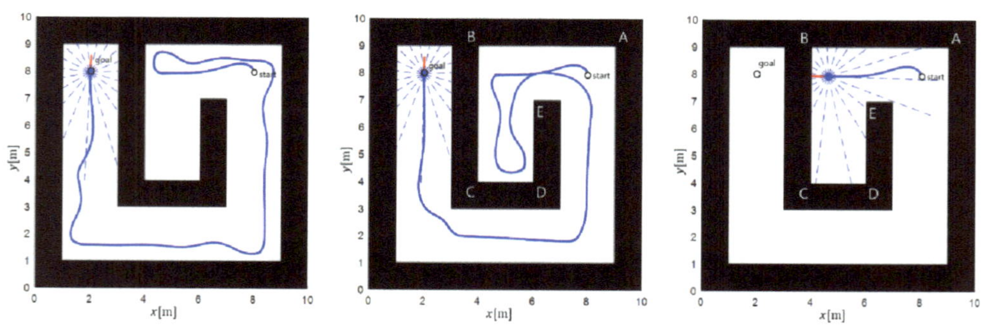

(a) (b) (c)

Figure 20. For a map that contains a dead end: (**a**) The driving path determined by *previous* $\mathbf{e}_b$. The path length is 29.41 [m]. (**b**) The driving path determined by *new* $\mathbf{e}_b$. The path length is 30.91 [m]. (**c**) The moment of the local minimum.

Table 1 shows how much the robot's mileage is reduced by the proposed algorithm. It represents the distance and ratio reduced by *new* $\mathbf{e}_b$ and the dead-end algorithm based on the path determined by *previous* $\mathbf{e}_b$ for all simulations shown in the paper. The driving distance is reduced by up to 68 [%] using *new* $\mathbf{e}_b$ and by up to 44 [%] using the dead-end algorithm. It is confirmed that the virtual hill algorithm is improved by the proposed method.

Table 1. Reduction in mileage according to algorithms.

	Mileage Using *Previous* $\mathbf{e}_b$ [m]	Mileage Using *New* $\mathbf{e}_b$ [m]	Mileage Using Dead-End Algorithm [m]	Reduction Proportion Using *New* $\mathbf{e}_b$ [%]	Reduction Proportion Using Dead-End Algorithm [%]
Figure 9	44.26	13.82	-	68.78	-
Figure 12	16.51	11	-	33.37	-
Figure 13	41.08	12.9	-	68.60	-
Figure 14	36.66	12.87	14.96	64.89	40.81
Figure 15	35.95	15.98	14.23	55.55	39.58
Figure 16	25.6	17.33	6.58	32.30	25.70
Figure 17	33.66	19.02	14.83	43.49	44.06
Figure 18	32.87	20.17	3.88	38.64	11.80
Figure 19	26.12	16.61	6.64	36.41	25.42
Figure 20	29.41	30.91	-	−5.10	-
Average [%]	-	-	-	43.69	31.23

5. Discussion

In order to compare with the previous virtual hill algorithm, simulations are performed on the same environment for both *previous* $\mathbf{e}_b$ and *new* $\mathbf{e}_b$. Also, the dead-end algorithm is applied when entering a blocked road, and how the path changes is checked. The circle mark is displayed at the position of $\mathbf{P}(t_0)$ for visual confirmation at the moment the blocked path is recognized. All simulations are saved as pictures, data files, and videos.

In Figures 12–19, each (a) is the result when *previous* $\mathbf{e}_b$ is applied, and each (b) is the result when *new* $\mathbf{e}_b$ is applied. They show that the moving distance can be significantly reduced when determining the driving direction by comparing the degree of openness of the two-way route regardless of the relative position of the robot and the goal. (c) Shows the result of applying the dead-end algorithm when a route including a blocked road is selected. It is possible to reduce the moving distance by more than about 4 m by coming back without entering deep into the dead end. In the service environment, if the mobile robot travels at a low speed of less than 0.2 m/s, it can save more than about 20 s. The saving distance and time can vary depending on the measurable distance of the sensor and control techniques. Very occasionally, the path along the *new* $\mathbf{e}_b$ is more inefficient, as shown in Figure 20. The maximum measurement distance of the sensor is 4 m. Since the distance from the robot position in Figure 20c to the obstacle CD is about 4 m, it is difficult for the robot to recognize the dead end road in the $(+)$ direction. Unless the robot has map information in advance or creates a map while moving, it may encounter dead ends when moving through an unknown space. The dead-end algorithm prevents the robot from having to drive unnecessarily deep into dead ends.

6. Conclusions

The virtual hill technique, which moves along a nearby obstacle when a robot is trapped in an LM, is easy to apply to an unknown variable environment. The virtual hill technique does not require information about space and obstacles, does not create a new the LM even in a complicated environment, and has the advantage of being easy to apply. However, there is a problem in that the mileage can vary greatly depending on which directional obstacle the robot moves along. In this study, the performance of the virtual hill technique is improved by compensating for this shortcoming using *new* $\mathbf{e}_b$ and the dead-end algorithm. The enhanced virtual hill algorithm makes it possible for a robot to reach its goal on a shorter path. Also, when a robot encounters blocked roads, it can come back without entering through the dead-end algorithm.

In future research, when the robot is trapped in the LM and extra potential is generated, it can be programmed to check in advance whether there is a dead end nearby and exclude that direction. In cases like Figure 20a, it is necessary to recognize in advance that if the robot moves in the $(+)$ direction, it will lead to a dead end. In order to recognize dead ends from a long distance, the sensor's obstacle measurement distance must be longer, and more sensor data must be processed for accuracy. It is unnecessary to process this much data in each control cycle. A method is needed to obtain detailed information about distant obstacles only when needed. To verify the proposed algorithm, additional simulations and experiments in diverse and dynamic environments are required. It is also necessary to establish indicators to quantitatively evaluate the stability and flexibility of robot driving. Through these, it is expected that the performance and limitations of the proposed algorithm will be analyzed, and shortcomings can be resolved in various conditions where static/dynamic obstacles exist in combination.

Author Contributions: Conceptualization, H.J.L. and M.C.L.; methodology, H.J.L.; software, H.J.L.; validation, H.J.L., M.-S.K. and M.C.L.; formal analysis, H.J.L.; investigation, H.J.L.; resources, H.J.L.; data curation, M.-S.K.; writing—original draft preparation, H.J.L.; writing—review and editing, M.-S.K. and M.C.L.; visualization, H.J.L.; supervision, M.C.L.; project administration, M.-S.K.; funding acquisition, M.-S.K. All authors have read and agreed to the published version of the manuscript.

Funding: This work was supported by the Development of High-Level Cognitive Prediction Sensor Technology for Autonomous Driving Mobility (RS-2022-00144500, Development of 3D Ultrasonic Sensor Technology for Vehicles based on Meta Structure) funded by the Ministry of Trade Industry & Energy (MOTIE, Republic of Korea).

Institutional Review Board Statement: Not applicable.

Informed Consent Statement: Not applicable.

Data Availability Statement: The original contributions presented in the study are included in the article, further inquiries can be directed to the corresponding authors.

Conflicts of Interest: The authors declare no conflicts of interest.

References

1. Rimon, E.; Koditschek, D.E. Exact robot navigation using cost functions: The case of distinct spherical boundaries in E^n. In Proceedings of the IEEE International Conference on Robotics and Automation, Philadelphia, PA, USA, 24–29 April 1988. [CrossRef]
2. Kim, J.-O.; Khosla, P. Real-time obstacle avoidance using harmonic potential functions. *IEEE Trans. Robot. Automat. Robot. Autom.* **1992**, *8*, 338–349. [CrossRef]
3. Feder, H.J.S.; Slotine, J.-J.E. Real-time path planning using harmonic potentials in dynamic environments. In Proceedings of the IEEE International Conference on Robotics and Automation, Albuquerque, NM, USA, 20–25 April 1997. [CrossRef]
4. Chengqing, L.; Ang, M.H.; Krishnan, H.; Yong, L.S. Virtual obstacle concept for local-minimum-recovery in potential-field based navigation. In Proceedings of the 2000 ICRA. Millennium Conference. IEEE International Conference on Robotics and Automation. Symposia Proceedings, San Francisco, CA, USA, 24–28 April 2000. [CrossRef]
5. Sfeir, J.; Saad, M.; Saliah-Hassane, H. An improved artificial potential field approach to real-time mobile robot path planning in an unknown environment. In Proceedings of the IEEE International Symposium on Robotic and Sensors Environments (ROSE), Montreal, QC, Canada, 17–18 September 2011. [CrossRef]
6. Chen, L. UUV path planning algorithm based on virtual obstacle. In Proceedings of the IEEE International Conference on Mechatronics and Automation, Tianjin, China, 3–6 August 2014. [CrossRef]
7. Hossain, M.A.; Ferdous, I. Autonomous robot path planning in dynamic environment using a new optimization technique inspired by bacterial foraging technique. *Robot. Auton. Syst.* **2015**, *64*, 137–141. [CrossRef]
8. Lin, P.; Choi, W.Y.; Lee, S.-H.; Chung, C.C. Model predictive path planning based on artificial potential field and its application to autonomous lane change. In Proceedings of the 20th International Conference on Control, Automation and Systems (ICCAS), Busan, Republic of Korea, 13–16 October 2020. [CrossRef]
9. Souza, R.M.J.A.; Lima, G.V.; Morais, A.S.; Oliveira-Lopes, L.C.; Ramos, D.C.; Tofoli, F.L. Modified artificial potential field for the path planning of aircraft swarms in three-dimensional environments. *Sensors* **2022**, *22*, 1558. [CrossRef] [PubMed]
10. Khatib, O. Real-time obstacle avoidance for manipulators and mobile robots. *Int. J. Robot. Res.* **1986**, *5*, 90–98. [CrossRef]
11. Khosla, P.; Volpe, R. Superquadric artificial potentials for obstacle avoidance and approach. In Proceedings of the IEEE International Conference on Robotics and Automation, Philadelphia, PA, USA, 24–29 April 1988. [CrossRef]
12. Volpe, R.; Khosla, P. Manipulator control with superquadric artificial potential functions: Theory and experiments. *IEEE Trans. Syst. Man Cybern.* **1990**, *20*, 1423–1436. [CrossRef]
13. Weerakoon, T.; Ishii, K.; Nassiraei, A.A.F. An artificial potential field based mobile robot navigation method to prevent from deadlock. *J. Artif. Intell. Soft Comput. Res.* **2015**, *5*, 189–203. [CrossRef]

14. Xia, X.; Li, T.; Sang, S.; Cheng, Y.; Ma, H.; Zhang, Q.; Yang, K. Path planning for obstacle avoidance of robot arm based on improved potential field method. *Sensors* **2023**, *23*, 3754. [CrossRef]
15. Li, X.; Li, G.; Bian, Z. Research on autonomous vehicle path planning algorithm based on improved RRT* algorithm and artificial potential field method. *Sensors* **2024**, *24*, 3899. [CrossRef]
16. De Medio, C.; Oriolo, G. Robot obstacle avoidance using vortex fields. In *Advances in Robot Kinematics*; Springer: Vienna, Austria, 1991. [CrossRef]
17. Szczepanski, R. Safe artificial potential field—Novel local path planning algorithm maintaining safe distance from obstacles. *IEEE Robot. Autom. Lett.* **2023**, *8*, 4823–4830. [CrossRef]
18. Melchiorre, M.; Scimmi, L.; Salamina, L.; Mauro, S.; Pastorelli, S. Robot collision avoidance based on artificial potential field with local attractors. In Proceedings of the 19th International Conference on Informatics in Control, Automation and Robotics, Lisbon, Portugal, 14–16 July 2022. [CrossRef]
19. Abdi, M.I.I.; Khan, M.U.; Güneş, A.; Mishra, D. Escaping local minima in path planning using a robust bacterial foraging algorithm. *Appl. Sci.* **2020**, *10*, 7905. [CrossRef]
20. Montiel, O.; Orozco-Rosas, U.; Sepúlveda, R. Path planning for mobile robots using Bacterial Potential Field for avoiding static and dynamic obstacles. *Expert Syst. Appl.* **2015**, *42*, 5177–5191. [CrossRef]
21. Lee, S.; Park, J. Cellular robotic collision-free path planning. In Proceedings of the Fifth International Conference on Advanced Robotics 'Robots in Unstructured Environments, Pisa, Italy, 19–22 June 1991. [CrossRef]
22. Li, Q.; Wang, L.; Chen, B.; Zhou, Z. An improved artificial potential field method for solving local minimum problem. In Proceedings of the 2nd International Conference on Intelligent Control and Information Processing, Harbin, China, 25–28 July 2011. [CrossRef]
23. Findi, A.H.; Marhaban, M.H.; Kamil, R.; Hassan, M.K. Collision prediction based genetic network programming-reinforcement learning for mobile robot navigation in unknown dynamic environments. *J. Electr. Eng. Technol.* **2017**, *12*, 890–903. [CrossRef]
24. Yao, Q.; Zheng, Z.; Qi, L.; Yuan, H.; Guo, X.; Zhao, M.; Liu, Z.; Yang, T. Path planning method with improved artificial potential field—A reinforcement learning perspective. *IEEE Access* **2020**, *8*, 135513–135523. [CrossRef]
25. Li, Q.; Ma, Q.; Weng, X. Dynamic path planning for mobile robots based on artificial potential field enhanced improved multiobjective snake optimization (APF-IMOSO). *J. Field Robot.* **2024**, *41*, 1843–1863. [CrossRef]
26. Kozhubaev, Y.; Yang, R. Simulation of dynamic path planning of symmetrical trajectory of mobile robots based on improved A* and artificial potential field fusion for natural resource exploration. *Symmetry* **2024**, *16*, 801. [CrossRef]
27. Wang, P.; Gao, S.; Li, L.; Sun, B.; Cheng, S. Obstacle avoidance path planning design for autonomous driving vehicles based on an improved artificial potential field algorithm. *Energies* **2019**, *12*, 2342. [CrossRef]
28. Lv, Q.; Hao, G.; Huang, Z.; Li, B.; Fu, D.; Zhao, H.; Chen, W.; Chen, S. Localized path planning for mobile robots based on a subarea-artificial potential field model. *Sensors* **2024**, *24*, 3604. [CrossRef]
29. Borenstein, J.; Koren, Y. Real-time obstacle avoidance for fast mobile robots. *IEEE Trans. Syst. Man Cybern.* **1989**, *19*, 1179–1187. [CrossRef]
30. Jung, J.-H.; Kim, D.-H. Local path planning of a mobile robot using a novel grid-based potential method. *Int. J. Fuzzy Log. Intell. Syst.* **2020**, *20*, 26–34. [CrossRef]
31. Dalai, S.; Irfan, M.; Singh, S.; Kishore, K.; Akbar, S.A. Heuristic guided artificial potential field for avoidance of small obstacles. In Proceedings of the 21st International Conference on Control, Automation and Systems (ICCAS), Jeju, Republic of Korea, 12–15 October 2021. [CrossRef]
32. Park, M.G.; Lee, M.C. A new technique to escape local minimum in artificial potential field based path planning. *KSME Int. J.* **2003**, *17*, 1876–1885. [CrossRef]
33. Ren, J.; McIsaac, K.A.; Patel, R.V.; Peters, T.M. A potential field model using generalized sigmoid functions. *IEEE Trans. Syst. Man Cybern. Part B Cybern.* **2007**, *37*, 477–484. [CrossRef] [PubMed]
34. Park, M.G.; Lee, M.C. Real-time path planning in unknown environment and a virtual hill concept to escape local minima. In Proceedings of the 30th Annual Conference of IEEE Industrial Electronics Society, IECON 2004, Busan, Republic of Korea, 2–6 November 2004. [CrossRef]

MDPI AG
Grosspeteranlage 5
4052 Basel
Switzerland
Tel.: +41 61 683 77 34

Applied Sciences Editorial Office
E-mail: applsci@mdpi.com
www.mdpi.com/journal/applsci

www.ingramcontent.com/pod-product-compliance
Lightning Source LLC
Chambersburg PA
CBHW040802250926
51850CB00007BA/1056
* 9 7 8 3 7 2 5 8 6 6 4 0 3 *